Manufacturing Process Planning

Manufacturing Process Planning

A Practical Approach for Mechanical Engineering

José V. Abellán-Nebot
Department of Industrial Systems Engineering and Design,
Universitat Jaume I, Castellón, Spain

Carlos Vila Pastor
Department of Mechanical Engineering and Materials,
Universitat Politècnica de València, Valencia, Spain

Héctor R. Siller
University of North Texas, Denton (Texas), USA

WILEY

Published by John Wiley & Sons, Inc., Hoboken, New Jersey.
Published simultaneously in Canada.

Library of Congress Cataloging-in-Publication Data Applied for:
Hardback ISBN: 9781394273508

Cover Design: Wiley
Cover Image: © Nordroden/Shutterstock

Set in 9.5/12.5pt STIXTwoText by Straive, Pondicherry, India

Contents

Foreword

This book is intended to be a reference textbook for teaching manufacturing process planning at undergraduate and graduate courses in the fields of design, manufacturing, and mechanical engineering. Process planning refers to all the activities that should be carried out in a manufacturing company to produce a product given the 3D model or technical drawing from the design office. The book sequentially illustrates all the necessary steps and provides practical examples at each stage of the process.

The book is organized as follows: the first chapters cover the fundamentals of process planning, the information that comes from the technical drawings, and the documents that process planners should create to communicate the required operations at the shop-floor level. Then, specific process planning tasks are explained and illustrated with practical examples. Some of these tasks are Selection of Manufacturing Processes, Process Planning in Sand Casting, Process Planning in Machining and Process Planning in Inspection. These four topics are detailed in different chapters and cover important activities performed in manufacturing companies. Fundamentals of casting, mold design, machining sequence identification, tool and parameter selection, geometric validation, CNC programming, and inspection reports are some of the key aspects that are reviewed throughout the book. Finally, manufacturing cost estimation and pricing are presented as the final activities performed by process planners to ensure a feasible and cost-effective production.

The book also includes, as supplementary files, five practical computer sessions relating to manufacturing process selection, casting simulation, pattern design, and CNC simulation. In addition, a project proposal is included for instructors that may be interested in teaching the course using a project-based approach. The project allows students to practice almost all of the concepts covered with in the book, using minimal workshop equipment found in most university facilities.

The authors of the book are recognized lecturers from the Universitat Jaume I (Spain), Universitat Politècnica de València (Spain), and the University of North Texas (United States), with more than 20 years of experience in teaching and research in the field of manufacturing.

The Spanish Society of Manufacturing Engineering (MES-SIF, Sociedad Ingeniería de Fabricación) acknowledges the fine work presented and recommends the book as a reference textbook for courses related to manufacturing process planning, especially for those who wish to study in detail sand casting and machining process planning. The activities presented in the proposed manufacturing process planning project and the computer sessions provide a good practical view of process planning and orient the instruction to a more active learning approach where learning by doing is a keystone.

Lorenzo Sevilla Hurtado
Chairman of the Spanish Society of Manufacturing Engineering
April 2024

Preface

We started to work on *Manufacturing Process Planning* in the 1990s in the field of concurrent and collaborative engineering. The thesis from Carlos Vila Pastor and Héctor R. Siller (Implementation Strategies of New Technologies in the Field of Concurrent Engineering 2001 and Collaborative Process Planning Model Applied to High Performance Machining of Hardened Steels Parts 2009) investigated process planning at different levels, paying special attention to machining process planning. In 2004, Abellan-Nebot joined the research team to work on aspects more related to intelligent machining systems and computer-aided manufacturing. Our research team has been always working on design and manufacturing practices, teaching topics such as integrated manufacturing, design for manufacturing, geometric product specifications, computer-aided manufacturing, manufacturing process planning and manufacturing process and materials selection at the Universitat Jaume I (Castellón, Spain), Universitat Politècnica de València (Valencia, Spain), Instituto Tecnológico y de Estudios Superiores de Monterrey (Monterrey, Mexico), and The University of North Texas (Texas, United States).

This book is mainly based on the material and notes prepared since the end of the 1990s for the course Manufacturing Process Planning, a graduate course for industrial engineering at the Universitat Jaume I and the Universitat Politècnica de València, which was later updated for a fourth-year course in mechanical engineering. Part of the material was first prepared by Professor Fernando Romero-Subirón, Pedro Rosado-Castellano, PhD, and Gracia M. Bruscas-Bellido, PhD, who also contributed to reviewing and providing useful feedback to improve the current manuscript. All this previous material was completed during the last eight years of teaching Manufacturing Process Planning by José V. Abellán-Nebot, PhD. During that period, active learning activities were prepared to motivate and engage the students in activities related to manufacturing process planning. A manufacturing process planning project was successfully applied and new material was prepared to cover CNC activities, inspection planning, computer sessions, and project activities. The positive performance and feedback from students were the spark that motivated us to prepare and share this book. With this book, the authors want to share the material on manufacturing process planning together with computer sessions and project activities with other engineers, students, and instructors around the world.

The result of the book is a self-contained manual that presents all the steps required in manufacturing process planning, starting with the identification of key aspects from a technical drawing, moving through manufacturing process selection, detailed process planning for casting, machining, and inspection, and finally, costing and pricing the product.

For the instructors, the book is organized to let them apply a project-based learning approach. Given a technical drawing and minimum laboratory equipment, the manufacturing process plan can be done by applying the contents learned in each chapter. For this purpose, the book provides as supplementary documentation with five computer sessions and one project description to let the instructors apply active learning practices in class.

The book is primarily intended to be useful to designers, mechanical and production engineers, and engineering students in general interested in learning the basics of process planning from a practical point of view.

The authors and publisher do not assume any responsibility with respect to the use of the information contained in this book.

April 2024 *José V. Abellán-Nebot, Carlos Vila Pastor, and Héctor R. Siller*

About the Companion Website

This book is accompanied by a companion website.

www.wiley.com/go/Manufacturing_Process_Planning_1e

This website includes:

1) Computer sessions 1-5
2) **Process Planning Project**

Part I

Fundamentals of Process Planning

1

Introduction to Process Planning

1.1 Introduction

Manufacturing process planning involves figuring out the best way to make a product or component, from choosing the right materials and methods to determining the most efficient sequence of steps. The main goal of manufacturing process planning is to optimize the production process, making it cost-effective while ensuring high-quality results. In small manufacturing companies, process planning is still conducted by manual methods that rely on experienced process planners who apply their knowledge in design and manufacturing to produce efficient and cost-effective process plans. More advanced manufacturing companies may utilize computer-aided process-planning (CAPP) systems to help in process-planning activities, saving money, and making more efficient process plans.

In this chapter, an overview of the manufacturing process-planning activities is presented, and the importance of these activities and the communication between design and manufacturing departments is highlighted to ensure competitive products in shorter lead times. The chapter briefly describes CAPP methods, whereas the manual methods are just schematically presented since this method is covered in detail throughout this book.

1.2 Process Planning Within Product Development Cycle

Process planning is the deliberate selection of the techniques needed to make a product affordable and competitive. Procedures, machine tools, and other equipment must be devised, chosen, and determined to transform raw materials into finished and assembled goods.

Process planning may be also considered as a bridge between product design and manufacturing, as illustrated in Figure 1.1. The initial steps of process planning coincide with the product design phase, involving decisions such as material selection and manufacturing methods such as casting, forging, or die casting. The formal endpoint for product design is marked by the release of all product data, usually in the form of three-dimensional (3D) models and engineering drawings, which specify the precise product details. At this juncture, process planning takes charge, outlining the comprehensive manufacturing plan for each component of the product.

Manufacturing Process Planning: A Practical Approach for Mechanical Engineering, First Edition.
José V. Abellán-Nebot, Carlos Vila Pastor, and Héctor R. Siller.

Companion website: www.wiley.com/go/Manufacturing_Process_Planning_1e

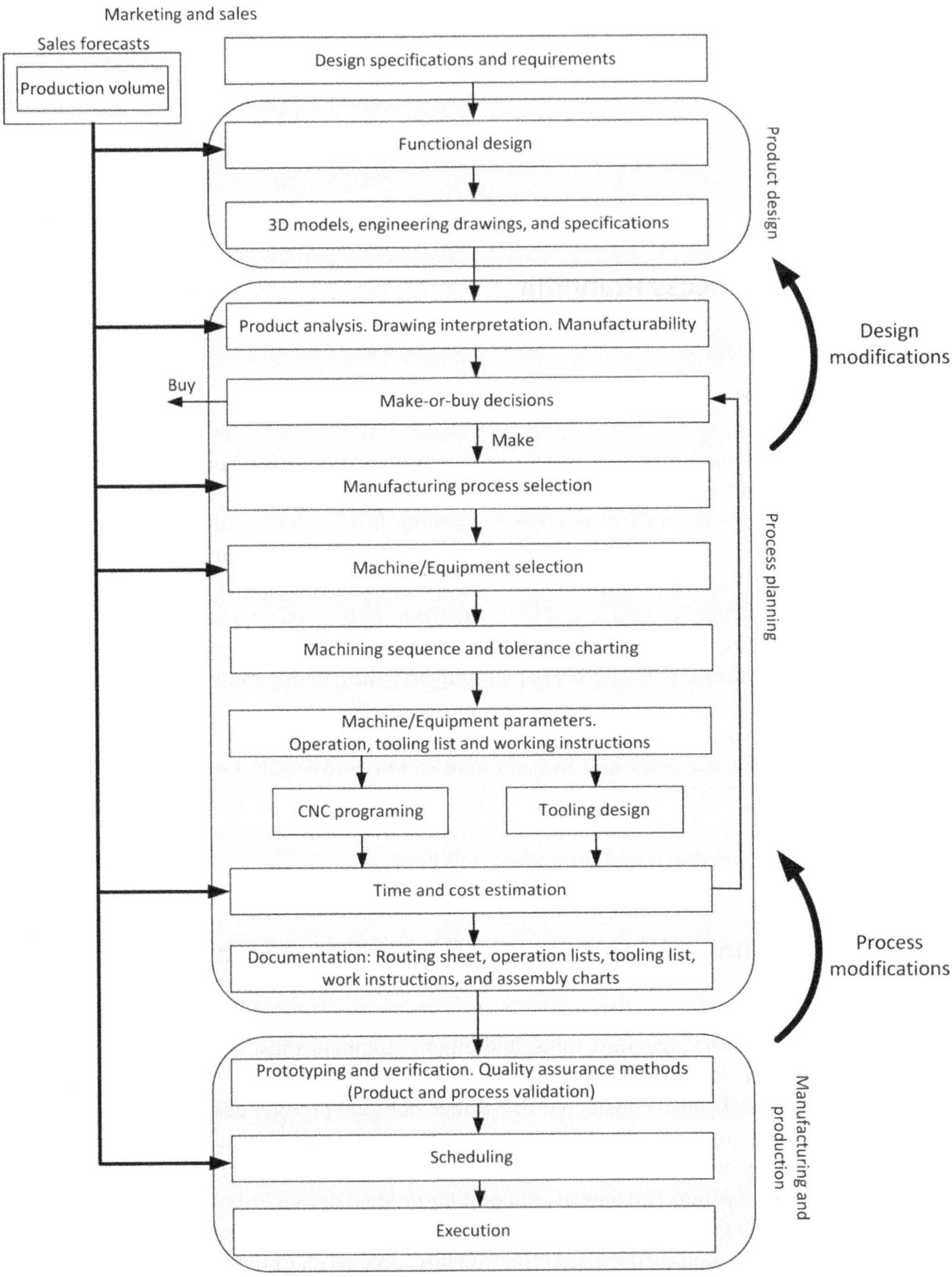

Figure 1.1 Traditional process planning deployment where machining operations are involved.

Process planning derives its input from 3D models or engineering drawings that detail what needs to be produced and in what quantity. These drawings are carefully scrutinized to determine the project's overall scope. For complex assembled products, considerable effort may be invested in breaking down the product into its constituent parts and subassemblies. Initial decisions regarding subassembly groupings, such as whether to manufacture certain parts or purchase them, as well as determining the general tooling requirements, may also be made. Subsequently, a detailed routing plan is developed for each part, requiring technical expertise in processes, machinery, and production economics. In essence, process planning translates the engineering drawing of a component into a group of manufacturing process steps, leading to the creation of a route sheet, which outlines the sequence of operations necessary for the component. The term "route sheet" is used because it also specifies the machines through which the part must pass to complete the sequence of operations (Kesavan 2009).

In the context of product design and manufacturing on a global scale, three key functions are involved: marketing and sales, design, and manufacturing. These functions traditionally follow a sequential process. Marketing assesses the contemporary market trends and requirements, creating new product concepts based on its findings. It also formulates specifications for the further enhancement of existing products, guided by market assessments. The design function utilizes these product concepts and specifications from marketing, providing a comprehensive overview of all parts and subassemblies required for the final product. This includes detailed 3D models, drawings, assembly plans, and bills of materials. The product requirements determined during the design phase are then assigned to the process-planning team. This information serves as the basis for creating detailed work instructions needed for production, defining, in case the part should be manufactured in-house, the manufacturing processes, the machine/equipment needed, tooling requirements, etc. These instructions are transferred to the manufacturing facility for production, as depicted in Figure 1.1.

Consequently, even though the design and manufacturing departments operate independently, the process-planning activity serves as a connection between them, following a sequential approach. This sequential approach, traditionally utilized in product design and manufacturing, assumes a lengthy lifespan for the product with a consistent demand over an extended period. However, in current competitive global markets, this assumption often does not hold true. As a result, manufacturing organizations have explored ways to reduce the time required for product design and manufacturing, commonly known as time to market. Some organizations have prioritized organizational changes with the intention of forcing all the stakeholders to be involved in the whole product development cycle and fostering cross-functional collaboration to expedite the design and manufacturing processes of their products. This method, based on the collaboration of cross-functional teams, is commonly known as concurrent engineering. However, it can be found with other terms such as simultaneous engineering or integrated product development.

Unlike a sequential approach, commonly referred to as "over-the-wall engineering" since no bidirectional communication exists to ensure product performance and manufacturability, concurrent engineering can be seen as a cooperative approach where members from all stages are in continuous communication to improve product design and facilitate the subsequent stages of production (Figure 1.2). Some activities that may be found under a concurrent engineering approach are listed in the following:

- Collaborate in the definition of product design specifications.
- Choose the best design and manufacturing techniques to employ.
- Propose modifications of geometrical specifications or relaxation of tolerances.
- Evaluate the design's manufacturability and modify it accordingly.
- Evaluate the assembly of the design to detect potential assembly issues and improvements.

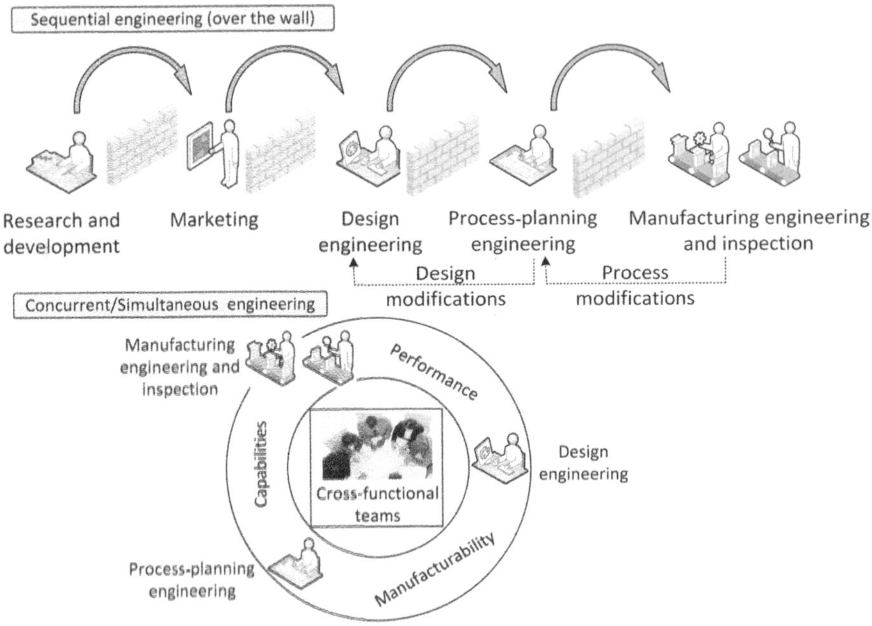

Figure 1.2 Sequential engineering versus concurrent engineering.

Besides reducing time to market, the main benefits of applying concurrent engineering are reducing cost and increasing product quality and productivity. However, managing concurrent engineering is challenging, and it requires the commitment of all stakeholders, and very often some of them are reticent to multidisciplinary collaboration.

Regarding product and production costs, it is important to note that approximately 70% of the total cost of a product is determined during the design phase (Ullman 1992). However, some studies reduce this impact to 47% on average (Ulrich and Pearson 1993). This cost allocation is primarily influenced by material selection and the associated manufacturing methods. The remaining 30% is attributed to decisions made regarding the manufacturing process, such as the utilization of production equipment and tooling. Additionally, it is widely recognized that any modifications made to the product during later stages have a greater impact on production costs compared to earlier stages, and the time required to implement such changes is longer. Figure 1.3 shows the impact of product design on product cost and production cost in common manufacturing companies.

Due to these reasons, concurrent engineering should be considered a widespread practice within manufacturing companies to minimize lead times and decrease the overall costs associated with the final product.

1.3 Process-Planning Levels of Detail and Activities

As stated earlier, process planning seeks to completely define the manufacturing processes, tools, process parameters, inspection methods, etc., required to manufacture a product according to the 3D models and technical drawings from the design office. Process planning can be conducted at different levels of detail defining four different types of process planning (ElMaraghy and Nassehi 2014): generic planning, macroplanning, detailed planning, and microplanning.

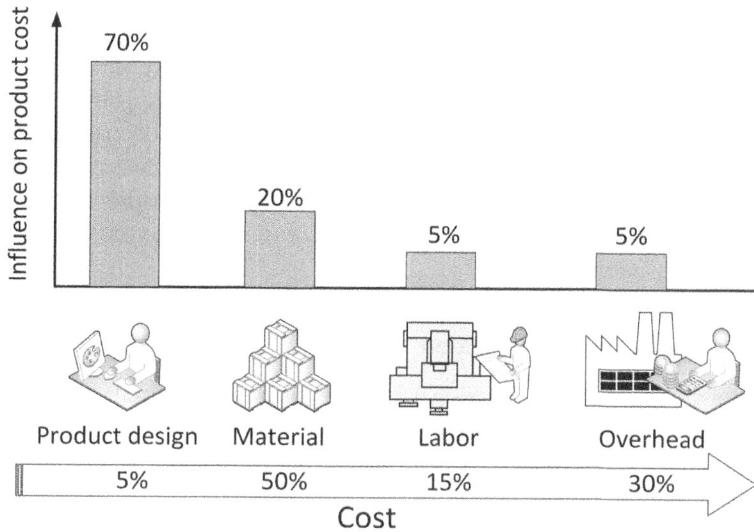

Figure 1.3 Impact of product design on product cost and production cost in common manufacturing companies. *Source:* Adapted from Munro and Associates (1989).

Generic planning refers to rough process planning where only manufacturing process selection is conducted according to material, production volume, process capabilities, and cost. Macroplanning details the equipment selection related to manufacturing, assembly, and inspection, and defines the routing sheets, a document that shows the route of each part with the stations and equipment involved at each step. Detailed planning refers to specific information at each station about the sequence of operations, tooling, fixtures, etc. Finally, microplanning describes process and operation parameters, NC programs, inspection procedures (conformity zones, sampling strategy, etc.), working instructions, calculation of overall times and costs, etc., which ends with the documentation of operation sheets. Although a first analysis for a new product development starts with generic planning, a final microplanning will be required before production.

Many different process-planning activities are required depending on the level of detail needed. For a complete process planning before production, the following activities are commonly conducted:

1.3.1 Interpretation of Product Design Data and Manufacturability

The initial stage in developing a process plan for any component involves referring to the product design data, mainly in the form of 3D models and engineering drawings. Interpreting the product data involves analyzing the part's geometry, dimensions, and correlated dimensional and geometric tolerances, as well as surface finish conditions and material specifications. Production information about lead time and production volume is also gathered.

A key analysis to be conducted at this step is a manufacturability analysis. Process planners should evaluate and propose design modifications that may reduce manufacturing costs and time. Practices related to design for manufacturing (DFM) and design for assembly (DFA) are applied for this purpose, and modification or relaxation of geometrical specifications and tolerances are also studied and shared with the design team.

1.3.2 Bill of Materials Analysis and Make-or-Buy Decision

Given the final 3D models and technical drawings, all components of the product are analyzed to decide which parts are going to be purchased from the market and which ones are going to be manufactured in-house. Typically, if the part can be produced in-house, a cost comparison between making versus buying is done. Assumptions are often required regarding factors such as expected capacity, demand patterns, waste, inventory levels, and fluctuations in material availability and affordability. After the make-or-purchase decisions are made, the bill of materials for the product is defined, including the make-or-buy nature.

1.3.3 Manufacturing Process Evaluation and Selection

After analyzing the technical drawings and deciding which parts are being manufactured in-house, process planners can start to evaluate which manufacturing processes are more appropriate for producing the parts.

At this stage, the part material highly determines the manufacturing process selection. Thus, the process planner may engage in meaningful discussions with design engineers to reconsider the material selection. For example, some alloys are well-suitable for casting or machining processes but are less adequate for forming processes. Note that these discussions are typically conducted at the design stage when cross-functional teams work together.

Manufacturing process selection requires identifying both primary processes (e.g., casting, forging, rolling, extrusion, drawing, and bending) and secondary processes (e.g., machining, surface finishing, and joining/assembly processes) that are required for reaching part specifications under the constraints of production volume and production rate identified. For example, every manufacturing process and equipment possesses inherent capabilities when it comes to achieving a certain surface quality. Therefore, any process or machinery that cannot meet the specified surface quality requirements would either be eliminated beforehand in the selection process, or additional finishing operations would need to be incorporated.

Therefore, after analyzing the technical drawings, materials, and production requirements, a list of potential processes can be compiled. Those candidates that can be applied to fulfill part and production requirements are then compared in terms of cost to ultimately identify the best manufacturing processes. At this step, early cost estimation tools are applied to assist in the final manufacturing process selection. After the selection of appropriate manufacturing processes, the raw material size for producing the part, including its size and any specific requirements such as heat treatment, are also defined.

1.3.4 Selection of Machines and Equipment

The next stage, once the process planner has decided which manufacturing methods to use, is to choose the right equipment, taking availability, process capability, production range, and cost into account. Factors such as the size and weight of the components being produced, the physical dimensions of the machinery, tooling availability, setup and production costs, and any limitations in terms of power and force are relevant aspects of the final equipment selection.

1.3.5 Machining Sequence and Tolerance Charting

After selecting manufacturing processes, machines, and equipment, the process sequence is defined. When secondary machining operations are needed, the machining sequence should be analyzed in detail to avoid excessive manufacturing tolerance accumulation that may produce parts out of specifications. Machining sequence diagrams and tolerance charts facilitate the definition of proper machining sequences to ensure part specifications.

1.3.6 Selection of Operation and Process Parameters

After selecting the machines and the sequence of operations to be conducted, it becomes necessary to define specific parameters for each operation on each machine. When dealing with casting or plastic deformation machinery, temperatures, cycle time, speeds, etc., are critical parameters. In the case of machining equipment, cutting tools should be selected depending on the geometry to be machined, the material of the workpiece, and the quality and capability requirements. Additionally, cutting parameters should be defined together with cutting strategies and cutting tool paths. Machining handbooks and cutting data recommendations from tool manufacturers are used for this purpose. It is possible to calculate the amount of nonvalue-added time (such as handling and setup) and operation time needed for each operation based on the process parameters. This facilitates the compilation of the total time needed for the job for a later cost estimation.

1.3.7 Tooling Design

Tools, dies/molds, and workholding systems are also defined for each processing step, and if specific fixtures are needed, a fixture design activity is conducted. For instance, in casting, patterns and molds must be designed. In the case of machining operations, the process planner analyzes the workpiece and the final part to determine the appropriate workholding devices. If a fixture or jig is required for a specific operation and the location and clamping conditions have been identified, this information is typically passed on to a specialized fixturing developer. Based on these requirements, the fixturing specialist generates detailed drawings and a general arrangement drawing of the workholding device to be designed and fabricated. In many cases, this may involve modifying an existing design or utilizing standardized components (Scallan 2003).

1.3.8 Time and Cost Estimation

Estimating cycle times and related expenses for the process plan is another responsibility of the process planners. This involves forecasting manufacturing costs and lead times associated with creating a component or product based on historical data or standard data from handbooks. Unlike early cost estimates at the manufacturing process selection step, a more detailed estimation is conducted now since data about the specific equipment used, the process and operation parameters, and inspection procedures have already been defined. At this step, an accurate manufacturing cost estimation is crucial for determining the unit cost and assessing the product's success. This estimation may result in changes in batch size considerations or whether to produce or purchase a particular product or component that was not clearly identified at the first steps of process planning.

1.3.9 Documentation and Communication

Documents are created and shared with the design and manufacturing departments as a part of the different preceding activities conducted. Basically, process-planning documentation presents the methods, sequence of tasks, parameters, time, and workforce required for manufacturing a product and they should clearly identify the version of the document, date, and the person in charge.

At the generic and macroplanning level, assembly charts, operation charts, and routing sheets are defined. The assembly and operation process charts graphically show the sequence of operations (assembly, manufacturing, or inspection operations) that are necessary to produce a product and it shows the equipment and time required at each operation. In a similar vein, routing sheets provide a tabular overview of the raw materials' journey through the manufacturing process. A list of the production tools and equipment that will be used is usually included. Routing sheets are typically used to accompany materials as they move around the production shop floor, giving operators pertinent information at every turn.

At detailed and microplanning levels, operation lists, tooling lists, work instruction sheets, and inspection sheets are defined. The operation list shows a detailed plan for every operation defined in the routing sheet, including information about operation and process parameters such as feeds and speeds in case of machine tools, setup procedures, and operation times. The tooling list serves the purpose of providing a comprehensive specification for all necessary tooling across various operations. For specific activities, work instruction sheets may be provided to offer supplementary information that is not covered in the operation or process sheets, providing additional details for the tasks at hand. Finally, inspection sheets and quality reports present comprehensive instructions and procedures for examining and measuring crucial quality attributes. If there is a need to collect and record specific quality data for statistical process control (SPC) analysis, an inspection or quality report is required.

1.3.10 Selection of Quality Assurance Methods

Process planners should determine the quality assurance methods at the product and process validation stage and at the production stage in collaboration with quality engineers. Inspection stages or in-process inspection measurements, sampling plans, the instruments required, and the conformity zones are defined for all critical product specifications. It is crucial to achieve a compromise between guaranteeing product quality and avoiding excessive inspection tasks, which would extend the processing time and increase the total manufacturing cost (Scallan 2003).

1.4 Technical Data Required for Process Planning

Process planning requires dealing with reliable information about product design, manufacturing capabilities, equipment capacities, manufacturing process parameters, etc. In general, the following sources of information should be available during process planning (Adithan 2007):

- Process plans of selected universal parts or other process plans from previous products that could serve as a good starting point.
- Product data and engineering drawings pertaining to each product component. For example, component drawings, product specifications, requirements, and a bill of materials outlining the quantity of each component used in the product.
- Design data handbooks (charts of limits, fits and tolerances, material properties, etc.).
- Manufacturing process capabilities and material–process compatibility matrices for process selection.
- Machinery data handbooks (tables of process parameters for different manufacturing conditions and different work materials).
- Cutting data from vendors of machining tools and inserts.
- Specifications of machines available on the shop floor (e.g., for machine tools: travel, spindle and motor power ratings, working dimensions, etc.).

- Machine hourly rate for all equipment available in the shop.
- Tables of standard costs related to manufacturing processes, materials, operators, etc.
- Tables of defining inspection sampling plans.

1.5 Process-Planning Methods

Process planning can be classified into two primary methods: manual process planning and computer assisted planning (CAPP). Expert engineers carry out manual process planning, taking their time to study and develop a workable plan based on the shop floor's existing tools and equipment. CAPP approaches can be grouped into variant CAPP or generative CAPP, and both seek to reduce or avoid relying on expert process planners, saving time and cost along process-planning activities. While manual process planning is most suitable for small shops that cannot afford the cost and implementation of CAPP systems, large companies with a wide variety of product geometries to be manufactured in-house can take advantage of using these computer-aided tools.

1.5.1 Manual Planning Methods

In manual process planning, all the activities shown in Figure 1.1 are conducted by experienced process planners. The most relevant tasks are listed in the following:

1) The process planner chooses the best manufacturing processes for producing the component by looking at the technical drawing, applying their manufacturing experience, and being aware of the available resources. They also choose the size and shape of the material to be utilized, frequently based on designs that have already been created for related elements.
2) The process planner determines the manufacturing parameters for each operation conducted at each manufacturing process. For machine tools operations, it is determined the tools, feeds, and speeds for a particular workpiece material. To ascertain the scheduled times for each handling and machining step involved, they also refer to manuals and standard data. The setup time and time per unit quantity for each activity are then determined using these time estimates.
3) All the relevant information from step 1 is documented in the routing sheet. This sheet provides a comprehensive list of all the operations. For each operation, an operations list is defined by specifying all the details mentioned in step 2.

The manual process-planning method is usually based on preset workflows for workpiece types that are often manufactured on the shop floor. After analyzing the drawing and identifying the required production processes, the process planner can choose and appropriately modify the corresponding preset sequence of activities.

1.5.2 Computer-Aided Process Planning

Process planning is still mostly a knowledge-intensive manual operation carried out by skilled engineers, but there have been numerous attempts to use computer tools to automate process-planning tasks. The primary application of CAPP systems is found in machining processes, where the most crucial tasks that significantly influence production cost, total time, and manufacturability are setup planning and sequencing of operations.

The CAPP systems let process planners generate manufacturing process plans with the aid of advanced computer software. CAPP systems are known to be more efficient than manual approaches and their main advantages are as follows:

- Automation and efficiency. Manufacturing plan generation time and effort can be greatly decreased by using CAPP technologies to automate process-planning operations. Moreover, the process planner's expertise and experience are less important to the process-planning process.
- Optimization and cost reduction. The software can perform optimization algorithms to identify process plans that minimize production time, reduce costs, or optimize other specified objectives. It considers various factors (e.g., time, cost, or quality) and constraints to find the best possible solution.
- Flexibility and adaptability. The software can easily adapt the process plans to changes in product designs or manufacturing requirements. It can quickly generate new process plans based on updated inputs, allowing manufacturers to respond to changes efficiently.

There are two principal approaches when dealing with CAPP systems: the variant approach and the generative approach.

1.5.2.1 Variant Approach

In CAPP systems, the variant approach, also known as the retrieval method, refers to CAPP systems that obtain a standard plan from a database and adapt it to fit a specific part. Typically, a coding system and a group technology (GT) classification are used to organize the parts based on their geometrical and manufacturing characteristics. This coding scheme uses a string of alphanumeric characters to stand for particular attributes shared by all parts or products in the part family.

The Opitz coding system (13 digits), MICLASS (12 digits), KK3 system (21 digits), and DCLASS (8 digits) are examples of typical GT coding systems. One of the most widely used GT coding schemes in manufacturing companies is the Opitz system, which consists of 13 numbers categorized into three groups. First to fifth digit codes are named form codes, and they are used to describe the geometrical attributes of the part. The sixth to ninth digit codes are supplementary codes that provide relevant information to manufacturing such as raw material shape, material, etc. The latter group of digits, tenth to thirteenth digits, are secondary codes that give information about production processes and production sequences. Figures 1.4 and 1.5 show the possible values and the corresponding explanation for the form and supplementary codes. As it can be observed, increasing the value of the digit in the form codes means a more complex part. If a component has several features in the same code digit, the value of the feature with higher complexity is chosen. For the secondary codes, they can be defined specifically according to the company's needs. An example of using the Opitz coding system is shown in Figure 1.6.

The variant process planning has two operational stages, a preparatory stage and a production stage (Chryssolouris 2013). The preparatory stage involves the following tasks (Figure 1.7):

1) Part classification. The first step is to apply GT to classify parts into groups based on their similarities in terms of shape, size, function, and manufacturing requirements.
2) Group formation. Once parts are classified into families and their process plans have been added to the system, the next step is to form groups based on the similarity of manufacturing processes required to produce those parts. Parts within the same family often share similar machining operations, tooling requirements, and production methods. By forming groups, CAPP can leverage the similarities to design standardized process plans for each group.

Digit 1		
Part Class		
0	Rotational parts	L/D < 0.5
1	Rotational parts	0.5 < L/D < 3
2	Rotational parts	L/D > 3
3	Rotational parts	With deviation L/D ≤ 2
4	Rotational parts	With deviation L/D > 2
5	Rotational parts	Special
6	Non-rotational parts	A/B ≤ 3 A/C ≥ 4
7	Non-rotational parts	A/B > 3
8	Non-rotational parts	A/B ≤ 3 A/C < 4
9	Non-rotational parts	Special

Digit 2			
External shape, external shape elements			
0	Smooth, no shape elements		
1	Stepped to one end		No Shape elements
2	Stepped to one end	or smoth	Thread
3	Stepped to one end	or smoth	Functional Groove
4	Stepped to both ends		No Shape elements
5	Stepped to both ends		Thread
6	Stepped to both ends		Functional Groove
7	Functional cone		
8	Operating thread		
9	All others		

Digit 3		
Internal shape, internal shape elements		
0	No hole, no breakthrough	
1	Smooth or stepped to one end	No Shape elements
2	Smooth or stepped to one end	Thread
3	Smooth or stepped to one end	Functional Groove
4	Stepped to both ends	No Shape elements
5	Stepped to both ends	Thread
6	Stepped to both ends	Functional Groove
7	Functional cone	
8	Operating thread	
9	All others	

Digit 4	
Plane-Surface Machining	
0	No Surface machining
1	Surface Plane and/or curved in one direction, external
2	External Plane Surface related by Graduation around a circle
3	External Groove and/or slot
4	External Spline (polygon)
5	Internal Plane/Surface and/or slot, external spline
6	Internal Plane/Surface and/or slot
7	Internal Spline (polygon)
8	Internal and External polygon, groove and/or slot
9	All others

Digit 5		
Auxiliary Holes and Gear teeth		
0	No gear Teeth	No auxiliary hole
1	No gear Teeth	Axial, not on pitch circle diameter
2	No gear Teeth	Axial, on pitch circle diameter
3	No gear Teeth	Radial, not on pitch circle diameter
4	No gear Teeth	Axial and/or radial and/or other direction
5	No gear Teeth	Axial and/or radial on PCD and/or other direction
6	With gear Teeth	Spur Gear Teeth
7	With gear Teeth	Bevel Gear Teeth
8	With gear Teeth	Other Gear Teeth
9	With gear Teeth	All others

Figure 1.4 Explanation of the form codes (digits 1–5) in the Opitz coding system. *Source:* Opitz (1970)/ with permission of Elsevier.

Digit 6		Digit 7		Digit 8		Digit 9	
Dimension D (rotational part) or part length A (nonrotational) (mm)		Material		Initial form		Accuracy in coded digits	
0	≤20	0	Cast iron	0	Round bar, black	0	No accuracy higher than IT7
1	>20 ≤50	1	Modullar cast iron Malleable cast iron	1	Round bar, bright drawn	1	2
2	>50 ≤100	2	Mild steel (not heat treated)	2	Bar: triangular, square, and hexagonal	2	3
3	>100 ≤160	3	Hard steel	3	Tubing	3	4
4	>160 ≤250	4	Mild and hard steel (heat treated)	4	Angle, U-, T-, and similar sections	4	5
5	>250 ≤400	5	Alloy steel	5	Sheet	5	2 and 3
6	>400 ≤600	6	Alloy steel heat treated	6	Plate and slab	6	2 and 4
7	>600 ≤1,000	7	Nonferrous metal	7	Cast and forge component	7	2 and 5
8	>1,000 ≤2,000	8	Light alloy	8	Welded assembly	8	3 and 4
9	>2,000	9	Other materials	9	Premachined components	9	2, 3, 4 and 5

Figure 1.5 Explanation of the supplementary codes in the Opitz coding system. Digit 9 identifies which form codes (digits 1–5) require higher accuracy than IT7. *Source:* Opitz (1970)/with permission of Elsevier.

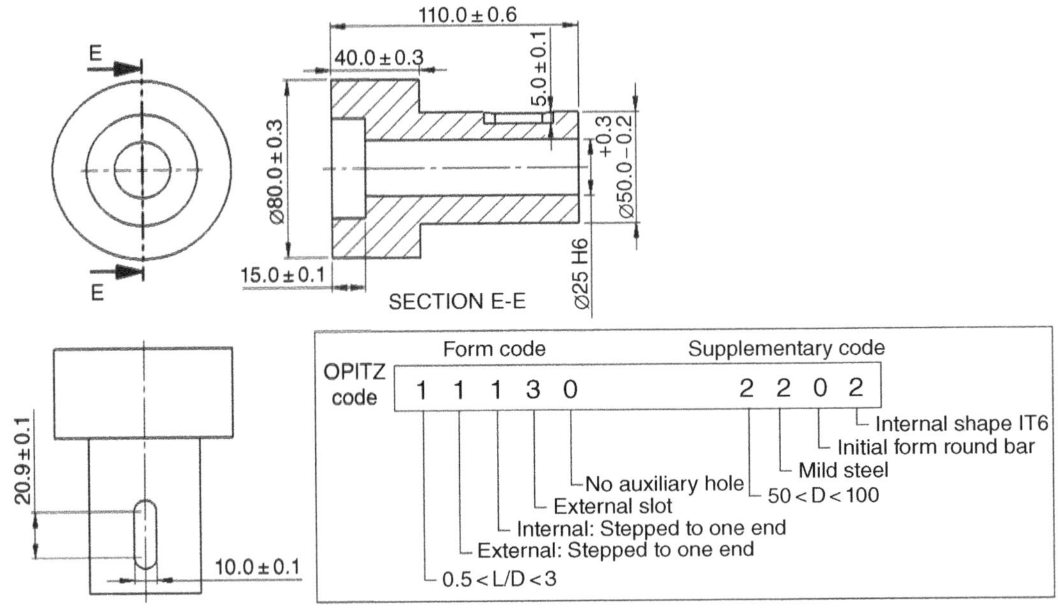

Note: Part made of mild steel from a round bar from rolling.

Figure 1.6 Example of application of the Opitz coding system.

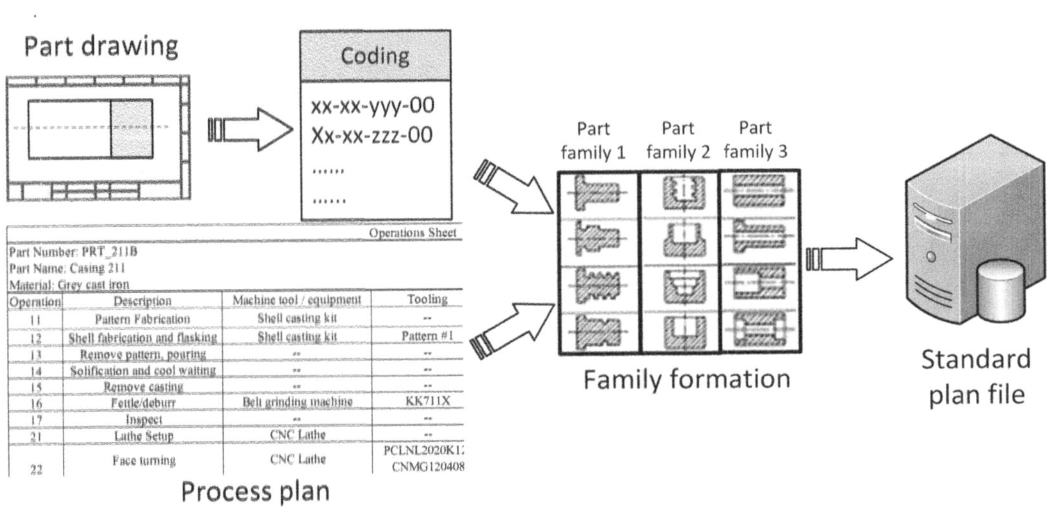

Figure 1.7 Preparatory stage of CAPP variant approach. *Source:* Adapted from Chryssolouris (2013).

3) Creation of process-planning templates. For each part group, process-planning templates or standard process plans are created based on previous process plans. These templates define the sequence of manufacturing operations, tooling requirements, machining parameters, inspection points, and any other relevant information needed to produce the parts in that group. The templates serve as a guide for generating specific process plans for individual parts within the group.

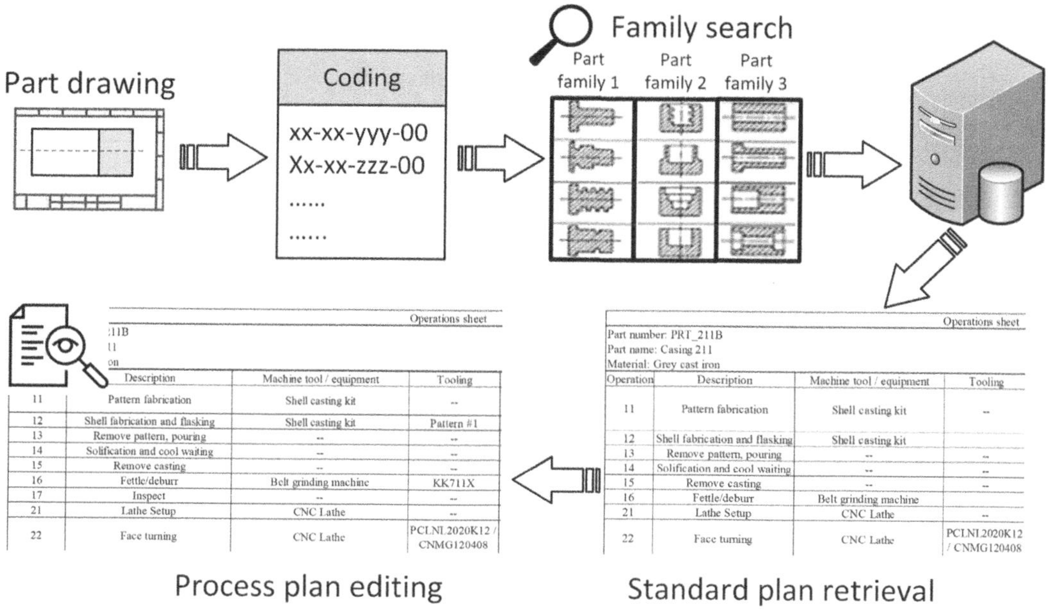

Figure 1.8 Production stage of CAPP variant approach. *Source:* Adapted from Chryssolouris (2013).

The second stage is named the production stage and it involves coding and classifying new components so the CAPP system identifies the part's family and selects the corresponding process-planning template. The system then customizes the template to the specific requirements of the individual part, considering any unique features or tolerances. This customization results in a variant process plan that is specific to the particular part while still benefiting from the standardized processes of its group (Figure 1.8).

The primary benefits of a CAPP based on the variant approach are that the process planner has control over the final process plan and that they are comparatively easier to use and comprehend than those based on the generative approach. They may be a very useful tool for helping to organize and support process-planning efforts. Variant CAPP systems, however, are limited to planning components that are very similar to those that have already been planned. Process planners still need to examine and modify the standard plan for the component.

1.5.2.2 Generative Approach

The foundation of this CAPP approach is starting from scratch with each component and developing a new process plan. Because of this, this system does not make use of items that have been already manufactured; instead, it combines product and process data to construct a process plan automatically for a new component. These plans are produced by computer software using formulas, algorithms, decision logic, and geometric analysis. CAPP systems use automatic feature identification algorithms to classify features and propose the best manufacturing operations. For instance, in the machining process planning different setups and machining sequences may be defined according to the features identified, their tool access directions, and the datum surfaces available. In accordance with machining practices, the CAPP system defines machining sequencing based on knowledge-focused reasoning rules and heuristics.

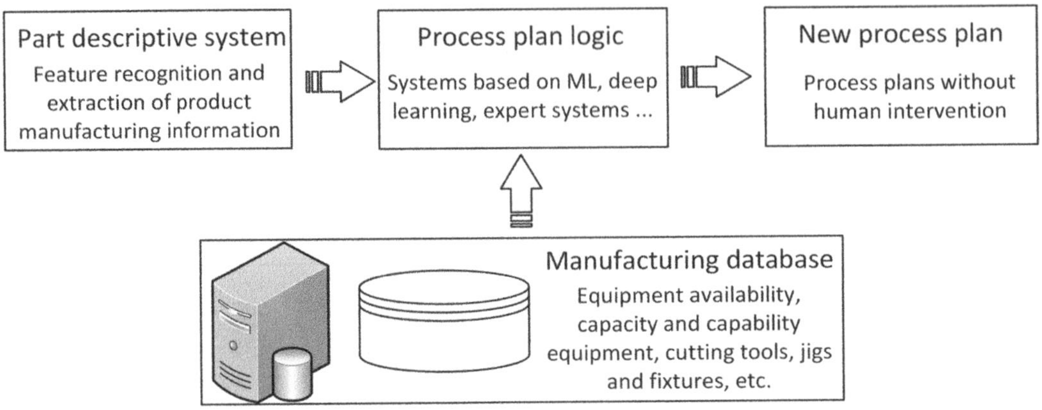

Figure 1.9 CAPP generative approach.

The major components that define a generative CAPP software are shown in Figure 1.9 and are defined as (Scallan 2003):

- A part-based system that defines the material and geometry, including geometric and dimensional tolerances and surface quality specifications. This system interprets the product manufacturing information (PMI) annotations from CAD models and conducts feature recognition processes to completely describe each component of the product.
- A database of information on the manufacturing machinery accessible and their capacities and capabilities.
- A subsystem that detects and sequences the operations and the required process parameters based on the part description and available equipment. This is the core of the generative CAPP system, and it relies on advanced algorithms based on expert systems or other techniques.
- A report generator to create the process plan.

Generative process planning often incorporates knowledge and rules from domain experts, which are encoded into the software. This CAPP approach applies advanced computer methods to automate and optimize the process-planning tasks. Theoretically, human intervention is not necessary for generative CAPP solutions. However, this is rarely the case because most systems need some kind of assistance. In fact, most CAPP works are not easily scalable to other processes, according to recent surveys, and no completely generative CAPP systems have been implemented yet (Al-wswasi et al. 2018).

The specific computer methods used in generative CAPP software solutions can vary and it depends on the complexity of the manufacturing processes, available data, desired optimization objectives, and the specific requirements of the industry. Some of these advanced computer methods are discussed as follows:

- Expert systems. Expert systems utilize rule-based reasoning to capture the knowledge and expertise of human process-planning experts. These systems encode domain-specific rules, best practices, and heuristics into an inference engine, which can generate process plans based on input parameters and constraints.
- Machine learning (ML). ML algorithms are commonly used in generative CAPP to analyze historical process-planning data and learn patterns, relationships, and rules from the data. This learned knowledge can then be used to generate new process plans based on similar input data.

- Deep learning approaches. Neural networks, particularly deep learning architectures, can be used in generative CAPP to analyze and interpret complex data, such as product designs or manufacturing parameters. Neural networks can learn from large datasets and generate process plans based on the learned patterns and relationships. They can be employed for tasks such as feature recognition, process optimization, or decision-making in generative CAPP.
- Constraint programming. Constraint programming is a technique that models and solves problems by defining constraints and finding solutions that satisfy those constraints. In generative CAPP, constraint programming can define and enforce various manufacturing constraints, such as material availability, tooling constraints, machine capabilities, and process sequencing requirements. It helps generate feasible process plans under different types of constraints.

Questions and Exercises

1 What is process planning?

2 What is the meaning of "concurrent engineering" and why is important in process planning?

3 Explain the impact of product design on production cost and the final product cost.

4 Enumerate the process-planning activities commonly carried out in the manufacturing industry.

5 Describe the different levels of process planning according to the details of the process planning.

6 What is the difference between the routing sheet and the operation list sheet?

7 Which are the most common process-planning methods? Describe them briefly.

8 Explain the main CAPP methods in industry.

9 Explain the main advantages of applying CAPP methods in industry.

References

Adithan, M. (2007). *Process Planning and Cost Estimation*. New Age International (P) Limited, Publishers.

Al-wswasi, M., Ivanov, A., and Makatsoris, H. (2018). A survey on smart automated computer-aided process planning (ACAPP) techniques. *International Journal of Advanced Manufacturing Technology* 97 (1): 809–832.

Chryssolouris, G. (2013). *Manufacturing Systems: Theory and Practice*. Springer Science & Business Media.

ElMaraghy, H., and Nassehi, A. (2014). Computer-aided process planning. In: *CIRP Encyclopedia of Production Engineering* (ed. L. Laperrière and G. Reinhart). Berlin, Heidelberg: Springer https://doi.org/10.1007/978-3-642-20617-7_6551.

Kesavan, R. (2009). *Process Planning and Cost Estimation*. New Age International.

Munro and Associates (1989). Who has the biggest shadow? https://www.leandesign.com/Newsletters/munroreport002.html (accessed April 2024).

Opitz, H. (1970). *A Classification System to Describe Workpieces*, Translated by Acton Taylor. New York: Pergamon Press.

Scallan, P. (2003). *Process Planning: The Design/Manufacture Interface*. Butterworth-Heinemann.

Ullman, D.G. (1992). *The Mechanical Design Process*, vol. 2. New York: McGraw-Hill.

Ulrich, K. T., and Pearson, S. A. (1993). Does product design really determine 80% of manufacturing cost? Massachusetts Institute of Technology.

2

Product Data and Manufacturability

2.1 Introduction

When the process planning department receives all data of a product, usually in the form of three-dimensional (3D) models and their associated engineering drawings, planners must analyze and ensure that any defined specification in the data will be fulfilled during the manufacturing stage. In manufacturing companies that are hired to produce a specific product or parts, all product data is treated as a binding agreement between the manufacturer and the customer, and this is the main reason why planners should carefully check and analyze the product data before a formal agreement is signed. The drawings (a typical engineering drawing is shown in Figure 2.1), created by the design department, represent specific functional requirements that have been established based on the part's intended role within the overall product. The part is precisely defined to ensure that, when combined with the entire mechanism, it performs its technical functions as intended (Halevi 2003). In addition to general details like the component's material, designation, coding, production quantity, or weight, the drawing provides essential information about the dimensions and tolerances (both dimensional and geometric) of the part.

This chapter reviews common information that manufacturing process planners may find in product data, such as 3D models, engineering drawings, and their corresponding bill of materials (BOMs). Dimensional and geometrical specifications according to the International Organization for Standardization/American Society of Mechanical Engineers (ISO/ASME) standards are briefly explained, paying special attention to the ambiguity issue in the design intent that usually exists if these standards are not correctly applied. This issue can be a critical problem when outsourcing manufacturing components. The importance of avoiding it with the correct use of ISO/ASME standards about geometric product specifications (GPS) is shown by providing some examples. Additionally, other notations in engineering drawings, such as general tolerances and surface roughness indications, are reviewed.

Manufacturing Process Planning: A Practical Approach for Mechanical Engineering, First Edition.
José V. Abellán-Nebot, Carlos Vila Pastor, and Héctor R. Siller.

Companion website: www.wiley.com/go/Manufacturing_Process_Planning_1e

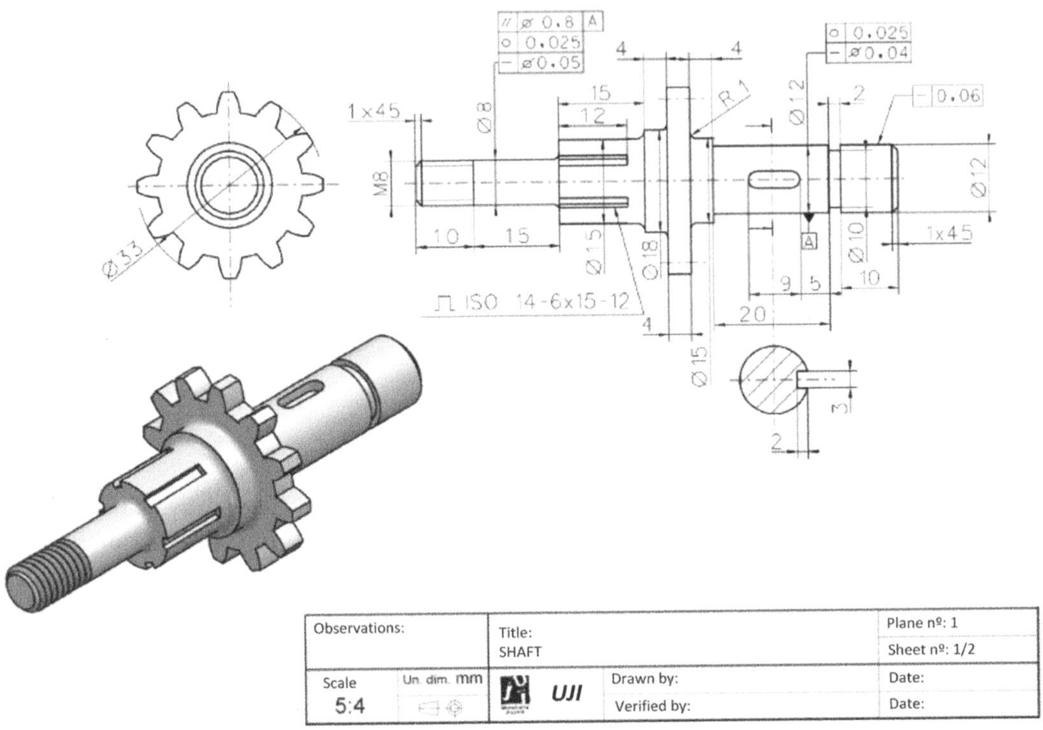

Figure 2.1 Engineering drawing of a mechanical part. *Source:* Company et al. (2007)/with permission of Universitat Jaume I.

2.2 Product Data

2.2.1 Product Structure and Bill of Materials

A product structure arranges the components that make up a product hierarchically, making it possible to depict the parts and their attributes. The materials, components, subassemblies, and other elements can be visually represented in a hierarchical format using the product structure. This depicts how the pieces are grouped together on an assembly drawing or how they converge at different points during the manufacturing process (Figure 2.2).

Engineers start the process of creating a new product by laying out a framework that lists the essential parts and systems needed to build the intended product. After that, each component is evaluated to see if a commercial component can be purchased to perform the required function or if custom parts should be designed. The engineer specifies the necessity for new pieces when appropriate existing alternatives cannot be found, which causes the product's structure to gradually expand. Product structure allows for the construction of complex, one-of-a-kind product structures as well as the development and reuse of multiple pieces and assemblies.

A graphic representation of a product called product 1 is provided in Figure 2.3. The product structure is graphically represented, and the number of items required to create each of the parent goods is indicated in parentheses. A BOM for product 1 is also included in the figure, where the total amount of each item's usage is compiled into a single list for the product.

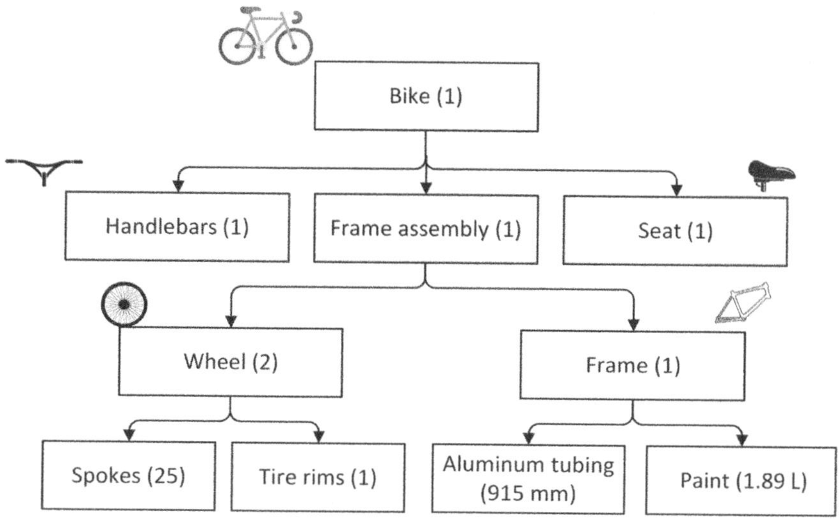

Figure 2.2 Example of a product structure.

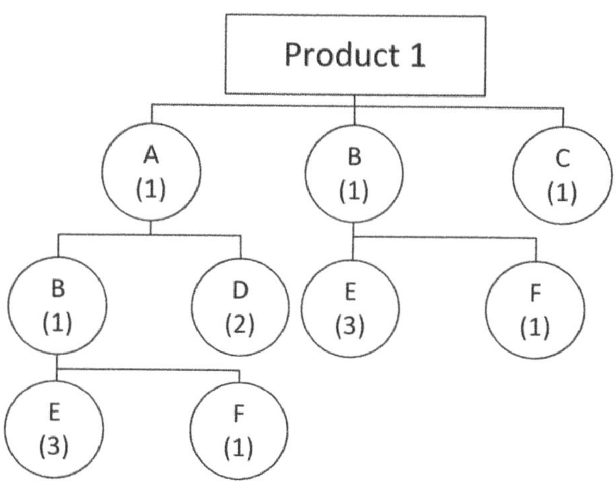

Bill of materials (BOM)

Part	Quantity
A	1
B	2
C	1
D	2
E	6
F	2

Figure 2.3 Product structure for product 1 and summarized bill of materials (BOM).

The main properties that are commonly listed in a BOM are the following (Adithan 2007):

- Name, material, and quantity of individual parts required to complete an assembly.
- If the individual components are going to be purchased or manufactured. It may be more cost-effective to buy the part rather than manufacturing it in-house, especially if the components are common commercial parts such as bolts, nuts, washers, bearings, or plastic-molded parts.
- The name, date, and signature of the engineer who prepared the BOM. This is typically found in the title block of drawings.

Figure 2.4 shows an example of a BOM as a part list in a drawing.

A product normally needs two different kinds of BOMs: an engineering BOM (eBOM, usually named directly as BOM) and a manufacturing BOM (mBOM). Items are mostly arranged in the eBOM according to how they relate to the parent product as shown in assembly drawings. In other

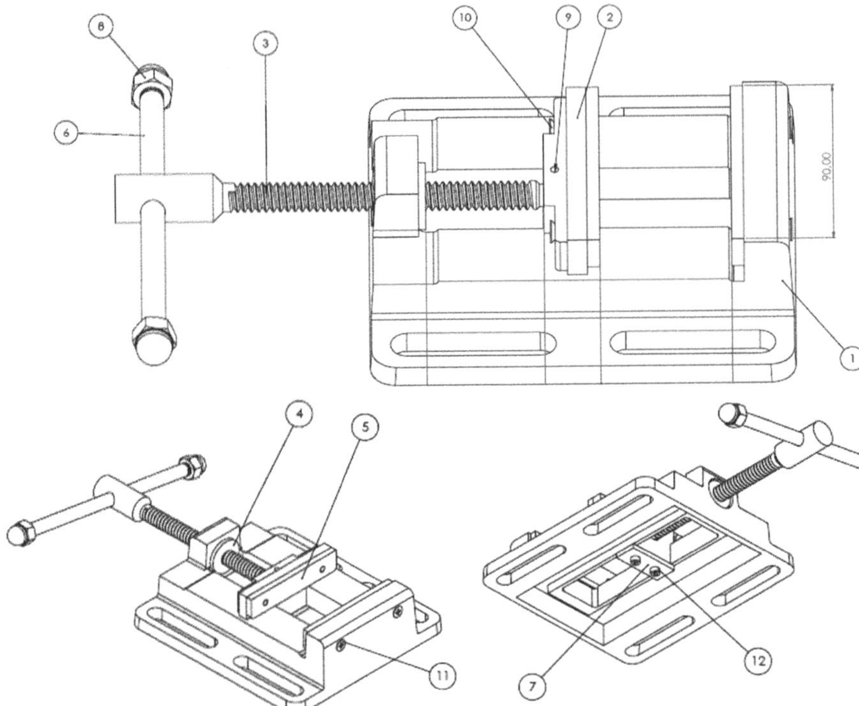

Nº	Part name	Units	Material description	Buy/Make	Nº	Part name	Units	Material description	Buy/Make
1	Fixed jaw	1	Grey iron	Make	7	Guide plate	1	Grey iron	Make
2	Movile jaw	1	Grey iron	Make	8	Nut	2	Steel SAE 1010	Buy
3	Spindle	1	Steel AISI 1045	Make	9	Set screw	1	Steel SAE 1010	Buy
4	Casing	1	Bronze UNS C22000	Buy	10	Bolt M5X20	2	Steel SAE 1010	Buy
5	Clamping plate	2	Steel AISI 1045	Make	11	Bolt M5X25	2	Steel SAE 1010	Buy
6	Lever	1	Steel AISI 1045	Make	12	Bolt M4X8	2	Steel SAE 1010	Buy

Figure 2.4 Example of an engineering BOMs (eBOMs). *Source:* Personat Martínez (2022).

words, the eBOM contains all of the product's components as well as technical information about them from a design perspective. Nevertheless, this method might not fully describe the grouping of parts at every stage of production or contain all the information required for procurement or manufacturing support. To achieve this, the eBOM is converted into the mBOM, which contains every item needed from an assembly and production perspective. Therefore, the mBOM specifies the processes that are required at the execution stage. Figure 2.5 shows the mBOM for the previous bicycle product (Figure 2.2).

The eBOM is used by engineers and process planners to identify which parts need process planning. The production planning and control department relies on the eBOM to decide on manufacturing and scheduling dates. On the contrary, the mBOM is used to estimate standard times as well as time allowances (cycle time) for manufacturing and assembly procedures and to guarantee that all required components are available during the assembly process.

2.2.2 3D Models and Engineering Drawings

Traditionally, the 3D design models elaborated by the design department are passed to the manufacturing department by transferring them to 2D engineering drawings, including the necessary

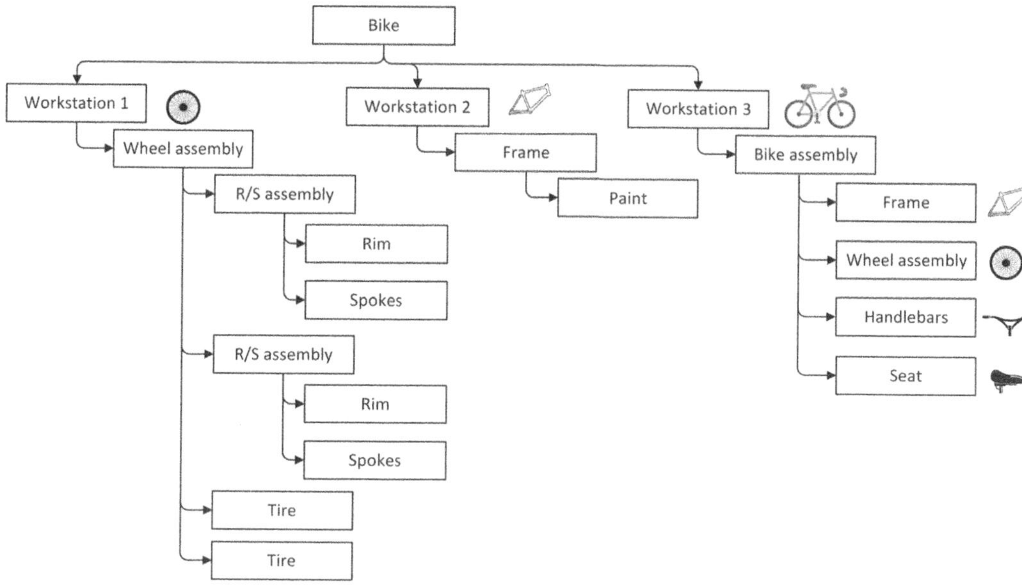

Figure 2.5 Example of manufacturing BOM (mBOM) for a bicycle product.

production information such as dimensional and geometrical tolerances, materials, surface roughness, welding information, etc. However, more recent practices apply the model-based definition (MBD) approach to work directly with the 3D models with all the manufacturing information embedded in the model itself eliminating the need for engineering drawings. Under this approach, the 3D models include all the required engineering data using 3D annotations that are interpretable among different software. Figure 2.6 shows an example of this type of 3D design model. All this engineering data is named as product manufacturing information (PMI), and they must comply with the ISO/ASME standards about product data representation and exchange (ASME Y14.41 and ISO 10303, ASME (2019), ISO (2024)). PMI data are created with 3D annotation tools in the computer-aided design (CAD) systems and can be imported and interpreted in other software to assist in different tasks such as tolerance analysis, generation of computer numerical controlled (CNC) programs for machining or coordinate measuring machine (CMM) programs for inspection. For example, computer-aided manufacturing (CAM) software can interpret the PMI in the 3D model and can adapt the cutting tool paths and conditions according to the information related to dimensional tolerances and surface quality. This approach saves significant time and cost in the product development process, and it also opens the door to automating other activities downstream.

The use of the MBD approach in the industry is gaining relevance since it presents important benefits such as the following:

- Reduces human-induced production and inspection errors, saving labor and time.
- Reduces product development cycles, which means faster times to market.
- Improves communication and makes an easier collaboration among stakeholders.
- Allows up-to-date information embedded and easily revisable.
- Reduces waste since there are no paper files to maintain and no misinterpreting 2D drawings, which cuts down on the waste from scraps and rework.

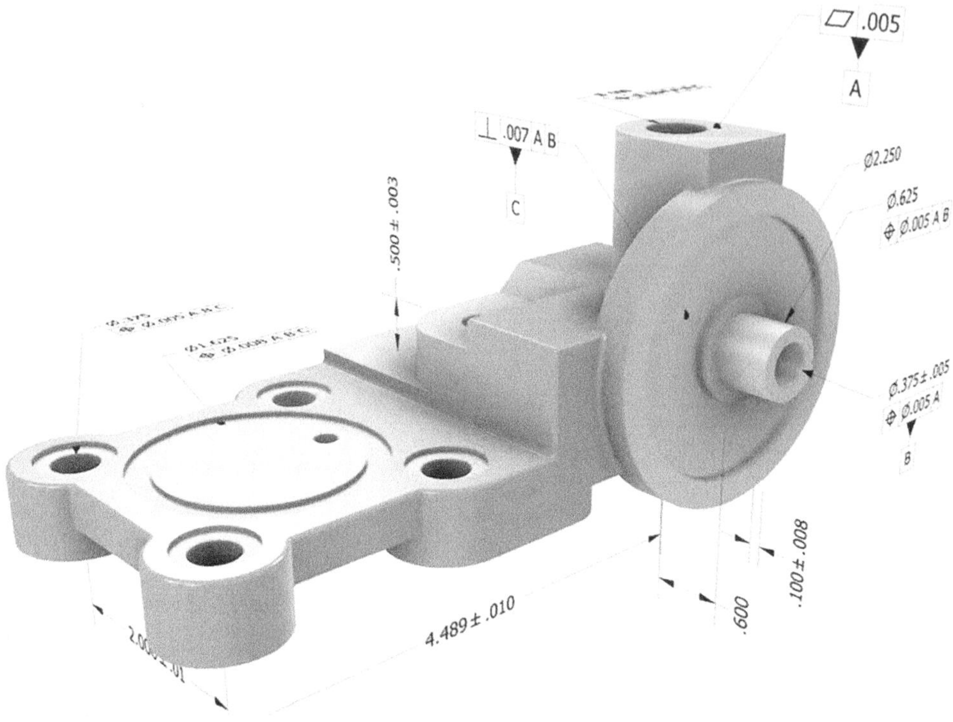

Figure 2.6 Example of a 3D model with product manufacturing information (PMI). *Source:* Courtesy of Autodesk, Inc.

2.2.3 Dimensional and Geometric Tolerances

2.2.3.1 Dimensional Tolerances

The dimensions of the parts in the design drawings or 3D models represent the basic or nominal dimensions of the part and its shape. These are the values that the designer would like to have for the part if it could be manufactured to an exact size without errors or variations in the manufacturing process. However, this is not possible, so tolerances are used to define the limits of permissible variation for the part to maintain its functionality. The parts that are outside these limits will be discarded or reprocessed.

Dimensional tolerances can be specified in different ways:

- Bilateral tolerances. The deviation from nominal values can be from both sides – positive and negative – for example, $13^{+0.020}_{-0.020}$; $13^{+0.020}_{-0.005}$.
- Unilateral tolerance. The deviation is only admissible in one direction, for example, $13^{+0.020}_{+0.005}$.
- Maximum and minimum dimensions, for example, 2.505–2.495.
- According to ISO format using IT indices and ISO positions, for example, 20h6; 18F9. See Appendix 1 for the ISO Tables according to "ISO 286-1: 2010 Geometrical product specifications (GPS) – ISO code system for tolerances on linear sizes – Part 1: Basis of tolerances, deviations and fits."

Figure 2.7 shows an example of dimensional tolerancing for a cylindrical part using maximum and minimum admissible dimensions.

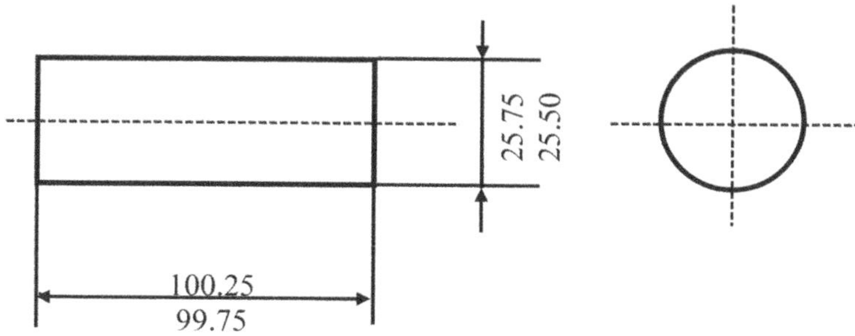

Figure 2.7 Dimensions for a cylindrical part. Units in mm.

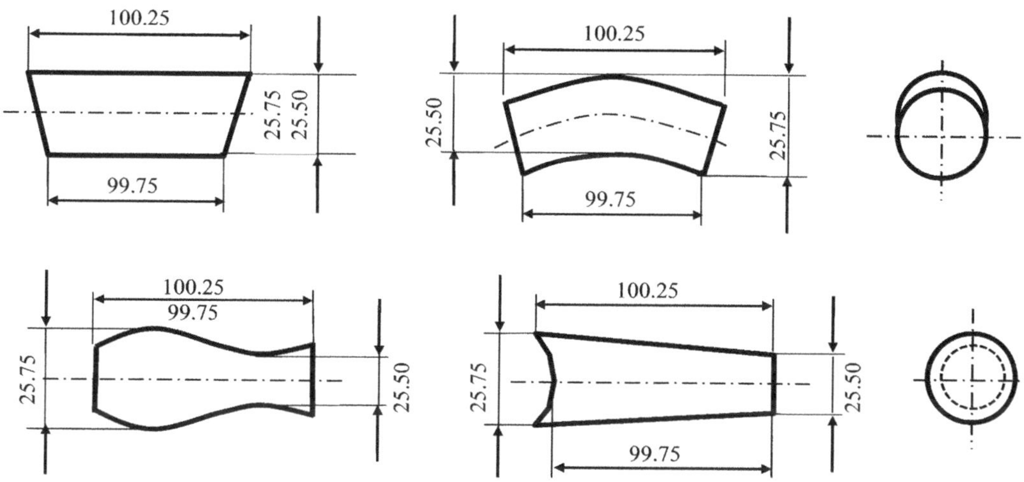

Figure 2.8 Four possible shapes that fit the previous dimensioned part. Units in mm.

2.2.3.2 Geometric Tolerances

Although the sizes of parts can be indicated quite accurately on the drawings and 3D models with dimensional tolerances, they often do not define the true geometric shape of the workpiece with sufficient accuracy. Consider, for example, the cylindrical part from Figure 2.7, dimensioned with considerable accuracy. With the specified dimensions, this part can be made of any of the forms shown in Figure 2.8 which must be accepted as good.

To limit these geometric shapes that may produce functional or interchangeability issues, the part of Figure 2.7 should specify a set of geometric characteristics to define the part more accurately. These indications are known as geometric tolerances. Figure 2.9 shows the initial part with the indications of geometric tolerances that uniquely define the acceptable geometry of the part.

Table 2.1 shows different geometric tolerances established by the ISO/ASME standards, which are classified as intrinsic (individual feature, unreferenced) or extrinsic (related features, referenced), the type of tolerance, and the associated symbol.

2.2.3.3 Datum Features and Datum Reference Frames

The extrinsic geometrical tolerances relate different features, for example, the perpendicularity tolerance relates the allowable perpendicular deviation between features. The information about the features' relationship is given in the feature control frame, which defines the type of geometric

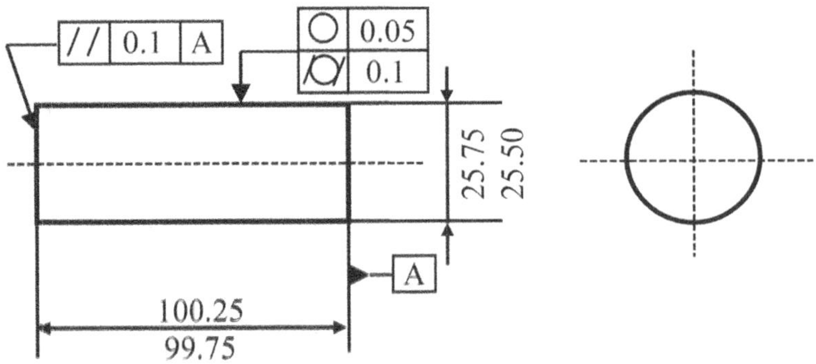

Figure 2.9 Geometric tolerancing of a cylindrical part. Units in mm.

Table 2.1 Classification and symbols of geometric tolerancing.

Application	Type	Characterization	Symbol
Individual features	Form	Straightness	⏤
		Flatness	⏥
		Circularity	○
		Cylindricity	⌭
Individual or related surfaces	Profile	Line profile	⌒
		Surface profile	⌓
Related features	Orientation	Angularity	∠
		Perpendicularity	⊥
		Parallelism	//
	Location	Position	⌖
		Concentricity	◎
		Symmetry	⌯
	Run-out	Circular run-out	↗
		Total run-out	⌰

tolerancing, the tolerance zone, and the datum references. The datum references indicate the datum features (surfaces or features of size such as holes or widths) that are used to locate the tolerancing zone. To correctly locate the tolerance zone, the feature control frame indicates the datum reference frame (DRF) that defines the coordinate system that locks the part's degrees of freedom. This is done by indicating in the feature control frame the sequence of datum features that apply to position and orient the part.

An important distinction should be made between the concept of datums and datum features. While datum features indicate the physical feature of the part that is used to reference the geometrical tolerance, a datum is a theoretical and perfect feature (plane, axis, or point) that represents the reference feature, that is, a datum is a geometrically exact representation of the datum feature. Unlike datums, datum features are real entities that are not exact and may have an irregular form. When conducting an inspection operation, measurement instruments physically touch the datum features, and the datum are generated by the measurement equipment (gauge pins, angle plates, V-blocks, etc.). In this sense, the instrument surfaces that touch the datum features are called datum simulators since they provide the exact representation of the datums.

Figures 2.10a, b and 2.11 illustrate previous concepts related to datums and DRFs.

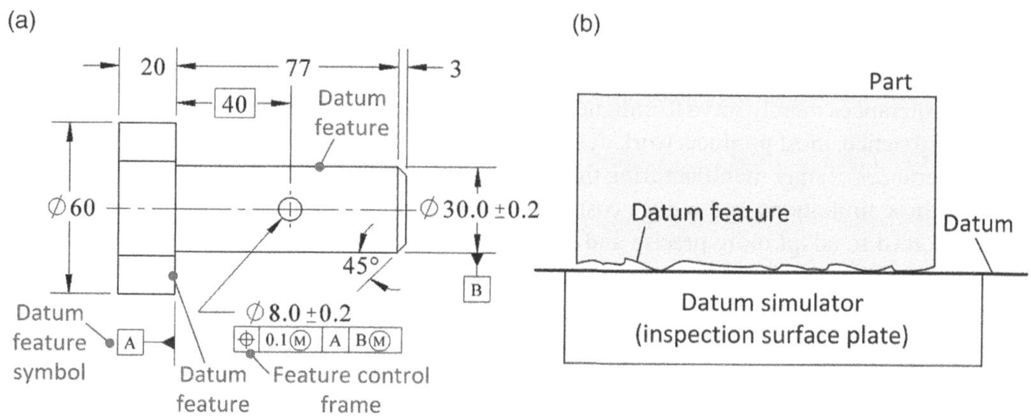

Figure 2.10 (a) Datum feature and feature control frame. Units in mm; (b) datum, datum feature, and datum simulator in an inspection process.

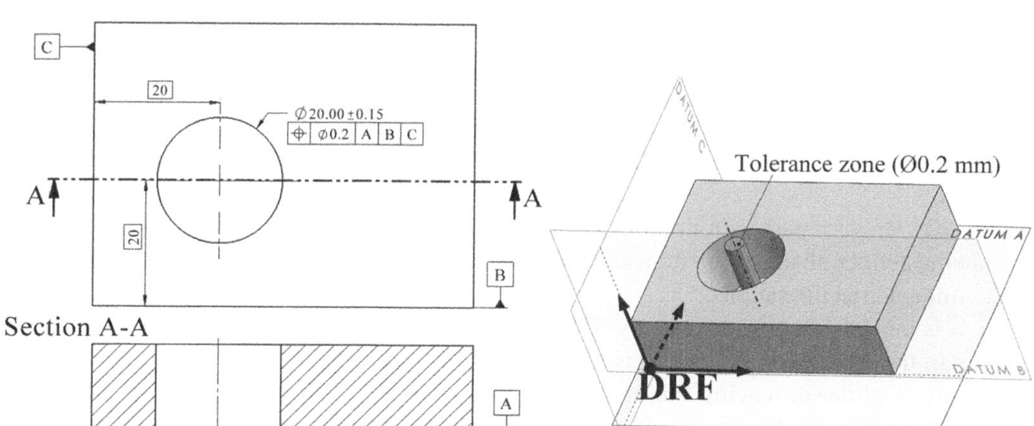

Figure 2.11 Datum reference frame (DRF) and tolerance zone according to the feature control frame and the defined DRF. Units in mm.

2.2.4 Traditional Tolerancing Versus GPS/Geometric Dimensioning and Tolerancing (GD&T)

Conventional or traditional tolerancing, commonly referred to as "direct tolerancing," is a method of tolerancing where only dimensional tolerances are given, usually in ± form, and no geometrical tolerances are provided. When manufacturing is outsourced, drawings based on traditional tolerancing usually result in a lot of issues because the part is not completely defined, and the supplier does not have a shared understanding between the customer's design department and his own workshop. It is crucial to keep in mind that when a drawing is given to another business, it either becomes a contract or a portion of one. As a result, the two parties' legally enforceable agreement is contained in the tolerancing notations.

The ambiguity that arises when applying traditional tolerancing is a fundamental issue. It raises some questions regarding just what must be produced, as it was presented in Figure 2.8. The specifications frequently contain so much ambiguity or uncertainty that even if the parts are manufactured to fulfill the tolerance requirements, they may not function as intended. In some cases, geometric tolerances are combined with conventional tolerancing to alleviate the ambiguity issue. This is called hybrid tolerancing, but the ambiguity problem is not entirely resolved because the geometric tolerances merely serve to enhance conventional tolerancing.

In actual practice, most products work despite this ambiguity issue, but it results in higher costs (tighter tolerances), larger manufacturing times to clarify design dimensions, higher waste, etc. To overcome these limitations and reduce costs and lead times, modern manufacturing companies are being forced to adopt more precise and unambiguous tolerancing approaches using the GPS approach from ISO standards (ISO 14638: 2015, ISO [2015] and related standards) or the equivalent GD&T approach from ASME standards (ASME Y14.5: 2018, ASME [2018]).

GPS (from now onward, GPS and GD&T are interchangeable) is required to guarantee that the drawing clearly conveys the design intent. 3D models and drawings are increasingly being used for production outsourcing, and any ambiguity in the geometry of the part will result in larger manufacturing costs and times. Conventional tolerancing is only effective in cases where the manufactured parts exhibit minimal form flaws and all angles are nearly faultless when compared to size tolerances. If this is not the case, GPS ought to be used.

The fundamental benefit of GPS is that it makes it possible to express tolerances with better precision and distinct meaning, especially in the presence of severe shape errors and angular variations in the part. The GPS tolerancing can very accurately imitate the part's operation since GPS sets the functional part tolerances. Experience has demonstrated that to make the part function; traditional tolerancing pushes the designer to use smaller tolerances – smaller tolerances than those GPS would have used in the drawing (Shepherd 1992). This is the financial fallout from conventional tolerancing, which frequently causes needless issues with production and inspection.

Figure 2.12 shows a part with traditional tolerancing bounded with dimensional tolerances ±. The drawing defines all necessary dimensions of the part. However, this tolerancing approach has several fundamental limitations.

First, it is impossible to know exactly from where to where the specified dimensions go. For example, in the dimension 15 ± 0.1, it is not known the measuring direction where a deviation of ± 0.1 mm is allowed, leaving the ambiguous part design in two alternatives, as shown in Figure 2.13a. Similarly, the distance 20 ± 0.15 mm from the hole center to the edge of the workpiece is also ambiguously defined since it is not known whether the specification refers to the distance between the median plane and the average circumference, the distance between the plane and the circle outside the material, or the distance between the closest points, as shown in Figure 2.13b.

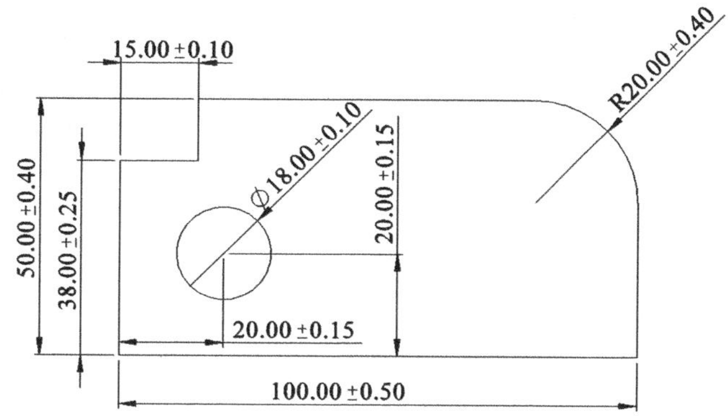

Figure 2.12 Part with traditional tolerancing. Units in mm. *Source:* Example adapted from Nielsen (2012).

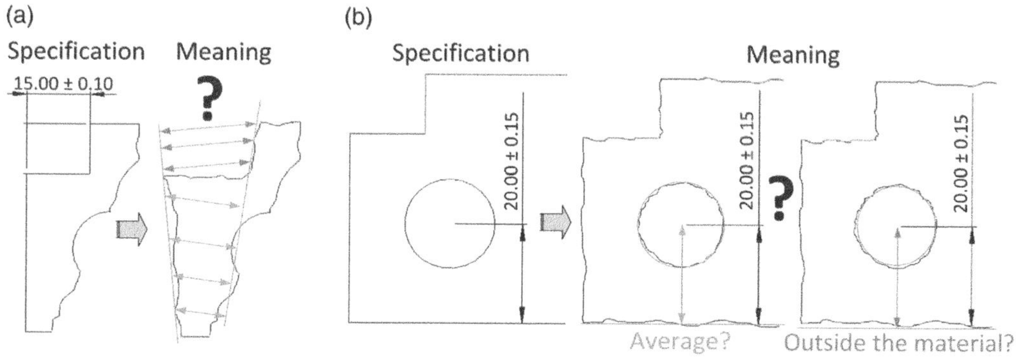

Figure 2.13 Ambiguity in traditional tolerancing: (a) plane position; (b) hole position. Units in mm. *Source:* Example adapted from Nielsen (2012).

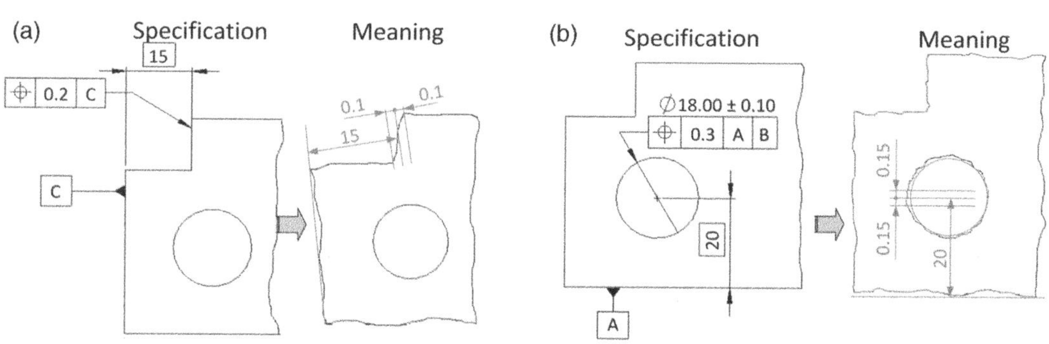

Figure 2.14 Unambiguous geometry definition using geometric product specifications (GPS): (a) plane position; (b) hole position. Units in mm. *Source:* Adapted from Nielsen (2012).

The tolerancing according to the ISO standard "Geometrical product specifications (GPS) – Geometrical tolerancing – Tolerances of form, orientation, location, and run-out (ISO 1101: 2013)" allows unambiguously define the geometries of the part described above. For example, the indication of a reference element and a positioning tolerance defines precisely what deviation is allowed in both cases, the lateral deviation of the surface, and the position of the hole with respect to the reference (Figure 2.14a, b).

Another important limitation of traditional tolerancing is that the dimensions cannot be assured that they are aligned with each other or perpendicular to each other; however, they might be in the drawing. For example, all dimensions may be within tolerances, although the generated parts present an important squaring error (Figure 2.15). A correct use of ISO GPS will allow the restriction of the orientation defects according to part functionality.

For the example shown in Figure 2.12, the GPS tolerancing approach can precisely limit the part's geometry without ambiguity. Figure 2.16 shows the proper GPS tolerancing approach for the example.

Another important limitation of the traditional tolerancing approach is that references are not specified, so the operator does not know which surface should be used as the datum surface for clamping, inspecting, and so on. Figures 2.17 and 2.18 show an example where the position of the two holes is not unequivocally defined, which may result in different interpretations for manufacturing or inspection.

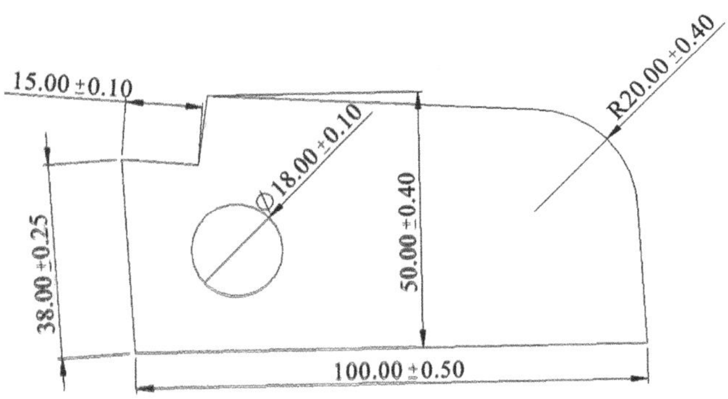

Figure 2.15 Orientation defects not limited to the traditional tolerancing approach. Units in mm. *Source:* Adapted from Nielsen (2012).

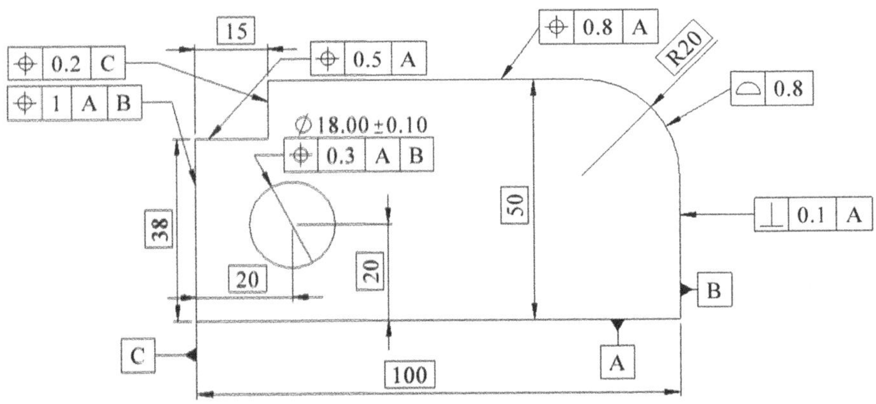

Figure 2.16 Part with GPS tolerancing approach. Units in mm.

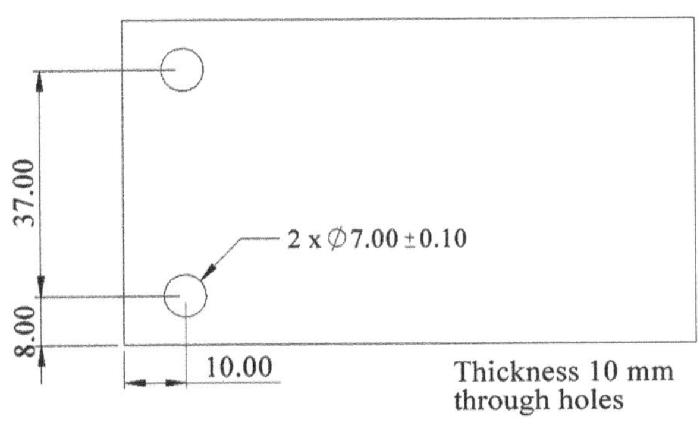

Figure 2.17 Traditional tolerancing example. References are not specified for the position of the hole. Units in mm.

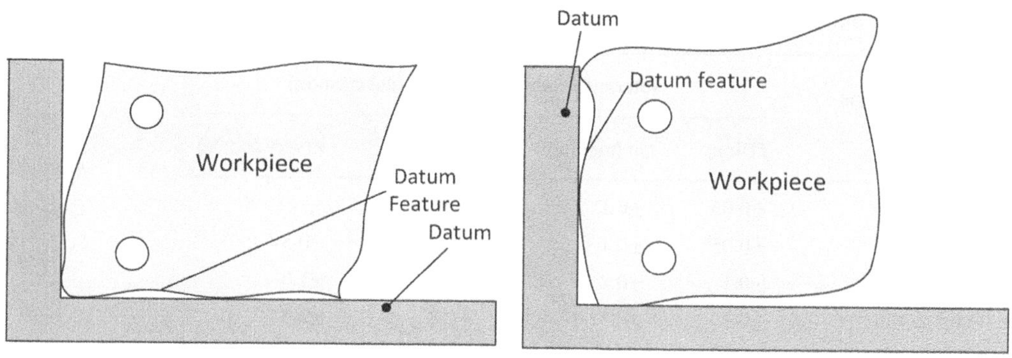

Figure 2.18 Different interpretations related to references for inspecting the previous part. Which assembly is adequate to inspect the previous part?

Traditional tolerancing can be used in GPS-based drawings but must be limited to geometric elements of size and geometric elements that define depth or length. For the reasons described above, traditional tolerancing should not be used to locate geometric elements.

2.2.5 General Tolerances

2.2.5.1 Dimensional and Geometrical Tolerances for Machined Parts

Standards "ISO 2768-1/2:1989 General tolerances – Part 1: Tolerances for linear and angular dimensions without individual tolerance indications" and "Part 2: Geometrical tolerances for features without individual tolerance indications" allow to facilitate the definition of generic tolerances in machined parts without having to indicate them one by one on the drawings. General tolerances apply to all dimensions or geometries of the part that do not show any tolerance indication.

Note that those dimensions or geometries that are functional and require a tight manufacturing tolerance should not be defined by general tolerances. Instead, they must be specified in the drawing by their respective dimensional and geometric limitations.

The ISO standard defines four classes of generic dimensional tolerances: fine quality, *f*; average quality, *m*; coarse quality, *c*; and very rude quality, *v*, These four classes are comparable to a quality index IT11, IT13, IT15, and IT17, respectively. According to the tolerance class and nominal value, the general dimensional tolerances related to dimensions and angles are defined in Tables 2.2 and 2.3.

For geometric tolerances, three types of tolerance classes are defined: H, K, and L going from more to less quality. Thus, according to general geometric tolerances, the geometric deviations related to straightness, flatness, perpendicularity, run-out, and symmetry are defined in Tables 2.4–2.7.

For circularity, the general circularity tolerance is numerically equal to the diameter tolerance, but in no case will it be higher than the corresponding value of tolerance given by the general circular run-out from Table 2.6.

2.2.5.2 Dimensional and Geometrical Tolerances for Metal Casting Parts

For casting parts, general tolerances can be found in norm ISO 8062, "Dimensional and geometrical tolerances for molded parts – Part 3: General dimensional and geometrical tolerances and machining allowances for castings." The norm defines the following terms:

- DCT. Dimensional casting tolerance.
- DCTG. Dimensional casting tolerance grade.
- GCT. Geometrical casting tolerance.
- GCTG. Geometrical casting tolerance grade.

Table 2.2 General dimensional tolerances (length).

Permissible deviations in mm for ranges in nominal lengths	Tolerance class designation (description)			
	f (fine)	*m* (medium)	*c* (coarse)	*v* (very coarse)
0.5 up to 3	±0.05	±0.1	±0.2	—
Over 3 up to 6	±0.05	±0.1	±0.3	±0.5
Over 6 up to 30	±0.1	±0.2	±0.5	±1.0
Over 30 up to 120	±0.15	±0.3	±0.8	±1.5
Over 120 up to 400	±0.2	±0.5	±1.2	±2.5
Over 400 up to 1,000	±0.3	±0.8	±2.0	±4.0
Over 1,000 up to 2,000	±0.5	±1.2	±3.0	±6.0
Over 2,000 up to 4,000	—	±2.0	±4.0	±8.0

Values in mm.
Source: Norm ISO 2768-1. Adapted from ISO (1989a).

Table 2.3 General dimensional tolerances (angles).

Permissible deviations in mm for ranges in nominal lengths	Tolerance class designation (description)			
	f (fine)	*m* (medium)	*c* (coarse)	*v* (very coarse)
Up to 10	±1°	±1°	±1° 30′	±3°
Over 10 up to 50	±0° 30′	±0° 30′	±1°	±2°
Over 50 up to 120	±0° 20′	±0° 20′	±0° 30′	±1°
Over 120 up to 400	±0° 10′	±0° 10′	±0° 20′	±0° 30′
Over 400	±0° 5′	±0° 5′	±0° 10′	±0° 20′

Source: Norm ISO 2768-1. Adapted from ISO (1989a).

Table 2.4 General geometric tolerances. Straightness and flatness.

Ranges of nominal lengths in mm	Tolerance class		
	H	*K*	*L*
Up to 10	0.02	0.05	0.1
>10–30	0.05	0.1	0.2
>30–100	0.1	0.2	0.4
>100–300	0.2	0.4	0.8
>300–1,000	0.3	0.6	1.2
>1,000–3,000	0.4	0.8	1.6

Values in mm.
Source: Norm ISO 2768-2. Adapted from ISO (1989b).

Table 2.5 General geometric tolerances. Perpendicularity.

Ranges of nominal lengths in mm	Tolerance class		
	H	*K*	*L*
Up to 100	0.2	0.4	0.6
>100–300	0.3	0.6	1.0
>300–1,000	0.4	0.8	1.5
>1,000–3,000	0.5	1.0	2.0

Values in mm.
Source: Norm ISO 2768-2. Adapted from ISO (1989b).

Table 2.6 General geometric tolerances. Circular run-out.

Any length or size	Tolerance class		
	H	*K*	*L*
	0.1	0.2	0.5

Values in mm.
Source: Norm ISO 2768-2. Adapted from ISO (1989b).

Table 2.7 General geometric tolerances. Symmetry.

Ranges of nominal lengths in mm	Tolerance class		
	H	*K*	*L*
Up to 100	0.5	0.6	0.6
>100–300	0.5	0.6	1
>300–1,000	0.5	0.8	1.5
>1,000–3,000	0.5	1	2

Values in mm.
Source: Norm ISO 2768-2. Adapted from ISO (1989b).

General dimensional and geometrical tolerances for roundness, parallelism, perpendicularity, symmetry, and flatness are defined as shown in Tables 2.8–2.10 (other tolerances, such as straightness and coaxiality, can be found in norm ISO 8062). As shown in the tables, the tolerances are defined according to 16 grades for dimensional tolerances and 7 grades for geometrical tolerances. According to the type of casting process, the norm indicates the most common general tolerance grades for each process (see Tables 2.11 and 2.12).

2.2.5.3 General Tolerance Indication in Engineering Drawings

When general tolerances are applied, they must be indicated in the drawing title block or nearby using the inscription of the standard and tolerance class. Thus, in the case of using a class m for dimensional and class K for geometric in machining parts, it is indicated "ISO 2768-mK."

Table 2.8 General dimensional tolerances for castings.

Nominal dimensions related to the molded part		Linear dimensional tolerance for dimensional casting tolerance grade (DCTG)[a] dimensions in mm															
		1	2	3	4	5	6	7	8	9	10	11	12	13	14	15	15wt[b]
—	≤10	0.09	0.13	0.18	0.26	0.36	0.52	0.74	1	1.5	2	2.8	4.2	—	—	—	—
>10	≤16	0.1	0.14	0.2	0.28	0.38	0.54	0.78	1.1	1.6	2.2	3	4.4	—	—	—	—
>16	≤25	0.11	0.15	0.22	0.3	0.42	0.58	0.82	1.2	1.7	2.4	3.2	4.6	6	8	10	12
>25	≤40	0.12	0.17	0.24	0.32	0.46	0.64	0.9	1.3	1.8	2.6	3.6	5	7	9	11	14
>40	≤63	0.13	0.18	0.26	0.36	0.5	0.7	1	1.4	2	2.8	4	5.6	8	10	12	16
>63	≤100	0.14	0.2	0.28	0.4	0.56	0.78	1.1	1.6	2.2	3.2	4.4	6	9	11	14	18
>100	≤160	0.15	0.22	0.3	0.44	0.62	0.88	1.2	1.8	2.5	3.6	5	7	10	12	16	20
>160	≤250	—	0.24	0.34	0.5	0.7	1	1.4	2	2.8	4	5.6	8	11	14	18	22
>250	≤400	—	—	0.4	0.56	0.78	1.1	1.6	2.2	3.2	4.4	6.2	9	12	16	20	25
>400	≤630	—	—	—	0.64	0.9	1.2	1.8	2.6	3.6	5	7	10	14	18	22	28
>630	≤1,000	—	—	—	—	1	1.4	2	2.8	4	6	8	11	16	20	25	32
>1,000	≤1,600	—	—	—	—	—	1.6	2.2	3.2	4.6	7	9	13	18	23	29	37

[a] For wall thicknesses in grades DCTG 1 to DCTG 15, one grade coarser applies.
[b] Grade DCTG 15wt (wall thickness) exists only for wall thicknesses of castings generally specified to DCTG 15.
Source: Norm ISO 8062-3 ISO (2009).

Table 2.9 General geometrical tolerances for castings. Roundness, parallelism, perpendicularity, and symmetry.

Longest nominal dimension of the feature		Tolerance for geometrical casting tolerance grade (GCTG) dimensions in mm						
		GCTG 2	GCTG 3	GCTG 4	GCTG 5	GCTG 6	GCTG 7	GCTG 8
—	≤10	0.18	0.27	0.4	0.6	0.9	1.4	2
>10	≤30	0.27	0.4	0.6	0.9	1.4	2	3
>30	≤100	0.4	0.6	0.9	1.4	2	3	4.5
>100	≤300	0.6	0.9	1.4	2	3	4.5	7
>300	≤1,000	0.9	1.4	2	3	4.5	7	10
>1,000	≤3,000	—	—	—	6	9	14	20
>3,000	≤6,000	—	—	—	12	18	28	40
>6,000	≤10,000	—	—	—	24	36	56	80

Source: Norm ISO 8062-3 ISO (2009).

Figures 2.19 and 2.20 show two examples of generic tolerances. Figures 2.21 and 2.22 show the constraints of the final part according to those general tolerances.

For the case of castings, the same procedure applies. When general orientation tolerances apply, the datum system should be indicated in the drawing using the term "ISO 8062-3 DS" followed by the datum surfaces. An example of a casting drawing with general tolerances is shown in Figure 2.23.

Table 2.10 General geometrical tolerances for castings. Flatness.

Longest nominal dimension of the feature		Flatness tolerance for geometrical casting tolerance grade (GCTG) dimensions in mm						
		GCTG 2	GCTG 3	GCTG 4	GCTG 5	GCTG 6	GCTG 7	GCTG 8
—	≤10	0.12	0.18	0.27	0.4	0.6	0.9	1.4
>10	≤30	0.18	0.27	0.4	0.6	0.9	1.4	2
>30	≤100	0.27	0.4	0.6	0.9	1.4	2	3
>100	≤300	0.4	0.6	0.9	1.4	2	3	4.5
>300	≤1,000	0.6	0.9	1.4	2	3	4.5	7
>1,000	≤3,000	—	—	—	4	6	9	14
>3,000	≤6,000	—	—	—	8	12	18	28
>6,000	≤10,000	—	—	—	16	24	36	56

Source: Norm ISO 8062-3 ISO (2009).

Table 2.11 DCTG values for different casting processes.

Method	Dimensional casting tolerance grade (DCTG) for casting material								
	Steel	Gray iron	Spheroidal graphite iron	Malleable iron	Copper alloys	Zinc alloys	Light metal alloys	Nickel-based alloys	Cobalt-based alloys
Sand cast and hand molding	11–14	11–14	11–14	11–14	10–13	10–13	9–12	11–14	11–14
Sand cast, machine molding, and shell molding	8–12	8–12	8–12	8–12	8–10	8–10	7–9	8–12	8–12
Metallic permanent mold (except pressure die casting)	—	7–9	7–9	7–9	7–9	7–9	6–8	—	—
Pressure die casting	—	—	—	—	6–8	3–6	[a]	—	—
Investment casting	[b]	[b]	[b]	—	[b]	—	[b]	[b]	[b]

Note: For complex castings, one grade coarser is recommended.

[a] The following apply depending on the largest overall dimension: ≤50 mm, DCTG 6; >50 mm ≤180 mm, DCTG 7; >180 mm ≤500 mm, DCTG 8; >500 mm, DCTG 9.

[b] For investment casting, the following apply depending on the largest overall dimension: ≤100 mm, DCTG 4–6; >100 mm ≤400 mm, DCTG 4–8; >400 mm, DCTG 4–9.

Source: Norm ISO 8062-3 ISO (2009).

Table 2.12 GCTG values for different casting processes.

	Geometrical casting tolerance grade (GCTG) for casting material								
Method	**Steel**	**Gray iron**	**Spheroidal graphite iron**	**Malleable iron**	**Copper alloys**	**Zinc alloys**	**Light metal alloys**	**Nickel-based alloys**	**Cobalt-based alloys**
Sand cast and hand molding	6–8	5–7	5–7	5–7	5–7	5–7	5–7	6–8	6–8
Sand cast, machine molding, and shell molding	5–7	4–6	4–6	4–6	4–6	4–6	4–6	5–7	5–7
Metallic permanent mold (except pressure die casting)	—	—	—	—	3–5	—	3–5	—	—
Pressure die casting[a]	—	—	—	—	2–4	2–4	2–4	—	—
Investment casting	[b]	3–5	3–5	3–5	3–5	2–4	3–5	[b]	[b]

Note: For complex castings, one grade coarser is recommended.
[a] For pressure die castings, the following apply: Grade GCTG 2: should only be used by special agreement; Grade GCTG 3: ordinary castings without side sliders for the contour; Grade GCTG 4: complex castings as well as castings with side sliders for the contour.
[b] For investment casting, the following apply depending on the largest overall dimension: ≤100 mm, GCTG 4–6; >100 mm ≤400 mm, GCTG 4–8; >400 mm, GCTG 4–9.
Source: Norm ISO 8062-3 ISO (2009).

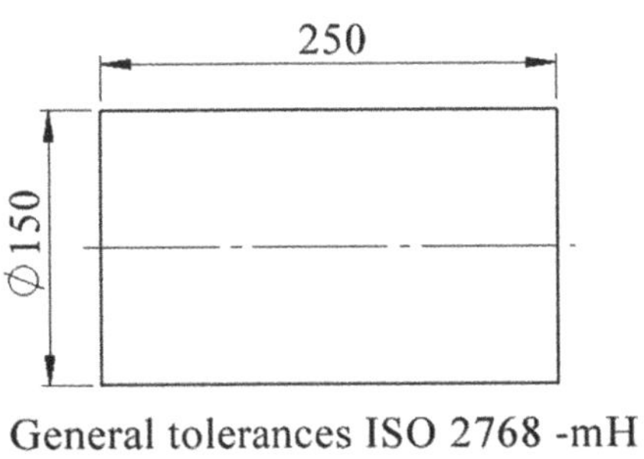

Figure 2.19 Example 1 – part design with general tolerances.

2.2.6 Surface Roughness and Surface Treatments

In engineering drawings, surface roughness is indicated using the symbol shown in Figure 2.24. The fields a, b, c, d, and e in this symbol can be used to indicate which manufacturing process should be employed, the level of surface roughness (Ra, Rz), the direction of cutting tool marks, and so on. According to norm ISO 1302–2002 (ISO 2002), the following characteristics may be indicated:

- Field a. Maximum surface roughness value in μm or maximum surface roughness index (N1–N12).
- Field a and b. If a maximum and minimum value is required, both values are indicated, the higher value over the lower value.
- Field c. Manufacturing process, surface treatment, and coating; for example, turned, plated, ground, etc.
- Field d. Direction of cutting tool marks.
- Field e. Machining allowance.

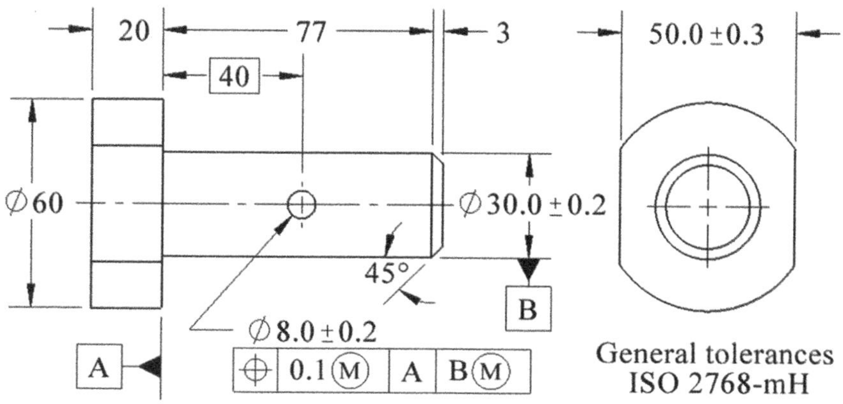

Figure 2.20 Example 2 – part design with general tolerances.

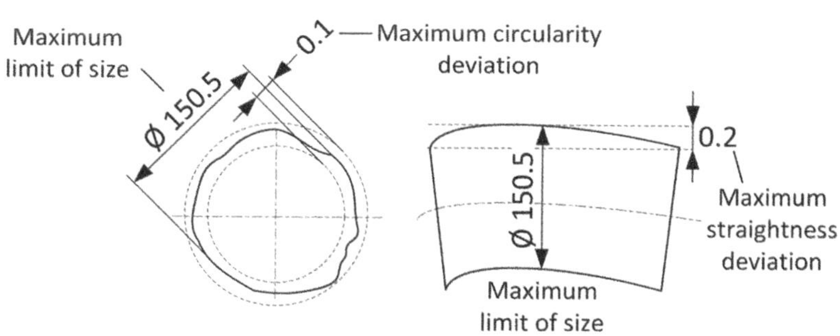

Figure 2.21 Interpretation of part design drawing – example 1.

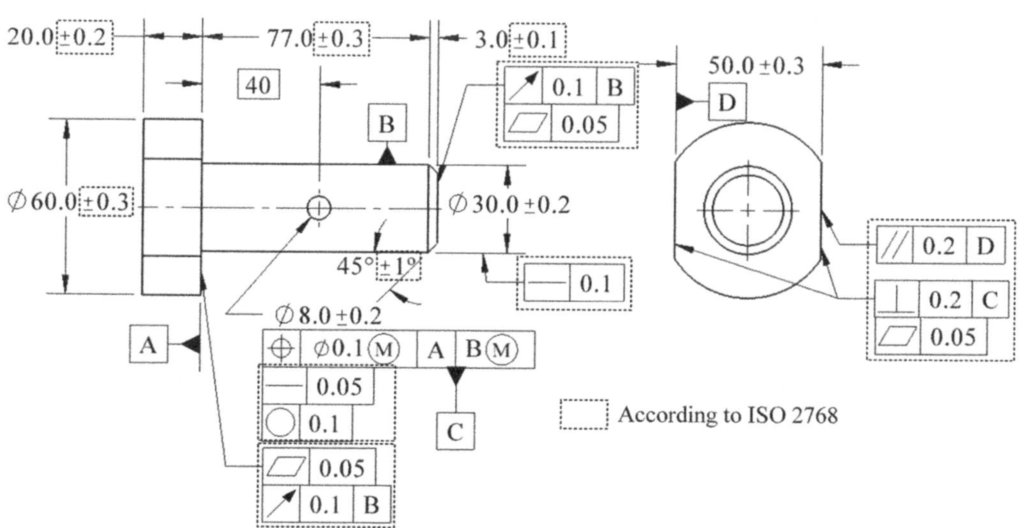

Figure 2.22 Interpretation of part design drawing – example 2.

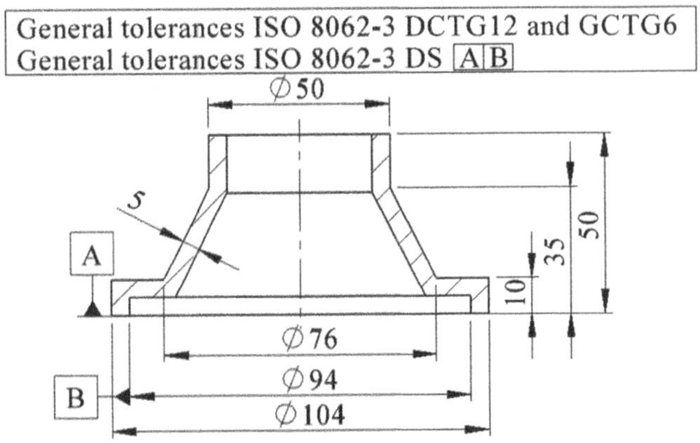

Figure 2.23 Casting drawing with general tolerances.

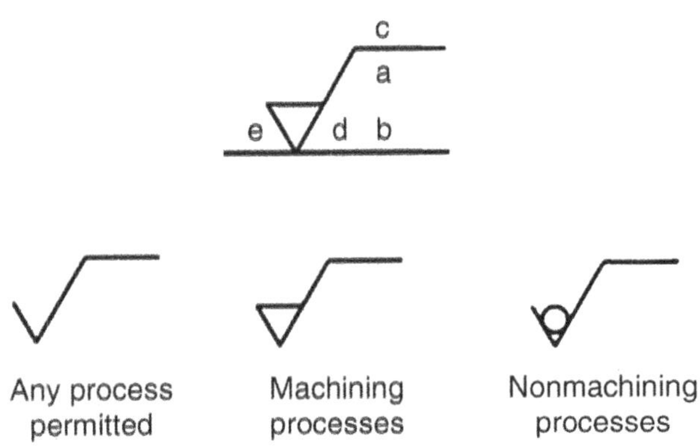

Figure 2.24 Symbols related to surface roughness.

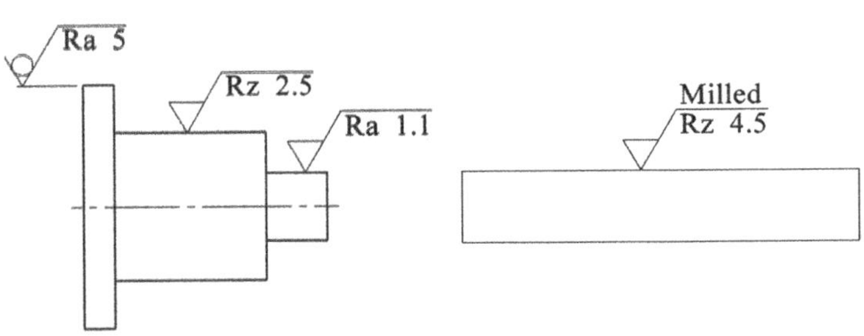

Figure 2.25 Example of surface roughness indication in a drawing.

Figure 2.25 shows an example of the application of surface roughness symbols. Table 2.13 shows the equivalence of Nx (N1–N12) classes from ISO 1302 norm, the average surface roughness, and common manufacturing processes able to reach that surface roughness.

It should be noted that allowing a high surface roughness value in a dimension where a tight dimensional tolerance is needed makes no sense since microgeometry defects in surface roughness may reach the same level as macrogeometry defects defined by the dimensional tolerance. Therefore, for a given dimensional tolerance, a minimal surface roughness requirement may be defined. Table 2.14 provides common values considered by process planners to relate dimensional tolerances and minimal surface roughness requirements.

Table 2.13 Surface roughness classes according to ISO 1302 and examples of potential manufacturing processes.

Roughness (μm)	Roughness class	Applications
50	N12	Forging, casting
25	N11	
12.5	N10	Roughing machining
6.3	N9	
3.2	N8	Finishing machining, broaching
1.6	N7	
0.8	N6	Grinding, boring
0.4	N5	
0.2	N4	Fine grinding, honing
0.1	N3	
0.05	N2	Lapping
0.025	N1	

Table 2.14 Relationships between dimensional tolerances and minimal surface roughness requirements.

Dimensional tolerance ± (mm)	Surface roughness Ra (μm)	Dimensional tolerance ± (mm)	Surface roughness Ra (μm)
<0.005	<0.20	0.100	2.5
0.010	0.32	0.150	3.75
0.020	0.80	0.2	5.0
0.040	1.32	0.25	6.25
0.060	1.8	0.6	12.5
0.080	2.12	1.0	25.0

Source: Halevi and Weill (1995), Chapter 6.

2.3 Manufacturability

Designers should apply design for manufacturing and assembly (DFMA) techniques when designing a product. There are many DFMA tools integrated into common CAD systems to assist designers in this process. Figure 2.26 illustrates an example of a DFMA tool integrated into a CAD system (SolidWorks). However, when process planners review the engineering drawings, design changes may also be recommended to improve the manufacturing process of the product and reduce manufacturing costs. Although design considerations related to manufacturing depend on the manufacturing process chosen and, at this stage, manufacturing processes have not been selected yet, process planners may expect the use of some types of processes according to materials and part geometry. In any case, when manufacturing process selection is conducted, additional recommendations may also be given to improve the part's manufacturability. It should be noted that proper

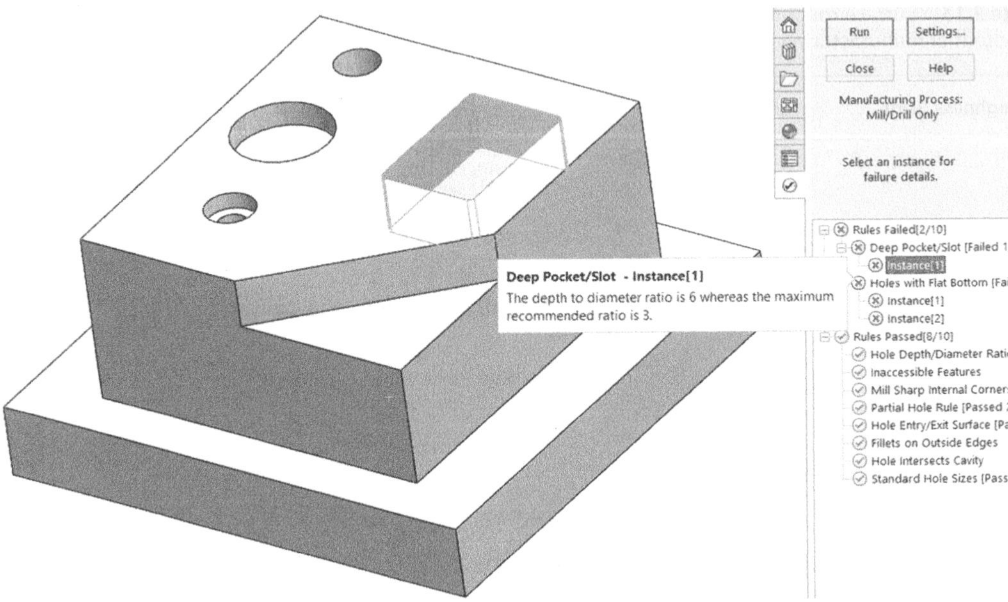

Figure 2.26 Design for manufacturing and assembly (DFMA) tools included in conventional CAD systems assist designers in applying good design for manufacturing practices.

DFMA practices lead to shorter manufacturing/assembly times, lower costs, increased product reliability, and elimination of process waste.

Some basic design considerations for metal parts to be produced by machining and casting are discussed in the following subsections. Readers interested in more detailed DFMA practices may refer to (Boothroyd et al. 2010; Bralla 1998; Trucks 1987).

2.3.1 Machining Parts

Important characteristics to be checked in machining designed parts are (Trucks 1987):

- Interior corner fillets. Avoid small internal corners in pockets or similar features that force the use of different cutting tools with different radii, increasing machining time and cost. As a guideline, the recommended radius is 1.3 times the cutting tool radius.
- Floor fillets. The radius of floor fillets should be compatible with standard bull or ball end mills. No floor fillets are easily machined by flat-end mills.
- Hole diameter-to-depth ratio. Due to issues with tool stiffness, this ratio should be less than 1:10, and ratios around 1:4 are recommended.
- Cavity width-to-depth ratio. Similar to the previous case, due to tool stiffness issues, cavity width-to-depth ratios are recommended to be less than 1:4.
- Wall thickness. Avoid machined walls with a thickness of less than 0.8 mm due to difficulties in reaching tight tolerances (deflection issues).

- Machined text. If text should be machined on a surface, use machine-engraved text instead of embossed text for lower costs. A text font of Sans-Serif or Arial and 20 points size or higher are recommended.
- Flat-bottomed holes. Avoid flat-bottomed holes if they are small and are expected to be drilled. Cone-bottomed holes with the angle of common drills for the applied material are recommended.
- Entry of holes. The axis of the hole should be perpendicular to the entrance and exit surfaces to avoid tool deviation, breakages, and burr removal issues.
- Thread depth. A minimum length for threads is 1.5 times the hole diameter although a factor of four is recommended. For blind holes, a no-thread zone with a length of 1.5 times the hole diameter is needed at the end of the hole.

2.3.2 Metal Casting Parts

Important characteristics to be checked in metal casting designed parts are (Trucks 1987):

- Uniform thickness. A good practice is to keep the number of varying thicknesses in the design to a minimum and always design smooth transitions between different thicknesses. Sharp corners and small radii in a change of section should be avoided since they may cause turbulence during casting.
- Fillets. Avoid sharp corners that can reduce tooling life. The applied fillets should be higher than 1 mm, but not too high to avoid the formation of casting defects such as pipes. A radius equivalent to one-third of the summation of the intersected walls is recommended.
- Joint design. Joints result in an area of metal with a higher volume than the rest of the junction, which may produce hot spots with solidification issues, such as pipes. Some of these joints can be slightly modified to reduce this problem by following some joint design rules.
- Ribs. The ribs section should be 25% of the section of the main member unless the resultant rib is considered too thin.
- Minimum wall thickness. Avoid wall thicknesses too thin that may produce early solidification issues. Table 2.15 shows an example of the minimum thickness for aluminum alloys in different casting processes.

Table 2.15 Minimum wall thickness for aluminum alloys in some casting processes.

	Minimum wall thickness (mm)			
Wall area (cm^2)	**Sand casting**	**Gravity casting**	**Die casting**	**Investment casting**
<25	3.15	2.35	1.00	1.25
>25–65	3.75	3.15	1.5	1.5
>65–125	4.7	3.9	2	1.75
>125	6.25	4.7	2.5	2.5

Source: Trucks (1987).

Questions and Exercises

1 Why is a correct drawing interpretation so important for process planning?

2 The ISO standard classifies the geometrical tolerances as intrinsic (individual features) and extrinsic (related features) tolerances. What is the difference? Indicate some examples of each type.

3 Which type of tolerance is perpendicularity, flatness, and position?

4 Explain the difference between datum and datum features.

5 Why is GPS important to be used in engineering drawings?

6 A cylindrical part has a length of 200 mm and a diameter of 80 mm. The notation in the title box is "general tolerances ISO 2768–cL." Which is the tolerance value of these dimensions, and which is the geometric tolerance related to the axis straightness?

7 From the figures depicted in the following, identify if there is any mistake or nonsense indication and propose a correction.

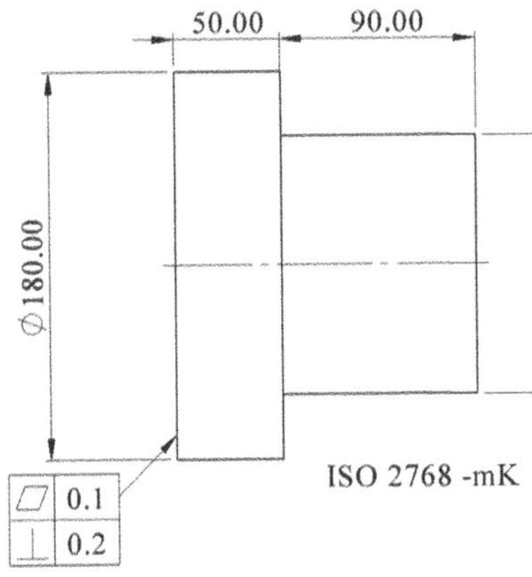

(a)

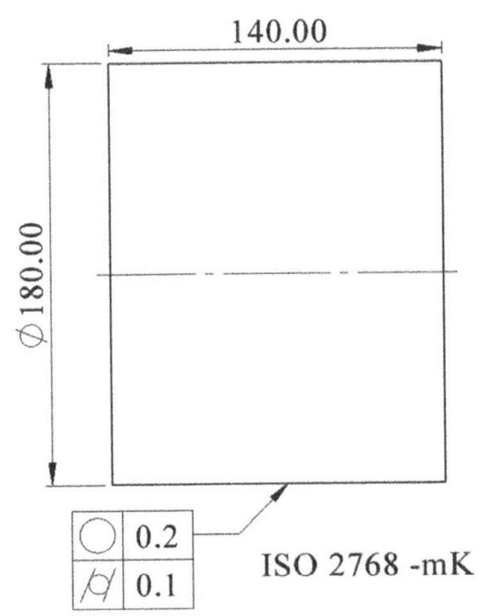

(b)

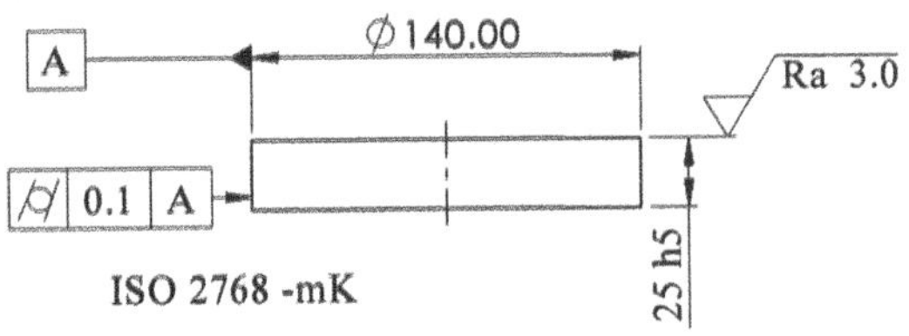

(c)

8 For the parts shown in the following, propose a design change to improve manufacturability. Assume that the parts are being produced using machining processes.

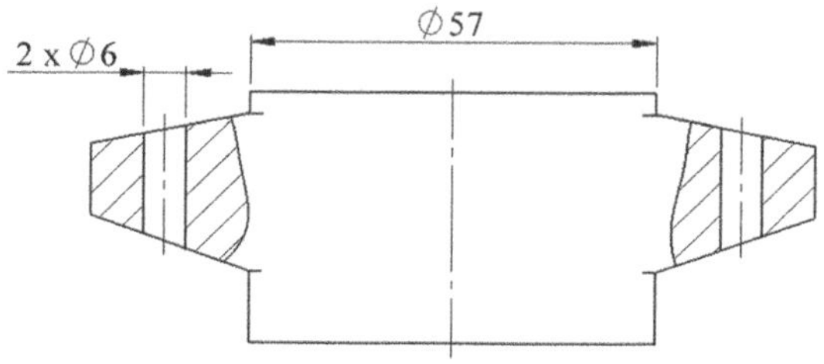

(a)

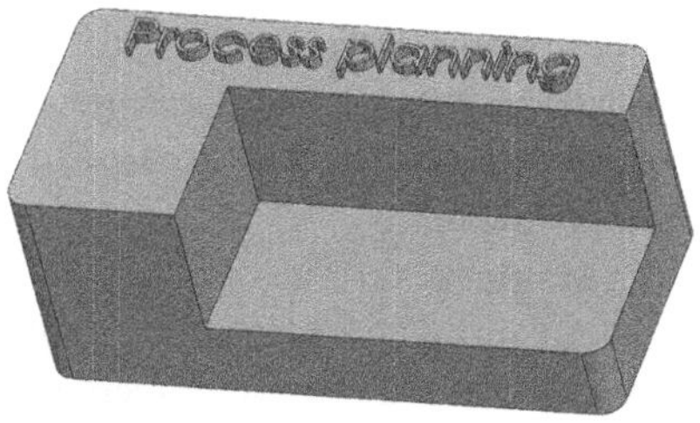

(b)

References

Adithan, M. (2007). *Process Planning and Cost Estimation*. New Age International (P) Limited, Publishers.

ASME. (2018). *ASME Y14.5 – Dimensioning and Tolerancing*. New York: The American Society of Mechanical Engineers.

ASME. (2019). *ASME Y14.41 – Digital Product Definition Data Practices*. New York: The American Society of Mechanical Engineers.

Boothroyd, G., Dewhurst, P., and Knight, W.A. (2010). *Product Design for Manufacture and Assembly*. New York: CRC Press.

Bralla, J.G. (1998). *Design for Manufacturability Handbook*, 2e. McGraw Hill.

Company, P., Vergara, M., and Mondragón, S. (2007). *Dibujo Industrial*. Col·lecció "Treballs d'Informàtica i Tecnologia," vol. 27. Publicacions Universitat Jaume I.

Halevi, G. (2003). *Process and Operation Planning: Revised Edition of The Principles of Process Planning: A Logical Approach*. Springer Science & Business Media.

Halevi, G., and Weill, R. (1995). *Principles of Process Planning: A Logical Approach*. Springer Science & Business Media.

ISO. (1989a). *ISO 2768-1: 1989. General Tolerances – Part 1: Tolerances for Linear and Angular Dimensions Without Individual Tolerance Indications*. Geneva, Switzerland: International Organization for Standardization.

ISO. (1989b). *ISO 2768-2: 1989. General Tolerances – Part 2: Geometrical Tolerances for Features Without Individual Tolerance Indications*. Geneva, Switzerland: International Organization for Standardization.

ISO. (2002). *ISO 1302: 2002. Geometrical Product Specifications (GPS) – Indication of Surface Texture in Technical Product Documentation*. Geneva, Switzerland: International Organization for Standardization.

ISO. (2009). *ISO 8062-3: 2009. Geometrical Product Specifications (GPS) – Dimensional and Geometrical Tolerances for Moulded Parts – Part 3: General Dimensional and Geometrical Tolerances and Machining Allowances for Castings*. Geneva, Switzerland: International Organization for Standardization.

ISO. (2015). *ISO 14638: 2015. Geometrical Product Specifications (GPS) – Matrix Model*. Geneva, Switzerland: International Organization for Standardization.

ISO. (2024). *ISO 10303: 2024. Industrial Automation Systems and Integration – Product Data Representation and Exchange. Part 1: Overview and Fundamental Principles*. Geneva, Switzerland: International Organization for Standardization.

Nielsen, H.S. (2012). *The ISO Geometrical Product Specifications Handbook. Find Your Yay in GPS*. Danish Standards.

Personat Martínez, K. (2022). Cálculo, diseño y planificación del proceso de fabricación de una mordaza para taladro. Mechanical engineering degree dissertation. Universitat Jaume I.

Shepherd, D.W. (1992). Are your tolerances too tight? *Quality* 31 (9): 51.

Trucks, H.E. (1987). *Designing for Economical Production*. Dearborn, MI: Society of Manufacturing Engineers.

3

Selection of Manufacturing Processes

3.1 Introduction

Choosing the right material for a part or product is a critical aspect of the design and manufacturing process. Designers are responsible for not only creating products that serve their intended purpose but also selecting the most suitable material. However, the selection of a specific material can limit the available manufacturing options. It is important to recognize that material and process choices are not limitless and are influenced by the organization's manufacturing capabilities and the range of materials and processes available.

During the design phase, manufacturing typically assesses the chosen material. In some cases, manufacturing may suggest an alternative material if the chosen one is not easily manufacturable or if it leads to high manufacturing costs. This prompts a meaningful discussion between manufacturing and design teams. Such discussions require manufacturing to have a thorough understanding of available materials, their properties, and their suitability for different manufacturing processes. This underscores the close connection between material selection and the choice of manufacturing processes (Scallan 2003).

The goal of this chapter is to introduce common materials and manufacturing processes and outline methods for selecting them.

3.2 Materials in Manufacturing

The variety and breadth of materials utilized in modern manufacturing are immense. Nonetheless, the materials utilized in manufacturing can be divided into four major types: composites, ceramics, metals, and polymers.

The most widely used materials in production are metals. Table 3.1 illustrates the further classification of metals into ferrous and nonferrous categories (Scallan 2003).

Ceramics typically consist of a combination of metallic and nonmetallic substances, and they often exhibit greater strength and hardness than metals when exposed to high temperatures. However, ceramics are more prone to brittleness compared to metals, which is a significant drawback associated with their use. Ceramics can be categorized into two main groups based on their composition: oxides and nonoxides. Nonoxide ceramics are also divided into carbides and nitrides ceramics (Table 3.2).

Manufacturing Process Planning: A Practical Approach for Mechanical Engineering, First Edition.
José V. Abellán-Nebot, Carlos Vila Pastor, and Héctor R. Siller.

Companion website: www.wiley.com/go/Manufacturing_Process_Planning_1e

Table 3.1 Common classification of metals.

Metals	Ferrous metals	Iron	Wrought iron	
			Cast iron	Gray cast iron, white cast iron, malleable iron, spheroidal graphite (SG) iron
		Steel	Carbon steel	Low-carbon steel, medium-carbon steel, and high-carbon steel
			Alloy steel	Low alloy
				High alloy: tool steel, high-speed steel, and stainless steel
	Nonferrous metals	Light alloys	Aluminum alloys, titanium alloys, and magnesium alloys	
		Heavy alloys	Copper alloys (copper–zinc, brass, and bronze)	
		Refractory metals	Molybdenum, niobium, tantalum, and tungsten	
		Precious metals	Platinum, gold, and silver	

Table 3.2 Common classification of ceramics and polymers.

Ceramics	Oxide ceramics		Alumina, magnesia, zirconia, and silica
	Nonoxide ceramics	Carbide ceramics	Tungsten carbide, titanium carbide, and silicon carbide
		Nitride ceramics	Cubic boron nitride, titanium nitride, and silicon nitride
Polymers	Thermoplastics		Nylons, acrylics, polyethylenes, and poly(vinylchloride)
	Thermosets		Epoxy resins, phenolics, polyesters, and polyurethanes
	Elastomers		Latex, neoprene, urethane, and silicone

In recent decades, the utilization of polymers in manufacturing, especially as substitutes for other materials, has gained significant importance. A mere observation of the average car is enough to understand the extensive application and significance of polymers as manufacturing materials. For example, take the widespread usage of polymers in bumpers and automotive interiors. The three different categories of polymers are elastomers, thermosets, and thermoplastics, as depicted in Table 3.2.

More recently, great advances in materials that consist of two or more distinct phases bonded together have impacted industries such as aeronautics, space, or automobiles. These materials are named composite materials to distinguish them from those dissolved in one another, such as metal alloys. The main benefits of composites are their high resistance to wear, corrosion, and high-temperature exposure. Furthermore, they can be fabricated in large, complex shapes that would be difficult or impossible to make with other materials and reduce machining, assembly, and fastener costs (Kutz 2015).

Composites are usually classified by the type of material used for the matrix. The three primary categories of composites are organic matrix composites (polymer and carbon matrix), metal matrix composites, and ceramic matrix composites. Furthermore, depending on the reinforcement type, composite materials can be classified into particulate composites, fiber-reinforced composites, and structural composites (Ibrahim et al. 2015). Table 3.3 shows the common classification of composite materials.

Table 3.3 Common classification of composites.

Composite materials (based on matrix)	Metal matrix composites	Matrix: aluminum, magnesium, titanium, lead Particles: silicon carbide, aluminum oxide, graphite, boron
	Organic matrix composites (polymer and carbon matrix composites)	Matrix: epoxy, polyester, phenolic, and graphite Fibers: carbon and glass
	Ceramic matrix composites	Matrix: silicon carbide, aluminum oxide, and silicon nitride Fibers: carbon, aluminum oxide, silicon oxide, and silicon carbide
Composite materials (based on reinforcement)	Particulate composites	Particles in flakes or powder shape (e.g., concrete and wood particles)
	Fiber-reinforced composites	Carbon fibers, glass fibers, aramid fibers, natural fibers, etc.
	Structural composites	Sandwich structures and monolithic structures

3.3 Manufacturing Processes

Manufacturing engineers have a wide range of options when it comes to selecting a suitable manufacturing process. Each process possesses its inherent capabilities in terms of surface finish, dimensional accuracy, and geometric accuracy within certain ranges. Moreover, some processes may be better suitable for producing specific shapes or features. Despite the intrinsic capacities of different manufacturing processes, it is important to note that more than one process can fulfill the manufacturing requirements. Hence, having a good understanding of the available process types is crucial for an optimal manufacturing process selection.

To produce a part, a sequence of manufacturing processes is required, and they are commonly classified as basic or primary processes, secondary processes, property-enhancing processes, and finishing processes (Figure 3.1). Basic or primary processes try to optimize the use of material and obtain a geometry very close to the final shape of the part. These processes are also called net or

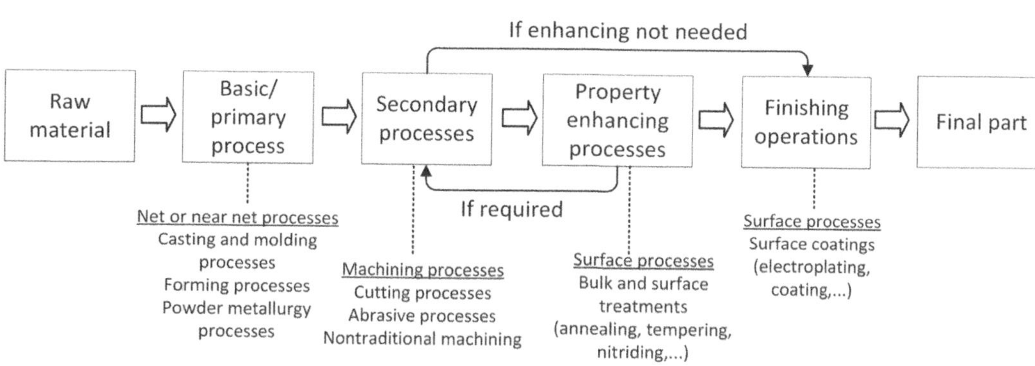

Figure 3.1 A common sequence of manufacturing processes to produce a part.

near net shape manufacturing processes (net processes do not require secondary operations to reach the final geometry) and are mainly casting, molding, forming, and powder metallurgy processes. Secondary processes transform the resultant part from basic processes into the final geometry or very close to it. For instance, after a basic process of casting, machining operations are common secondary processes; after flat rolling processes, stamping operations such as punching or bending may correspond to secondary operations. Processes to enhance material properties may be needed after secondary operations, for instance, heat-treating operations. After processes such as heat treatment, it may be necessary to conduct additional secondary processes to reach the required tolerances. Finally, finishing operations such as powder coating, painting, electroplating, etc., may be conducted at the end of the process.

Process planners should know the main manufacturing processes, their limitations, capabilities, and typical applications to guide them in proper manufacturing process selection. The next subsections provide basic information about the five main groups in manufacturing processes (joining processes are omitted here): casting and molding, forming, powder metallurgy, machining, and surface processes. For a more detailed explanation, readers may refer to the relevant books given in the references (Black and Kohser 2008; Groover 2020; Kalpakjian and Schmid 2009).

3.3.1 Casting and Molding Processes

The wide variety of casting processes enables the manufacturing of parts or products in almost any desired shape or size. Casting processes are known for their efficiency in transforming raw metals into finished parts, making them a rapid manufacturing method. Furthermore, utilizing the appropriate casting process can minimize the requirement for additional postprocessing steps. On the contrary, molding processes refer to similar processes applied to plastic materials.

In casting, the molten metal is poured into a mold cavity that represents the final product. The molten metal cools while heat is transferred through the mold, solidifying into the required shape. Casting processes can be divided into two main categories: expendable mold processes and permanent mold processes.

When using expendable mold methods, the casting is removed by destroying the molds. Sand, plaster, and ceramics combined with a bonding agent are common mold materials in these cases. On the contrary, since the mold is reusable in permanent mold techniques, it needs to be made such that the casting may be removed with ease. Permanent molds are usually composed of metals and refractory coatings (e.g., tool steel alloys) that are strong at high temperatures. Generally, casting processes are employed because they can process materials that would be difficult to handle using other methods. Furthermore, they allow the production of complex shapes with internal cavities or hollow sections, and are often the most cost-effective method of production.

3.3.1.1 General Characteristics of Casting and Molding Processes

Table 3.4 briefly describes each casting and molding process and some applications (Wick et al. 1984). Figure 3.2 shows some characteristics of the main casting processes. As it can be noted, casting/molding methods are versatile in terms of the range of part sizes that may be accomplished and the intricacy of geometric shapes that can be produced. They also show versatility in terms of the materials they can work with, such as ceramics, polymers, and metals. Castability is the degree to which a material can be properly cast while still meeting quality standards. The fluidity and reactivity of the molten material are the main elements impacting castability, while casting techniques can also have an impact. The capacity of the molten material to flow and fill the entire mold cavity is referred to as fluidity.

Table 3.4 Classification, description, and use of different casting processes.

Type	Process	Description	Applications
Expendable mold processes	Sand casting	Mold is made by ramming sand around wood or metal patterns. Sand bonded with clay or chemicals.	Engine blocks, transmission cases, valve casings, and gear blanks.
	Vacuum casting	Like sand casting but instead of clay or chemicals, vacuum is used to attain mechanical bonds.	Airframe components, bulkheads, and hip and knee implant prostheses.
	Shell casting	Sand coated with thermosetting resin is poured over a heated metal pattern to form a shell. Halves are stripped and assembled.	Cylinder heads, connecting rods.
	Plaster casting	Molds are made by casting plaster around patterns. Molds are baked after pattern removal.	Gears, valves, ornaments, fittings, and tooling.
	Ceramic casting	A slurry of ceramic aggregate and binder poured over a pattern and chemically set. The resultant mold is stabilized and cured at high temperature.	Cast steel dies and tools for extrusion, forging, and die casting. Aerospace components.
	Investment casting	A wax pattern is coated with a refractory shell by casting or dipping. The assembly is then heated to melt out the wax pattern. Metal is cast into the ceramic mold.	Turbine blades, armament parts, pipe fittings, lock parts, hand tools, jewelry, and prostheses.
	Full-mold process (lost foam)	The pattern is made from foamed polystyrene. Sand is rammed around the pattern. When metal is poured into the mold, the pattern vaporizes.	Automotive components such as inlet manifolds, heat exchangers, cylinder heads, and blocks.
Permanent mold processes	Low-pressure casting	Gas pressure is used to inject molten metal into a permanent mold.	Suspension, steering, and engine components.
	Die casting	Molten metal is forced into a metallic die under high pressure. In hot-chamber machines (for zinc), metal is pumped into the mold. In cold–chamber machines (for aluminum), metal is ladled into a shot chamber.	Engine components, pump components, appliance housing.
	Gravity casting	Metal is poured into permanent molds made from metal or graphite.	Automotive pistons, gears, gear housings, and wheels.
	Centrifugal casting	A permanent mold for a symmetrical part, such as a pipe, is rotated rapidly around its longitudinal axis. Centrifugal forces distribute molten metal evenly along the mold.	Pipes, wheels, pulleys, and nozzles.
	[a]Injection molding	Polymer is molten and injected into a permanent mold.	Housings, containers, caps, fittings, and plastic casings.
	[a]Compression molding	The polymer is placed into an open, heated mold. The mold closes and compresses the material. The polymer cures in the heated mold.	Hoods, fenders, scoops, and spoilers.

[a] Processes used for polymer processing (molding processes). The rest of processes corresponds to metal processing.
Source: Adapted from Wick et al. (1984).

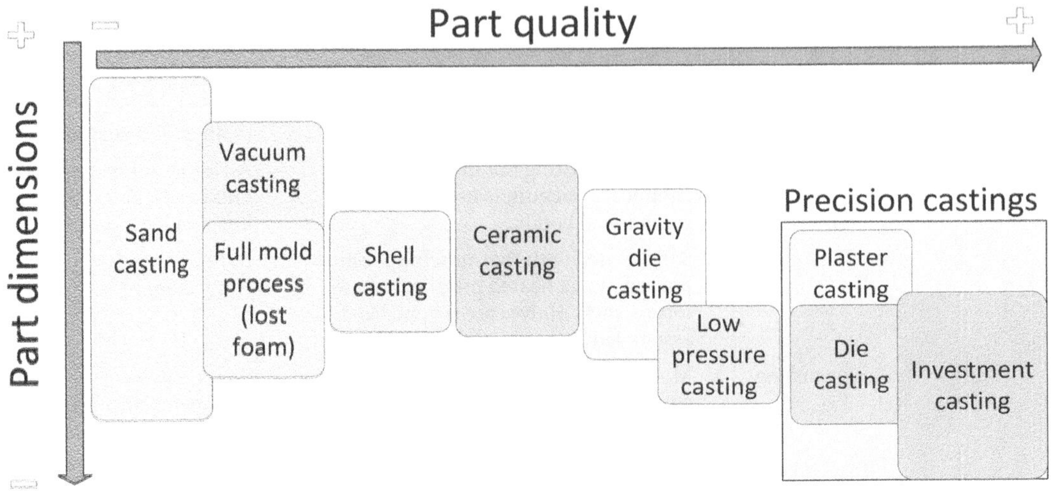

Figure 3.2 Casting processes classification according to part dimensions and part quality. *Source:* Adapted from ECS Staff Report (2006).

Choosing the suitable casting process relies on several considerations, such as the material properties, dimensions, weight, and intricacy of the geometry, as well as labor, equipment, and tooling expenses. Other factors include the desired tolerances and surface finish, required strength, quantity, production rate, and overall quality standards. These characteristics are mentioned for a few of the more often used casting techniques in Tables 3.5 and 3.6 (Scallan 2003).

Previous casting processes apply for specific production volumes, named the economic batch quantity (EBQ), which is defined by different cost terms such as labor/operating cost, equipment/capital cost, and tooling cost. Table 3.6 shows typical ranges for production volume, production rate, and costs for common casting processes. The table also shows common tool life values of their corresponding molds or patterns in the case of expendable mold processes.

3.3.2 Forming Processes

In forming, a solid billet of material is typically shaped through the application of force to achieve the desired shape. While forming can produce intricate shapes, it is not as versatile as casting when it comes to geometric possibilities. However, forming has the advantage of yielding superior microstructure that enhances mechanical properties.

Forming processes can be divided into two main categories: bulk forming and sheet forming. The following lists some typical processes for both categories:

- Bulk-forming. Rolling, drawing, extrusion, and forging.
- Sheet forming. Sheet–metal forming, vacuum forming, blow molding, superplastic forming, and contact molding.

Bulk-forming processes are commonly referred to as deformation processes in metals. In these processes, a tool's compressive loading causes most of the shaping's deformation. The material is typically heated at a temperature higher than its recrystallization temperature during bulk forming. At this temperature, there are no negative effects on the material's mechanical characteristics, and higher deformation ratios can be applied in most metals. If bulk forming is performed at

Table 3.5 Characteristics of casting processes.

Process	Material	Geometry (shape)	Weight (kg)	Minimum thickness (mm)	Maximum thickness (mm)	Typical tolerance (mm)	Surface roughness (μm)
Sand casting	Any metal, but steel difficult	P-C, P-N, 3D-S, 3D-H	0.01–10,000	2.5	No limit	±1.0	8–25
Vacuum casting	Any metal	P-C, P-N, 3D-S, 3D-H	0.3–180	1.5	3	±0.2	1.6–3.2
Shell casting	Any metal except Ti alloys and refractory alloys	P-C, P-N, 3D-S, 3D-H	0.3–10	2.0	50	±0.2	0.8–6.3
Plaster casting	Nonferrous metals (Al, Mg, Cu, Zn)	P-C, P-N, 3D-S, 3D-H	0.02–25	1.0	No limit	±0.05	0.8–3.2
Ceramic casting	Primary stainless steels, tool steels and superalloys	P-C, P-N, 3D-S, 3D-H	0.1–50	1.2	No limit	±0.2	1–6
Investment casting	Any metal	P-C, P-N, 3D-S, 3D-H	0.005–100	1	75	±0.05	0.4–3.2
Full-mold process (lost foam)	Any metal except Ti alloys and refractory alloys	P-C, P-N, 3D-S, 3D-H	0.5–25	3	100	±1.0	4–20
Low-pressure casting	Primary aluminum alloys	P-C, P-N, 3D-S, 3D-H	5–25	4	10	±0.2	1–6
Die casting	Nonferrous metals (Al, Mg, Zn, and Cu)	P-C, P-N, 3D-S, 3D-H	0.05–300	0.5	12	±0.05	0.4–3.2
Gravity die casting	Nonferrous metals (Al, Mg, Zn, and Cu). Some iron.	P-C, P-N, 3D-S, 3D-H	0.5–300	2	50	±0.1	1–6
Centrifugal casting	Any metal and some ceramics	P-C, P-N, 3D-S, 3D-H	1–5,000	2	100	±0.2	1.6–12
[a]Injection molding	Most thermoplastics, but also some thermosets, composites, and elastomers	P-C, P-N, 3D-S, 3D-H	6–25	0.4	13	±0.05	0.2–0.8
[a]Compression molding	Most thermoplastics, but also some thermosets, composites, and elastomers	3D-S; F-S; F-D	0.05–15	0.25	25	±0.05	0.2–0.8

3D-H, three-dimensional (hollow); 3D-S, three-dimensional (solid); P-C, prismatic (circular); P-NC, prismatic (noncircular); S-D, sheet (dished); S-F, sheet (flat).

[a] Processes used for polymer processing (molding processes). The rest of processes corresponds to metal processing.

Source: Adapted from Kalpakjian and Schmid (2009) and Scallan (2003).

Table 3.6 Casting process. Economic considerations.

Process	Economic batch quantity (EBQ)	Production rate (parts/h)	Operating cost ($/h)	Capital cost ($)	Tooling cost ($)	Tool life (units)
Sand casting	1–100,000	1–20	40–140	1,500–8,000	150–1,500	100–1,000
Vacuum casting	1–100	1–10	40–140	1,700–9,000	170–1,000	1–5
Shell casting	1,000–100,000	5–50	40–140	1,700–17,000	1,700–9,000	1,000–10,000
Plaster casting	50–5,000	1–10	40–140	1,700–9,000	340–2,000	500–5,000
Ceramic casting	50–1,000	1–10	40–140	900–9,000	150–2,000	50–500
Investment casting	1–10,000	1–200	60–200	300–1,500	50–400	1–10
Full-mold process (lost foam)	1–1,000	1–10	40–140	500–3,000	50–300	1–15
Low-pressure die casting	2,000–100,000	5–50	80–250	15,000–150,000	8,000–23,000	30,000–100,000
Die casting	10,000–1,000,000	2–200	80–250	150,000–800,000	6,000–90,000	20,000–1,000,000
Gravity die casting	1,000–10,000	5–50	80–250	15,000–60,000	5,000–20,000	10,000–100,000
Centrifugal casting	10–10,000	1–20	80–250	2,000–20,000	600–6,000	100–500
Injection molding	10,000–1,000,000	60–1,000	40–110	30,000–600,000	3,000–30,000	10,000–1,000,000
Compression molding	500–2,000,000	10–100	40–110	15,000–80,000	1,500–15,000	2,000–200,000

Source: Kalpakjian and Schmid (2009) and Scallan (2003).

temperatures below recrystallization, only ductile materials can be processed. Rolling, drawing, extrusion, and forging are examples of bulk processing, hot rolling being the most popular method.

As the name implies, sheet forming is used as raw material sheets with high surface area to thickness ratios. Bending, stretching, or shearing loads are typically applied to produce the form change. Similar to bulk-forming techniques, successful sheet forming depends on the ductility of the material being used. Vacuum forming, blow molding, and sheet–metal forming (including bending, deep drawing, and stretch forming) are all examples of sheet-forming techniques.

3.3.2.1 General Characteristics of Bulk Forming

The selection of a suitable forming process will depend on several factors, including the material, size, weight, complexity of the geometry, labor expenses, tooling costs, necessary tolerances and surface finishes, strength, quantity, and production rate. Tables 3.7 and 3.8 list these parameters for a few of the more popular bulk-forming methods (Scallan 2003). For general characteristics in sheet forming, please refer to Kalpakjian and Schmid (2009) and Scallan (2003).

3.3.3 Powder Metallurgy Processes

Powder processing, also known as powder metallurgy, is the process of molding parts made of metal powders by applying heat and pressure, sometimes at the same time. The powder metallurgy process usually consists of four main phases, with extra processing added if needed. First, the metal powder must be combined with any necessary additives, including binding agents or lubricants. Then, the powder mixture is compressed. In hot compacting, heat, and pressure are applied simultaneously, while in cold compacting, the metal powder is just subjected to pressure. When it comes to heat compacting, the procedure is finished unless further processing is required. However, with cold compacting, after the compacting is complete, the part goes through a procedure called

Table 3.7 Characteristics of bulk-forming processes.

Process	Material	Geometry (shape)	Weight (kg)	Minimum Thickness (mm)	Maximum thickness (mm)	Typical tolerance (mm)	Surface roughness (μm)
Shape rolling	Most metals	P-C; P-NC	10–1,000	0.5	No limit	±0.5	1–25
Shape drawing	Steel, Al, and Cu alloys (ductile metals)	P-C; P-NC	10–1,000	0.1	10	±0.1	0.2–0.8
Cold heading and impact extrusion	Steel, Al, and Cu alloys (ductile metals)	P-C; P-NC; 3D-S	0.1–50	1	No limit	±0.1	0.4–3.2
Hot extrusion	Most metals	P-C; P-NC	1–500	1	No limit	±0.1	1–25
Hot forging (open die)	Most metals	P-C; P-NC; 3D-S	0.1–200,000	5	No limit	±0.5	1–25
Hot forging (impression die)	Most metals	P-C; P-NC; 3D-S	0.1–100	0.1	No limit	±0.5	1–25

3D-H, three-dimensional (hollow); 3D-S, three-dimensional (solid); P-C, prismatic (circular); P-NC, prismatic (noncircular); S-D, sheet (dished); S-F, sheet (flat).

Source: Adapted from Scallan (2003).

Table 3.8 Bulk-forming process selection table. Economic considerations.

Process	EBQ	Production rate	Operating cost ($/h)	Capital cost ($)	Tooling cost ($)	Tool life (m or parts)
Shape rolling	100–10,000 m	2–600 m/h	60–180	200,000–2,000,000	5,000–30,000	10,000–1,000,000
Shape drawing	>1,000 m	100–60,000 m/h	30–90	9,000–90,000	2,000–20,000	1,000–100,000
Cold heading	20,000–10,000,000 parts	1,000–10,000 parts/h	30–100	200,000–1,000,000	500–10,000	10,000–100,000
Impact extrusion	5,000–10,000,000 parts	100–10,000 parts/h	40–120	200,000–2,000,000	2,000–9,000	10,000–100,000
Hot extrusion	50–10,000 m	1–300 m/h	60–200	20,000–2,000,000	1,000–5,000	100–100,000
Hot forging (open die)	1–10,000 parts	1–50 parts/h	60–250	300,000–900,000	100–1,000	1,000–10,000
Hot forging (impression die)	100–10,000,000 parts	10–300 parts/h	60–200	400,000–900,000	7,000–20,000	1,000–1,000,000

Source: Adapted from Kalpakjian and Schmid (2009) and Scallan (2003).

sintering. Sintering involves the controlled application of heat to facilitate the formation of metallurgical bonds among the powdered metal particles. The temperatures employed typically reach around 70–80% of the melting point of the material. Following sintering, additional processing might be necessary. The powder metallurgy process is employed for the mass production of small, intricate geometries that are challenging to fabricate using alternative methods, often eliminating the need for secondary processing. For general characteristics in powder metallurgy processes, please refer to Kalpakjian and Schmid (2009) and Scallan (2003).

3.3.4 Machining Processes

A range of operations falls under the heading of machining procedures, which entail removing material from the workpiece. These can be divided into three major subcategories, cutting processes, abrasive processes, and nontraditional machining processes (Figure 3.3).

The great majority of machining procedures employed are classified as cutting procedures because they remove material from the workpiece in the form of chips, leaving the completed geometry behind. The part quality in these processes ranges from IT6 in boring operations to IT13-12 in drilling or shaping operations. For better dimensional and surface roughness quality, abrasive processes are required. Under these processes, IT5-01 and surface roughness up to 0.01 μm can be reached. Nontraditional machining processes are applied under special circumstances when cutting processes cannot be applied or may be uneconomical. For instance, machining very deep and small holes in tool steels requires electrodischarge machining or machining at micro- or nanoscale requires chemical machining.

Machining may be the most cost-effective way of manufacturing when producing small batches. Additionally, there are several reasons why machining operations are used, including the necessity for improved surface quality that cannot be achieved by casting or forming processes alone and the

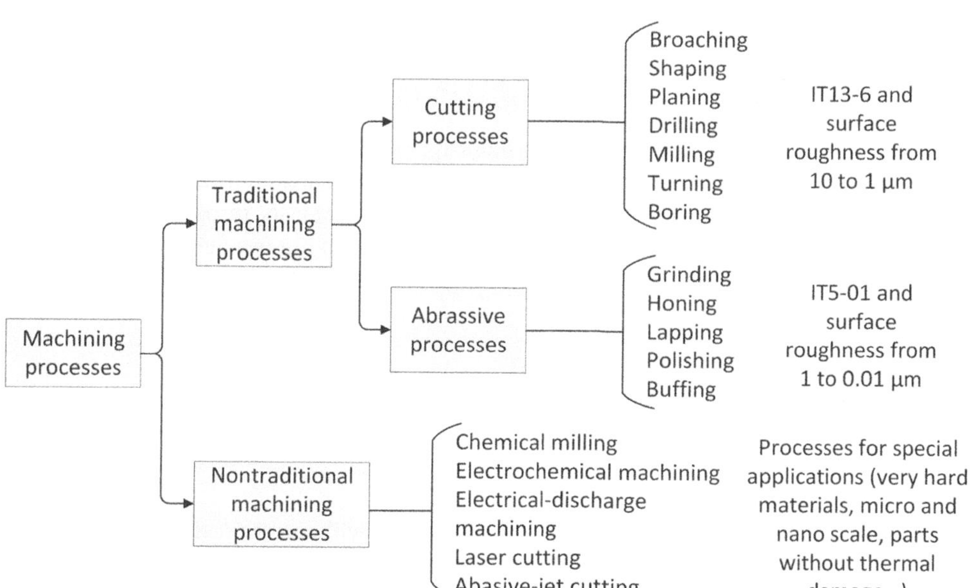

Figure 3.3 Machining processes classification with dimensional quality (IT) and surface roughness ranges.

need for tighter dimensional precision. However, there are several clear drawbacks to using machining operations. By their own nature, they produce waste material; they demand more capital, energy, and labor per unit of output than casting and forming operations; and they often take longer than casting and forming procedures. Despite these drawbacks, machining techniques are prevalent and play an essential role in manufacturing.

3.3.4.1 General Characteristics of Machining Processes

A multitude of parameters will determine which machining process is best. Among them, the most relevant are the geometrical quality achievable and the costs (labor, equipment, and tooling costs). The surface roughness and main geometrical tolerances for most common machining processes are shown in Table 3.9. In terms of economy, the processes are listed in order from the cheapest to the most expensive one. Figure 3.4 also shows graphically common surface roughness ranges of operation and relative costs of machining processes and other manufacturing processes.

Furthermore, machining processes require a minimum machining allowance for proper material removal. Table 3.10 shows typical machining allowances required for cutting and abrasive processes.

For nontraditional machining processes, the main characteristics related to typical tolerances, surface roughness, costs, and production rates can be found in Table 3.11.

3.3.5 Surface Processes

Surface processing is usually done after all manufacturing processing is complete, apart from heat treatments. Bulk treatments, surface treatments, surface coatings, and cleaning procedures are the most popular surface procedures (Table 3.12).

Bulk treatments such as annealing, precipitation hardening, etc., alter the microstructure and subsequently the properties of the majority of the material being treated, whereas surface treatments (e.g., anodizing, nitriding, etc.) only alter the mechanical properties of the material's surface, usually corrosion resistance and hardness. Table 3.13 presents some types of bulk and surface treatments and their effects on the material.

Similarly, surface coatings are used to improve mechanical properties at the surface level. To increase corrosion resistance, impact resistance, and thermal barrier properties, metal surfaces are coated with materials such as carbides, powdered metals and alloys, and ceramics. Lastly, throughout processing, a number of parts and/or goods may pick up surface imperfections. To remove these surface impurities, cleaning techniques such as solvent cleaning, pickling, and vapor degreasing can be used.

3.4 Manufacturing Process Selection

3.4.1 Factors in Process Selection

The critical factors to be considered in manufacturing process selection are:

- Shape or geometry considerations. The geometry of the part will constrain the manufacturing process that can be used. For instance, hollow parts cannot be manufactured by forging processes, or very thin walls are not possible to be cast in sand casting processes. The size and weight of the parts are also critical for process selection.

Table 3.9 Characteristics of traditional/conventional machining processes.

Process	Surface roughness (μm)	Dimensional tolerances		Geometrical tolerances (mm)			
		(mm)	International tolerance (IT) grade	Cylindricity	Flatness/ straightness	Parallelism/ perpendicularity	Concentricity
Drilling	0.4–6	±0.1	13–9	0.05–0.1	0.04	0.1–0.2	0.1
Turning	0.1–12	±0.02	13–7	0.01	0.01	0.01–0.02	0.01
Milling	0.5–20	±0.05	13–8	0.02	0.02	0.02	—
Boring	0.1–2	±0.02	9–6	0.005–0.01	0.005	0.005–0.01	0.005–0.01
Broaching	0.4–6	±0.02	9–6	0.01	0.01	0.01–0.02	—
Grinding	0.05–2	±0.005	6–4	0.004	0.004	0.002–0.005	0.002–0.005
Honing	0.025–0.5	±0.002	5–2	0.002	0.002	0.002	0.002
Lapping	0.01–0.4	±0.002	3–01	0.001	0.001	0.002	0.002

Source: Adapted from Scallan (2003), Chapter 4 and Halevi and Weill (1995), Chapter 6.

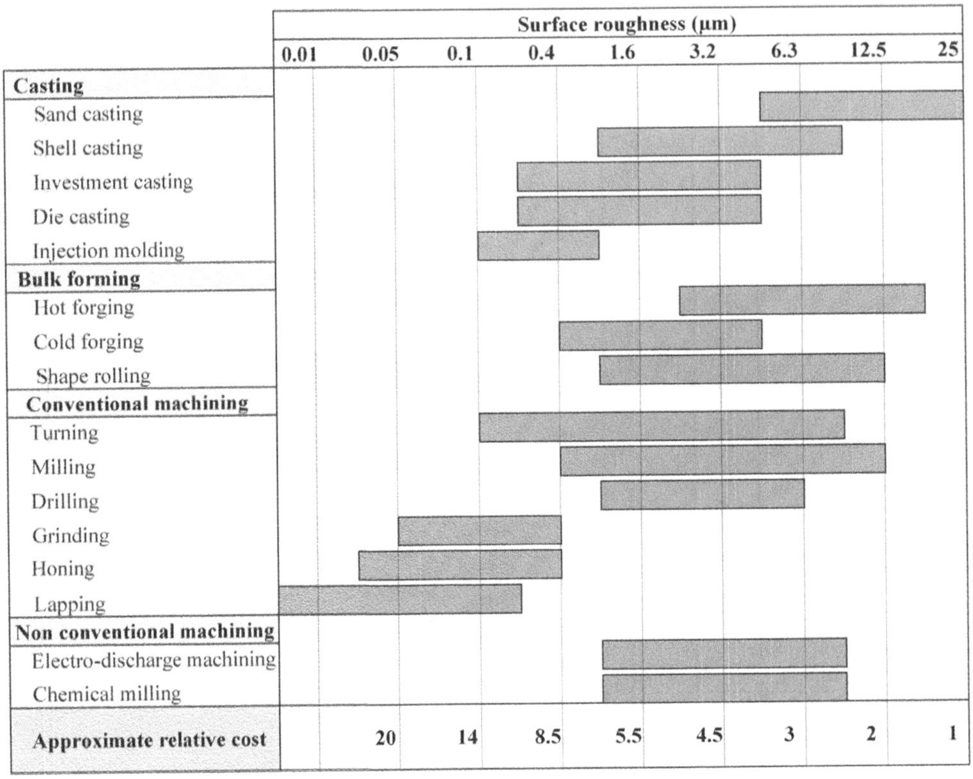

Figure 3.4 Surface finishes for some common manufacturing processes. *Source:* Adapted from Black and Kohser (2008), Chapter 35.

Table 3.10 Minimum machining allowance for different machining processes.

Type of machining process		Minimum allowance (mm)
Parts with cutting operations (milling and turning)	Finishing operations	0.25–0.15
Parts with abrasive operations	Grinding	0.2–0.1
	Honing	0.15–0.1
	Superhoning	Around 0.05

Source: Adapted from Zamanillo (1993).

- Material and manufacturing process compatibility. Since not all materials are suitable for all manufacturing processes, the material itself will restrict the techniques that can be employed. For instance, steel cannot be cast in plaster molding since plaster would be burnt at steel pouring temperatures.
- Dimensional and tolerancing compatibility. Dimensional and geometric tolerances will limit the potential selection of manufacturing processes. The capability of manufacturing processes has to match the quality specifications of the parts, or if not, the global selection of primary and secondary processes should match. The same applies to surface roughness specifications.

Table 3.11 Characteristics of nontraditional machining processes.

Process	Material	Weight/size	Typical tolerance (mm)	Surface finish (μm)	Relative cost	Production rate
Chemical milling	Most materials	No limit	±0.005–0.3	0.4–6	Low	Low
Electrochemical machining	Any electrical conductive metal	No limit	±0.01–0.25	0.2–6	Medium–high	Medium
Electrodischarge machining	Any electrical conductive metal	No limit	±0.005–0.08	0.4–12	Medium–high	Low
Laser cutting	Any metal	Up to 40-mm steel thickness[a]	±0.03–0.08	0.8–6	High	High
Ultrasonic machining	Any material	No limit	±0.005	0.4–6	Medium	Low

[a] For a 20-kW fiber laser cutting machine.
Source: Adapted from Scallan (2003).

Table 3.12 General classification of surface processes.

Surface processes	Bulk treatments	Solution treatments	
		Precipitation hardening, Induction hardening	
		Annealing	Process annealing
			Stress–relief annealing
			Tempering
	Surface treatments	Anodizing, carburizing, nitriding, carbonitriding, and boronizing	
		Flame hardening	
		Shot peening	
		Burnishing	
	Surface coatings	Thermal spraying and ceramic coating	
		Electroplating and electroless plating	
		Physical vapor deposition and chemical vapor deposition	
	Cleaning processes	Mechanical cleaning	Wire brushing and belt sanding, buffing and tumbling
		Chemical cleaning	Alkaline cleaning, solvent cleaning, vapor degreasing, pickling, and ultrasonic cleaning

Source: Adapted from Scallan (2003).

- Economic factors. The production cost is highly dependent on the type of manufacturing process used. For example, depending on the production volume, some manufacturing processes may be more cost-effective than others. Therefore, it is important to estimate manufacturing costs in advance to select the optimal process. Key factors to consider include the required manufacturing volume, production rates, and economic batch size.

Table 3.13 Selected treatments and their effects.

Treatment	Description	Effects	Materials
Precipitation hardening	Metal that has been heated to a moderate temperature and maintained there	Enhances strength and hardness	Al and Cu alloys Some stainless steels
Annealing	The air cools after the metal is heated to a predetermined temperature	Enhances ductility	Low-carbon steels
Tempering	Metal that has been heated to a certain temperature and cooled gradually	Decreases brittleness, increases ductility, and toughness	Most steels
Anodizing	Using metal submerged in an acidic solution as an anode to oxidize its surface	Enhances corrosion resistance	Al, Mg, Zn, and Ti alloys
Carburizing	Metal that has been heated with carbon dioxide and then cooled	Enhances hardness	Low-carbon steels, alloy steels
Nitriding	Ammonia gas-heated material that is subsequently cooled	Enhances hardness	Most steels
Electroplating	Metals coated with metals by using them as cathodes in electrolyte solutions	Differs on plating materials (increases corrosion resistance, strength, etc.)	Most metals

Source: Adapted from Scallan (2003).

3.4.2 Process Selection Method

Process selection requires the use of a process selection method to make the process more methodical and reduce the complexity of this activity. The proposed process selection has been modified from (Scallan 2003) to help practitioners comprehend the intricacies of the process while also giving them a methodical approach to follow. When employing the presented methodology, the following two fundamental assumptions are made:

- The material was chosen first, and it has been indicated at the design stage.
- Every piece of information found in the design documents – parts lists, drawings, and other materials – is complete, and interpreting the engineering drawings can yield all the information needed for production (production volume, production rates, etc.).

There are six steps to follow in the proposed process selection method:

1) Drawing interpretation:
   a) Identify material.
   b) Identify geometry class.
   c) List general quality characteristics (general tolerances and surface roughness).
   d) List specific key product characteristics (specific features that require tighten tolerances or roughness requirements).
   e) List special features, that is, features that may require nonconventional operations (electrodischarge machining operations, etc.).
   f) List surface processes required, that is, heat treatments, coatings, etc.

2) Material and compatibility processes:
   a) List the potential processes according to material and production volume.
3) Geometry and process compatibility:
   a) Discard noncompatible processes due to geometry constraints.
4) Manufacturing processes technically feasible:
   a) Reduce the previous list of potential processes by considering:
      i) Geometry:
         1) Size.
         2) Weight.
         3) Thickness.
      ii) Tolerances and surface roughness.
5) Manufacturing processes economically feasible:
   a) Reduce the previous list of potential processes by considering:
      i) EBQ.
6) Identifying a process:
   a) From the remaining manufacturing process candidates, select the most economical one considering the following features:
      i) Operating costs.
      ii) Capital costs.
      iii) Tooling costs and tool life.
      iv) Secondary processes costs.

3.4.2.1 Drawing Interpretation

As specified in Chapter 2, the three-dimensional (3D) model/drawing interpretation initiates process selection. After identifying the material, the geometry of the part should be identified according to the geometry classification shown in Figure 3.5. According to Ashby (1999), the following shapes can be distinguished: prismatic-circular (P-C); prismatic noncircular (P-NC); sheet-flat (S-F); sheet-dished (S-D); 3D solid (3D-S), and 3D hollow (3D-H).

Then, the drawing is analyzed to complete the report shown in Table 3.14. Information that should be included is surface finish, dimensional and geometric tolerances, special features that require nontraditional manufacturing methods, etc.

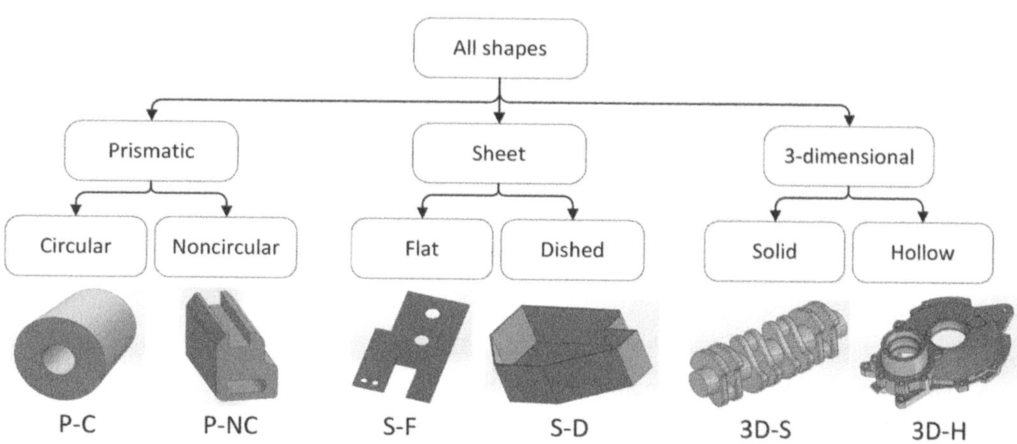

Figure 3.5 Geometry classification. *Source:* Adapted from Ashby (1999).

Table 3.14 Report for drawing interpretation.

Part name	**–**
Material	—
Geometry class	—
General tolerance	—
General surface roughness	—
Specific key product characteristics (dimensional and geometric tolerances)	—
Specific key product characteristics (surface roughness)	—
Special features (features that require nonconventional operations)	—
Surface processes (heat treatment, coating, etc.)	—

3.4.2.2 Material and Process Compatibility

Once the drawing has been analyzed, the next step is to study the compatibility between manufacturing processes and materials. Although the information about materials applied in different manufacturing processes has been reviewed above previously, a more straightforward view of material and process compatibility can be conducted using Tables 3.15 and 3.16 if only casting/molding, bulk-forming, and sheet-forming processes are considered.

3.4.2.3 Geometry and Process Compatibility

From the previous list of candidate processes, the geometry of the product will constrain the potential use of a specific manufacturing process. The geometry compatibility for each manufacturing process is shown in the tables from Section 3.3; however, Table 3.17 can be used for a more straightforward selection of geometry and process compatibility. The report shown in Table 3.18 can be used to discard the noncompatible processes and narrow the list of candidates.

3.4.2.4 Manufacturing Processes Technically Feasible

Then, for the remaining list of manufacturing processes, the following critical factors are checked:

- Geometry (size, weight, and thickness).
- Tolerances (dimensional tolerances, geometric tolerances, and surface roughness).

The template shown in Table 3.19 can be used for this selection process. The tables in Section 3.3 provide the information about the process capability in terms of weight, size, tolerances, surface roughness, etc.

3.4.2.5 Manufacturing Processes Economically Feasible

At this step, only technically feasible manufacturing processes are kept in the list of candidates to analyze their economic feasibility. The item to be studied is the EBQ. Although the analysis of the economic feasibility of a manufacturing process depends on many factors, an

Table 3.15 Compatibility between materials and manufacturing processes. Casting and molding processes.

Process	Cast iron	Carbon steel	Alloy steel	Stainless steel	Al alloys	Cu alloys	Zn alloys	Mg alloys	Ti alloys	Ni alloys	Thermo-plastics	Thermoset plastics
Sand casting	●	●	●	●	●	●	—	●	—	●	X	X
Investment casting	—	●	●	●	●	●	—	—	—	●	X	X
Die casting	X	X	X	X	●	—	●	●	X	X	X	X
Vacuum casting	●	●	●	●	●	●	—	●	—	●	X	X
Shell casting	●	●	—	—	●	●	—	—	—	●	X	X
Plaster casting	X	X	X	X	●	●	●	●	X	X	X	X
Ceramic casting	—	●	●	●	●	●	—	—	—	●	X	X
Full-mold process	●	●	●	●	●	●	—	●	—	●	X	X
Low-pressure casting	X	X	X	X	●	—	—	●	X	X	X	X
Gravity die casting	—	X	X	X	●	●	●	●	X	X	X	X
Centrifugal casting	●	●	●	●	●	●	—	●	—	●	X	X
Injection molding	X	X	X	X	X	X	X	X	X	X	●	—
Compression molding	X	X	X	X	X	X	X	X	X	X	●	●

●, normal practice; —, less-common practice; X, not applicable
Source: Adapted from Ashby (1999).

Table 3.16 Compatibility between materials and manufacturing processes. Bulk-forming and sheet-forming processes.

Process	Cast iron	Carbon steel	Alloy steel	Stainless steel	Al alloys	Cu alloys	Zn alloys	Mg alloys	Ti alloys	Ni alloys	Thermo-plastics	Thermoset plastics
Hot extrusion	X	●	—	—	●	●	X	●	—	—	X	X
Hot forging (impression die)	X	●	●	●	●	●	X	●	—	—	X	X
Cold forging and extrusion	X	●	●	—	●	●	—	—	X	—	X	X
Cold heading	X	●	●	●	●	●	—	—	—	—	X	X
Impact Extrusion	X	—	—	—	●	—	—	—	—	—	X	X
Hot forging (open die)	X	●	●	●	●	●	—	●	●	●	X	X
Shape rolling	X	●	●	●	●	●	—	—	—	—	X	X
Shape drawing	X	●	●	●	●	●	—	—	—	—	X	X
Spinning	X	●	●	●	●	●	●	●	●	●	X	X
Sheet–metal shearing	X	●	●	●	●	●	●	●	●	●	X	X
Sheet–metal forming	X	●	●	●	●	●	●	●	●	●	X	X

●, normal practice; —, less-common practice; X, not applicable.
Source: Adapted from Ashby (1999).

Table 3.17 Geometry and process compatibility.

Process	Circular prismatic (P-C)	Noncircular prismatic (P-NC)	Flat sheet (S-F)	Dished sheet (S-D)	Three-dimensional solid (3D-S)	Three-dimensional hollow (3D-H)
Sand casting	●	●	—	—	●	●
Investment casting	●	●	—	—	●	●
Die casting	●	●	—	—	●	●
Vacuum casting	●	●	—	—	●	●
Shell casting	●	●	—	—	●	●
Plaster casting	●	●	—	—	●	●
Ceramic casting	●	●	—	—	●	●
Full-mold process	●	●	—	—	●	●
Low-pressure casting	●	●	—	—	●	●
Gravity die casting	●	●	—	—	●	●
Centrifugal casting	●	●	—	—	●	●
Injection molding	●	●	●	●	●	●
Compression molding	—	—	●	●	●	
Hot extrusion	●	●	—	—	—	—
Hot forging (impression die)	●	●	—	—	●	—
Cold forging and extrusion	●	●	—	●	●	—
Hot forging (open die)	●	●	—	—	●	—
Shape rolling	●	●	—	—	—	—
Shape drawing	●	●	—	—	—	—
Spinning	—	—	●	●	—	—
Sheet–metal shearing	—	—	●	●	—	—
Sheet–metal forming	—	—	●	●	—	—

●, normal practice.
Source: Adapted from Ashby (1999).

Table 3.18 Report for geometry compatibility of manufacturing processes.

	Part	Process 1	Process 2	Process 3	Process 4
Process name	—	—	—	—	—
Geometry compatible with the manufacturing process					
Geometry class	—	—	—	—	—
Is the process adequate?	—	—	—	—	—

Table 3.19 Report for technical feasibility of manufacturing processes.

	Part	Process 1	Process 2	Process 3	Process 4
Process name	—	—	—	—	—
Process capability					
Size	—	—	—	—	—
Weight	—	—	—	—	—
Thickness (maximum)	—	—	—	—	—
Thickness (minimum)	—	—	—	—	—
Dimensional tolerance	—	—	—	—	—
Surface roughness	—	—	—	—	—
Is the process adequate?	—	—	—	—	—

Table 3.20 Report for the economic feasibility of manufacturing processes.

	Part	Process 1	Process 2	Process 3	Process 4
Process name	—	—	—	—	—
Economic capability					
Production volume and EBQ	—	—	—	—	—
Is the process adequate?	—	—	—	—	—

easy way for process selection at a very early stage is to discard the processes that do not match their EBQ with the required production volume (Ashby 1999). Template from Table 3.20 can be used at this step.

3.4.2.6 Identifying a Process

After previous steps, a unique manufacturing process candidate may have been reached. However, in many cases, there are several manufacturing processes that can fulfill part specifications within the given constraints. To select one of these process candidates based on the production volume, further economic analysis in terms of labor, equipment, and tooling costs must be considered. By estimating these costs, a break-even analysis can be conducted to choose between different manufacturing processes (Figure 3.6).

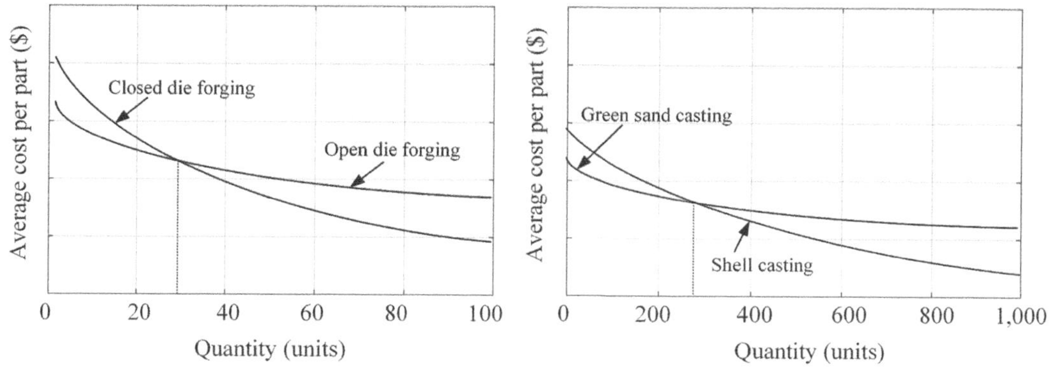

Figure 3.6 Examples of break-even analysis for different manufacturing processes applied to a specific part design. *Source:* Adapted from Ashby (1999).

Given a required batch quantity, n, and the manufacturing process costs (i.e., operating cost, capital cost, and tooling cost), a general part cost estimation can be obtained as follows:

$$\text{Cost per unit} = \text{Material cost} + \frac{\text{Tooling cost}}{\min(\text{Tool life}, n)} + \frac{\text{Operating cost}}{\text{Production rate}} + \frac{\text{Capital cost} \times a}{n}$$

where a is the amortization of the machinery during the production volume. Table 3.21 shows material cost for different alloys although this cost may be omitted for comparing different manufacturing processes if all compared processes are net or near net processes.

It should be noted that, depending on the primary process selected, the number of secondary operations may change. For instance, a more accurate casting process, such as investment casting with a manufacturing tolerance of ±0.05 mm, will require fewer secondary machining operations than shell molding processes where manufacturing tolerances are around ±0.2 mm. Therefore, for cost comparison, the secondary machining operations should be roughly estimated at this stage to include them in the cost per part. A straightforward way for early cost estimation of machining is to calculate the material volume to be removed, a general material removal rate according to the workpiece material (Table 3.22 shows some reference values for conventional machining), and the hourly rate. Hourly rates in machining may range from $60 to 90/hour for common machining operations.

Considering the cost from the above equation and estimating the cost of secondary processes, the template shown in Table 3.23 can be filled to conduct a final process selection based on a rough cost estimation per unit.

Table 3.21 Density and costs of different metal alloys. Data gathered from different databases and website sources.

Generic alloy	Density (kg/m^3)	Cost ($/kg)	Generic alloy	Density (kg/m^3)	Cost ($/kg)
Ductile iron	7,100	2–3.5	Aluminum 6061	2,710	2.0–4.0
Gray iron	6,920	1.5–2.5	Aluminum 7075	2,710	7.0
Low carbon steel	7,830	0.6–1.0	Magnesium alloys	1,800	4.0–10.0
High carbon steel	7,830	1.0–2.5	Bronce	8,800	5.0–15.0
Austenitic stainless steels	7,750	2.5–5.0	Ferritic stainless steels	7,750	3.0–6.0

Table 3.22 Common material removal rates in conventional machining.

Generic alloy	Material removal rate (cm^3/min)	Generic alloy	Material removal rate (cm^3/min)
Aluminum and magnesium (soft)	1,300	Cast iron (hard)	300
Aluminum and magnesium (hard)	650	Titanium (soft)	460
Brass and bronze (soft)	630	Titanium (hard)	260
Brass and bronze (hard)	300	Steel (150 BHN)	250
Cast iron (soft)	540	Steel (600 BHN)	100

Source: Adapted from Kennametal Inc.

Table 3.23 Report for final process selection.

		Process 1	Process 2	Process 3	Process 4
Process name		—	—	—	—
Processing costs					
Primary process	Material cost	—	—	—	—
	Production rate	—	—	—	—
	Operating cost	—	—	—	—
	Capital cost	—	—	—	—
	Tooling cost	—	—	—	—
	Tool life	—	—	—	—
	Cost per unit	—	—	—	—
Secondary processes	Removed volume	—	—	—	—
	Material removal rate	—	—	—	—
	Hourly rate	—	—	—	—
	Cost per unit	—	—	—	—
	Total cost per unit	—	—	—	—

3.4.3 Practical Example

Identify suitable manufacturing processes for the casing part shown in Figure 3.7.

a) Drawing interpretation
The material of the part is gray iron, and considering the geometry of the part in Figure 3.7, the most suitable classification is prismatic circular (P-C). The general dimensional tolerance is ±0.75 mm, and the general surface specification is 12.5 μm. Lastly, the production volume is 3,000 units in 3 months.

Analyzing the critical product characteristics, Table 3.24 is completed.

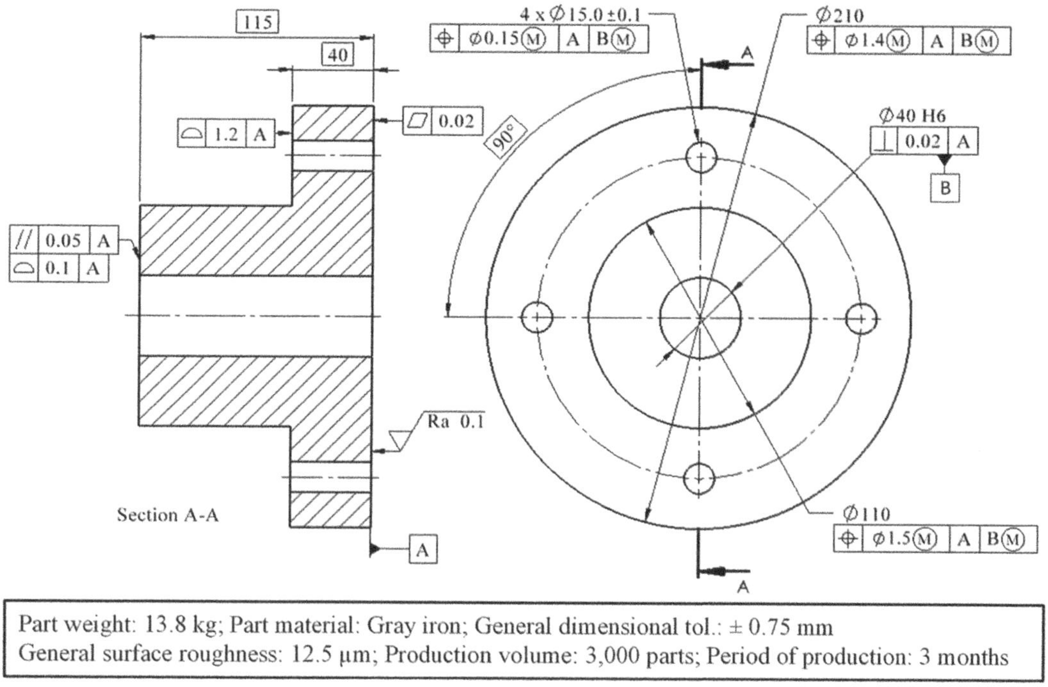

Figure 3.7 Casing for the case of study and part and production characteristics.

Table 3.24 Report for drawing interpretation. Case study.

Part name	**Casing**
Material	Gray iron
Geometry class	P-C
General tolerance	±0.75 mm
General surface roughness	12.5 μm
Specific key product Characteristics (dimensional and geometric tolerances)	• Ø40H6 mm with 0.02 mm perpendicularity tolerance. Requires machining. • Ø15 ± 0.1 mm and position tolerance of 0.15 mm may require drilling. Depends on the casting precision. • Surface with parallelism tolerance of 0.05 mm and profile of 0.1 mm requires machining. • Surface with flatness of 0.02 mm requires machining.
Specific key product characteristics (surface roughness)	• The base of the casing, with a surface finish of 0.1 μm. Requires turning/milling and grinding.
Special features (features that require nonconventional operations)	• None
Surface processes (heat treatment, coating, etc.)	• None

b) Material and process compatibility

According to the material and process compatibility (Tables 3.15 and Table 3.16) and limiting the example to casting processes, the following processes are potential candidates.

- Sand casting.
- Vacuum casting.
- Shell molding.
- Gravity die casting.
- Centrifugal casting.
- Investment casting.
- Ceramic mold casting.
- Full-mold process.

c) Geometry and process compatibility

From the previous list, all candidates are adequate from the geometrical point of view of the part as shown in Table 3.25.

d) Manufacturing processes technically feasible

From the previous list, the process capability is analyzed, filling the technical information (Table 3.26). As a result, there are six possible processes that could be used, and two processes – sand casting and full-mold processes – are discarded.

e) Manufacturing processes economically feasible

Considering the production volume and the EBQ, vacuum casting and ceramic mold casting are also discarded, as shown in Table 3.27.

Therefore, shell molding, gravity die casting (less common in practice for this material), centrifugal casting and investment casting are the final potential manufacturing processes for manufacturing the part.

Table 3.25 Report for geometry compatibility of manufacturing processes. Case study.

	Part	Process 1	Process 2	Process 3	Process 4	Process 5	Process 6	Process 7	Process 8
Process name	--------	Sand casting	Vacuum casting	Shell molding	Gravity die casting[a]	Centrifugal casting	Invest. casting	Ceramic mold casting	Full-mold process
Geometry compatibility									
Geometry class	P-C	[b]	[b]	[b]	[b]	[b]	[b]	[b]	[b]
Is the process adequate?	--------	Yes	Yes	Yes	Yes	Yes	Yes	Yes	Yes

[a] Less common practice.

[b] P-C, P-NC, 3D-S, 3D-H.

Table 3.26 Report for technical feasibility of manufacturing processes. Case study.

	Part	Process 1	Process 2	Process 3	Process 4	Process 5	Process 6	Process 7	Process 8
Process name	--------	Sand casting	Vacuum casting	Shell molding	Gravity die casting	Centrifugal casting	Invest. casting	Ceramic mold casting	Full-mold process
Process capability									
Size	Ø210×115 mm	—	—	—	—	—	—	—	—
Weight (kg)	13.8	0.05–No limit	0.3–180	0.05–100	0.05–300	1–5,000	0.005–100	0.1–50	0.5–25
Thickness (mm) (maximum)	50	No limit	3	50	50	100	75	—	100
Thickness (mm) (minimum)	40	2.5	1.5	2.0	2	2	1	1.2	3
Dimensional Tol (mm)	±0.75	±1	±0.2	±0.2	±0.1	±0.2	±0.05	±0.2	±1.0
Surface roughness (μm)	12.5	3.2–25	1.6–3.2	0.8–6.3	1–6	1.6–12.5	0.4–3.2	1–6	4–20
Is the process adequate?	--------	No	Yes	Yes	Yes	Yes	Yes	Yes	No

Table 3.27 Report for economic feasibility of manufacturing processes. Case study.

	Part	Process 1	Process 2	Process 3	Process 4	Process 5	Process 6	Process 7	Process 8
Process name	--------	Sand casting	Vacuum casting	Shell molding	Gravity die casting	Centrifugal casting	Invest. casting	Ceramic mold casting	Full-mold process
Economic capability									
Production volume and EBQ	3,000	—	1–100	1,000–100,000	1,000–10,000	10–10,000	1–10,000	50–1,000	—
Is the process adequate?	—	—	No	Yes	Yes	Yes	Yes	No	—

f) Identifying a process

From the previous list of candidates, a detailed cost analysis is required to select the most adequate process since four casting processes are technically and economically feasible. Ideally, process planners should consider the main economic characteristics of the equipment that the facility would purchase or is already installed to conduct a fair estimation of part manufacturing cost. In those cases, the operating cost, capital cost, and tooling cost may be approximately known and the optimal process in terms of cost can be obtained. However, at the early stage of product development, this specific information may not be available, and only ranges of common operating costs, capital investment, and tooling costs are known. In these cases, a rough estimation of these terms of cost can be obtained by selecting the operating cost, capital cost, and tooling cost proportional to the position of the production volume with respect to the EBQ. The reasoning is that higher production volume usually means bigger machinery, more complex and automatic tooling, and so on. Although this may not hold in all circumstances, the reasoning may provide a rough estimate that can help in the final decision if a more detailed economic analysis is not available. Based on this approach, Table 3.28 shows the ranges of production rate, operating cost, capital cost, tooling cost, production rate, and tool life, and the proportional point within those ranges. For instance, gravity die casting presents an EBQ range of 1,000–10,000, and the production volume for the case study is 3,000, Thus, the operating point within the EBQ range is 22.2%. Using this proportion, the costs according to each range are: operating cost \$117.7/hour, capital cost \$25,000, tooling cost \$8,333, and tool life 30,000 units. Additionally, using the same reasoning, the production rate is 15 units/hour. Assuming an amortization coefficient of 5% for all processes, the rough estimation of cost per unit for gravity die casting is

$$\text{Cost per unit} = C_{\text{material}} + \frac{\text{Tooling cost}}{\min(\text{Tool life}, n)} + \frac{\text{Operating cost}}{\text{Production rate}} + \frac{(\text{Capital cost}) \times a}{n}$$

$$\text{Cost per unit} = 13.8 \times \frac{\$2}{\text{kg}} + \frac{8{,}333}{\min(30{,}000;3{,}000)} + \frac{117.7}{15} + \frac{(25{,}000) \times 0.05}{3{,}000}$$

$$\text{Cost per unit} = 27.6 + 2.78 + 7.85 + 0.42 = \$38.6$$

Following the same procedure, Table 3.28 shows the final cost per unit for all the remaining candidates.

An additional calculation can be conducted here to check if the production volume can be produced within the defined period of time stated in the drawing. In this case, given production of 3,000 parts in 3 months, and assuming 1 month to conduct all process planning activities and production setup, the minimum production rate is estimated as

$$\text{Minimum production rate} = \frac{3{,}000}{(2 \times 5 \times 4 \times 2 \times 7)} = 5.4 \text{ parts / hour}$$

where it is assumed 2 shifts and 5 working days per week, 2 months of production, 4 weeks/month, and 7 hours per shift. Comparing with the expected production rates of all manufacturing process candidates, all of them can reach the required production volume.

Table 3.28 Report for final process selection (primary processing costs). Case study.

		Process 1	Process 2	Process 3	Process 4	Process 5	Process 6	Process 7	Process 8
Process name		Sand casting	Vacuum casting	Shell molding	Gravity die casting	Centrifugal casting	Invest. casting	Ceramic mold casting	Full-mold process
Processing costs									
Primary process	Material cost ($)	—	—	27.6	27.6	27.6	27.6	—	—
	Production rate (parts/h)	—	—	5–50 (5.9)	5–50 (15)	1–20 (6.7)	1–200 (60.7)	—	—
	Operating Cost ($/h)	—	—	40–140 (42.02)	80–250 (117.07)	80–250 (130.9)	60–200 (102)	—	—
	Capital Cost ($)	—	—	1,700–17,000 (2,009)	15,000–60,000 (25,000)	2,000-20,000 (7,387)	300–1,500 (659.9)	—	—
	Tooling Cost ($)	—	—	1,700–9,000 (1,847)	5,000–20,000 (8,333)	600–6,000 (2,216)	100–800 (155)	—	—
	Tool life (units)	—	—	1,000–10,000 (1,181)	10,000–100,000 (30,000)	100–500 (219.7)	1–10 (3.7)	—	—
	Cost per unit ($)	—	—	36.3	38.6	57.3	71.2	—	—

Furthermore, secondary processes are required for the key product characteristics that are not met through the primary manufacturing processes. For this case study, according to dimensional and geometrical tolerances, surface roughness, and part shape, the appropriate secondary processes could be the following:

- Turning to produce
  - The Ø40H6 mm with 0.02 mm perpendicularity tolerance. To reach IT6, a boring operation is required.
  - The base and top surfaces of the part that requires a surface roughness of 0.1 μm and parallelism of 0.05 mm, respectively. The base surface requires a subsequent grinding operation to reach that roughness value.
- Drilling to produce the 4×Ø15±0.1 mm and position tolerance of 0.15 mm.
- Grinding to finish the base of the part that requires a surface roughness of 0.1 μm and 0.02 mm flatness.

Table 3.29 shows the additional cost due to these secondary operations assuming a 2-mm machining allowance, 400-cm^3/min material removal rate, and an hourly rate of $90/hour. Note that the investment casting process can reach the quality of the holes of Ø15, and thus, drilling is not necessary after this primary process.

According to these results, shell molding seems to be the most adequate process under the assumptions considered in the example. This cost can be compared to a manufacturing process plan where the part is obtained only by machining operations with a round bar as raw material. Assuming bar dimensions of Ø220 and 120-mm in length, the total cost per unit considering the material cost ($31.56), the removed material (2,608 cm^3), the hourly rate ($90/hour), and the material removal rate (400 cm^3/min) is $41.34. As it can be seen, for this case study the shell molding process still seems to be the most appropriate process, with a cost per unit of $36.71.

3.5 Use of Software for Process Selection

Material and manufacturing process selection is a common problem in the development of a new product. The previous methodology assumes that the material is already selected and shows different steps for the best manufacturing process selection, for example, material and process compatibility, geometry compatibility, technical and economic criteria, etc. The information provided above permit designers to select the best casting and/or bulk-forming processes, as well as machining processes in terms of geometrical quality and costs. However, only basic casting and machining processes are considered, and there are many more primary and secondary processes that could be considered as candidates for a given application.

To consider more processes and conduct a more friendly manufacturing process selection, there are some commercial software systems available to assist designers in a complete search and selection. Examples of this type of software are Ansys Granta Edupack® and DFM Concurrent Costing®, software systems that include a large number of databases to correctly select both materials and processes and obtain an early cost estimation for proper product development. Under this software, the user imposes limits on material, shape, mass, section, tolerance, and so on. The result is a subset of processes that meet all the constraints, and then the economics are investigated by considering the desired batch size on an economic batch size chart.

Figure 3.8 shows an example of the use of Ansys Granta Edupack for the process selection of the casing example in Figure 3.7. In the first step, the user defines the characteristics of the part under study, such as material, shape, mass, section, tolerance, and so on. The result is a subset of processes

Table 3.29 Report for final process selection (secondary and total cost per unit). Case study.

		Process 1	Process 2	Process 3	Process 4	Process 5	Process 6	Process 7	Process 8
Process name		Sand casting	Vacuum casting	Shell mold.	Gravity die casting	Centrifugal casting	Invest. casting	Ceramic mold casting	Full-mold process
Processing costs (continuation from Table 3.28)									
Secondary processes	Removed volume (cm^3)	—	—	111.4	111.4	111.4	110.7	—	—
	Material removal rate (cm^3/min)	—	—	400	400	400	400	—	—
	Hourly rate ($/h)	—	—	90	90	90	90	—	—
	Cost per unit ($)	—	—	0.41	0.41	0.41	0.40	—	—
	Total cost per unit ($)	—	—	36.71	39.01	57.71	71.6	—	—

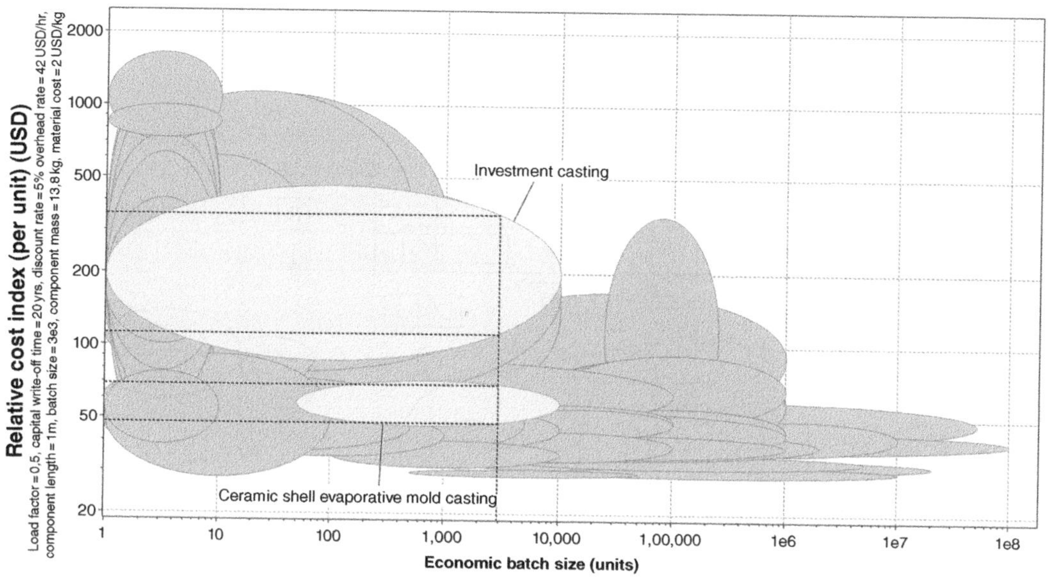

Figure 3.8 Relative cost index with respect to batch size given default cost parameters from Ansys Granta Edupack.

that meets all the constraints: the investment casting and the shell molding process. In this case, centrifugal casting is not included in the software database, and gravity die casting is restricted to nonferrous metals. From the resulting subset, the economics are then investigated by putting the desired batch size on the economic batch size chart. Keeping the default parameters of the cost models from the software, the range of costs for shell mold is lower than investment casting, a range of $32–85/part with respect to $100–300/part.

Questions and Exercises

1 Which is the most general classification of materials in manufacturing?

2 Which are the most common primary processes in manufacturing?

3 Which are the main factors to be considered in manufacturing process selection?

4 Name the main steps of the proposed method for manufacturing process selection.

5 Sort from highest to lowest surface finish the following processes: sand casting, turning, hot forging, grinding, and lapping.

6 Why is important to consider the compatibility geometry–material–manufacturing process?

7 For the mechanical drawings shown in Figures 3.9 and 3.10, apply the manufacturing process selection methodology to select the most appropriate manufacturing process.

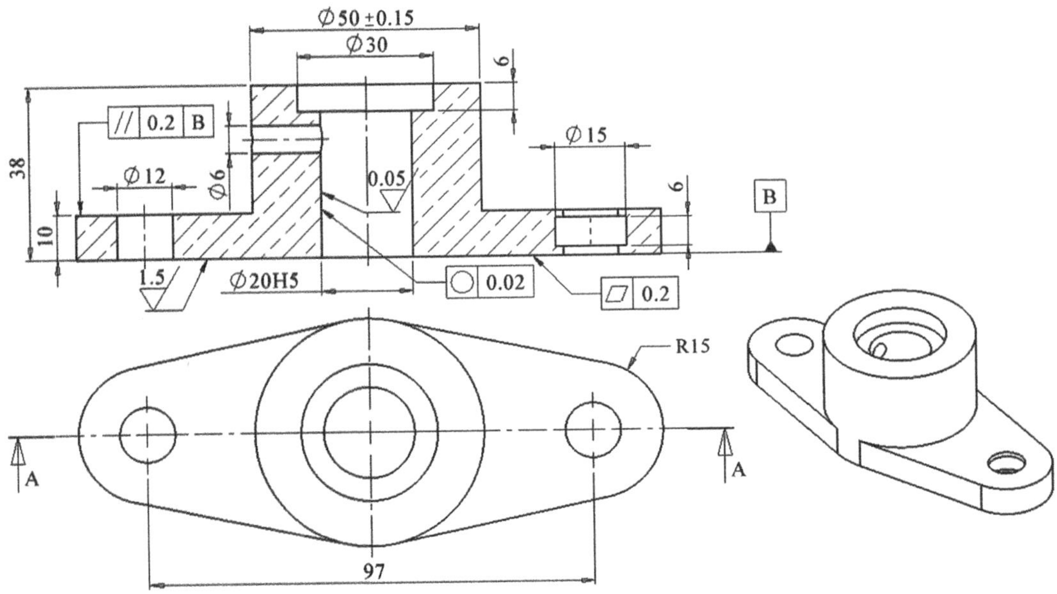

Figure 3.9 Part 1 – Support. N10 (ISO 1302); General tolerance ISO 2768 – cH. 900 units in 2 months. Material AISI D3, weight 737 g.

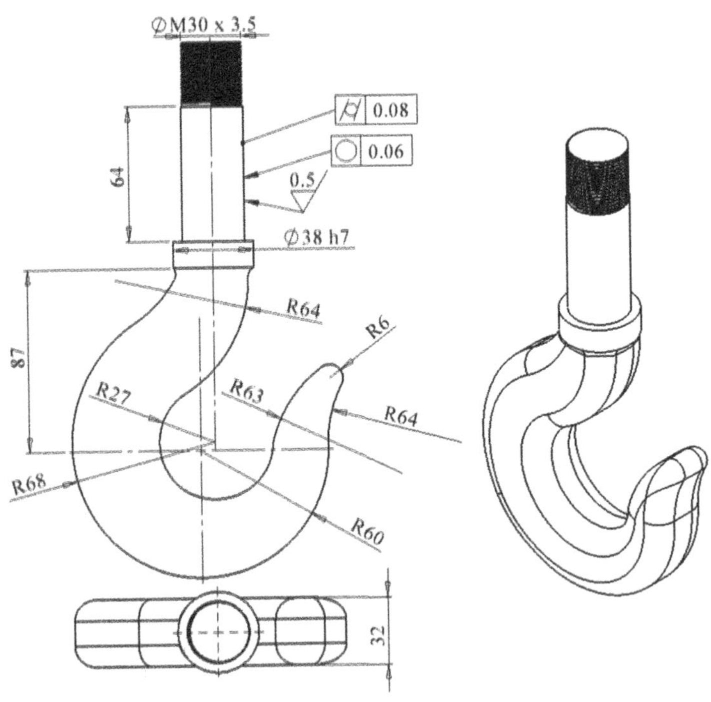

Figure 3.10 Part 2 – Hook. N10 (ISO 1302); General tolerance ISO 2768 – cH. 120,000 units per year and one year of production. Material DIN C67. Weight 2,786 g.

References

Ashby, M.F. (1999). *Material Selection in Mechanical Design*. Butterworth-Heinemann.

Black, J.T., and Kohser, R.A. (2008). *DeGarmo's Materials and Processes in Manufacturing*. Westford: Wiley.

ECS Staff Report. (2006). *Guide to Casting and Molding Processes*, 9–15. Engineered Casting Solutions, Casting Source Directory.

Groover, M.P. (2020). *Fundamentals of Modern Manufacturing: Materials, Processes, and Systems*. Wiley.

Halevi, G., and Weill, R. (1995). *Principles of Process Planning: A Logical Approach*. Springer Science & Business Media.

Ibrahim, I.D., Jamiru, T., Sadiku, R.E. et al. (2015). The use of polypropylene in bamboo fibre composites and their mechanical properties – a review. *Journal of Reinforced Plastics and Composites* 34 (16): 1347–1356.

Kalpakjian, S., and Schmid, S.R. (2009). *Manufacturing Engineering Technology*, 568–571. London, United Kingdom: Prentice Hall.

Kutz, M. (2015). Composite Materials. In: *Mechanical Engineers' Handbook: Materials and Mechanical Design*. Wiley.

Scallan, P. (2003). *Process Planning: The Design/Manufacture Interface*. Butterworth-Heinemann.

Wick, C., Benedict, J.T., and Veilleux, R.F. (1984). *Tool and Manufacturing Engineering Handbook*, 4 Charles Wick, John T. Benedict, and Raymon F. Veilleux, Volume II Forming, Chapter 16 "Casting," Society of Manufacturing Engineers, Dearbon, MI 16–1 to 16–66.

Zamanillo, J. (1993). *Planificación de procesos*. Universidad Politécnica de Valencia.

4

Process-Planning Documentation

4.1 Introduction

One of the main activities that process planners must conduct is the documenting and communicating of all the steps needed to be conducted to produce a product. Process planners must define the sequence of manufacturing processes, the equipment, the setups, process parameters, frequency of inspection, key product specifications to be inspected, etc. All the information should be available at shop-floor level for operators and technicians, and clear and understandable documents should be provided for this purpose.

This chapter describes some common documents generated by process planners to clearly define the manufacturing process plans. Although presented in this chapter, these documents are generated after detailing the manufacturing processes to be used, equipment, setup procedures, parameters, etc. In the next chapters, detailed information and guidelines are given to correctly fill in these documents when dealing with sand casting, machining processes, and inspection processes.

4.2 Process Flow Charts

The process flow chart provides a high-level graphical overview of the complete manufacturing process for a given component or assembly. This chart utilizes five basic symbols to depict operations, transportation, inspection, delays, and storage stages within the process.

- ●: Operation. The item is intentionally changed in one of its characteristics. Examples are milling a part, bending a metal sheet, casting a part, cleaning a casting part, etc.
- ➔: Transportation. The item is moved from one place to another (except when the movement occurs as an integral part of an operation or inspection). Examples are moving an item on a conveyor belt between operations, moving a part from a machine tool to another, or to storage.
- ■: Inspection. The item at an inspection point is compared with the quality standard established for that point.
- ◗: Delay. The next planned action does not take place, causing a delay in the process.
- ▼: Storage. An item is stored in a place/warehouse until the item is withdrawn.

Manufacturing Process Planning: A Practical Approach for Mechanical Engineering, First Edition.
José V. Abellán-Nebot, Carlos Vila Pastor, and Héctor R. Siller.

Companion website: www.wiley.com/go/Manufacturing_Process_Planning_1e

Flow process chart								
Item: Support ZDY-29	**Planner:** J.V. Abellán-Nebot	**Date:** 01-03-23						
Description: Flow process chart for the support ZDY-29, from blank at department store to inspection and storage								
Step	Details of method	Operation	Transportation	Inspection	Delay	Storage	Distance (m)	Time (min)
1	Blanks in department stores	○	⇨	□	D	▼	-	-
2	To lathe queue	○	➡	□	D	▽	30	0.75
3	At lathe in queue	○	⇨	□	◗	▽	-	-
4	Into station	○	➡	□	D	▽	-	-
5	Clamping and rough turn	●	⇨	□	D	▽	-	3.2
6	To milling queue	○	➡	□	D	▽	15	0.4
7	At milling in queue	○	⇨	□	◗	▽	-	-
8	Into station	○	➡	□	D	▽	-	-
9	Clamping and milling	●	⇨	□	D	▽	-	10
10	To bench inspection	○	➡	□	D	▽	5	0.1
11	Inspect finished part	○	⇨	■	D	▽	-	3
12	Place on cart	●	⇨	□	D	▽	-	0.1
13	Remain on cart until loaded	○	⇨	□	◗	▽	-	-
14	Full cart departs	○	➡	□	D	▽	80	0.5

Figure 4.1 Example of flow process chart.

Unlike operation and routing sheets that generally show value-added activities, the process flow chart also shows the nonvalue-added steps such as transportation and storage. Figure 4.1 illustrates a sample process flow chart. In this sense, the process flow chart is an efficient tool for process improvement, and it is commonly used in continuous process improvement activities.

4.3 Assembly Charts and Operation Process Charts

In the planning stages of developing a production system, the layouts of machines and the plant are not yet known. One can only visualize, from parts and product designs, the operations that are necessary and the sequence in which they must be performed. Assembly charts and operation process charts are the graphical representations of such methods. An assembly chart gives a broad overview of how several parts manufactured separately are to be assembled to make the final product. Figure 4.2 shows an example of an assembly chart of a bench vice where "SA" refers to "subassembly" and "A" refers to "assembly."

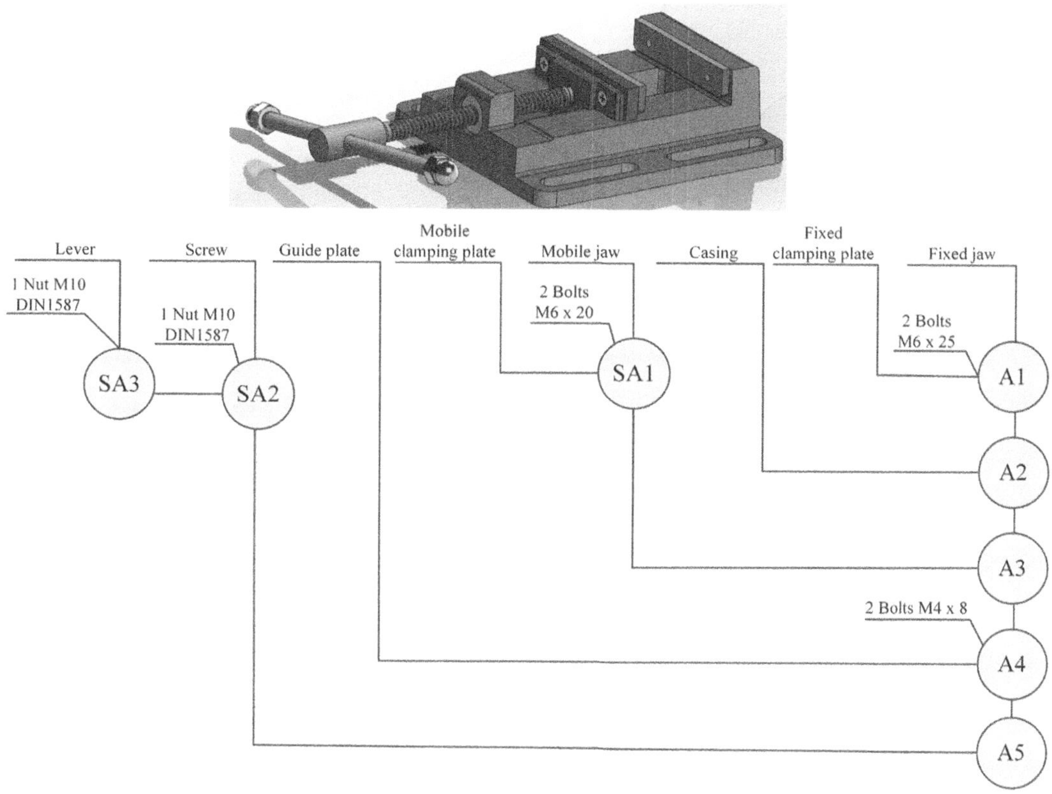

Figure 4.2 Example of assembly chart for a bench vice. *Source:* Personat Martínez (2022).

On the other hand, the operation process chart essentially refers to the task of detailing the assembly chart, showing every operation and inspection that each part needs as it progresses from raw material to the final assembly. Figure 4.3 shows the operation process chart of the previous example, a bench vice (Personat Martínez 2022).

4.4 Routing Sheets

The routing sheet, also known as process sheet or process planning sheet, is a basic document of process planning where it indicates the route that the raw material should follow along the production process, identifying the required production equipment, times, special tooling, etc. The information that is commonly included in the routing sheet is:

- Information about the product manufactured, for example, name and number of the part.
- Operations required to manufacture the product in the proper sequence.
- Machinery and tooling required at each operation.
- Setup, handling, operation, and cycle times.
- Quality control activities.

The header of a routing sheet contains basic product or component part information. The routing sheet's primary body lists the operation numbers in increments of 10 and describes the process.

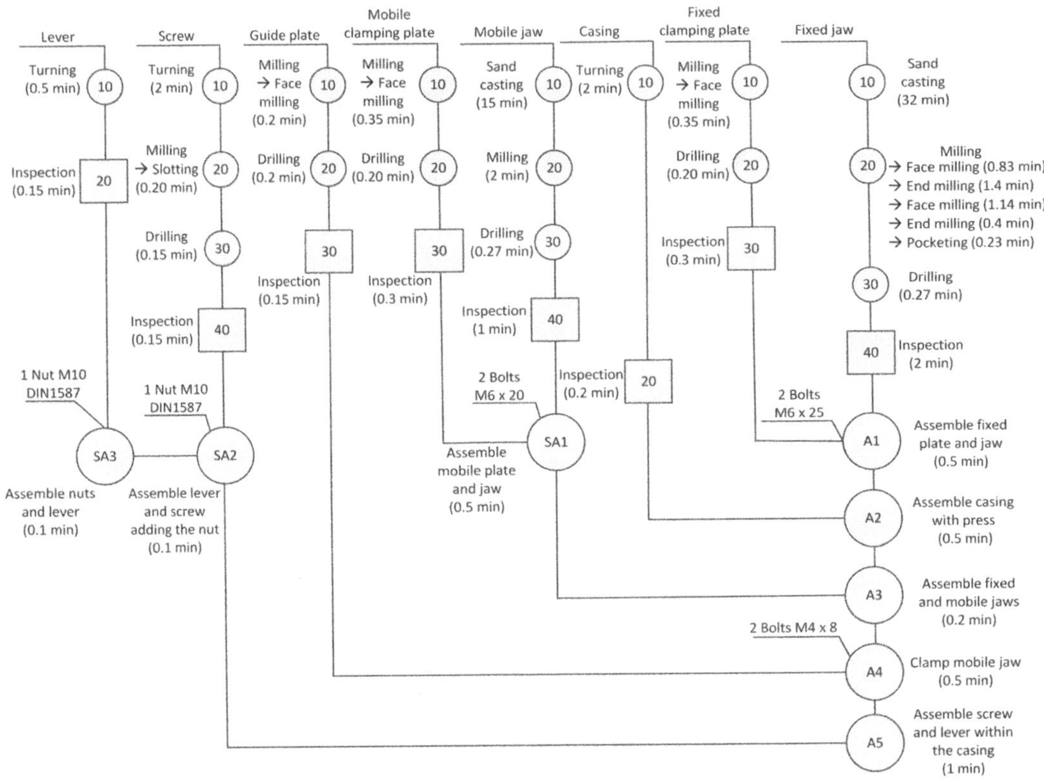

Figure 4.3 Example of operation process chart for a bench vice. *Source:* Personat Martínez (2022).

The department and equipment involved are then identified to indicate where the material must go for processing and what production equipment is required.

To illustrate the information presented in a routing sheet, the casing part shown in Figure 4.4 is used as a part example and the corresponding routing sheet is shown in Figure 4.5.

4.5 Operations Lists

The operations list or operations sheet includes a detailed description of all operations for all processes listed in the routing sheet. This list guides the workers on the activities to be done at each operation, indicating specific processing parameters, working instructions, etc. Unlike routing sheets that provide an overview of the production process, operation sheets are focused on each specific operation within the process, giving a more detailed and operative description.

The header of the operations list contains product/component part information whereas the main body of the document contains the operation numbers considering the numeration done in the routing sheet. For instance, in the routing sheet shown in Figure 4.5, casting the starting geometry is recorded as Operation number 10. Following that, all operations mentioned in the operations list that are related to this process are listed in increments of one prefixed by 10, that is, 11, 12, etc., as shown in Figure 4.6. It then provides a description of the operation, the machine tool that will be utilized, the tooling that will be used, and the process parameters that are appropriate for the operation. The final column is for comments to clarify any part of the procedure.

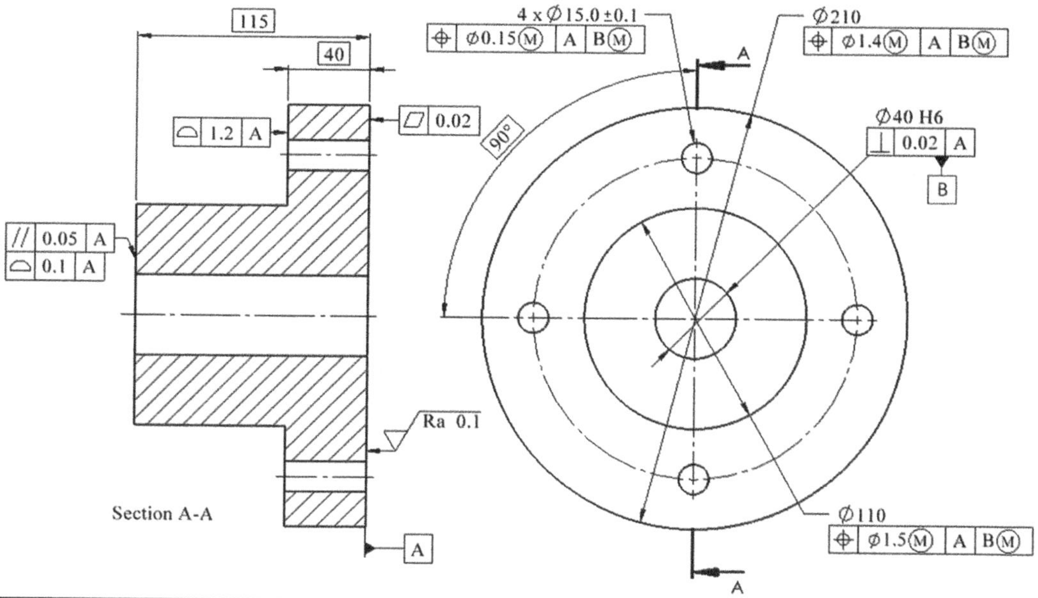

Part weight: 13.8 kg; Part material: Gray iron; General dimensional tol.: ± 0.75 mm
General surface roughness: 12.5 μm; Production volume: 3,000 parts; Period of production: 3 months

Figure 4.4 A casing part. Technical drawing.

Routing sheet							
Part number: PRT_211B Part name: Casing 211 Material: Gray cast iron				Process planner: J.V. Abellán Release date: 25/8/2023 Version: 4			
Operation	Description	Department/area	Machine tool/equipment	Setup time	Operation time	Parts	Operators
10	Shell casting	Foundry	Shell casting kit	10 min	49 min	4	1
20	Turning operations	Shop-floor	Lathe	25 min	8 min	1	1
30	Drilling operations	Shop-floor	Vertical drill machine	10 min	8 min	1	1
40	Grinding operations	Shop-floor	Plane grinder	10 min	5 min	1	1
50	Inspection	Quality department	Metrology devices	--	12 min	1/25	1

Figure 4.5 Routing sheet for the example given (casing part).

Operations sheet								
Part number: PRT_211B Part name: Casing 211 Material: Gray cast iron							Process planner: J.V. Abellán Release date: 25/3/2023 Version: 3	
Operation	Description	Machine tool/equipment	Tooling	Speed	Feed	Setup time	Operation time	Observations
11	Pattern fabrication	Shell casting kit	--	--	--	--	5 days	Tool life 2,000 uses. External fabrication
12	Shell fabrication and flasking	Shell casting kit	Pattern #1	--	--	10 min	15 min	--
13	Remove pattern, pouring	--	--	--	--	--	5 min	--
14	Solification and cool waiting	--	--	--	--	--	20 min	--
15	Remove casting	--	--	--	--	--	3 min	--
16	Fettle/deburr	Belt grinding machine	KK711X	2,800 rpm	--	--	3 min	--
17	Inspect	--	--	--	--	--	3 min	Visual inspection
21	Lathe setup	CNC lathe	--	--	--	25 min	--	Setup every batch size
22	Face turning	CNC lathe	PCLNL2020K12 / CNMG120408	180 m/min	0.1 mm/rev	--	3 min	--
23	Boring	CNC lathe	S12M-STFCL-11 / TCMT110202	180 m/min	0.05 mm/rev	--	4 min	--
24	Face turning	CNC lathe	PCLNL2020K12 / CNMG120408	180 m/min	0.1 mm/rev	--	2 min	--
31	Drilling setup	Vertical drilling machine	--	--	--	10 min	--	Setup every batch size
32	Drilling	Vertical drilling machine	D8	995 rpm	140 mm/min	--	3 min	--
33	Manual tool change	Vertical drilling machine	--	--	--	--	2 min	
34	Drilling	Vertical drilling machine	D15	500 rpm	100 mm/min	--	3 min	--
41	Plane grinding setup	Plane grinder	--	--	--	10 min	--	Setup every batch size
42	Plane grinding	Plane grinder	C60J11V18	9,000 rpm	1,500 mm/min	--	5 min	--
51	Inspection, plane tolerance	Dial indicator, metrology table	--	--	--	--	5 min	Sampling 1/25
52	Inspection, surface roughness	Profilometer	--	--	--	--	2 min	Sampling 1/25
53	Inspection, drilled holes position	Inspection gage #12	--	--	--	--	2 min	Sampling 1/25
54	Inspection, central hole dimension	Internal micrometer, 3-point contact	--	--	--	--	3 min	Sampling 1/25

Figure 4.6 Operations list for the example given.

Tooling list						
Part number: PRT_211B Part name: Casing 211 Material: Gray cast iron				Process planner: J.V. Abellán Release date: 25/3/2023 Version: 3		
Operation	Operative description	Tooling	Tool position	Tool description	Tool material	Observations
16	Fettle/deburr	KK711X	--	Abrasive belt	Al_2O_3	--
22	Face turning	PCLNL2020K12 / CNMG120408	T2 D5	Left-handed face/straight turning. Lead angle 95°, insert angle 80°	WC and TiN	--
23	Boring	S12M-STFCL-11 / TCMT110202	T6 D11	Left-handed boring tool. 90° lead angle and 60° insert angle	WC and TiAlN	--
24	Face turning	PCLNL2020K12 /CNMG120408	T2 D5	Left-handed face/straight turning. Lead angle 95°, insert angle 80°	WC and TiN	--
32	Drilling	D8	--	Ø8 carbide twist drill	Carbide tool	--
34	Drilling	D15	--	Ø15 carbide indexable drill	Carbide insert	--
42	Plane grinding	C60J11V18	--	300 x 40 x 127 abrasive wheel	Silicon carbide	--

Figure 4.7 Tooling list for the example given.

4.6 Tooling List

The operations list identifies the tooling to be used without including tooling details. To complete the information related to the tooling needed at each operation, a tooling list (Figure 4.7) is used. The operation numbers on the tooling sheet are related to those in the operations list.

4.7 CNC Programs

Computer numerical control (CNC) programs are required in most of the machine tools used in industry. Programs should be elaborated and simulated using specific computer-aided manufacturing (CAM) software to ensure no collisions and proper cutting tool path trajectories. These programs include all the information given in the operations list such as the type of cutting tools needed, fixturing, and cutting conditions (cutting speed, depth of cuts, feed rates, etc.). Figure 4.8 shows an example of a CNC program and its simulation using specific software.

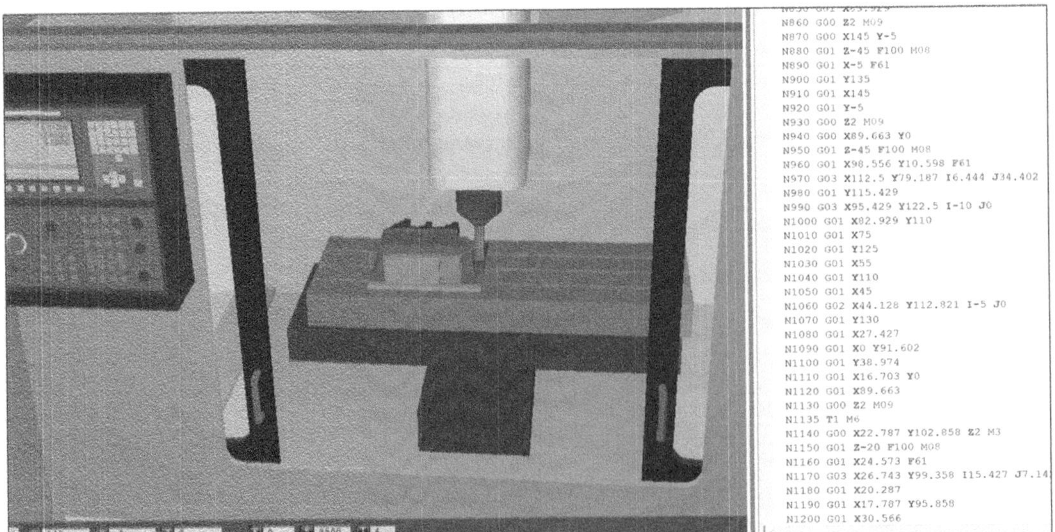

Figure 4.8 Example of CNC program under a simulation software.

Work instructions sheet						
Part #: 123-0012		**Part name:** Pump component A		**Date:** 2/2/2024		**Prepared by:** J.V. Abellan
Step	**Operation description**	**Key point**	**Sampling**	**Tool**	**Time**	**Sketch**
1	Get a work and remove burrs	Use gloves and security goggles		Sand belt	1 min	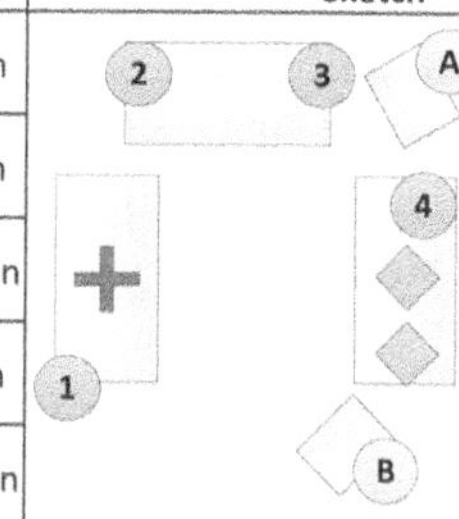
2	Load work into leak test machine	With both left and right hand			3 min	
3	Check leakage results. If fails, discard to container A				0.8 min	
4	Inspect	Measure inner hole (20 ± 0.1)	1/25	Micrometer	1 min	
5	Check appearance and put into container B	Check both sides			0.2 min	
				Total	**6 min**	Safety Quality

Figure 4.9 Example of a work instructions sheet.

4.8 Work Instructions Sheet

A work instructions sheet contains additional information not found in operation or process sheets. These instructions are specific to the duties that influence the equipment operator or personnel assigned to a certain activity. Instructions may include procedures, pictures, a sequence of steps to follow during the operation, and examples of poor quality or performance. Another critical component of work instruction is the specific procedures that an operator must take in the event of nonconforming parts (Rufe 2013). Therefore, work instructions may cover the following details:

- Equipment setup instructions.
- Equipment maintenance procedures.
- Material handling standards.
- Safety alerts for the employees.
- Reaction strategies for nonconforming parts.
- Any additional information.

A typical work instructions sheet is shown in Figure 4.9. The main columns indicate the operation description, describing each step in the order of work, and the key points, which refer to the main aspects that the workers should consider performing their work accurately, safely, conscientiously as well as quickly. As key points, the instructions indicate how the procedures should be performed paying special attention to three conditions: (i) a success/failure condition, something that controls if the work is defective; (ii) a safety condition, a behavior that may cause injury during performing the work; (iii) a work easier condition, something that makes the work easier and is reported to be followed by the worker.

4.9 Inspection Sheets and Reports

Inspection sheets contain step-by-step instructions for inspecting all the important quality characteristics of the product/component. If specific quality data needs to be collected and recorded for product and process validation, statistical process control (SPC), confirmation of safety conformance, or verification of government or regulatory compliance, an inspection report may be necessary.

Although it may depend on the type of inspection report, the main information that an inspection sheet should include is as listed in the following:

- Identification of the specifications of the part to be inspected, usually using balloons on the technical drawing.
- Identification of the type of measurement instrument used.
- Measurement uncertainty and conformity zones.
- Frequency of inspection and number of samples inspected.
- A field to add any observation that the operator may want to document.

One important report about the inspection is the so-called first article inspection (FAI) report (SAE AS9102 2000). FAI report allows suppliers to formally report measurements of the initial part manufactured in cooperation with a customer who reviews and approves it. Customers expect verification of a project's specifications, and manufacturer's need to document their work to set the standard for production runs. Therefore, the FAI report is documentation that proves a product has been manufactured to the required specifications. Note that FAI is a requirement in the aerospace and defense industry and other industries (e.g., automotive industry) use similar procedures for product and process validation. Figure 4.10 shows an example of an inspection report that is part of a more extended FAI report.

Questions and Exercises

1 Draw the assembly chart related to the assembly shown in Figure 4.11. Assume that R9 and R7 are initially assembled to create a subassembly for the left and right bracket, and there is an inspection operation after the product is totally assembled. After inspection, a part number tag and a box packing are added to finish the product.

2 Draw the assembly chart for the product shown in Figure 4.12.

3 A part is manufactured in a company as follows:

- From sand casting, the part is fabricated using a sand core. Both molds and cores are baked.
- After casting, the part is manually cleaned to remove any burnt sand or burrs.
- The part is moved to a special fixture where critical surfaces are inspected after casting.
- The part is moved to a mill machine, where a face milling operation is conducted using a vice as a fixture.
- The part is moved to a drilling machine, where two holes are drilled using a coated high-speed steel (HSS) drill. A three-jaw chuck is used to hold the part.
- The part is moved to a special fixture where critical surfaces are inspected.
- The machined part is moved to a finishing machine to polish the main surface.

According to this descriptive plan, write the following process planning documents (please add any additional information you may think is required in this process):

- Flow process chart.
- Routing sheet.
- Operations list.

5 Explain what a FAI report is and in which cases this report is mandatory.

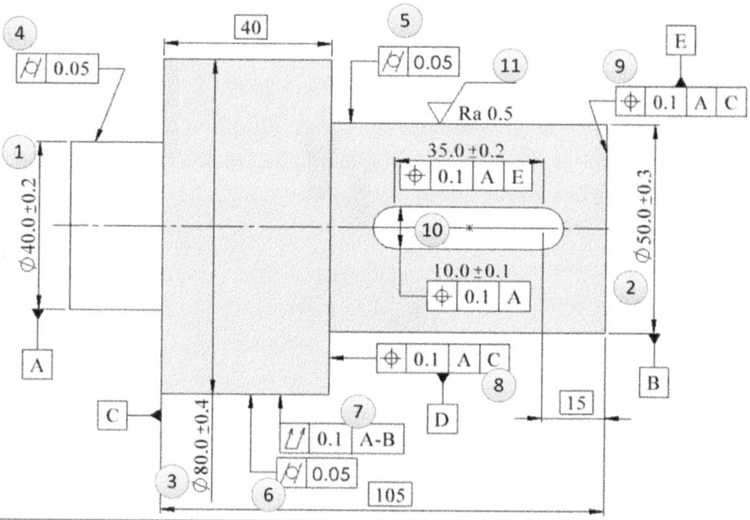

Part number: PRT_321A
Part name: Shaft

QA inspector: J.V. Abellán
QA inspection date: 25/3/2023
Units: mm
Inspection frequency: 3

#	Dimension	Conformity zone	Measurement method	Measurement uncertainty	Part 1	Part 2	Part 3	Decision	Observations
1	Ø40 ± 0.2	[40.18, 39.82]	Caliper	±0.02	40.02	40.05	40.00	Pass	---
2	Ø50 ± 0.3	[50.28, 49.72]	Caliper	±0.02	50.20	50.15	50.22	Pass	---
3	Ø80 ± 0.4	[80.38, 79.62]	Caliper	±0.02	79.88	79.82	79.90	Pass	---
4	⌭ 0.05	[0, 0.04]	Between centers and dial indicator	±0.01	0.02	0.02	0.03	Pass	---
5	⌭ 0.05	[0, 0.04]	Between centers and dial indicator	±0.01	0.01	0.02	0.01	Pass	---
6	⌭ 0.05	[0, 0.04]	Between centers and dial indicator	±0.01	0.03	0.04	0.03	Pass	---
7	⌰ 0.1 A-B	[0, 0.09]	Between centers and dial indicator	±0.01	0.05	0.07	0.04	Pass	---
8	⌖ 0.1 A C	[39.96, 40.04]	Height gauge	±0.01	39.99	40.01	40.01	Pass	---
9	⌖ 0.1 A C	[104.96, 105.04]	Height gauge	±0.01	104.98	105	105.00	Pass	---
10	⌖ 0.1 A E ⌖ 0.1 A	Pass	Inspection gauge A	-	Pass	Pass	Pass	Pass	---
11	√ Ra 0.5	Ra < 0.5 μm	Profilometer	-	0.28	0.35	0.25	Pass	---

Figure 4.10 Inspection report example. Units in mm.

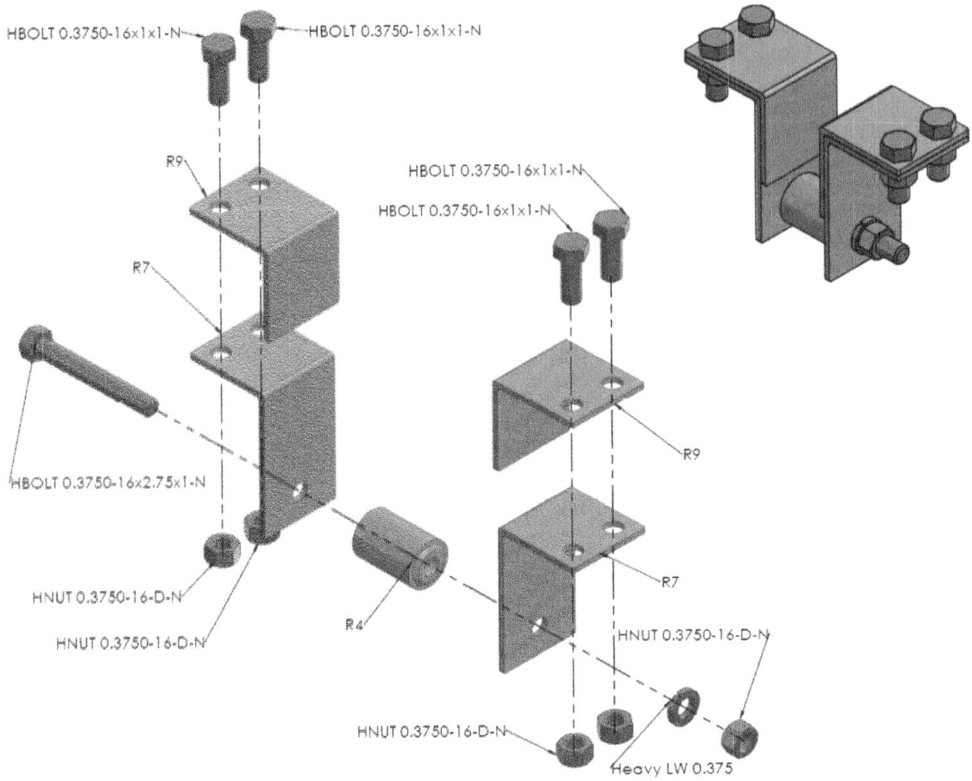

Figure 4.11 Assembly for Exercise 1.

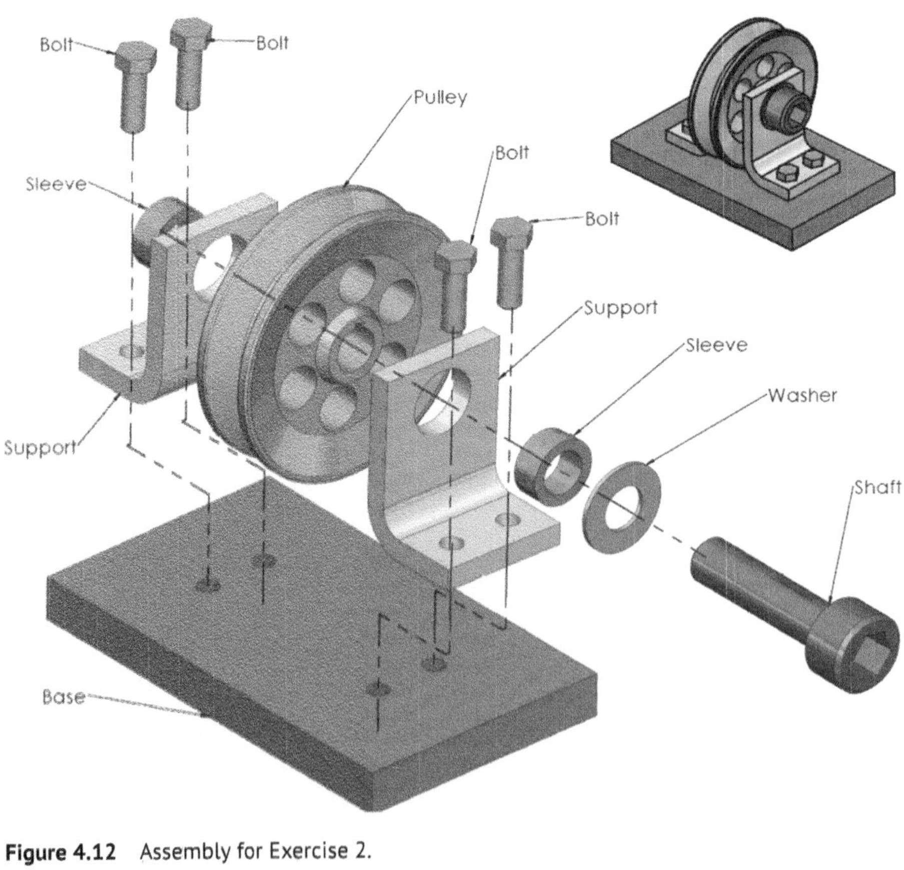

Figure 4.12 Assembly for Exercise 2.

References

Personat Martínez, K. (2022). Cálculo, diseño y planificación del proceso de fabricación de una mordaza para taladro. Mechanical engineering degree dissertation. Universitat Jaume I.

Rufe, P.D. (2013). *Fundamentals of Manufacturing*. Society of Manufacturing Engineers.

SAE AS9102. (2000). Aerospace first article inspection requirement.

Part II

Process Planning in Casting

5

Fundamentals of Casting

5.1 Introduction

Foundries produce castings, often known as near-net shape components, that are very similar to the shape of the finished product. Molten metal is poured into molds to form castings, with cores used to make hollow internal portions. The casting is extracted from the mold once the metal has cooled sufficiently and is then subjected to the required cleaning and finishing operations.

Today, metal casting parts are manufactured at a worldwide level. Figure 5.1 gives a snapshot of the global metal casting market in 2019, with China, India, and the United States being the most important countries in the production of metal castings. In the automobile industry, for instance, many components are manufactured using manufacturing casting processes (Figure 5.2).

If continuous casting is overlooked (almost any steel component is manufactured from a billet initially generated from continuous casting), approximately 90% of the metal castings in the world are manufactured by sand-casting (Black and Kohser 2020, Chapter 12). It is an important manufacturing technique for the manufacturing of automotive and heavy machinery components. Therefore, the importance of sand casting is evident, and process-planning activities related to casting may often include sand-casting processes.

This chapter reviews the fundamentals of sand casting, with special attention to the solidification process, which is, indeed, the key factor in achieving good casting parts. This chapter presents the basics that are later applied in the next chapter, where process planning in sand casting is covered in detail.

5.2 Sand Casting

This ancient method is used to cast metals in sand molds. The steps involved in sand casting are shown in Figure 5.3 and are listed in the following: (i) pattern and mold design; (ii) mold construction (imprinting a pattern and gating system in the sand); (iii) pouring molten metal into the mold cavity; (iv) permitting the metal to cool until it solidifies; (v) separating the sand mold; (vi) cleaning and finishing; and (vii) casting inspection.

Manufacturing Process Planning: A Practical Approach for Mechanical Engineering, First Edition.
José V. Abellán-Nebot, Carlos Vila Pastor, and Héctor R. Siller.

Companion website: www.wiley.com/go/Manufacturing_Process_Planning_1e

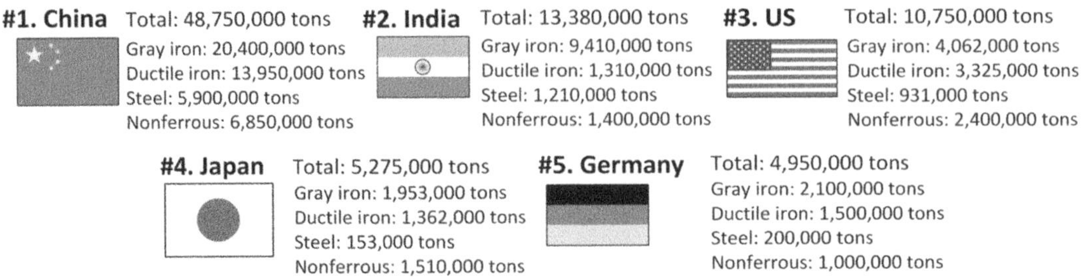

Figure 5.1 Top five casting producers in 2019. Total casting tons produced and production for the main four metal alloys. *Source:* Adapted from The World Foundry Organization (2020).

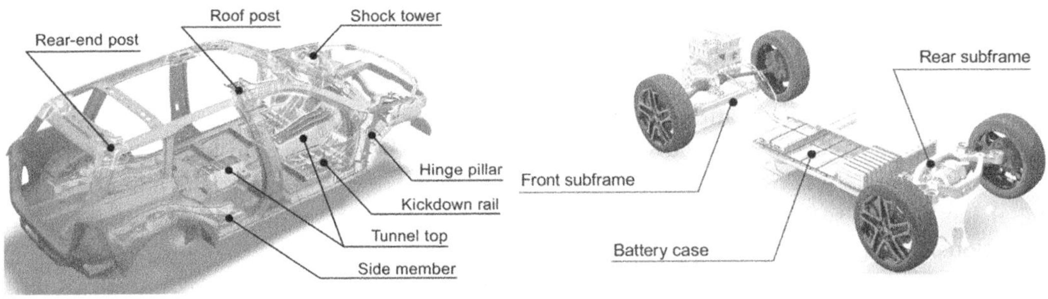

Figure 5.2 Aluminum die casting parts for an electrical vehicle. *Source:* Courtesy of Ryobi Company.

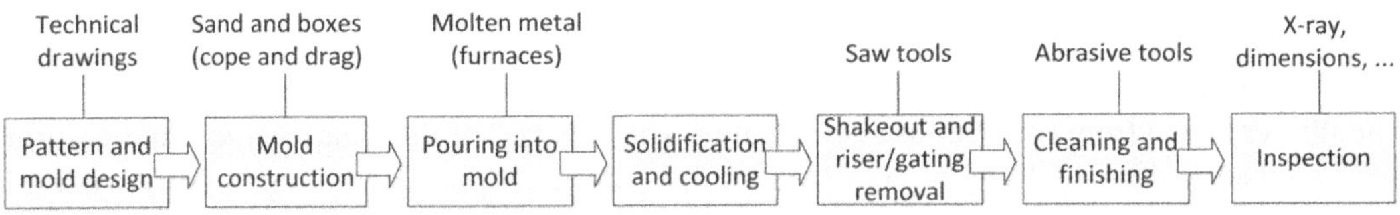

Figure 5.3 Outline of production steps in a typical sand-casting process.

The following are the main parts of a sand-casting mold (Figure 5.4):

- Flask. The flask holds the mold in place and is divided into two parts – the cope and the drag. The parting line separates the cope on top and the drag on bottom.
- Pouring basin. The pouring basin is the top area of the mold where the molten metal is poured.
- Sprue. The sprue is a vertical passage through which the molten metal is introduced into the runner system for filling the mold cavity. At the bottom part of the sprue, a sprue well is added to retain impurities.
- Runner system. The runner system is a horizontal passage that transfers the molten metal from the sprue to the mold cavity through channels.
- Gates. Gates are the part of the runner system that connects to the mold cavity.
- Risers. Risers are deposits of molten metal that supply material as the casting solidifies and shrinks.
- Cores. Cores are used to avoid molten metal filling specific zones of the cavity to create internal surfaces in the casting. Cores are made of sand and are later removed during cleaning operations.
- Vents. Vents are placed within molds to expel the gases produced as the molten metal fills the mold.

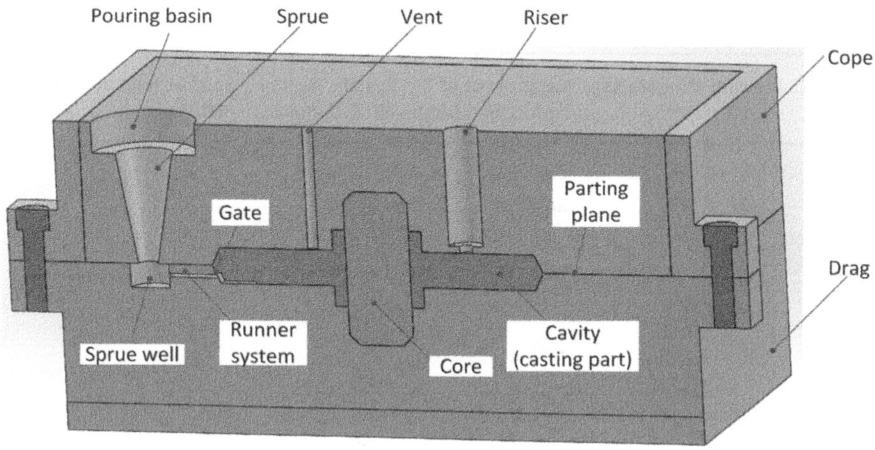

Figure 5.4 Features of sand-casting molds. Cross-section of typical sand-casting mold.

5.3 Heating the Metal

Different types of heating furnaces may be used to heat the metal to the molten temperature that is appropriate for casting. The heat required to melt the material, the heat of fusion needed to transform it into a liquid, and the heat needed to raise the molten metal's temperature to the right pouring temperature are all included in the total heat energy required. This can be expressed as (Groover 2007) follows:

$$H = \rho \times V \times \left[C_s \times (T_m - T_0) + H_f + C_l \times (T_p - T_m)\right]$$

where H is the total heat required to raise the temperature of the metal to the pouring temperature; ρ is the density of the metal; C_s is the specific heat for the solid metal; T_m is the melting temperature of the metal; T_0 is the starting temperature – usually room temperature; H_f is the heat of fusion; C_l is the specific heat of the liquid metal; T_p is the pouring temperature; and V is the volume of metal being heated. Some values of these thermophysical properties for different alloys are shown in Table 5.1. The specific heat at the liquid state of metal alloys is generally not provided in technical publications, but it usually tends to be higher compared to the solid state, around 20–40% higher, according to the values reported in Table 5.1.

It should be noted that the previous equation has limited application and is mainly used for academic purposes. Heat loss during the heating process, variations of the specific heat with temperature, change of solid to liquid state within a temperature range for most alloys, unknown specific heat at liquid state, etc., make accurate estimation of the heat needed at heating furnaces difficult (Groover 2007).

5.4 Pouring the Molten Metal

5.4.1 Flow Velocity and Reynolds Number

After heating, the metal is ready to be poured. A crucial step in the casting process is introducing the molten metal into the mold. The metal must flow into every area of the mold before it solidifies without eroding the mold cavity or generating gases and oxides.

Table 5.1 Different thermophysical properties for some metals and metal alloys.

Metal	Density, ρ (kg/m^3)	Melting temperature, T_m, (°C)	Latent heat of fusion, H_f (kJ/kg)	Specific heat at 20 °C, C_s (kJ/kg)	Specific heat at above melting point, C_l (kJ/kg)
Aluminum	2,710	660	397	0.903	1.17
Aluminum 6061	2,710	650	400	0.897	—
Iron	7,800	1,538	272	0.449	0.83
Gray cast iron	6,920	1,150	140	0.460	—
Low carbon Steel	7,830	1,370	250	0.486	—
Copper	8,850	1,085	205	0.385	0.51
Zinc	7,130	420	114	0.387	0.48
Tin	7,265	230	59	0.227	—
Brass (Cu–Zn)	8,890	910	210	0.380	—
Bronze (Cu–Sn)	8,800	880	200	0.377	—

Source: https://webbook.nist.gov/, https://www.energy.gov, and other website databases.

Pouring temperature, pouring rate, and fluid flow are the main three factors that affect the quality of the pouring process.

The temperature of the molten metal when it is poured into the mold is known as the pouring temperature. The volumetric rate at which the molten metal is poured into the mold is referred to as the pouring rate. Both pouring temperature and pouring rate are crucial parameters to avoid an incomplete filling of the mold cavity or flow turbulences. In the first case, the metal will cool and freeze before filling the cavity if the rate is too slow or the temperature is too low. In the second case, turbulence may arise if the pouring rate is too high. As opposed to laminar flow, which is a smooth and streamlined flow, turbulent flow results in an agitated and uneven flow. Turbulence generates metal oxides that may get stuck during solidification and lower the casting's quality. Furthermore, mold erosion is increased and may become a problem. For these reasons, turbulent flow should be avoided during pouring.

The Reynolds number, Re, is used to quantify this aspect of the fluid flow, representing the ratio of the inertia to the viscous forces in the fluid flow. The Reynolds number is defined as (Kalpakjian and Schmid 2009) follows:

$$\mathrm{Re} = \frac{v \times D \times \rho}{\eta},$$

where v is the velocity of the liquid, D is the diameter of the channel, and ρ and η are the density and viscosity of the liquid, respectively. For noncircular channels, D is the equivalent diameter defined as $4S/P$ (Murdock 1993, Chapter 4), where S is the cross-section of the channel and P is the perimeter. Representative values of density and viscosity of liquid elements and alloys are reported in Table 5.2.

The likelihood of a turbulent flow increases with Reynolds number. Re values in typical feed casting systems range from 2,000 to 20,000. Laminar flow is represented by a number of up to 2,000; a blend of laminar and turbulent flow is represented by a value between 2,000 and 20,000.

Table 5.2 Viscosity and density values of different elements and alloys in the liquid state.

Element/alloy	T_0 (°C)	ρ (kg/m^3) at T_0	η (mPa·s) at T_0	T_1 (°C)	η (mPa·s) at T_1
Al	660	2,380	1.38	776	1.08
Ti	1,600	4,130	5.1	1,666	3.9
Mg	650	1,590	—	923	1.25
Fe	1,535	7,030	5.4	1,926	3.19
Zn	419	6,580	3.50	—	—
Cu	1,083	8,000	4.38	—	—
Ni	1,460	7,900	4.58	—	—
Ti-6Al-4V	1,676	4,100	4.5	1,776	4.1
Al alloy A201	706	—	1.21	808	1.02
Gray cast iron	1,288	—	8.3	—	—
White cast iron	1,288	—	7.1	—	—

Source: Dinsdale and Quested (2004), Nakagawa et al. (1968), Turchi (2018), and Wang and Overfelt (2002).

In the feed systems, this kind of mixture is typically regarded as harmless. Re values above 20,000, on the contrary, indicate extreme turbulence in lost mold operations (sand casting, shell casting, etc.), which causes trapped air and the creation of slag (an oxide layer on the surface of molten metal) as a result of the reaction between the liquid metal and other gases and air (Kalpakjian and Schmid 2009, Chapter 10). In some technical publications, the Re values are limited according to the region of the flow rate analyzed within the casting mold and the material poured. For instance, Re values in aluminum sand casting are limited to 7,000 for sprues, 1,100 for gates, and 780 for mold cavities (Zych et al. 2014).

To estimate the velocity of the liquid during pouring, Bernoulli's theorem should be considered. This theorem establishes that the sum of the energies (head, pressure, kinetic, and friction) at any two places in a moving liquid is equal. This statement can be written in the following form (Beddoes and Bibby 1999, Chapter 2):

$$h_1 + \frac{P_1}{g \times \rho} + \frac{v_1^2}{2 \times g} + F_1 = h_2 + \frac{P_2}{g \times \rho} + \frac{v_2^2}{2 \times g} + F_{1-2},$$

where h is the height, P is the pressure in the liquid, ρ is the density of the molten metal, v is the velocity of the liquid, g is the constant of gravitational acceleration, and F is the load losses due to friction. Subscripts 1 and 2 indicate any two points in the flow of the liquid.

There are several ways to simplify the Bernoulli equation. If the system is assumed to be at atmospheric pressure throughout, and friction losses are disregarded (friction will, of course, alter the liquid flow through a sand mold), the equation can be simplified to

$$h_1 + \frac{v_1^2}{2 \times g} = h_2 + \frac{v_2^2}{2 \times g}$$

The initial speed of the molten metal at the base of the pouring sprue can be calculated using this expression.

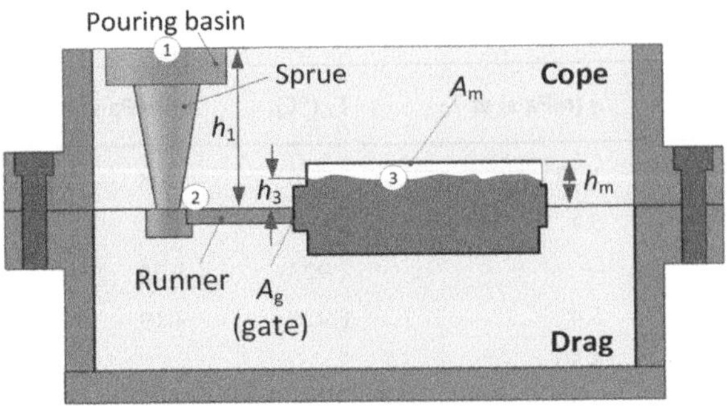

Figure 5.5 Filling process of a mold cavity in sand casting.

Molten metal is poured into the pouring basin at point 1, and the molten metal enters into the runner system at point 2, see Figure 5.5. If point 2 is used as a reference, then h_1 is the height of the cope, and point 2's height is zero. The metal's initial velocity at the top is zero when it is poured into the pouring cup and flows downward. Then, the equation becomes simplified to

$$h_1 = \frac{v_2^2}{2 \times g}$$

which can be reordered to express the flow velocity as

$$v_2 = \sqrt{2 \times g \times h_1}$$

Therefore, the molten metal flows along the runner with the v_2 velocity. The volumetric flow rate is equal to the flow velocity multiplied by the cross-sectional area of the runner. Thus, the volumetric flow rate at the runner is

$$Q = v_2 \times A_2$$

where A_2 is the cross-sectional area of the runner. According to the continuity law, which states that the volume rate of flow remains constant throughout the liquid, the flow velocity at the gate is obtained as

$$v_g = \frac{v_2 \times A_2}{A_g}$$

where A_g and v_g refer to the cross-sectional area and the flow velocity at the gate, respectively. The Reynolds value of the flow velocity at the gate should be checked as indicated previously to avoid turbulent flow.

5.4.2 Filling Time

Given the volumetric flow rate, the time of filling the cavity can be roughly estimated as

$$t_f = \frac{V_c}{Q}$$

where V_c is the volume of the mold cavity to be filled. However, as the mold starts to be filled above the parting plane, the molten metal speed at the gate decreases since this point is not under atmospheric pressure anymore. Thus, the previous equation underestimates the actual filling time. At point 3 (see Figure 5.5), the speed at the gate of the mold is calculated as

$$v_g = \sqrt{2 \times g \times (h_1 - h_3)}$$

Therefore, a more approximate calculation of filling time is obtained as

$$t_f = t_{fB} + t_{fA}$$

where t_{fB} is the filling time of the mold cavity below the parting plane, which can be calculated by the previous equation using V_c as the cavity volume below the parting plane, and t_{fA} is the filling time of the cavity above the parting plane, which is estimated as (Ghosh and Mallik 1986) follows:

$$t_{fA} = \frac{A_m}{A_g} \times \frac{1}{\sqrt{2 \times g}} \times 2 \times \left(\sqrt{h_1} - \sqrt{h_1 - h_m}\right)$$

where A_m and h_m is the cross-sectional area and height of the mold cavity, respectively (Figure 5.5). For not straight geometries, an approximate filling time can be obtained by using the projected area and the equivalent height for A_m and h_m, respectively.

5.4.3 Ceramic Filters

To prevent slag, oxides, and impurities from entering the cavity, ceramic filters may be used (Figure 5.6). Filters are most effective when placed in the gating system as close to the casting as possible (Chastain 2004, Chapter 5). Especially in the fields of small- and medium-sized castings such as automobiles, high-speed railways, machine tools, and hydraulic parts, the purification effect of the filter has been generally recognized by the majority of foundry practitioners, and the method of use has become increasingly mature.

5.5 Solidification and Cooling

5.5.1 Solidification Time

The time needed for the casting to harden after pouring is known as the solidification time. Chvorinov's rule – an empirical expression that indicates that the solidification time depends on the casting's size and shape – is usually applied for time estimation, and it is defined as (Groover 2007) follows:

$$T_s = C_m \times \left(\frac{V}{A}\right)^n$$

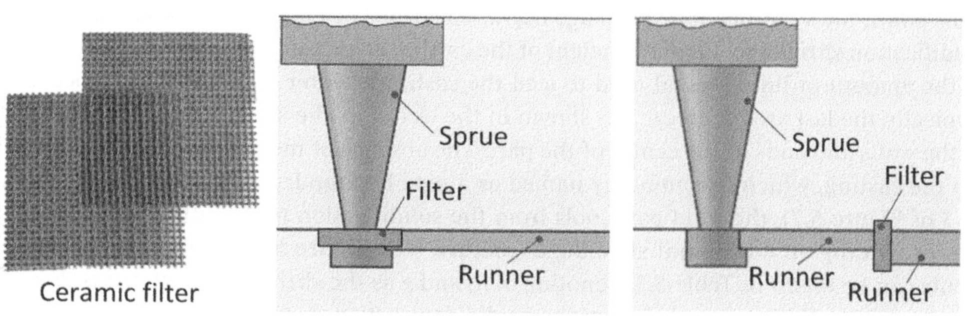

Figure 5.6 Example of ceramic filter and placement in a sand-casting mold.

where T_s is the total solidification time, V is the volume of the casting, A is the surface area of the casting, n is an exponent (n usually has a value of 2 in most sand-casting processes), and C_m is affected by the specifics of the casting process, such as the properties of the mold material (e.g., thermal conductivity), the thermal characteristics of the cast metal (e.g., the heat of fusion, specific heat, or thermal conductivity), and the pouring temperature in relation to the metal's melting point. The C_m coefficient can be estimated as (Jelinek and Elbel 2010; Marathe and Quadros 2021) follows:

$$C_m = \left[\frac{\pi}{4} \times \left(\frac{\rho_c \times H_f}{T_m - T_0} \right)^2 \times \frac{1}{k_m \times \rho_m \times C_h} \right]$$

where T_m is the melting temperature (°C), T_0 the initial mold temperature (°C), ρ_m the density of the mold (kg/m^3), C_h the specific heat of the mold (J/kg·K), k_m the thermal conductivity of the mold (W/m·K), ρ_c the density of the casting metal (kg/m^3) and H_f the latent heat of fusion of the casting metal (kJ/kg). Chvorinov's rule predicts that cast iron will cool and solidify more slowly if the ratio of its volume to surface area is higher. This idea is useful in both the riser and the mold's design as is shown in detail in Chapter 6.

For estimating the total time for solidification and cooling, the C_m parameter can be modified as shown in the following equation:

$$C_m = \left[\frac{\pi}{4} \times \left(\frac{\rho_c \times \left(H_f + C_s \times \left(T_m - T_b\right)\right)}{T_m - T_0} \right)^2 \times \frac{1}{k_m \times \rho_m \times C_h} \right]$$

where C_s is the specific heat of the casting metal at its solid state, and T_b is the temperature at which the cooling ends, that is, the temperature of mold breakage. For sand-casting processes, the following data can be used related to sand molds: $\rho_m = 1{,}500\,\text{kg/m}^3$, $C_h = 1{,}029\,\text{J/kg·K}$, and $k_m = 0.6\,\text{W/m·K}$.

5.5.2 Shrinkage

Three stages of shrinkage take place in a casting process (Groover 2007): liquid contraction during cooling before solidification, solidification shrinkage during the transition from liquid to solid, and thermal contraction of the solidified casting during cooling to ambient temperature. Figure 5.7, which depicts a cylindrical casting formed in an open mold, can be used to describe the three steps. Picture 1 of Figure 5.7 displays the molten metal right after pouring. The liquid metal contracts as it cools from pouring temperature to freezing temperature, lowering the liquid's height from its starting point. This liquid contraction is usually around 0.5% in most metals. At this point, the solidification shrinkage begins (picture 2 of Figure 5.7). Two effects result from solidification shrinkage. First, the height of the casting decreases as a result of contraction. Second, the amount of liquid metal used to feed the casting's center region is limited because this is typically the last area to freeze. As shown in the example, the solidification (nucleation) starts at the walls and ends in the center of the part. The absence of metal after shrinking leaves a hole in the casting, which is commonly named as a pipe by foundrymen. After solidification (picture 3 of Figure 5.7), the solid part cools from the solidification temperature to room temperature, producing an additional shrinkage (picture 4 of Figure 5.7). Common shrinkage coefficients can be found in Table 5.3, denoting α, β, and γ as the shrinkage coefficients during liquid contraction, solidification shrinkage, and solid contraction, respectively.

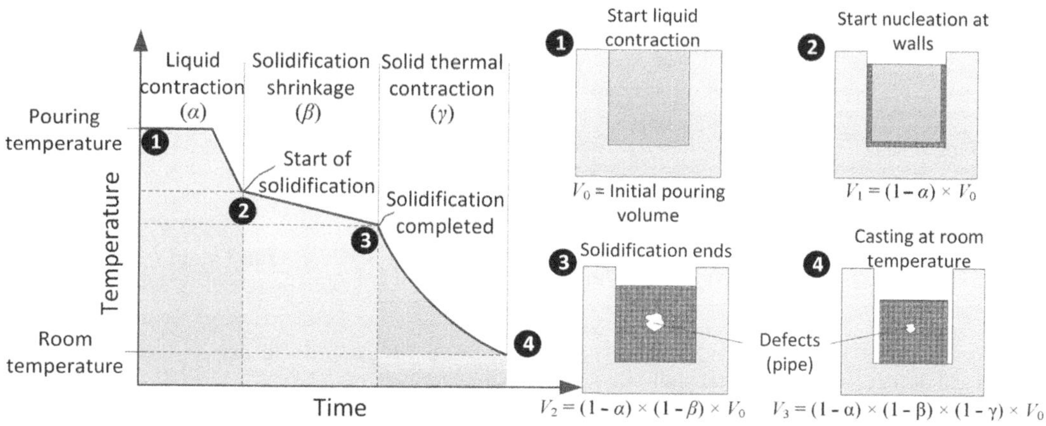

Figure 5.7 Cylindrical casting shrinkage during solidification and cooling: (1) starting level of molten metal immediately after pouring; (2) liquid contraction during cooling; (3) height reduction and pipe generation due to solidification shrinkage; and (4) thermal contraction during cooling of the solid metal. *Source:* Adapted from Groover (2007).

Table 5.3 Volumetric contraction for different casting metals. Liquid contraction, α, is usually around 0.5% in most metals.

Metal	Solidification shrinkage, β (%)	Solid thermal contraction, γ (%)
Aluminum	7.0	5.6
Aluminum alloy	7.0	5.0
Gray cast iron	1.8	3.0
Low carbon steel	3.0	7.2
Copper	4.5	7.5
Bronze (Cu–Sn)	5.5	6.0

Source: Adapted from Groover (2007).

5.5.3 Directional Solidification

As a good casting practice, the areas of the casting farthest from the liquid metal supply have to freeze first. From these areas, the solidification of the part should continue toward the riser(s) to limit the harmful effects of shrinkage. This will ensure that molten metal is always accessible from the risers, preventing shrinkage voids (i.e., pipes) during freezing. This strategy is called directional solidification, and it is commonly applied in the sand-casting mold design. By applying Chvorinov's rule in the design of the casting itself, a good directional solidification can be obtained. For instance, by placing parts of the casting with lower V/A ratios away from the riser, freezing will happen first in these areas, and the supply of liquid metal for the rest of the casting will remain open until the bulkier parts freeze (Groover 2007).

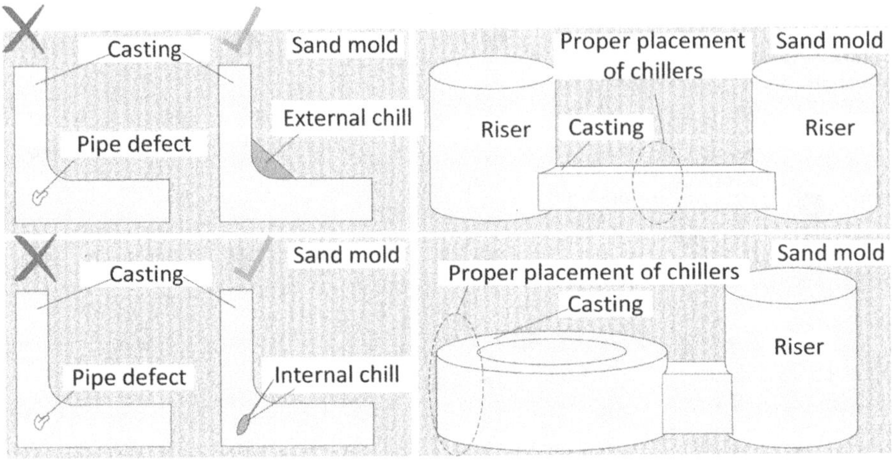

Figure 5.8 (Left top) External chill and (left bottom) internal chill to promote rapid freezing and likely defects if chills are not used. (Right) Examples of proper placement of chillers in castings.

The use of chills – internal or external heat sinks that quickly cause the freezing of some areas of the casting – is another method for promoting directional solidification. Small metal components called internal chills are inserted into the cavity before pouring, so the molten metal will solidify first around these components. Making the internal chill from the same metal as the casting itself will make it easier to obtain the desired chemical similarity between the internal chill and the metal being poured.

External chills are metal inserts in the mold cavity's walls that can quickly remove heat from the liquid metal to accelerate solidification. They are frequently employed successfully in areas of the casting where feeding liquid metal is challenging, promoting quick freezing in these areas. In contrast, the link to liquid metal is still open. Well-designed chills can locally reduce the modulus of the area where they are placed by up to 50%, significantly reducing the solidification time (Rio Tinto 2000). Refer to Chapter 6 for the concept of modulus in sand-casting molds.

Figure 5.8 illustrates a possible application of external and internal chills and the likely result in the casting if chills were not used. On a circular casting, a properly placed chill at the furthest point from the riser will promote directional solidification. The use of chills at specific locations can control the sequential solidification of castings and increase the feeding distance of the riser (Chastain 2004, Chapter 6).

5.6 Casting Quality

There are some flaws that all casting techniques share. The following is a description of the most typical flaws in casting processes (Rao 2007, Chapter 7):

- Incomplete casting. This defect results when an inadequate metal volume is poured or when the mold cavity is not properly filled due to misruns (caused by early solidification) or runouts (resulting from metal being lost from the mold after pouring).
- Hot tearing. Sometimes referred to as hot cracking, hot tearing occurs when the casting cannot shrink due to a hard mold during the latter phases of solidification or the first cooling phase following solidification.

- Cold shuts. This defect is generated when two molten metal flows merged in the cavity are not properly fused together due to premature freezing.
- Pipes or shrinkage cavities. This defect is caused by solidification shrinkage and results in interior voids in the casting that may result in mechanical weakness.
- Surface depressions. Similar to pipes, surface depressions occur by solidification shrinkage, but instead of internal voids, the defects appear at the superficial level in the form of depressions.
- Microporosity. This defect is presented in the form of a network of tiny spaces locally dispersed and is caused by a localized solidification shrinkage of the final liquid metal in the dendritic structure of the casting.

Additionally, some of the following flaws are exclusive to sand castings since they are caused by the usage of sand molds:

- Sand blow is a flaw that appears as a balloon-shaped gas cavity and is brought on by the release of mold gases during the pouring process.
- Penetration is a surface flaw that develops when a liquid metal has high fluidity and permeates the sand mold or sand core. As a result, the casting surface is made up of a combination of metal and sand grains.
- Mold shift describes flaws brought on by the relative displacement of the cope and drag in a mold, which causes a step to appear at the separating line in the cast product.
- Core shift is comparable to mold shift, but it involves the movement of the core.
- Mold crack happens when the mold's strength is insufficient, causing a crack that can allow the liquid metal to enter.

5.7 Cleaning Operations

After the casting hardens and is removed from the mold, a few more steps are usually required. These procedures consist of trimming, removing the core, cleaning the surface, inspecting, and, if necessary, heat treating. All these operations in foundry facilities are typically referred to as cleaning operations.

Trimming is the process of removing superfluous metal from the cast item, including sprues, runners, risers, parting-line flash, chaplets, etc. The removal of any cores that were used in the part's casting is also conducted as cleaning operations. They can occasionally be eliminated by mechanically or manually shaking the casting using a hammer or press.

During surface cleaning, sand is removed from the casting's surface. The surface can be cleaned by wire brushing, chemical pickling, tumble drying, and airblasting with a metal shot or coarse sand grit.

Castings may include flaws, and inspection is necessary to find them. After inspection, castings are frequently heat-treated to improve their mechanical properties and reduce internal tensions.

Questions and Exercises

1 Explain the purpose of the following parts in a sand-casting mold: (i) core; (ii) riser; (iii) sprue; and (iv) gate.

2 During sand casting, turbulence may produce impurities and mold erosion. What are the main factors that influence molten metal turbulence?

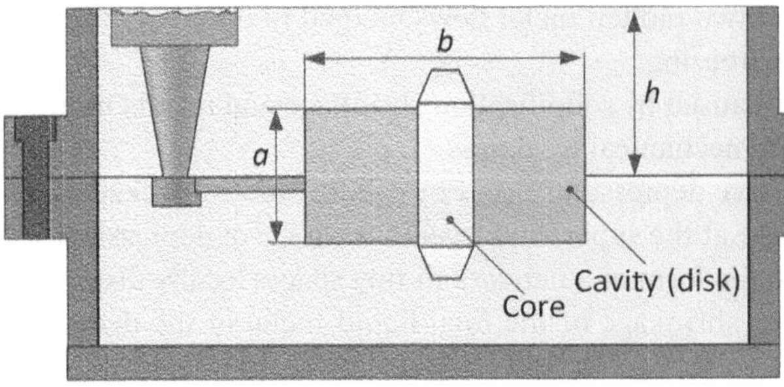

Figure 5.9 Sand mold for Exercise 7. Material: gray cast iron.

3 Explain the continuity law applied in sand casting.

4 Write Chvorinov's rule and explain the purpose of its use in sand casting.

5 Explain the meaning of a pipe as a casting defect. How is it produced, and how can it be avoided?

6 Enumerate four common sand-casting defects in the industry.

7 Given the sand-casting mold shown in Figure 5.9 (produces a disk of gray cast iron), where $a = 160$ mm; $b = 320$ mm; core diameter 50 mm; runner and gate cross-section 30×30 mm, and $h = 190$ mm, estimate the following:
- Heat required. Assume a desired pouring temperature of 1,200 °C. Assume a specific heat for a liquid state 30% higher than the one at a solid state and 25 °C for the initial temperature of the metal.
- Reynolds number at the gate.
- Filling time.
- Solidification time.
- Total cooling time. Assume that the breakage is done when casting cools and reaches 50 °C.
- Final dimensions of the part.

References

Beddoes, J., and Bibby, M. (1999). *Principles of Metal Manufacturing Processes.* Butterworth-Heinemann.

Black, J.T., and Kohser, R.A. (2020). *DeGarmo's Materials and Processes in Manufacturing*. Wiley.

Chastain, S., (2004). *Metal Casting: A Sand Casting Manual for the Small Foundry*, vol. 2. Jacksonville, FL.

Dinsdale, A.T., and Quested, P.N. (2004). The viscosity of aluminium and its alloys – a review of data and models. *Journal of Materials Science* 39: 7221–7228.

Ghosh, A., and Mallik, A.K. (1986). *Manufacturing Science*. Wiley Eastern.

Groover, M.P. (2007). *Fundamentals of Modern Manufacturing: Materials Processes, and Systems*. Wiley.

Jelinek, P., and Elbel, T. (2010). Chvorinov's rule and determination of coefficient of heat accumulation of moulds with non-quartz base sands. *Archives of Foundry Engineering* 10 (4): 77–82.

Kalpakjian, S., and Schmid, S.R. (2009). *Manufacturing Engineering and Technology*, 6e. London, UK: Prentice Hall.

Marathe, S., and Quadros, C. (2021). Thermal hot spot prediction in high pressure die casting by determination of Chvorinov's rule shape constant. *Materials Today: Proceedings* 47: 5607–5617.

Murdock, J.W. (1993). *Fundamental Fluid Mechanics for the Practicing Engineer*. CRC Press.

Nakagawa, Y., Suzuki, K., and Momose, A. (1968). Study on the viscosity of molten iron. *Transactions of the Japan Institute of Metals* 9 (1): 67–72.

Rao, T.R. (2007). *Metal Casting: Principles and Practice*. New Age International.

The World Foundry Organization. (2020). WFO Global Foundry Report 2020. Actual situation of the worldwide casting industry. October 2020.

Tinto, R. (2000). *Ductile Iron, the Essentials of Gating and Risering System Design*. Rio Tinto Iron & Titanium Inc.

Turchi, P.A. (2018). Viscosity and surface tension of metals. Lawrence Livermore National Laboratory. Document No. LLNL-TR-750486, Livermore, CA (United States).

Wang, D., and Overfelt, R.A. (2002). Oscillating cup viscosity measurements of aluminum alloys: A201, A319, and A356. *International Journal of Thermophysics* 23: 1063–1076.

Zych, J., Wójcik, M., Kolczyk, J., and Snopkiewicz, T. (2014). Metallostatic pressure function in the quality of Al-Si alloy castings made in sand moulds. *Metallurgy and Foundry Engineering* 40 (3): 191–202. https://doi.org/10.7494/mafe.2014.40.3.191.

6

Process Planning in Sand Casting

6.1 Introduction

Figure 6.1 shows a representation of a standard sand-casting mold. This mold is enclosed within a two-part flask: the upper part referred to as the cope, and the lower part known as the drag. The interface between the cope and drag is termed the parting plane, which can also be referred to as the parting line when observing the casting in a two-dimensional context.

The vertical pathway through which molten metal is introduced into the casting is named as the sprue. Within the parting plane, there are horizontal passages known as runners, responsible for distributing the molten metal. The linkages connecting these runners to the cavities in the casting are named as gates or ingates. Additional components of the casting, strategically placed to supply molten metal as the casting solidifies and contracts, are termed risers or feeders. Occasionally, metal fragments are positioned within the mold to accelerate solidification at specific points; these fragments are referred to as chills.

The process of designing the casting mold initiates with the identification of the parting plane. Subsequently, determining the optimal gate placements to ensure even distribution during casting follows. Once gate locations are established, an appropriate design for the runner system and sprue is chosen. The cavity dimensions are defined considering the part's geometry and employing basic design principles. Shrinkage, machining allowance, and drafts are taken into account at this step. Often, multiple iterations of mold designs are necessary before achieving a successful and defect-free casting.

This chapter presents general guidelines extracted from different references (Beddoes and Bibby 1999; Chastain 2004; Guleyupoglu 1998; Rao 2006) that process planners can follow for a proper design of sand-casting molds.

Manufacturing Process Planning: A Practical Approach for Mechanical Engineering, First Edition.
José V. Abellán-Nebot, Carlos Vila Pastor, and Héctor R. Siller.

Companion website: www.wiley.com/go/Manufacturing_Process_Planning_1e

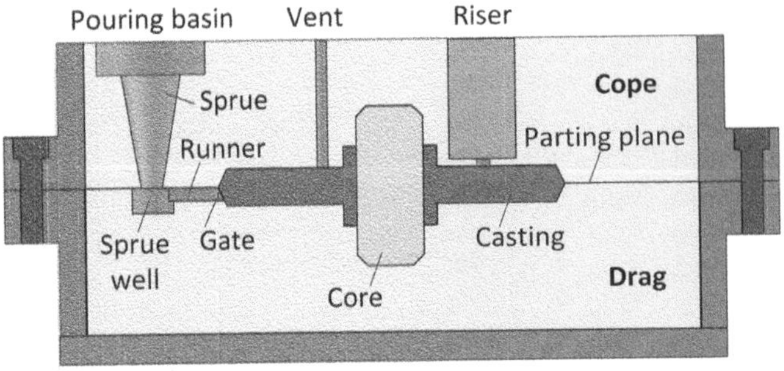

Figure 6.1 Parts of a generic sand-casting mold.

6.2 Mold-Making Guidelines

6.2.1 Part Orientation Rules

Typically, the casting design decision that needs to be decided first is the part orientation. The part orientation will define the orientation of the cavity with respect to the gravity action and will determine the dimensions of the mold. The rules to define part orientation are:

1) Minimize the height of the casting.
2) Place open spaces down.
3) Place the casting such that top risers can be placed on high points on the casting for the heavy sections.

6.2.2 Parting Plane Rules

The parting plane depends on the geometry of the part to be cast and will define both pattern and mold geometry. Some guidelines for parting plane definition are explained in Section 6.3 Pattern Fabrication Guidelines.

6.2.3 Gating System

The gating system is usually referred to as the system composed of the sprue, the runner, and the gate. The guidelines for designing these three components in a sand-casting mold are reviewed as follows.

6.2.3.1 Sprue Rules

The sprue goes from the pouring basin to the runner system, where the molten metal is distributed to reach the cavity. A proper sprue design is of great importance to prevent turbulence during filling. Some standard guidelines employed in the design of sprues are shown in the following:

- The minimum height of the sprue is defined by the casting itself and the height of the top riser.
- The dimensions of the sprue should be calculated to avoid turbulence at the runner system. Oversized sprues may result in elevated flow rates, contributing to turbulent flow.
- Rectangular cross-section sprues are preferable compared to circular ones of the equivalent cross-sectional area due to the lower critical velocity for inducing turbulence. Additionally, circular cross-section sprues tend to be more prone to forming vortices. However, small round sprues with limited height and radius pose minimal vortex-related problems, are simpler to create, and hence, prove more cost-effective for smaller castings.

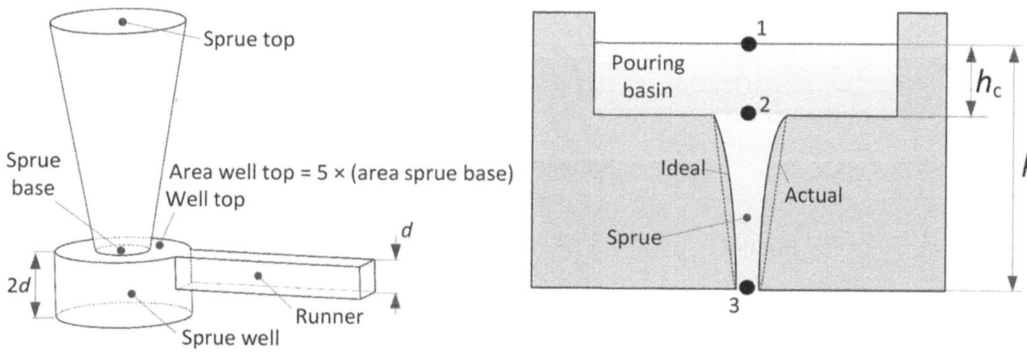

Figure 6.2 Sprue well and pouring basin design. *Source:* Beddoes and Bibby (1999) and Chastain (2004).

- The sprue should direct the molten metal into a standardized well area to diminish the kinetic energy of the metal. The well's top area is usually recommended to be five times that of the sprue exit area, while the well's depth is twice that of the runner depth (Figure 6.2).
- The ideal placement for the sprue is as distant as feasible from the gates. This is mainly explained by the common generation of turbulent flow at the end of the sprue. To avoid or minimize its effect, an elongated path to the gate and the inclusion of a filter at the sprue base is recommended.
- A minimum taper of about 5% should be applied to the sprues to prevent air aspiration and uncontrolled descent of the metal (Figure 6.2). Slightly exceeding the required taper for the sprues offers an additional margin of safety against air aspiration. The theoretical relationship between the sprue top area (A_2), the sprue base area (A_3), the pouring basin height (h_c), and the total sprue height (h_t) to avoid air aspiration is (Beddoes and Bibby 1999) as follows:

$$\frac{A_2}{A_3} = \sqrt{\frac{h_t}{h_c}}$$

6.2.3.2 Runner Rules

Runners serve as passageways that transport molten metal from the sprue well to the gates, facilitating the entry of metal into the mold cavity. The established principles applied to the design of runners are:

1) Employ standardized dimensions and configurations for runners.
2) In sand castings, rectangular cross-sectional profiles are recommended.
3) Steering clear of abrupt alterations in the runner direction is advisable. If the directional shift exceeds approximately 15°, filleting is required at the junction.
4) For elongated parts, runners traverse along the component where different gates are placed. It is considered sound practice to progressively reduce the runner's cross-sectional area after it passes through each gate, adjusting for the gate area (Figure 6.3). This practice generates a balanced gating system that ensures a consistent metal velocity at every entry point into the mold cavity.
5) Extensions, also known as blind ends, are commonly incorporated into runners to trap the first part of the molten metal flow that tends to concentrate impurities and oxides.
6) The area of the runner is maintained at three to ten times the cross-sectional area of the sprue exit, as prescribed by the gating ratio (see Section 6.2.3.3).

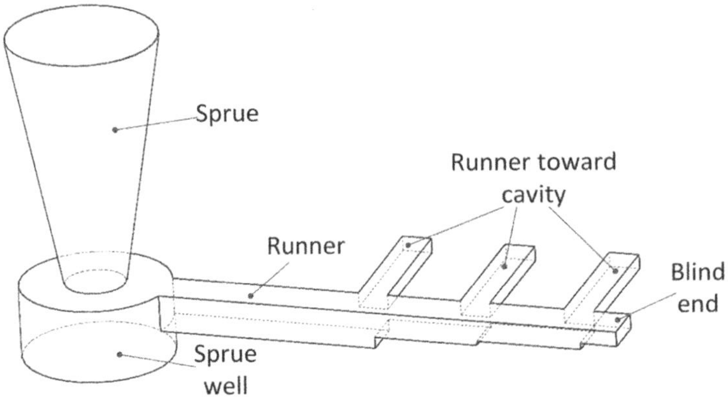

Figure 6.3 Balanced gating system. *Source:* Adapted from Chastain (2004).

6.2.3.3 Gate Rules

Gates serve as pathways connecting runners to the mold cavity that reproduces the part. The subsequent guidelines for gate design are commonly applied:

1) Direct the gate toward thicker sections.
2) Adopt standardized configurations and dimensions for gates. Rectangular gates are the most prevalent choice.
3) Position gates strategically to minimize disturbance and prevent erosion of the sand mold due to the flow of molten metal.
4) Keep the appropriate gating ratio. Appropriate ratios among the cross-sectional areas of the sprue, runners, and gates should be kept to achieve optimal filling conditions. Some examples of common gating ratios are provided in Table 6.1. For instance, if a 4:3:2 ratio is adopted, the combined area of the runners should be three-fourths the area of the sprue base, while the collective area of the gates should be two-thirds the area of the runners.
5) The incorporation of fillets between the gates and the casting is beneficial.
6) Unless the casting is of a small and simple nature, multiple gates should be utilized to evenly distribute metal throughout the mold. This approach facilitates quicker filling, reduces the risk of hot spots, and offers the advantage of lower pouring temperatures, which enhances the metallurgical structure and mitigates temperature gradients within the casting.

Table 6.1 Gating ratios and densities for different alloys.

Material	Density, kg/m^3	Gating ratio (sprue base: runner: gate)
Aluminum	2,710	1:2:1, 1:3:3, 1:4:4, 1:2:4
Copper	8,850	2:8:1, 3:9:1
Gray cast iron	6,900–7,800	2:1.5:1, 2:3:1
Ductile iron	6,900–7,300	1.15:1.1:1, 1.25:1.13:1
Magnesium	1,740	1:2:2, 1:4:4
Steels	7,830	1:1:7, 1:2:1, 1:2:1.5
Brass	8,890	1:1:1, 1:2:3

Source: Rao (2006).

7) A gate length of 10–20 mm is generally adequate for small castings while 30–60 mm is common in medium/large castings.
8) For rectangular cross-section gates, the width-to-thickness ratio is usually around 3–6 to avoid junction hot spots. Furthermore, the thickness is related to the adjacent wall thickness of the casting and should be less than 0.5–0.8 times this thickness.

6.2.3.4 Steps for Gating System Design

The first step in designing a gating system is to determine the casting weight. From here it can be determined the optimal pouring rate. Optimal pouring rates depend on several factors that include the type of alloy, the fluidity of the metal, the pouring temperature, and the weight of the casting. In the study by Chastain (2004), it is shown how to obtain the pouring rate for casting iron, aluminum, brass, and bronze. As a rough guide, Table 6.2 from the study by Rao (2006) shows typical pouring rates for different materials and casting weights.

After determining the pouring rate, the pouring time or filling time of the mold is calculated by dividing the casting weight (W) by the pouring rate (P_r):

$$t_{\text{filling}} = \frac{W}{P_r}$$

Then, the area of the sprue base is calculated. For this purpose, the effective sprue height H according to the placement of the pattern in the mold is obtained as follows:

$$H = h - \frac{p^2}{2c}$$

where h is the height of the sprue, c is the total height of the mold cavity and p is the height of the mold cavity in the cope (Figure 6.4). Next, the area of the sprue base is calculated as follows:

$$A_s = \frac{W}{d \times \rho \times t_{\text{filling}} \times \sqrt{2g \times H}}$$

where ρ is the density of the molten metal and d is the discharge coefficient for the sprue (0.88 for round tapered, 0.47 for round straight sprue, 0.74 for squared tapered sprue).

Finally, using the adequate gating ratio (Table 6.1), the cross-sectional areas of the runner and gates are calculated.

Table 6.2 Pouring rates in kg/s for castings.

	Pouring rate in kg/s and casting weight			
Metal	**Up to 10 kg**	**10–50 kg**	**50–100 kg**	**100–500 kg**
Pig iron	1.1	1.5–2.0	3.0–4.0	3.5–6.0
Steel	1.2–1.4	1.9–2.5	4.0–5.0	4.5–7.0
Aluminum alloys	0.25–0.3	0.5–0.7	1.0–1.3	1.2–2.0

Source: Rao (2006).

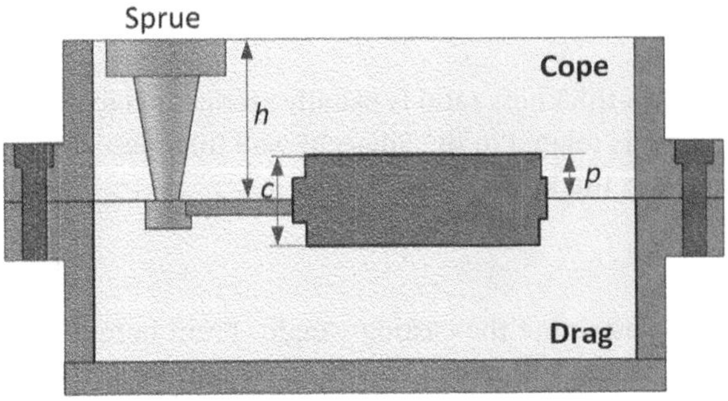

Figure 6.4 Parameters for the effective sprue height (H) calculation.

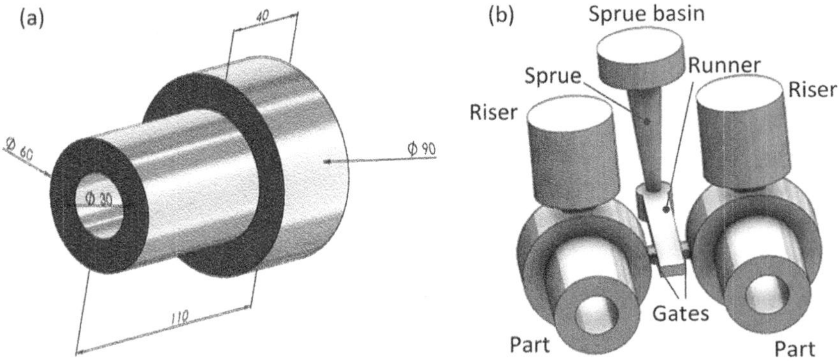

Figure 6.5 Aluminum part (a) and (b) complete casting result.

6.2.3.5 Practical Example

Figure 6.5a show an aluminum part and Figure 6.5b shows the resulting casting from a sand-casting process. The mold design has two cavities, a sprue height of 91 mm, and a pouring basin height of 20 mm. For each cavity, a cylindrical riser of 60-mm height and 60-mm diameter is designed. For this mold, it is desired to evaluate the main dimensions of the sprue, runner, and gates. The weight of each part after casting and trimming is 1 kg whereas the weight of each riser is 0.46 kg.

The weight of the casting (without considering runners and sprue) is $(1.46)\times 2 = 2.92$ kg. Thus, from Table 6.2 the proper pouring rate is 0.3 kg/s. The pouring time is

$$t_{\text{filling}} = \frac{W}{P_r} = \frac{2.92}{0.3} = 9.7\,\text{s}$$

The effective sprue height is

$$H = (91+20) - \frac{45^2}{2\times 45} = 88.5\,\text{mm}$$

Thus, the area of the sprue base is

$$A_s = \frac{W}{c\times \rho \times t_{\text{filling}} \times \sqrt{2g\times H}} = \frac{2.92}{0.88\times 2.375\times 10^{-6}\times 9.7\times \sqrt{2\times 9{,}810\times 88.5}}$$

$$= \frac{2.92}{0.0267} = 109.3\,\text{mm}^2$$

where c is 0.88 (sprue round tapered) and the density of molten aluminum is $2.375 \cdot 10^{-6}\,\text{kg/mm}^3$. The diameter at the sprue base is $11.8 \approx 11.5\,\text{mm}$ ($A_s = 103.9\,\text{mm}^2$).

To calculate runner area and dimensions, the gating ratio 1:3:3 is used (Table 6.1). Then, the runner area is

$$A_r = 3 \times \frac{\pi D^2}{4} = 3 \times \pi \times \frac{11.5^2}{4} = 311.6\,\text{mm}^2$$

Assuming a rectangular cross-section, the runner dimensions would be approximately $25 \times 12.5\,\text{mm}$ if an aspect ratio of 2 (the width-to-height ratio) is considered. The gates area, since there are two cavities, is $0.5 \times 311.6 = 155.8\,\text{mm}^2$ per gate and the dimensions of each one are approximately $21 \times 7\,\text{mm}$ if a width-to-thickness ratio of 3 is applied. The length of the gate is set to 10 mm.

The area of the sprue well is 5 times the area of the sprue base, so the diameter of the sprue well is

$$A_{\text{well}} = 5 \times A_s \rightarrow 5 \times \frac{\pi \times 11.5^2}{4} = \frac{\pi \times D^2}{4} \rightarrow D = 25.7\,\text{mm}$$

The well depth is 2 times the runner depth, so $2 \times 12.5 = 25\,\text{mm}$.

Finally, the area at the top of the sprue, considering the pouring basin height of 20 mm, is

$$\frac{A_{\text{top}}}{A_{\text{base}}} = \sqrt{\frac{h_t}{h_c}} \rightarrow \frac{A_{\text{top}}}{103.9} = \sqrt{\frac{91}{20}} \rightarrow A_{\text{top}} = 221.6\,\text{mm}^2$$

Then, the sprue top diameter is around 17 mm.

6.2.4 Riser's Rules

Risers are molten metal reservoirs used to feed the casting when it solidifies. If more molten metal cannot be fed to a mold cavity region, the shrinkage that occurs during solidification results in voids. Risers are intended to remove casting voids caused by shrinkage and for this purpose, they should solidify last. Risers also function as pressure heads to feed thin portions and as exits for gases and dross trapped in the metal. Typical riser designs are top risers and side risers depending on where the riser is placed, and open risers and blind risers if they are opened to the atmosphere or not. Figure 6.6 shows some examples of different riser designs.

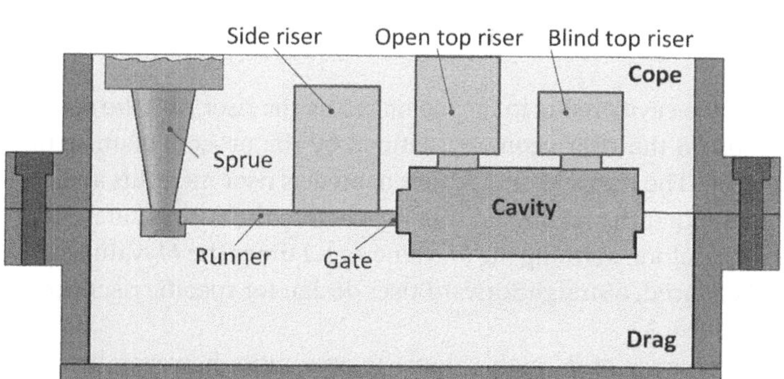

Figure 6.6 Top risers, side risers, blind and open risers.

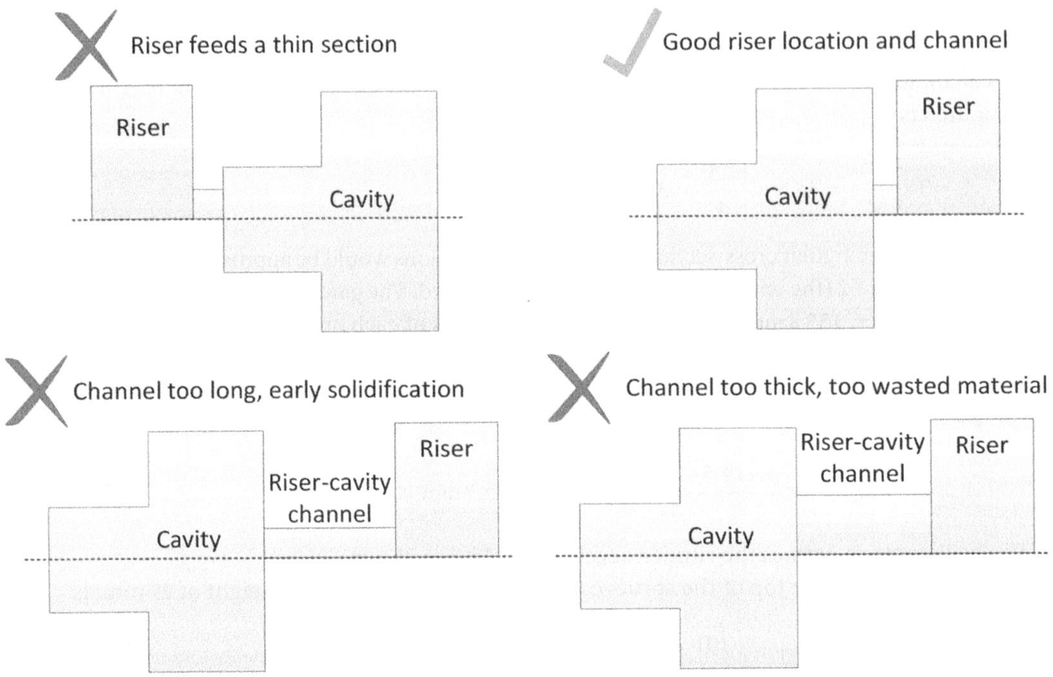

Figure 6.7 Correct and incorrect riser's locations and channel dimensions.

The general rules for riser design are:

1) Risers are situated close to areas of the casting that are thick (Figure 6.7).
2) Riser and casting connections must be designed to avoid early freezing (Figure 6.7). Thin and long channels will freeze before the heavy section has entirely solidified.
3) Side risers are usually placed close to the gates. Thin-walled castings commonly employ side risers.
4) If bosses are present away from the gates, top risers can be used to minimize shrinkage problems.
5) Open risers are simpler to produce, remove, and clean up after manufacturing than blind risers.
6) To ensure that the region fed by the riser solidifies earlier, risers should have higher volume to area ratios than the region fed. According to Chvorinov's rule, the riser should ensure:

$$\frac{V_{\text{riser}}}{A_{\text{riser}}} > \frac{V_{\text{cavity}}}{A_{\text{cavity}}} \rightarrow M_{\text{r}} > M_{\text{c}}$$

where the volume to area ratios of the cavity refers to the region fed by the riser, and the volume to area ratios of the riser depends on the riser geometry defined by the process planner that should ensure the above inequation. The terms M_{r} and M_{c} are named as riser modulus and cavity modulus, respectively. To ensure that the region fed solidifies before the riser, a safety coefficient of 1.2 is usually adopted. Therefore, defining the M_{r} value as 1.2 times the M_{c} value of the region where a riser is going to be placed, a straightforward riser design for specific riser geometries can be done according to Table 6.3.
7) The ideal riser shape is a sphere because of its high volume-to-area ratio, however, working with spherical risers is challenging. Since they have the second-best volume-to-area ratio, round cylinders are widely employed as risers. Practical foundry experience has demonstrated that

Table 6.3 Volume, area, and modulus for different riser geometries.

	Riser geometry		
	Spherical	**Cylindrical**	**Prismatic**
	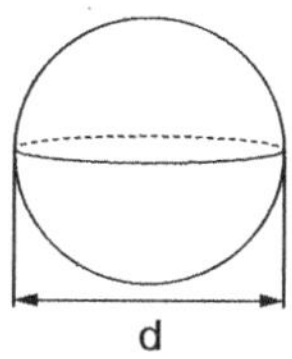	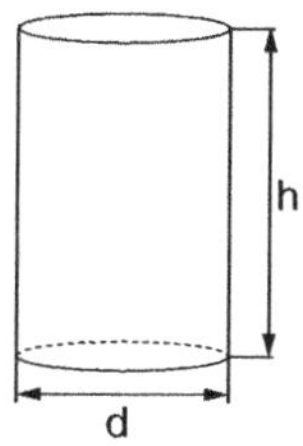	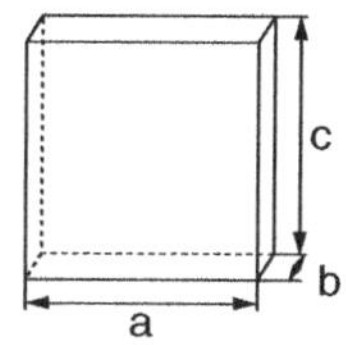
V_r	$\frac{1}{6} \times \pi \times d^3$	$\frac{1}{4} \times \pi \times d^2 \times h$	$a \times b \times c$
A_r	$\pi \times d^2$	$\frac{1}{2} \times \pi \times d \times h + \pi \times d \times h$	$2 \times a \times b + 2 \times a \times c + 2 \times b \times c$
M_r	$\frac{1}{6} \times d$	$\frac{d \times h}{2 \times (d + 2 \times h)}$	$\frac{a \times b \times c}{2 \times a \times b + 2 \times a \times c + 2 \times b \times c}$

under this shape, height values of 1.5 times its diameter maximize feeding while using the least quantity of metal (Kaushish 2010, Chapter 5).

8) The connection of risers with the cavity, named the neck risers, needs special attention. If the neck is too small, it freezes off and cannot supply feed metal to the casting. If it is too large, it becomes difficult to remove. Figure 6.8 shows some recommendations from sand-casting manuals (Chastain 2004; Wang et al. 2010).

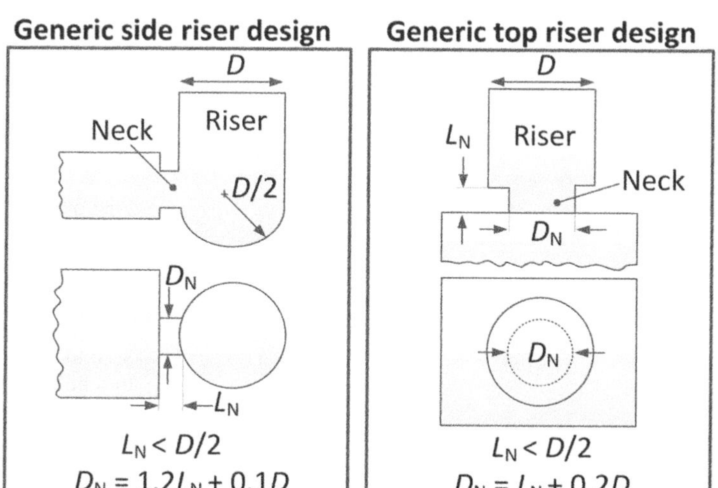

Figure 6.8 Neck riser recommendations. *Source:* Chastain (2004) and Wang et al. (2010).

6.2.4.1 Practical Example

The 3D model given in Figure 6.9 represents a support that is going to be produced by sand casting. The gate is placed in Section 1 (the thicker region of the part), and to analyze the riser design, the casting is divided into three sections where the modulus (V/A) of each section is given in Table 6.4.

According to Table 6.4, the middle section of the part presents the lower modulus value, which means that it will solidify earlier. Therefore, both sides of the part remain disconnected from the molted metal that enters from the runner system into Section 1. To avoid pipe shrinkages on both sides of the part, two risers are needed to provide molten metal to both sides while solidification is produced. Another possibility would be to add a chiller to force that Section 3 solidifies earlier than Section 2. Under this situation, only one riser at Section 1 would be needed since a directional solidification would occur. Assuming the use of two risers, each one is designed to have a modulus

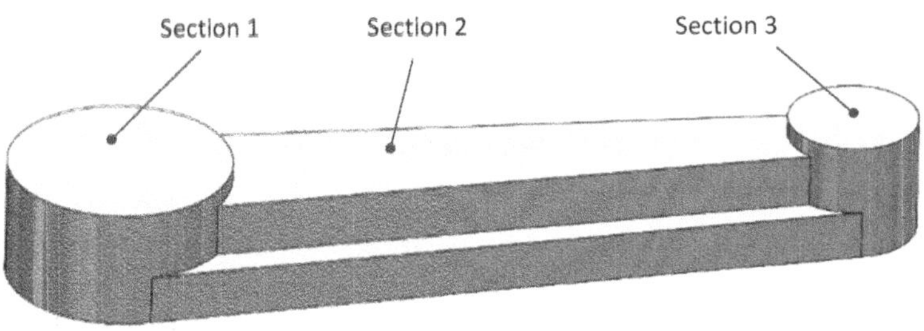

Figure 6.9 CAD part to be cast.

Table 6.4 Modulus of different sections in the part to be cast.

Casting section	Volume (mm^3)	Contact surface (mm^2)	Modulus (mm)	Observation
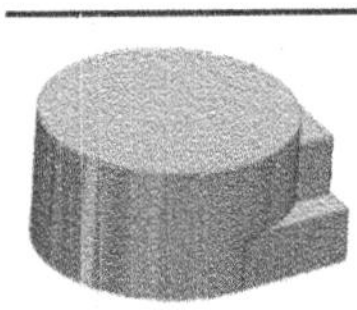 Section 1	102,612	10,760	9.53	Last part to solidify
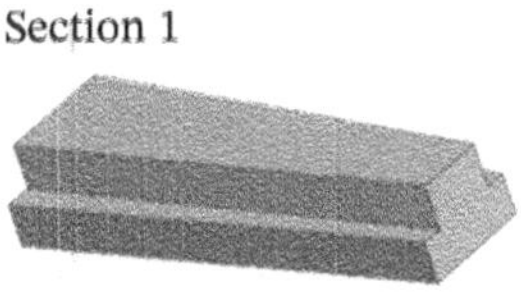 Section 2	179,520	28,429	6.31	First part to solidify
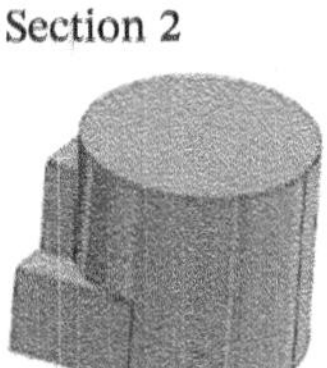 Section 3	34,578	4,706	7.34	Second part to solidify

higher than 1.2 times the modulus of the section that it feeds. Therefore, the designs of the two top risers are as listed as follows:

- Riser 1. $M_r > 1.2 \times 9.53 = 11.43$. Assuming a cylindrical riser with a height 1.5 times the diameter, the riser modulus is

$$M_r = \frac{d \times h}{2 \times (d + 2 \times h)} = \frac{1.5 \times d^2}{2 \times (d + 3 \times d)} = \frac{1.5 \times d}{8}$$

  Thus, the diameter and height of the cylindrical riser are

$$d = 16 \times \frac{M_r}{3} = 16 \times \frac{11.43}{3} \approx 61\,\text{mm}$$

$$h = 1.5 \times d = 91.5\,\text{mm}$$

  The maximum value of the riser neck is $d/2$. Considering a value of $d/6$, the neck is $61/6 \approx 10\,\text{mm}$, and the neck diameter is $= 10 + 0.2 \times 61 \approx 22\,\text{mm}$.
- Riser 2. $M_r > 1.2 \times 7.34 = 8.8$. Assuming the same height-to-diameter ratio, the dimensions of the riser are

$$d = 16 \times \frac{M_r}{3} = 16 \times \frac{8.8}{3} \approx 47\,\text{mm}$$

$$h = 1.5 \times d = 70.5\,\text{mm}$$

  The maximum value of the riser neck is $d/2$. Considering a value of $d/6$, the neck is $47/6 \approx 8\,\text{mm}$, and the neck diameter is $= 8 + 0.2 \times 47 \approx 17\,\text{mm}$.

As a result, the cast part with the two risers is shown in Figure 6.10. Figure 6.11a shows the result of the casting process with, and Figure 6.11b shows the result of the casting process without, the addition of the risers. As it can be seen, the risers avoid pipe shrinkages and porosities on both sides of the part.

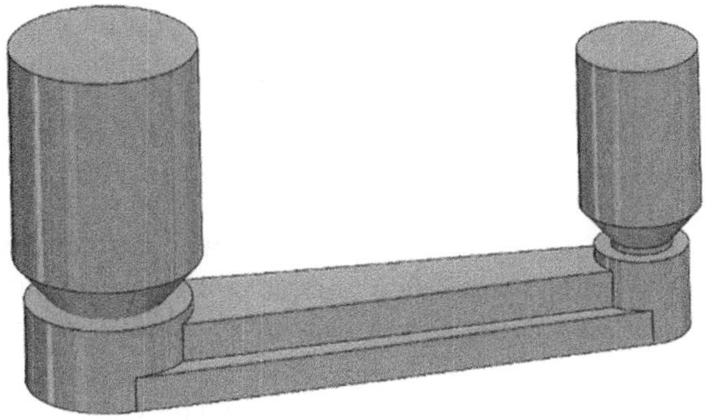

Figure 6.10 Risers included in the design to improve part quality.

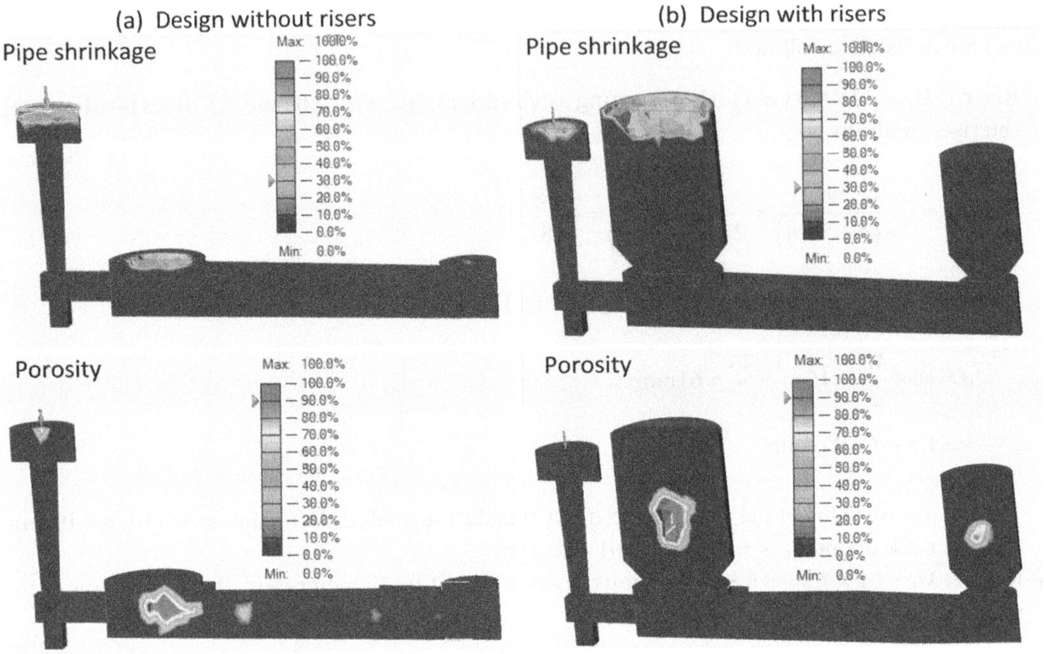

Figure 6.11 Simulation of the casting process without (a) and with (b) risers. Risers will retain the shrinkage and porosity, improving part quality.

6.3 Pattern Fabrication Guidelines

In the process of designing and crafting a pattern, many aspects should be defined such as follows:

- Determining the optimal location for the parting line to facilitate foundry operations.
- Selecting the methods to achieve the specified external and internal shape of the casting most effectively (e.g., core design, geometry simplification for pattern design, etc.).
- Determining the flow of molten metal during filling for correct pattern and core design (e.g., mold erosion, core stability, etc.).
- Determining how the casting will contract as it solidifies and cools to room temperature to set shrinkage allowances.
- Deciding where and to what extent draft angles should be applied.
- Identifying the locations and amounts of extra material needed for machining.
- Selecting the appropriate pattern type, material, and fabrication process.

As shown in Figure 6.12, sand-casting tool fabrication usually requires a 3D computer-aided design (CAD) model to conduct casting simulations and iterate different patterns and core designs. When casting analysis results in adequate casting quality, the fabrication of the pattern is conducted by computer numerical control (CNC) machining or additive manufacturing if low-volume production or rapid fabrication applies. Ultimately, the choice of pattern fabrication method hinges on a delicate balance between lead time, cost, production volume, and precision.

The following guidelines can be used for a correct pattern fabrication process.

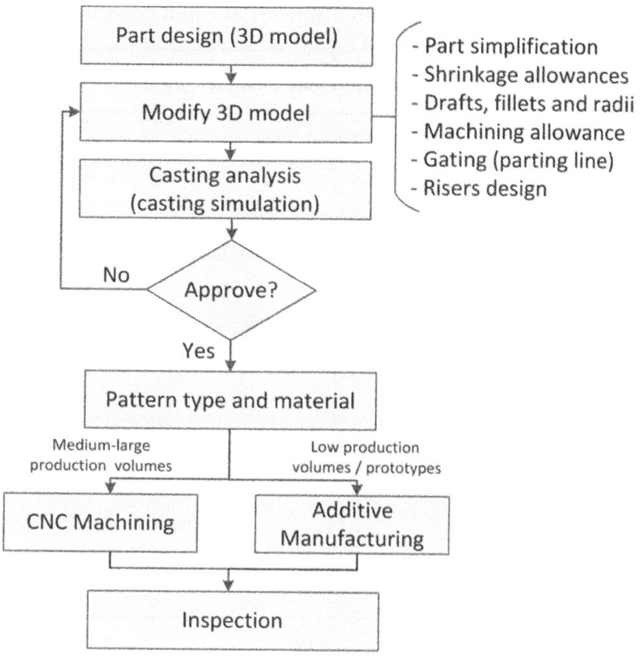

Figure 6.12 Tool fabrication process for sand casting.

6.3.1 Geometry Simplification

Sand casting cannot produce very small holes or fine details on casting surfaces. In these circumstances, it is preferable to streamline the casting process by removing such elements and incorporating them into secondary operations such as turning, milling, or drilling. Additional geometric simplifications depend on the level of accuracy required, the casting process capabilities, and the mold construction technique used. Figure 6.13 shows an example of geometric simplification of a casting part. Note that the internal hole of Ø16 mm and the slots are fine details that need machining (Figure 6.13b) while the casting (Figure 6.13a) quality is enough for the rest of the geometry.

In relation to hole simplification, it should be noted that small holes are more economical to be produced by machining than by creating and using a core in the mold cavity. The minimum

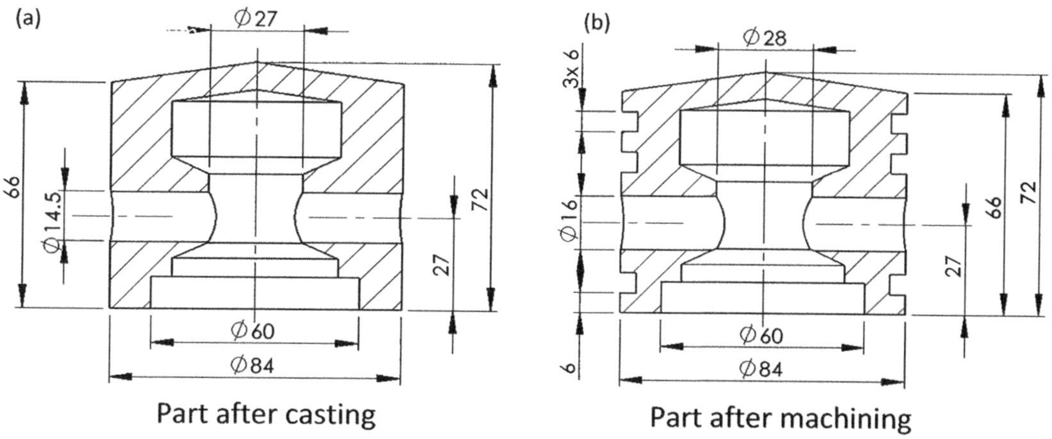

Figure 6.13 Simplified version for casting (a) and actual component after machining (b).

Table 6.5 Ranges of minimum diameters of cored holes in the casting.

Casting method	Minimum hole diameter (mm)
Sand casting	4.8–6.3
Permanent mold	4.8–6.3
Shell mold	3.2–6.3
Die casting	Cu alloys, 4.8 mm; Al alloys, 2.4 mm; Mg alloys, 2.4 mm; Zn alloys, 0.8 mm
Plaster mold	12.7
Investment casting	0.5–1.3

Source: Adapted from Beeley (2001).

Table 6.6 Minimum thickness for different casting processes and materials.

	Minimum section thickness (mm)		
Casting method	Aluminum	Magnesium	Steel
Sand casting	3.18	3.96	4.75
Permanent mold	2.36	3.18	—
Die casting	1.57	2.36	—
Investment casting	1.57	1.57	2.36

Source: Adapted from Black and Kohser (2020).

recommended hole that can be produced by cores in casting depends on the metal cast, the casting process, the thickness of the section where the hole is located, and the length of the hole. In gray irons by sand-casting processes, holes below 8 mm are difficult to produce; in aluminum die-cast parts, the minimum hole to be cast is usually 3 mm (Ravi 2005, Chapter 3). Table 6.5 shows some ranges of minimum diameters of cored holes in casting (Beeley 2001).

Thickness should be also studied in detail before patternmaking. For instance, walls thinner than recommended minimum values should be avoided due to filling and mold-making issues. Table 6.6 shows some thickness limits for different casting processes (Black and Kohser 2020, Chapter 11).

6.3.2 Shrinkage Allowance

In Chapter 5, three volumetric contractions were identified during the casting process: liquid contraction, solidification shrinkage, and solid thermal contraction. The first two shrinkages reduce the volume of molten metal in the mold, but this volume is compensated by a proper design of the mold or/and risers, that provide a correct directional solidification. However, once the cavity is solidified, the cooling process shrinks the casting part. This shrinkage must be taken into account and it should be incorporated as a corrective factor into the pattern design. This factor is commonly named as the patternmaker's shrinkage factor by foundrymen. Therefore, the pattern is typically crafted with slightly larger dimensions than the final part requires to compensate for the solid thermal contraction. This oversizing ensures that, once the casting reaches room temperature, its dimensions align with the design intent before secondary operations are placed, staying within acceptable tolerances (Wang et al. 2010).

The magnitude of the shrinkage allowance depends on the specific casting alloy employed, as different metals contract at varying rates during cooling. Table 6.7 shows an average amount of shrinkage for important cast metals. Note that the percentage of shrinkage listed in the table refers to dimensional shrinkage instead of volumetric shrinkages reported in Chapter 5. Furthermore, casting shrinkage depends on shape, thickness, casting temperature, mold temperature, and mold strength, in addition to the material used. Therefore, it is preferable to estimate the quantity of shrinkage based on historical data gathered from previous experiences in similar castings.

Another crucial aspect to consider in shrinkage allowances is the resistance of the mold to the casting's contraction, influenced by factors such as the casting's shape and the material of the mold itself. Take, for example, the casting illustrated in Figure 6.14. In this case, dimension *X* experiences minimal constraints on its ability to contract because the casting will naturally separate from the mold walls as it shrinks. Consequently, the full value of the patternmaker's allowance would be applied to this dimension. Conversely, the contraction of diameter *Y* is significantly restricted, depending on the rigidity of the core used to create the hole. Therefore, dimension *Y* would receive little to no patternmaker's allowance. This example underscores that the application of the patternmaker's allowance heavily relies on judgment and experience. Additionally, the shape of the casting can profoundly influence the extent of local shrinkage. Various mold characteristics, such as mold rigidity, risers, gates, and so forth, can also impact casting shrinkage.

Finally, in cases where precise tolerances are essential for a specific dimension, an initially larger-than-normal shrinkage allowance may be employed. Subsequently, the pattern is fine-tuned through iterative adjustments based on precise measurements of the actual shrinkage that occurs, with machining modifications made accordingly.

Table 6.7 Typical patternmaker's shrinkage for important casting metals.

Metal/alloy	Amount of shrinkage (%)	Metal/alloy	Amount of shrinkage (%)
Gray cast iron	0.83–1.3	Brasses	1.30–1.6
White cast iron	2.1	Aluminum bronze	2.1
Malleable cast iron	0.78–1.0	Phosphor bronze	1.0–1.6
Steels	2.0–2.6	Aluminum alloys	1.3
Magnesium alloys	1.3	Zinc and zinc alloys	2.6

Source: Adapted from Hurst (1996) and Kalpakjian and Schmid (2009).

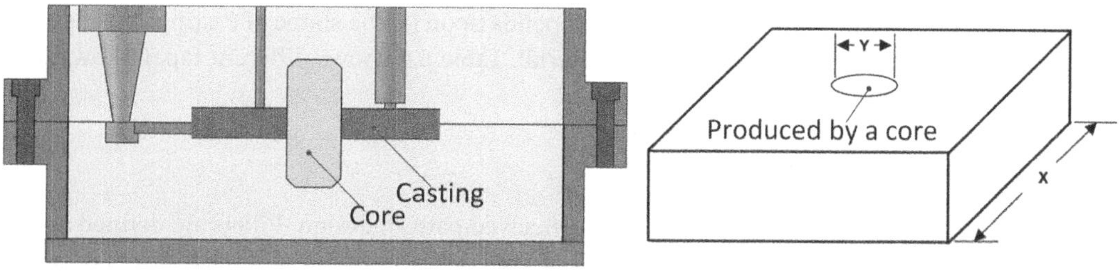

Figure 6.14 Two contraction situations. *Y*-Dimension cannot shrink (constrained by a core), *X*-Dimension can freely shrink.

Table 6.8 General machining allowance for different metals.

	Dimension (mm)		
Metal/alloy	**<150**	**150–300**	**300–500**
Cast irons	2.4	3.2	4.0
Steels	3.2	4.8	6.4
Nonferrous materials	1.6	1.6	2.4

Source: Adapted from Beeley 2001, Chapter 7.

6.3.3 Machining Allowance

Casting dimensions are larger than final design dimensions to provide stock for machining. This material is added when higher dimensional quality is needed after casting, so secondary operations such as turning or milling are required. This additional material is called machining allowance and it should be considered in the pattern design. The following elements affect pattern enlargement caused by machining allowance:

1) The machining technique (such as turning, grinding, drilling, etc.). For instance, less metal is removed by grinding than by turning.
2) Metal properties (ferrous or nonferrous, soft or hard, and easily machinable or difficult). For instance, iron oxidizes, and thus, a higher allowance may be needed for iron than for aluminum or magnesium.
3) The casting technique employed. For centrifugal casting, a larger allowance is needed on the inner side where most of the impurities is retained. Sand castings require more machining allowance while precision casting such as investment casting, plaster casting and die casting need less.
4) The casting's dimensions and shape. For instance, long castings have increased warpage, and thus, larger machining allowance is needed to ensure part quality after machining.

It should be remarked that the amount of metal remaining for machining must be kept to a minimum while still being sufficient to allow cutters to fully remove the casting's scale, impurities, and imperfections. Common machining allowances for different dimensions and alloys are shown in Table 6.8.

6.3.4 Draft

When pattern's vertical faces are tapered or angled (known as draft), the pattern is easier to be removed from the sand mold, as shown in Figure 6.15. No matter what kind of pattern is employed, draft must be used. The amount of draft needed depends upon (i) the shape of casting; (ii) depth of casting; (iii) casting process; and (iv) casting material. Table 6.9 shows different taper allowances used for different casting processes.

6.3.5 Fillets and Radii

Fillets and radii are key characteristics of a well-conceived pattern design. Fillets are defined as the rounding-out of internal corners of the pattern. Radii are the rounding-out of the external edges of the pattern. Both enhance the visual appeal of the casting while also aiding in the distribution of strains and the reduction of inherent stresses that arise during casting.

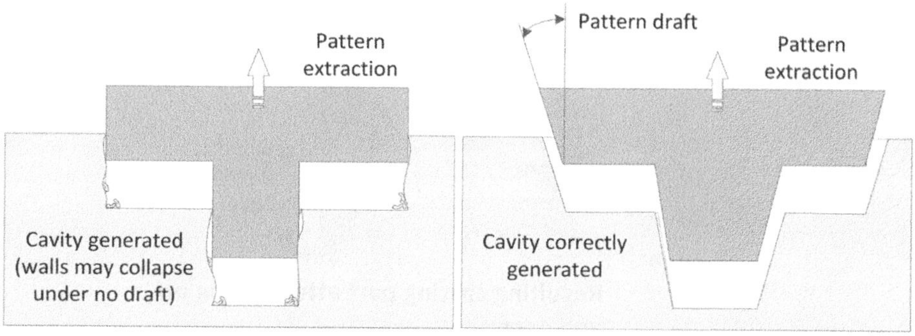

Figure 6.15 Example of taper allowance.

Table 6.9 Common draft values for different casting processes. The values may be reduced if the height of the pattern is higher than 50 mm.

Draft representation	Material	Recommended draft angle
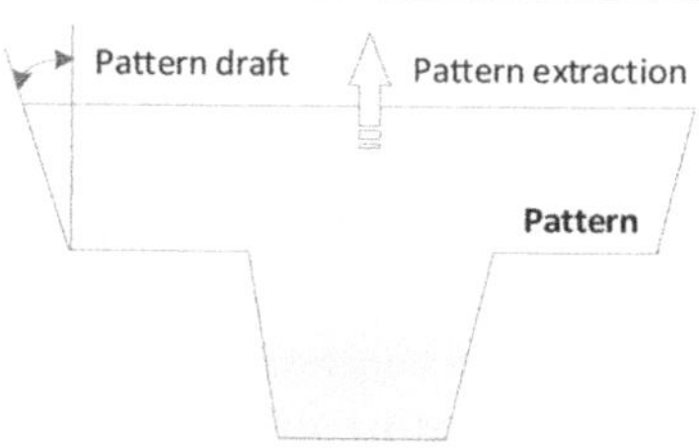	Die casting	1–3°
	Investment casting[a]	None
	Permanent mold	2–3°
	Shell mold	0.5°
	Ceramic mold	1°
	Plaster mold	0.5–1°
	Sand casting	1–5°
	Lost foam casting[a]	None

[a] Although 0–2° drafts may be required to extract wax/polystyrene patterns from molds.
Source: Data from Black and Kohser (2020).

Internal corners may produce localized areas that solidify more slowly than the rest of the part, which may result in many casting defects such as hot tears, cracking, or shrinkage. External corners rapidly lose heat to the surrounding mold material, which may generate additional defects such as misruns (Kay 2002). Additionally, any abrupt change along the pathway of the molten metal may induce turbulence. This turbulence encourages the mingling of the molten metal with air in the casting cavity, resulting in the formation of oxide inclusions. The presence of sufficient fillets/radii facilitates the smooth flow of molten metal, ensuring thorough cavity filling and minimizing the likelihood of turbulent flow. As a general practice, fillets/radii with a radius of 4–5 mm are commonly utilized.

6.3.6 Parting Line

In sand casting, the mold is typically split into two halves, the cope and the drag, to allow for the removal of the pattern. The surface where the two mold halves meet is known as parting plane or parting line if a 2D perspective is considered. It is generally preferable to have the parting line

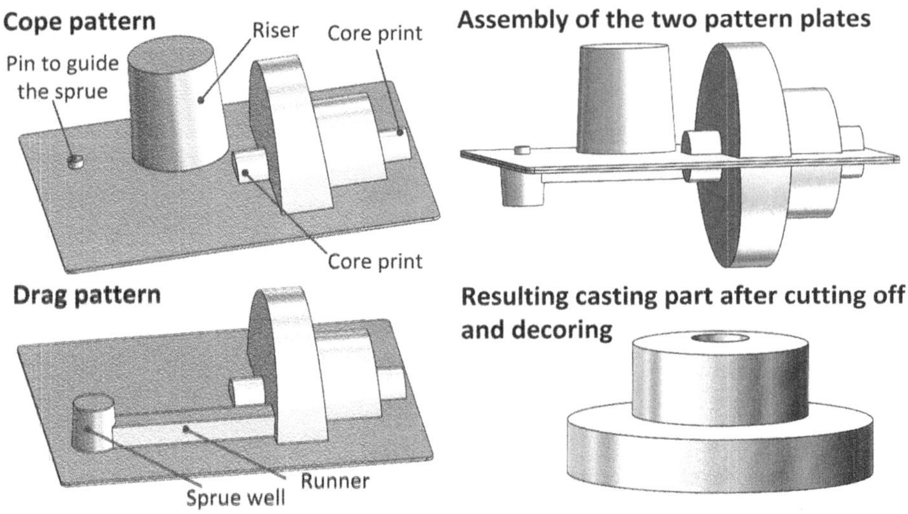

Figure 6.16 Split patterns and resulting casting part.

in a single plane because it simplifies pattern design, pattern creation, mold construction, and assembly. Additionally, patterns with straight parting lines tend to be more cost-effective compared to those with irregular or offset parting lines.

A key rule in the placement of the parting plane is to select the casting's biggest cross-section to facilitate the removal of the pattern. Following this rule, when a casting design exhibits symmetry around a central axis or plane, determining the position of the parting line is straightforward. Furthermore, the runners, gates, and sprue well are typically positioned in the drag so that the top of these channels or cavities is formed by the separating plane. An example of a split pattern is shown in Figure 6.16, where both cope and drag patterns are illustrated.

6.3.7 Core Prints and Chaplets

Whenever there is a need to incorporate a core to shape the inner contours of a casting, it becomes crucial to establish the correct positioning and support for this core within the mold. To address this requirement, core prints are extensions or projections intentionally integrated into the pattern. These core prints are carefully designed to resist the upward pressure exerted on the core by the molten metal once it is poured into the mold. Additionally, core prints are designed to prevent any displacement of the core when the molten metal flows into the mold cavity.

The buoyant force in casting is estimated as:

$$P = V \times (\rho - d)$$

where P is the buoyant force, V is the volume of the core in the mold cavity, ρ is the density of the liquid metal, and d is the density of the core material. For common sand cores, a density of 1,387 kg/m^3 is assumed. A more straightforward estimation can be done by expressing the buoyant force as a proportion of the weight of the core according to Table 6.10.

Table 6.10 Ratios of buoyant force to core weight.

Liquid metal	Density (kg/m^3)	Ratio of buoyant force to core weight
Aluminum	2,710	0.66
Brass	8,890	4.25
Copper	8,850	4.50
Cast iron	6,900–7,800	3.50
Steel	7,830	3.90
Zinc	7,130	3.38

Source: Adapted from Brown (2000).

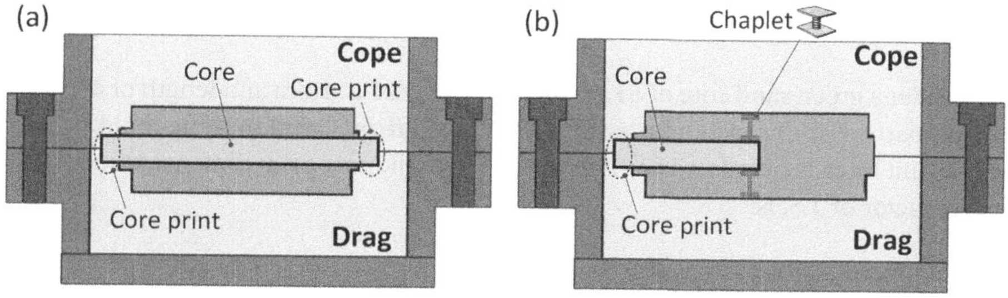

Figure 6.17 (a) Core prints to support core; (b) chaplets added to support the core.

To keep the core in position without distorting the mold, it is empirically considered that a core print has to be able to support a load of 3.5 N/cm^2 (Rao 2006). Therefore, the core print area should satisfy:

$$A > \frac{P}{0.035}$$

where A is the area of the core print in mm^2 and P is the buoyancy force exerted on the core and it is expressed in N. For facilitating core assembly, a minimum core print length of 10 mm is recommended. In case the position of the core (Figure 6.17a) cannot be ensured only with the core prints, chaplets (Figure 6.17b) can be added. Table 6.11 shows recommended core prints for horizontal and vertical cores (h_1 and v_1, respectively), according to Rao (2006).

Additionally, the core itself may collapse due to the buoyant forces. For green sand cores, it can be assumed shear and tensile strength ranges are within 100–150 kPa while for baked cores, 1,000–1,500 kPa is a reasonable range. When it is expected that cores will need to resist higher values, chaplets should be used.

Cores are commonly supported from one side (the other side is free) or from both sides (Figure 6.17). For cylindrical cores with one side support, the maximum shear stress is F/A (F is the buoyant force and A is the core cross-section) whereas for two sides supported cores, the maximum shear stress is $F/(2A)$. In the last case, the maximum tensile tension is given at the middle of the core due to the bending moment and its maximum value is estimated as $\left[8 \times F \times \left(\frac{L}{2}\right)\right] / \left(\pi \times d^3\right)$, where L is the core length and d is the core diameter.

Table 6.11 Recommended core prints for horizontal and vertical cores.

	Core length (mm)							
	<50		51–150		151–300		301–500	
Core diameter (mm)	v_1	h_1	v_1	h_1	v_1	h_1	v_1	h_1
Up to 25	20	15	25	25	—	40	—	—
26–50	20	20	40	35	60	45	70	60
51–100	25	25	35	40	50	50	70	70
101–200	30	30	30	50	40	55	60	80

Source: Rao (2006).

For instance, consider a green sand core of 5 kg with a diameter of 100 mm and length of 460 mm for casting gray iron parts. For this example, assume that the core is placed into the mold by two sides. Then, the buoyant force is calculated as $3.5 \times 5 = 17.5$ kgf. The core print dimension required, considering a safety factor of 1.5, is:

$$A > \frac{17.5 \times 9.81 \times 1.5}{0.035} \rightarrow A > 7{,}357\,\text{mm}^2$$

The core print projected area is $D \times L$, and since there are two core prints (two sides), the minimum length for each core print is

$$7{,}357 \leq 2 \times 100 \times L \rightarrow L > 37\,\text{mm}$$

After designing the core print, the core resistance is analyzed. In this example, since the core is supported by two sides, the maximum tensile stress is

$$\sigma = \frac{\left(8 \times F \times \frac{L}{2}\right)}{\pi \times d^3} = \frac{\left(8 \times 17.5 \times 9.81 \times \frac{460}{2}\right)}{\pi \times 100^3} = 0.1\,\text{MPa} = 100\,\text{kPa}$$

Since the tensile stress at the middle of the core is lower than 100–150 kPa for green sand cores, a chaplet is not required.

6.3.8 Opening Forces on Molds

The force of the molten metal due to metallostatic pressure will force open the boxes and result in runouts if a mold is not properly clamped. Therefore, it is crucial to quantify the opening force to utilize the appropriate weighting or clamping devices.

The metallostatic force, F_m, can be estimated by taking the area of the cross-section of the casting on which this is acting. The projected area of the casting in the parting plane, A_p, is the area on which the metal pressure will be acting. As shown in Figure 6.18, the metallostatic pressure depends on the height given by $h - c$. Hence, the force F_m is calculated as

$$F_m = A_p \times \rho \times (h - c)$$

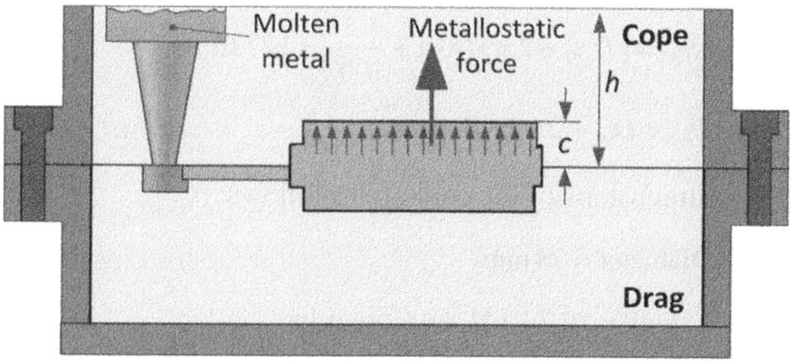

Figure 6.18 Parameters h and c related to the opening force of the cope.

Note that when the metal enters the mold cavity at a high flow rate, the loss of momentum would exert much higher pressure than that accounted for the aforementioned static condition. It is a general practice to increase 50% the estimated force to account for this dynamic effect (Brown 2000, Chapter 1).

6.3.9 Practical Example

Define the pattern and core required for a casting process to manufacture the cylindrical parts shown in Figure 6.19. The material is aluminum alloy and the machining allowance is defined as 3 mm.

First, the machining and shrinkage allowance are added to the model. The shrinkage for aluminum alloys is 1.3% (Table 6.7) and the machining allowance is given in the example and its value is 3 mm. Thus, the pattern is defined by the envelope of the final cast part with the following values:

$$D_{180} = \left(D_{180} + 2 \times 3\right) \times \left(1 + \frac{1.3}{100}\right) = 188.4\,\text{mm}$$

$$D_{50} = \left(D_{50} + 2 \times 3\right) \times \left(1 + \frac{1.3}{100}\right) = 56.7\,\text{mm}$$

$$D_{140} = D_{140} \times \left(1 + \frac{1.3}{100}\right) = 141.8\,\text{mm}$$

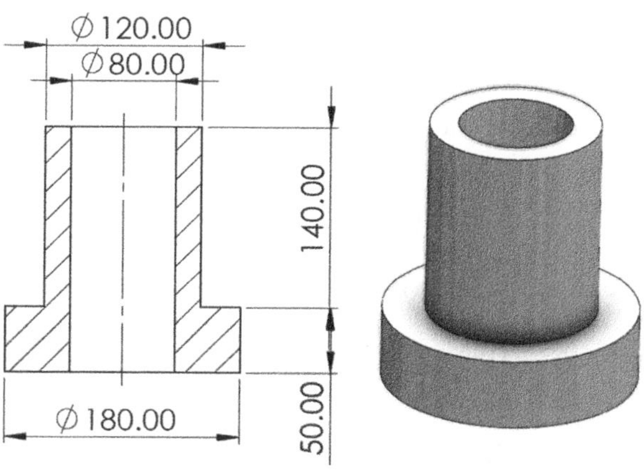

Figure 6.19 Cylindrical part to be cast. Dimensions in mm.

$$D_{120} = (D_{120} + 2\times3)\times\left(1+\frac{1.3}{100}\right) = 127.6\,\text{mm}$$

$$D_{80} = D_{80} - 2\times3 = 74\,\text{mm}$$

The dimensions of the core are as follows:

$$\text{Diameter} = 74\,\text{mm}$$

$$\text{Length} = 56.7 + 141.8 + \text{Core print}$$

The pattern will present a core print. The length of the core print is defined to ensure that the core resists the buoyant force and stays in position without distorting the mold. The buoyant force is expressed as

$$F_b = 0.66\times W$$

where W is the weight of the core within the cavity, which is expressed as

$$W = \rho\times V = 1{,}387\times\pi\times\frac{0.074^2}{4}\times(0.0567+0.1418) = 1.18\,\text{kg}$$

and the core sand density is assumed to be 1,387 kg/m^3. Thus,

$$F_b = 0.66\times W = 0.66\times1.18 = 0.78\,\text{kgf}$$

The core print required assuming a safety factor of 1.5 is

$$A > \frac{0.78\times9.81\times1.5}{0.035} \rightarrow A > 328\,\text{mm}^2$$

The core print projected area is $D\times L$, and since there are two print cores (two sides), the minimum core print length required is

$$328 = 2\times74\times L \rightarrow L = 2.2\,\text{mm}$$

This minimum core print value will ensure that the mold has no distortion. In this case, the minimum value is too small to be practical, and thus, a recommended core print value of 50 mm from Table 6.11 is chosen.

Next, the core resistance should be analyzed. For a two-side support, the maximum tensile tension is

$$\sigma = \frac{\left(8\times F\times\frac{L}{2}\right)}{\pi\times d^3} = \frac{\left(8\times0.78\times9.81\times\frac{198.5}{2}\right)}{\pi\times74^3} = 0.0048\,\text{MPa} = 4.8\,\text{kPa}$$

Since the tensile tension is less than 100–150 kPa, the core will resist the buoyant forces.

The opening force in this mold is

$$F_m = A_p\times\rho\times(h-c)$$

Assuming a sprue height of 150 mm,

$$F_{m1} = (188.4\times56.7)\times2.65\times10^{-6}\times\left(150-\frac{3}{4}\times\left(\frac{188.4}{2}\right)\right) = 2.25\,\text{kgf}$$

$$F_{m2} = (141.8\times127.6)\times2.65\times10^{-6}\times\left(150-\frac{3}{4}\times\left(\frac{127.6}{2}\right)\right) = 4.90\,\text{kgf}$$

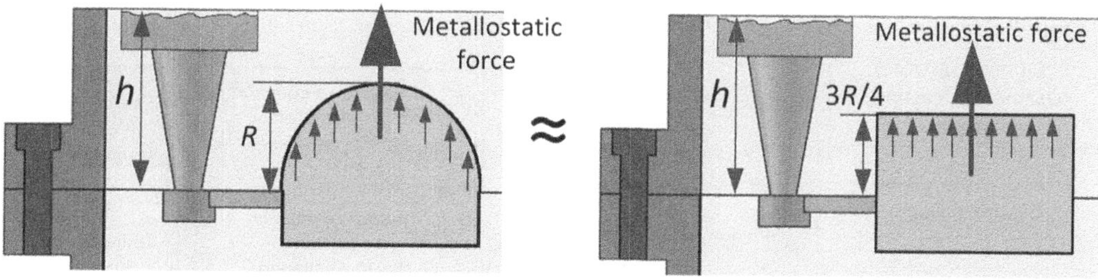

Figure 6.20 Metallostatic force in a cylindrical cavity and equivalent flat cavity.

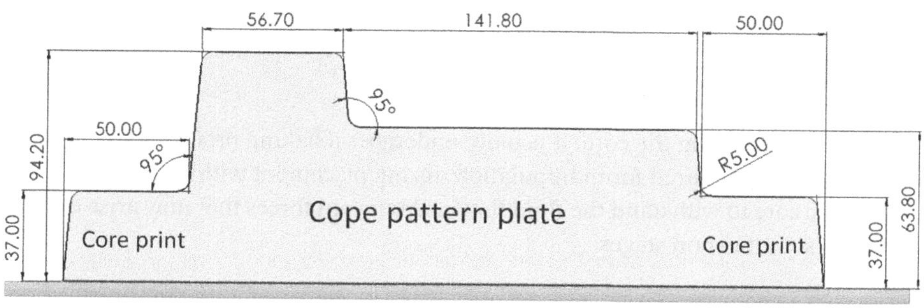

Figure 6.21 Cross-section of the cope pattern with core prints. Dimensions in mm.

Note that since the cavity is cylindrical, the value of c depends on the point analyzed within the cavity and it is not constant. An average from each cylindrical cavity is considered as three-fourth of the radius as shown in Figure 6.20, since it presents the same volume of sand above the cavity. Thus, the final opening force is

$$F_m = F_{m1} + F_{m2} = 7.15\,kgf$$

If the weight of the cope with the sand is lower than this value, the cope and drag should be clamped together during pouring to avoid lifting. At this step, a safety factor of 1.5 is usually applied. Finally, all surfaces have a generous draft of 5° since the process is a sand-casting process (Table 6.9). The fillets/radii of all corners are 5-mm radius. Figure 6.21 shows the final cope pattern plate dimensions.

6.4 Core and Core Box Fabrication

Cores are sand inserts used to shape the interior surfaces of a casting that cannot be formed by the pattern. Patternmaker must determine the quantity and configuration of required cores and subsequently devise and construct the corresponding core boxes. Each core must possess a shape that core makers can readily produce, and that the foundry can easily handle. Furthermore, these cores must be assembled into the mold without causing interference with other cores or mold components. A crucial consideration is ensuring the accurate positioning and secure retention of the cores during the filling and solidification steps.

Typically, cores are crafted using a blend of sand and organic binders, which are favored over clay due to their ability to disintegrate when exposed to the high temperatures of molten alloys. This characteristic facilitates their removal during the shakeout process. Cores are fabricated by compacting the raw sand mixture into a core box. In some cases, metal rods are employed to

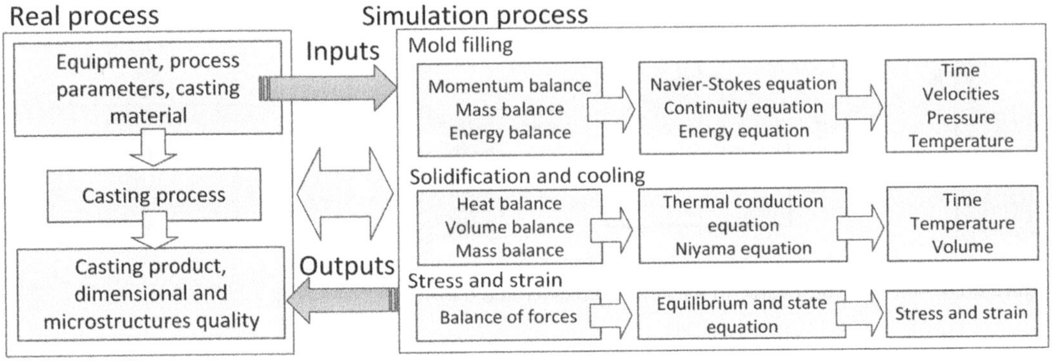

Figure 6.22 Different analysis, equations, and variables typically handled by casting simulation software. *Source:* Adapted from Khan and Sheikh (2018).

reinforce the core. After compacting the core, it usually undergoes a baking process in an oven to impart the necessary strength required for manipulation during placement within the mold. This strength also enables the core to withstand the fluid flow and buoyant forces that may arise during the casting's filling and solidification stages.

6.5 Casting Simulation

Some popular casting simulation programs are AutoCAST-X, MAGMA, ProCAST, SolidCAST, and Altair Inspire Cast. The main inputs and outputs of casting simulation software are shown in Figure 6.22. In general, these programs analyze mold filling (erosion, inclusions, cold shuts, and misruns), the solidification process (macro and microporosity), the cooling process (residual stresses, cracks, and tears), and microstructure (grain size, mechanical properties). Figure 6.23 shows an example of use of Altair Inspire Cast for validating a casting design of a valve component. Therefore, casting simulations let process planners easily validate riser designs and gate placement, detect issues in flow direction and thermal gradients, and predict defects such as pipes and microporosity.

6.6 Melting and Casting

Depending on the metal to be cast, different furnace characteristics and melting requirements must be satisfied. An estimation of the energy required to be provided by the furnace for melting can be done given the type of material and mass, as shown in Chapter 5.

During casting, tools such as ladles are used to extract molten metal from the furnace that is then poured into the molds. It is important to consider that the molten metal temperature will decrease due to heat loss through the environment and the final pouring temperature will be lower than that from the furnace. Since pouring temperature is a critical factor in avoiding incomplete castings, fluidity issues, etc., a fair estimation of this temperature drop is needed to set up the furnace temperature. Table 6.12 provides the time for different temperature losses in uncovered ladles from 50 kg to 500 kg in iron foundries. For steady-state cooling, the cooling-rate for 25 kg uncovered ladles is defined as 30 °C/min; for 500 kg uncovered ladles, the value drops to 6 °C/min (BCIRA 1980). Furthermore, Table 6.13 shows recommended pouring temperatures for different metals and thicknesses. Note that higher pouring temperatures lead to higher costs and, in some cases, it also leads to reactivity issues with atmospheric gases and casting defects such as flash or penetration issues.

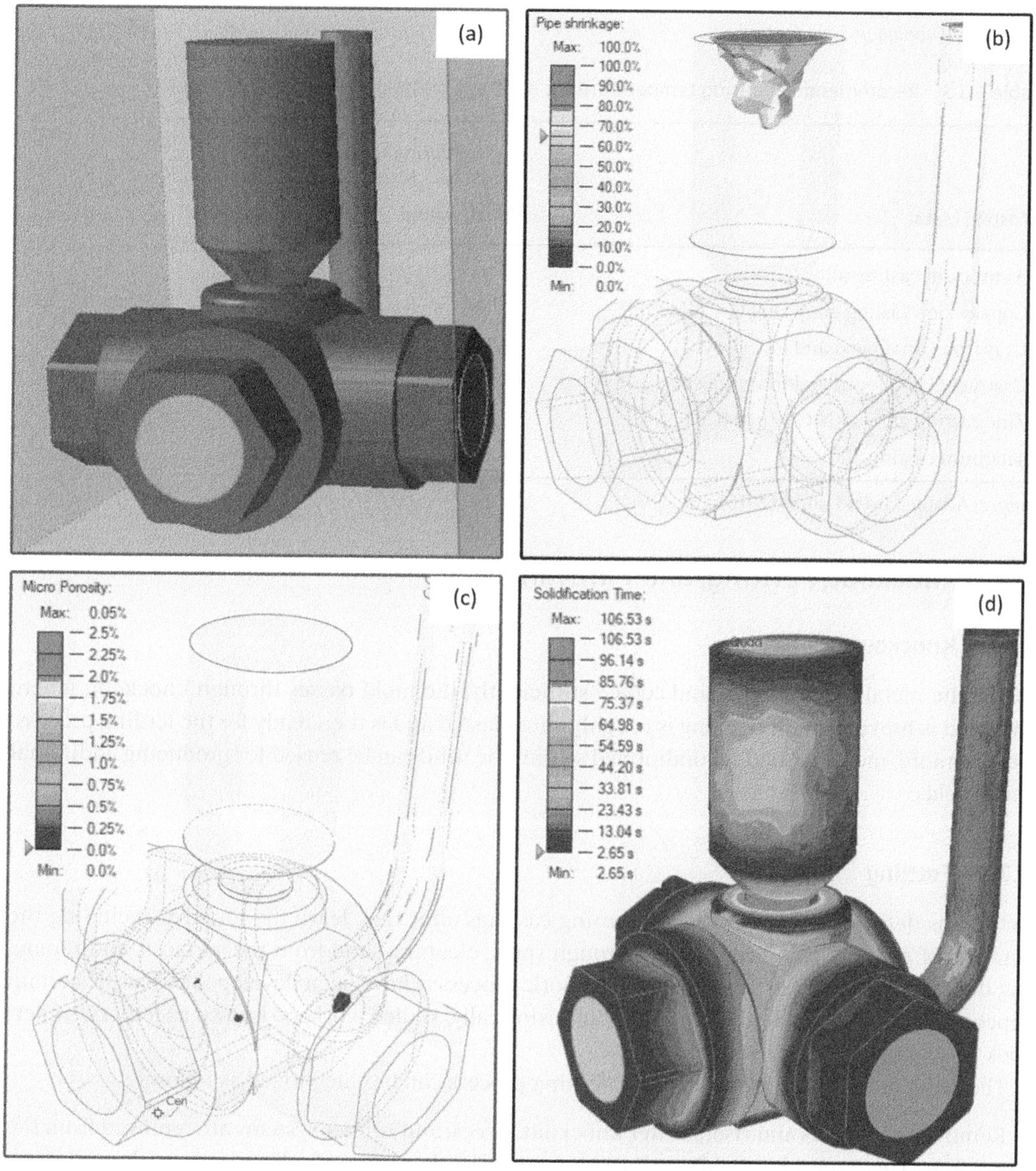

Figure 6.23 Analysis of a valve in casting simulation software (Altair Inspire Cast). (a) Valve component; (b) simulation results about pipe shrinkage; (c) microporosity; and (d) solidification time.

Table 6.12 Time for different temperature losses in uncovered ladles from 50 to 500 kg in iron foundries.

	Time (min) for the following temperature loss				
Ladle capacity, kg	**50 °C**	**100 °C**	**150 °C**	**200 °C**	**250 °C**
50	2.3	4.5	6.8	9.0	11
150	4.4	8.8	13	17	22
300	5.9	11	17	23	29
500	6.1	12	18	24	30

Source: Adapted from BCIRA (1980).

Table 6.13 Recommended pouring temperature for different materials and part thickness.

	Pouring temperature (°C)	
Casting metal	**Thick walled**	**Thin walled**
Aluminum casting alloy	680	750
Copper-zinc casting alloy DIN G-CuZn33Pb	1,040	1,160
Gray iron (unalloyed and low alloyed)	1,340	1,480
Cast steel (unalloyed and low alloyed)	1,560	1,700
Zinc casting alloy DIN GDZnAl4Cu1	420	440
Titanium casting alloy	1,700	1,800

Source: Adapted from Franke (2015).

6.7 Knockout, Fettling, and Finishing

6.7.1 Knockout

When the metal has solidified and cooled sufficiently, the mold passes through knockout, where the mold is broken and the casting is roughly shot blasted and is then ready for the fettling process. Furthermore, molding sand is conditioned so that the sand can be reused for producing additional sand molds.

6.7.2 Fettling

Fettling is defined as the process of cleaning castings after they leave the foundry, including the removal of runners/gates, risers, filling rough spots, cleaning sand from the surfaces, and removing the cores. In recent times, automated robotic processes have been developed to handle certain repetitive aspects of the fettling process, but historically, skilled workers, known as fettlers, undertook this labor-intensive task manually.

The sequence of tasks executed in the fettling process can be categorized as follows:

- Removal of runners and risers. After knockout, risers, and runner systems are removed from the castings. The removal is conducted using a metal hand saw for soft alloys such as aluminum, or more powerful saw-cutting machines for steels although other methods can be applied such as abrasive cut-off wheels, flame cutting, and arc cutting.
- Decoring. Decoring, as the name suggests, is merely the removal of cores from castings or their knocking out. Decoring tools such as core vibrators, water-jet core knockout units, shot-blasting, and many more can be used to remove cores.
- Burnt sand and scale removal. Manual shot blasting is commonly applied at this step to remove scale and burnt sand. For mass-produced small castings, the outer surfaces are cleaned using tumbling barrels with periodic action. This cleaning process involves the castings rubbing against white iron picks within a tumbling barrel.
- Chipping and grinding of castings. To eliminate excess metal and flash from the completed casting, chipping, and grinding are employed. Chipping involves removing flash using chipping hammers and chisels. Grinding, on the contrary, is a smoothing process used to correct irregularities and remove excessive metal.

6.7.3 Finishing

The latter phase of the cleaning process after casting is known as finishing. Further surface refinement can be achieved through various methods such as machining, chemical treatments, polishing, buffing, blasting, and painting. Certain types of steel and malleable iron may undergo heat treatment after cleaning, which can impact the casting's surface. Therefore, these metals often require specialized finishing processes. Nonferrous castings can undergo finishing through chemical, electrolytic, and mechanical procedures. Another aspect of finishing that can be applied to castings involves the removal of impurities such as dust, dirt, scales, sand, and grit, achieved through a salt-bath cleaning process.

In some cases, casting parts present defects that will not pass quality control. When these defects are not critical and can be corrected, a rectification process also named salvage operation is conducted. Methods employed for rectifying defects in castings include electric welding, metal spraying, and gas welding. Whether salvage is feasible or not depends on factors such as the defect's size, the weight of the casting, the casting's alloy, the complexity of the repair, the cost of salvage, and the availability of the necessary equipment.

6.8 Inspection

Casting inspection is required for a correct assessment of the quality of castings, which ultimately leads to their acceptance or rejection. Different types of inspection can be conducted in casting: visual, dimensional, mechanical, and metallurgical.

- Visual inspection. Visual inspection involves the examination of completed castings using the naked eye to assess potential casting defects (dirt, misruns, cold shuts, blowholes, surface dross, etc.). Visual inspection is typically conducted both before the cleaning and annealing processes and after the fettling operation. For tiny cracks, pores, or other surface glitches that are hard to detect by the human eye, dye penetrant testing is used for both ferrous and nonferrous materials.
- Dimensional inspection. Dimensional inspection entails the measurement of the finished casting's dimensions. Measurement instruments such as calipers and gauges are used to analyze if dimensions are within the admissible values to move the casting to subsequent operations.
- Mechanical inspection. Mechanical tests such as hardness testing, tensile/impact testing and load testing can be also conducted.
- Metallurgical inspection. Chemical composition analysis, evaluation of casting integrity, and other related casting tests are conducted to confirm the internal quality of the casting part. Chemical analysis often involves taking a sample of molten metal poured into a special mold that is analyzed by spectrographic atomic absorption. For internal inspections, ultrasonic testing or radiographic inspection is applied to detect internal imperfections such as pipes or inclusions.

6.9 Operations List in Sand Casting

A typical operations list in sand casting may look like the one shown in Figure 6.24. As can be seen, all required operations for sand casting are indicated with the estimated operation time required. Setup times, specific tooling and any relevant information are also indicated in the casting operations list.

Operations sheet								
Part number: PRT_211B Part name: Casing 211 Material: Gray cast iron							Process planner: J.V. Abellán Release date: 25/3/2023 Version: 3	
Operation	Description	Machine tool/equipment	Tooling	Speed	Feed	Setup time	Operation time	Observations
11	Pattern fabrication	Shell casting kit	--	--	--	--	5 d	Tool life 2,000 uses. External fabrication
12	Shell fabrication and flasking	Shell casting kit	Pattern #1	--	--	10 min	15 min	--
13	Remove pattern, pouring	--	--	--	--	--	5 min	--
14	Solification and cool waiting	--	--	--	--	--	20 min	--
15	Remove casting	--	--	--	--	--	3 min	--
16	Fettle/deburr	Belt grinding machine	KK711X	2,800 rpm	--	--	3 min	--
17	Inspect	--	--	--	--	--	3 min	Visual inspection

Figure 6.24 Typical operations list in sand casting.

Questions and Exercises

1 Explain the factors to be considered for pattern design in sand casting.

2 Design the gating system for casting the mild steel part shown in Figure 6.25. Define reasonable values for the equipment required (size of the flask [cope; drag], etc.).

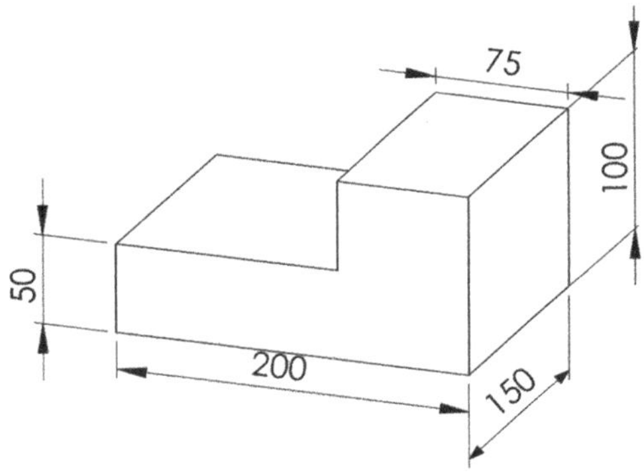

Figure 6.25 Part Exercise 2. Dimensions in mm.

3 Calculate the proper riser and its placement for casting the part shown in Figure 6.26.

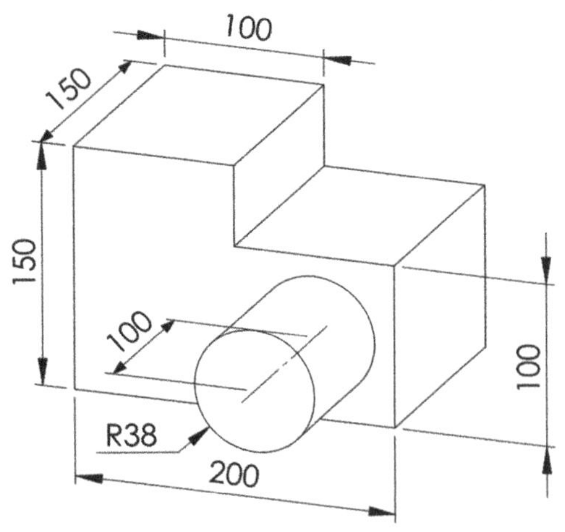

Figure 6.26 Part Exercise 3. Dimensions in mm.

4 A core of 15-mm diameter and 100 mm is held by two sides in a mold cavity. Calculate the buoyancy force that the core should resist and if chaplets are required. Assume cast iron as the casting material.

5 Calculate the opening force due to the metallostatic pressure of the molten metal in the mold shown in Figure 6.27.

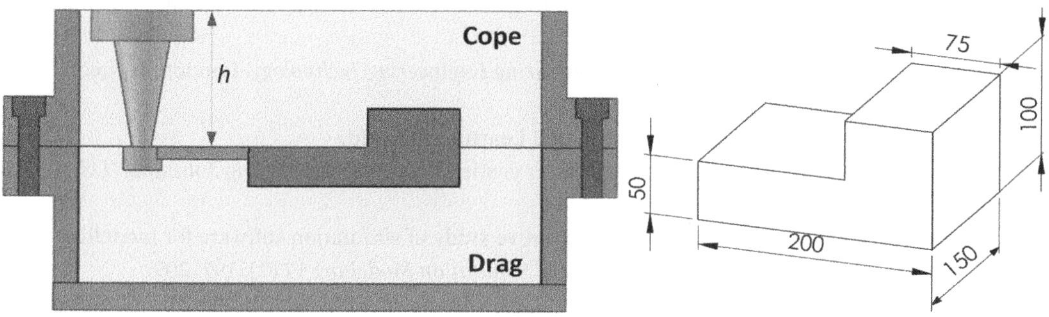

Figure 6.27 Part Exercise 5. Cavity dimensions in mm.

6 Figure 6.28 shows a casting part after machining. If the machining allowance is 2 mm and the material is cast iron. Calculate the dimensions of the pattern to create the mold cavity.

Figure 6.28 Part Exercise 6. Dimensions in mm.

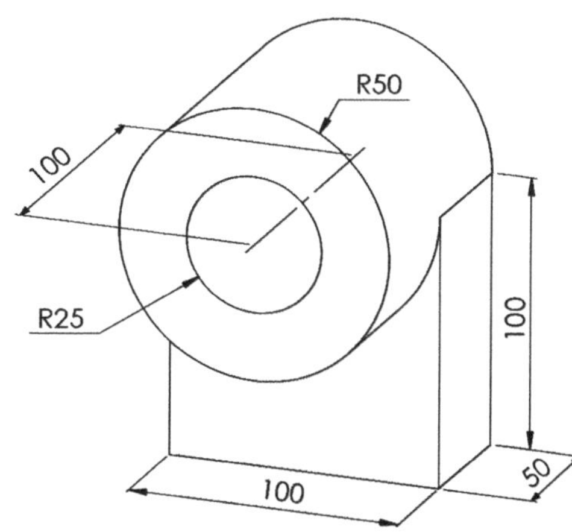

References

BCIRA (1980). Cooling-rates and heat losses in iron foundry ladles. BCIRA Broadsheet, No. 182, 2 pp.

Beddoes, J., and Bibby, M. (1999). *Principles of Metal Manufacturing Processes*. Butterworth-Heinemann.

Beeley, P. (2001). *Foundry Technology*. Elsevier.

Black, J.T., and Kohser, R.A. (2020). *DeGarmo's Materials and Processes in Manufacturing*. Wiley.

Brown, J.R. (2000). Sand and green sand. In: *Foseco Ferrous Foundryman's Handbook*, 152–153. Butterworth-Heinemann.

Chastain, S. (2004). *Metal Casting: a Sand Casting Manual for the Small Foundry*, vol. 2. Jacksonville, FL: Stephen Chastain.

Franke, S. (2015). *Pocket Guide Foundry 2019: Giesserei Praxis*. Berlin, Germany: Schiele and Schön GmbH.

Guleyupoglu, S. (1998). *Casting Process Design Guidelines*, 869–876. Transactions-American Foundrymens Society.

Hurst, S. (1996). *Metal Casting: Appropriate Technology in the Small Foundry*. Intermediate Technology Publications.

Kalpakjian, S., and Schmid, S.R. (2009). *Manufacturing Engineering Technology*. London, United Kingdom: Prentice Hall.

Kaushish, J.P. (2010). *Manufacturing Processes*. PHI Learning Pvt. Ltd.

Kay, I.M. (2002). Patternmaking "Tricks" for better castings. *Engineered Casting Solutions (USA)* 4 (1): 48–49.

Khan, M.A.A., and Sheikh, A.K. (2018). A comparative study of simulation software for modelling metal casting processes. *International Journal of Simulation Modelling* 17 (2): 197–209.

Rao, P.N. (2006). *Manufacturing Technology Foundry, Forming and Welding*. McGraw Hill Education.

Ravi, B. (2005). *Metal Casting: Computer-aided Design and Analysis*. PHI Learning Pvt. Ltd.

Wang, W., Stoll, H.W., and Conley, J.G. (2010). Tool design and construction for sand casting. In: *Rapid Tooling Guidelines for Sand Casting*, 11–36. US: Springer.

Part III

Process Planning in Machining

7

Definition and Selection of Workholding Systems

7.1 Introduction

A workholding system refers to any mechanism utilized to secure a workpiece and prevent it from shifting during manufacturing operations, typically machining operations. Furthermore, workholding systems are designed to reduce operator labor by simplifying the tasks of locating and clamping, ultimately leading to an improved production rate (Nee and Tao 2004).

Workholding devices can be general-purpose workholders (such as clamps, vises, and chucks), jigs, or fixtures. Jigs are commonly used for guiding cutting tools during the machining operation, specially to avoid deflection and issues during cutting entries. The most common types of jigs are drill jigs, which serve various purposes, including drilling, tapping, reaming, chamfering, counterboring, countersinking, and similar operations. On the contrary, fixtures offer a broader range of applications compared to jigs, and they do not guide the cutting tools but locate and hold the workpiece during machining. These specialized workholders are specifically designed for situations where complex structures should be built to position and orient a part precisely. Figure 7.1a,b shows examples for both jigs and fixtures in machining processes. The left side illustrates a commercial drill jig employing a V-block to secure cylindrical parts before the drilling process. The right side illustrates a permanent workholder used for machining a pedal of a freight vehicle.

Workholding design can be a tedious task and it may entail a large budget in industries such as automobile or aircraft where complex permanent fixtures are required. Process planners should take into account many factors for a successful workholding design, such as identifying reference surfaces or datums, machining capability in single setups, cutting and clamping forces estimation, ways of ease workpiece loading and unloading, prevention of cutting tool path interferences, etc.

In this chapter, the main principles of workholding for proper workholding systems design are presented. Locating and clamping guidelines are reviewed, and the main general-purpose workholders applied in machining are described. The chapter also presents a standard to formally indicate the configuration of the locating and clamping systems in technical drawings. The type of contact, the movements locked, and other characteristics of the workholding system can be graphically represented following the standard.

Manufacturing Process Planning: A Practical Approach for Mechanical Engineering, First Edition.
José V. Abellán-Nebot, Carlos Vila Pastor, and Héctor R. Siller.

Companion website: www.wiley.com/go/Manufacturing_Process_Planning_1e

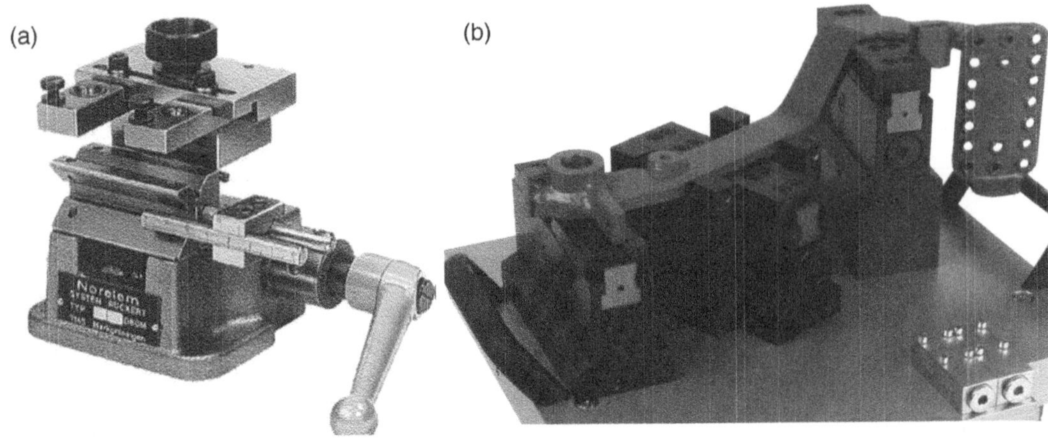

Figure 7.1 (a) Drill jig. Courtesy of Norelem GmbH & Co. (b) Permanent workholder used for machining a pedal of a freight vehicle. *Source:* Courtesy of Römheld GmbH.

7.2 Principles of Workholding

Locating and clamping are the fundamental functions of any workholding system. As such, the fundamental principles of locating and clamping must be thoroughly understood and are briefly reviewed in the following sections.

7.2.1 Locating and Clamping Principles

A part needs to be held in a setup that ensures a specific location and orientation to be machined properly. Throughout the production run, this arrangement must be reproducible. Additionally, the workholding mechanism must maintain the part's position despite cutting forces and vibrations.

Understanding the coordinate system that represents a solid object in three-space and its degrees of freedom is the first step in defining its location. Any solid object in three-space has six degrees of freedom, which corresponds to three translation movements along the x, y, and z axes and three rotations around each of these three axes. However, for fixturing design purposes, it is more recommendable to consider the positive and negative directions of these movements, which in practice defines 12 directions of movements. These 12 movements correspond to six translation movements along the x, y, and z axes (i.e., $\pm Tx$, $\pm Ty$, and $\pm Tz$, which refer to the movements from #1 to #6), and six rotations around each of these three axes (i.e., $\pm Rx$, $\pm Ry$, and $\pm Rz$, which refer to the movements from #7 to #12). A workpiece may move in any of these 12 directions (Figure 7.2).

The locators are the tools that accurately position the workpiece in the space, that is, they provide a secure stop. When pressed up against the stop, the workpiece is immobile. Additionally, clamps are the tools that exert a clamping force to secure the workpiece to the locators.

7.2.1.1 Forms of Location

Location can take one of three general shapes: plane, concentric, or radial, and in many cases, locating is performed by a combination of them (Carrlane Manufacturing Co. 2023). Workpieces are typically located on surfaces that may be flat, curved, or irregularly contoured. Figure 7.3a

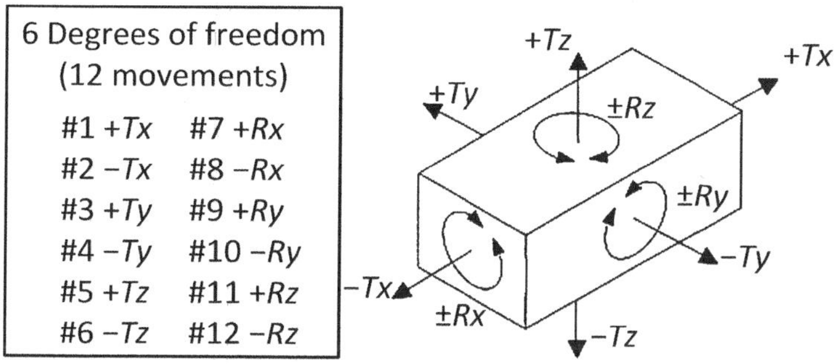

Figure 7.2 The 12 movements of a part in the three-space.

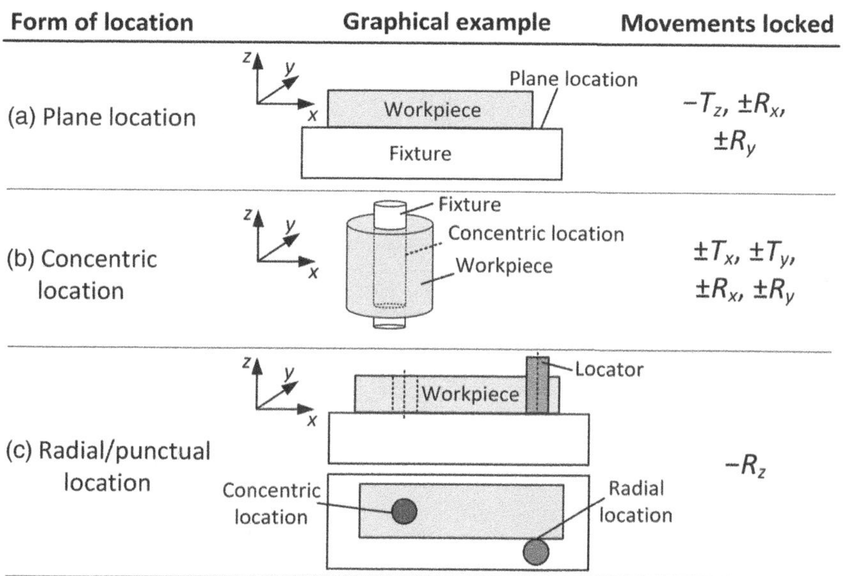

Figure 7.3 The three forms of location: plane, concentric, and radial. *Source:* Adapted from Carrlane Manufacturing Co. (2023).

shows how plane-locating devices often locate a part via its exterior surfaces, for instance, on a machine tool table or a plane surface from a fixture. A plane location will lock five movements, one translation ($-Tz$), and four rotation movements ($\pm Rx$, $\pm Ry$).

Another form of location is based on concentric location, usually locating the workpiece from a central axis. This axis might or might not be in the workpiece's center. A locating pin that is inserted into a hole is the most typical sort of concentric location (Figure 7.3b). A concentric location will lock four translations ($\pm Tx$, $\pm Ty$) and four rotation movements ($\pm Rx$, $\pm Ry$). Finally, to prevent the movement of a workpiece around a concentric locator, a radial locator is applied, as shown in Figure 7.3c. Therefore, one additional rotation movement is constrained ($-Rz$).

7.2.1.2 Locating from External Surfaces

External flat surfaces are frequently employed as workpiece features for positioning purposes. A flat surface is usually positioned by locators or supports that can be fixed or adjustable in height, as depicted in Figure 7.4.

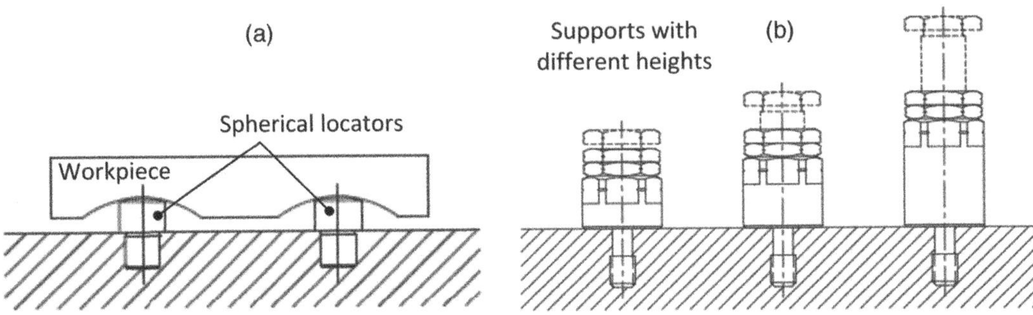

Figure 7.4 Spherical fixed locator (a) and adjustable supports to locate/support a workpiece (b).

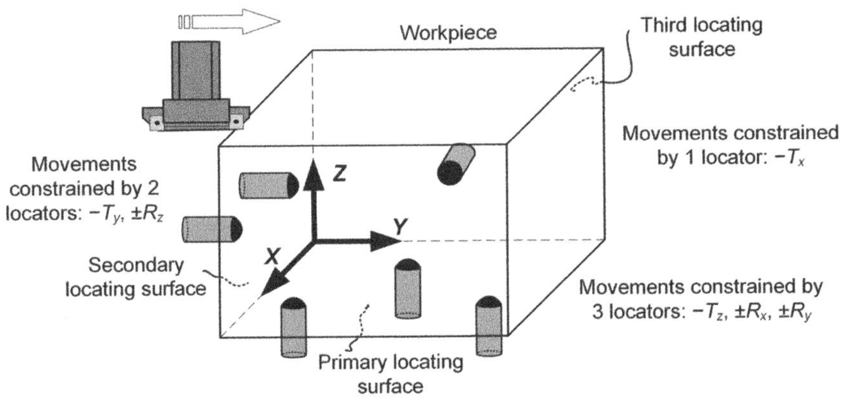

Figure 7.5 The 3-2-1 locating scheme and movements constrained by locators in a face-milling operation. Clamps are not shown.

Fixed supports, also known as solid supports, are locators with a predetermined height. They accurately position a surface along a single axis. Adjustable supports, on the contrary, offer variable height settings. Similar to solid supports, they excel at precise alignment along a single axis. Adjustable supports are particularly useful when workpiece variations necessitate adaptable support to accommodate varying heights. They are predominantly employed for cast or forged workpieces that exhibit uneven or irregular mounting surfaces.

The most widely adopted method for positioning a workpiece using its external profile is the 3-2-1 locating scheme. In this approach, six individual locators (typically spherical locator pins) are employed to establish references and constraints on the workpiece's movement. The 3-2-1 locating scheme is illustrated in Figure 7.5, and it is defined as follows. First, three locators are situated on the primary locating surface, typically at the bottom part of the workpiece. These locators restrict axial movement downward, along the $-z$ axis (movement #6), as well as the rotational movement about the x (#7 and #8) and y (#9 and #10) axes. Collectively, these three locators constrain five movements. The next two locators are placed on the secondary locating surface, constraining axial movement along the $-y$ axis (#4) and rotational movement about the z (#11 and #12) axis, constraining three additional movements. The final locator is positioned at the third locating surface, constraining axial movement in one direction along the $-x$ axis (#2). Together, these six locators collectively lock nine out of 12 movements. The remaining three movements (#1, #3, and #5) are typically constrained by clamping forces.

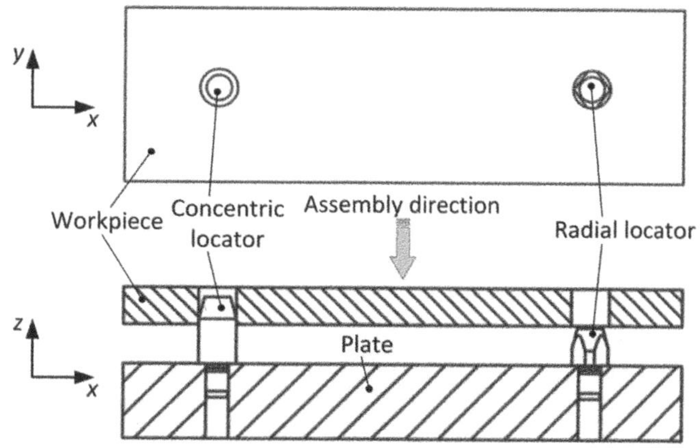

Figure 7.6 Two locating pins mounted on a plate lock 11-out-of-12 possible movements.

7.2.1.3 Locating from Internal Surfaces

Positioning a workpiece using internal surfaces represents an efficient form of placement, and it is commonly applied in industry, for instance, in machining engine blocks after casting. This method primarily relies on specific holes or hole patterns as key features for positioning. Two types of locators are typically employed for internal positioning: concentric locators, also referred to as round pins, and radial locators, also referred to as diamond pins.

A simple example to illustrate these two types of locators using internal surfaces is shown in Figure 7.6. The workpiece is placed over a plate that effectively constrains five movements (i.e., plane location), preventing any axial movement in the $-z$ (#6) direction and rotations around the x and y axes (from #7 to #10). The workpiece presents two holes, where a concentric and a radial locator are used for positioning. The concentric locator ensures that there is no axial movement along the x (#1, #2) and y (#3, #4) axes. Collectively, the plate and the concentric locator lock nine out of 12 movements. Finally, the radial locator restricts two movements by preventing rotational movement around the z (#11 and #12) axis. Combined, the plate and locators lock eleven out of 12 movements. The remaining movement in the $+z$ direction is typically constrained using a clamp.

7.2.1.4 Analyzing Machining Forces

When designing fixtures, a key aspect is to understand the direction and magnitude of the forces generated during the machining process. Knowing the magnitude of these cutting forces helps determine clamping requirements and potential workpiece deflections. In cases where workpiece deflection needs to be controlled, additional supports can be incorporated (see Section 7.2.3).

A correct workholding design ensures that the workpiece is held against the cutting forces by locators and not by the clamps. Following this basic design rule, the necessary clamping pressure is highly reduced. In other words, clamps are not intended to withstand the primary cutting forces; their sole purpose is to maintain the workpiece's position against the locators and counteract the secondary cutting forces.

The relationship between locators and clamps can be illustrated using a plain vice for milling operations. In Figure 7.7a, the plain vice combines both locating and clamping elements. The solid jaw and the vice body serve as the locators, while the movable jaw acts as the clamp. The recommended orientation places the locators in opposition to the cutting forces, and thus, the cutting forces are directed toward the solid jaw and vice body. Then, the position of the workpiece is ensured by the solid jaw, whereas the movable jaw, much like other clamps, merely secures the workpiece's position against the locators. A similar reasoning can be conducted in Figure 7.7b,

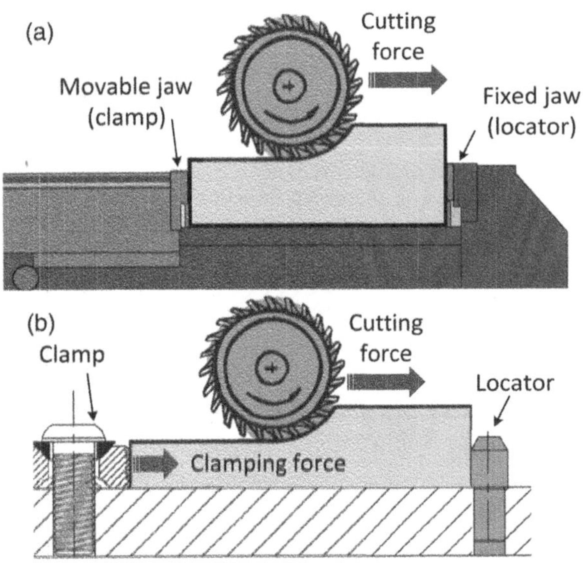

Figure 7.7 Cutting forces in a milling operation should be directed into the locating component: (a) plain vice; (b) locator and clamp mounted on a machine tool table.

where a locator and a clamp are used for holding a workpiece, and the cutting forces are directed toward the locator.

7.2.2 Locating Guidelines

There is no one location method or locator type that will be effective for all workholders. Each locator needs to be carefully planned into the fixture layout to carry out the required location in a proper manner. The following are some rules to follow when selecting and using locators.

7.2.2.1 Positioning Locators

Any locator's primary objective is to create a workpiece reference point to ensure position and orientation repeatability. To avoid significant variation from part to part, locators should, whenever possible, be positioned to make contact with a machined surface rather than a casting or forming surface. In addition to improving repeatability, a machined surface usually offers a more stable positioning method. To create a sufficient contact surface, the entire surface may sometimes be machined, but other times—especially when working with castings—only certain sections may be machined.

To ensure repeatability, another crucial consideration in locator placement is the spacing between the locators themselves. As a general guideline, locators should be spaced as far apart as is practically possible. This principle is exemplified in Figure 7.8. All the locating schemes shown in the figure employ the 3-2-1 locating method, with the only distinction being the locator spacing.

As it can be seen, only the first locating scheme can position the workpiece with good repeatability since a longer distance between locators compensates for the irregularities in either the locators or the workpiece surfaces and enhances overall stability. However, the other schemes are not adequate because locators are too close to each other or, in the case of the last scheme, the three locators are aligned, which prevents the stability of the workpiece, that is, the part is under-constrained. It should be noted that if locators are too close to each other, any small locator error would be magnified, producing a larger workpiece position deviation or even a slight rotation of the part.

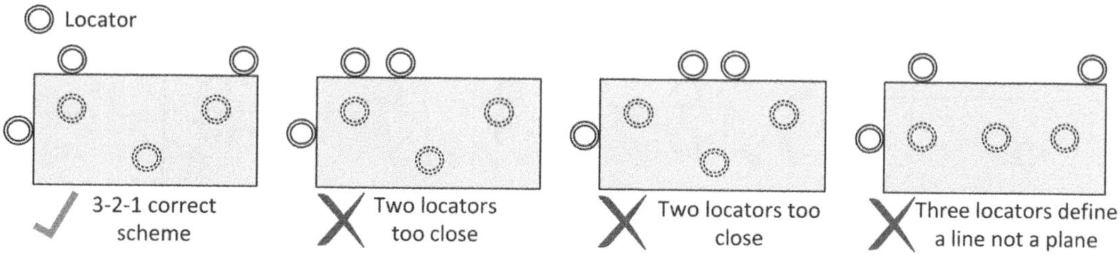

Figure 7.8 Correct and incorrect locating schemes for machining. Locators should be spaced as far apart as practical to compensate slight irregularities and ensure maximum stability.

Instead of external planar surfaces, machined holes are usually the preferred surfaces for positioning, as they offer the most comprehensive and stable form of positioning with the fewest number of locators. In this case, the same reasoning applies, and longer distances between radial and concentric locators will minimize the impact of positioning errors of the locators, ensuring a repeatability position of the workpiece.

Finally, the positioning of locators should also consider the cutting tool path and the loading and unloading movements of the workpiece to ensure easy and fast movements without interferences (Joshi 2003).

7.2.2.2 Avoiding Over-Constrained Workpieces

Workholding must ensure the repeatability of the workpiece position. For this purpose, the part must be constrained properly, avoiding parts to be over-constrained. As shown previously, under-constrained occurs when there is at least one direction of motion unconstrained, and thus, the part will not be stable during machining. On the contrary, over-constrained workpieces are given when different locators try to constrain the same degrees of freedom, and then, depending on the part and locator tolerances, the location of the part will vary from part to part, and the repeatability of part location is compromised.

Therefore, fixture design must follow the principle of isostatism. The concept of isostatism means that the workholding provides a unique location in space, and there is no ambiguity about the location of the part. To reach isostatic positioning, locators that constrain the same degrees of freedom (named redundant locators) should be avoided. The problem of redundant locators is illustrated in Figure 7.9. In case Figure 7.9a, it is observed that four locators are constraining the same part movements (up and down movement and part rotation). Given that workpiece dimensions may vary within their tolerances, the probability that all parts simultaneously rest on both surfaces is minimal. Thus, the position of the part may vary, as shown in option 1 and option 2, which refers to two possible workpiece positions. In Figure 7.9b, two locators and both concentric and radial locators constrain the same part movements (up and down movements and part rotations). Due to workpiece variability, the final position will depend on which surface touches the locators, as illustrated in the examples shown in option 1 and option 2. As shown in Figure 7.9a, b, the repeatability of the workpiece position is not ensured due to the existence of redundant locators.

Note that the same problem is given when locating surfaces instead of locators are applied, and those that are redundant and prevent repeatability of workpiece position should be avoided. The most straightforward way to eliminate redundant locations is by examining the blueprint of the part to determine which workpiece surface serves as the reference surface. Often, the part's dimensional specifications indicate which surfaces or features are critical and should be regarded as the main locating surfaces.

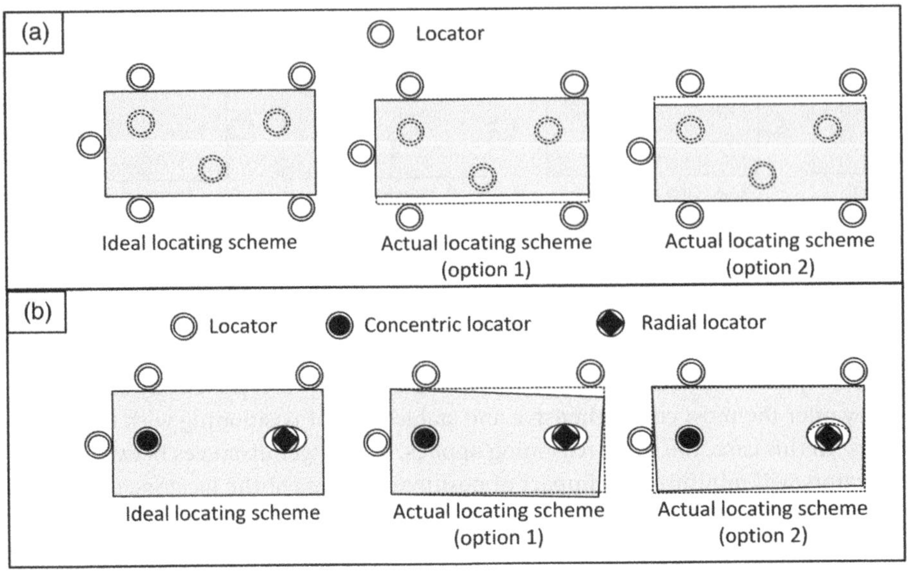

Figure 7.9 Examples of redundant locators that produce a non-repeatable location (options 1 and 2): (a) a 3-2-1 scheme with 2 redundant locators; (b) a concentric and radial scheme with 3 redundant locators.

7.2.2.3 Designing Foolproofing Layouts

Foolproofing designs avoid operator's mistakes when locating parts on workholding devices and can be a good strategy to avoid costly errors at the machining stage. The complexity of the foolproofing design depends on the workpiece design, but in some cases, the foolproofing design can be easily included in the workholding device, as shown in Figure 7.10.

7.2.2.4 Evaluating Locating Accuracy

As shown in Figure 7.8, the position of the locators, that is, the fixture layout, influences the accuracy of workpiece location. For instance, if locators are too close to each other, any small locator error would be magnified, producing a larger workpiece position deviation or even a slight rotation of the part. The accuracy of a fixture can be mathematically estimated by analyzing the variability of the fixture coordinate system given the fixture layout and the locators' tolerances. For a 3-2-1

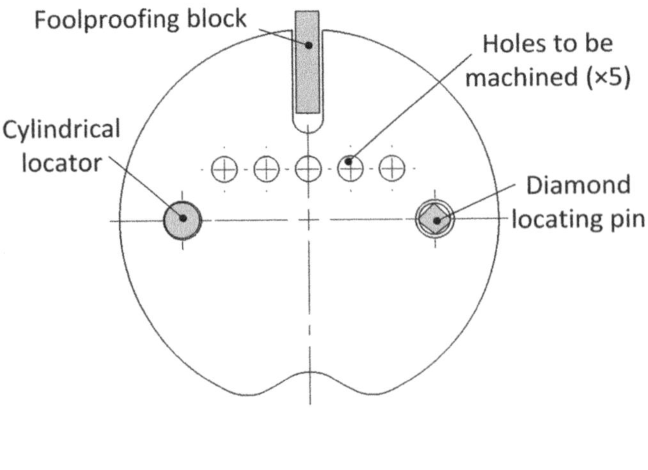

Figure 7.10 Example of foolproofing workholding design to ensure that components are loaded properly.

fixture layout, the accuracy of the fixture can be estimated considering the accurateness matrix of localization from Wang (2002) defined as follows:

$$\mathbf{A} = \sum_{i=1}^{6} a_i \cdot \begin{bmatrix} \mathbf{n}_i \cdot \mathbf{n}_i^T & \mathbf{n}_i \cdot (\mathbf{r}_i \times \mathbf{n}_i)^T \\ (\mathbf{r}_i \times \mathbf{n}_i) \cdot \mathbf{n}_i^T & (\mathbf{r}_i \times \mathbf{n}_i) \cdot (\mathbf{r}_i \times \mathbf{n}_i)^T \end{bmatrix}$$

where $\mathbf{n}_i$ is a vector that defines the workpiece surface normal at the ith contact point ($i = 1, \ldots, 6$); $\mathbf{r}_i$ is the position vector of each ith locator with respect to the fixture coordinate system; and a_i is named as the locator accurateness and is defined as $1/s_i$, where s_i is Tol/*36*, being "Tol" the locator tolerance or estimated locator error. Note that the superscript T refers to the transpose of the vector, and "×" refers to the cross-product vector.

The accuracy of the fixture layout is related to the determinant of matrix **A**. Maximizing this matrix means minimizing the overall variance of the fixture coordinate system that is used to locate the workpiece. Therefore, for comparing different layouts for the same workpiece, analyzing and selecting the one with the higher determinant of matrix **A** results in the best fixture layout configuration.

Figure 7.11 shows an example to illustrate the calculation process. In this example, three different 3-2-1 fixture layouts are compared by evaluating the determinant of matrix **A**, that is, the accurateness matrix of localization, and assuming that the tolerance of all locators is the same, 0.01 mm. As expected, case 3 presents the highest determinant value, which indicates that it is less sensitive to locators' inaccuracies, and thus, the most accurate fixture layout.

Note that this accuracy analysis does not consider aspects such as fixture rigidity, and it only analyzes the layout of locators in a 3-2-1 fixture scheme. It should also be noted that the required accuracy in a workholding design is defined by the final part specifications. Generally, 20–50% of part tolerance can be attributed to the fixturing device (Hoffman 2012).

7.2.3 Clamping Guidelines

Besides locating the workpiece accurately and repeatedly, a workholding system should clamp it sufficiently to prevent movement during the machining operation. Clamps are the primary tools

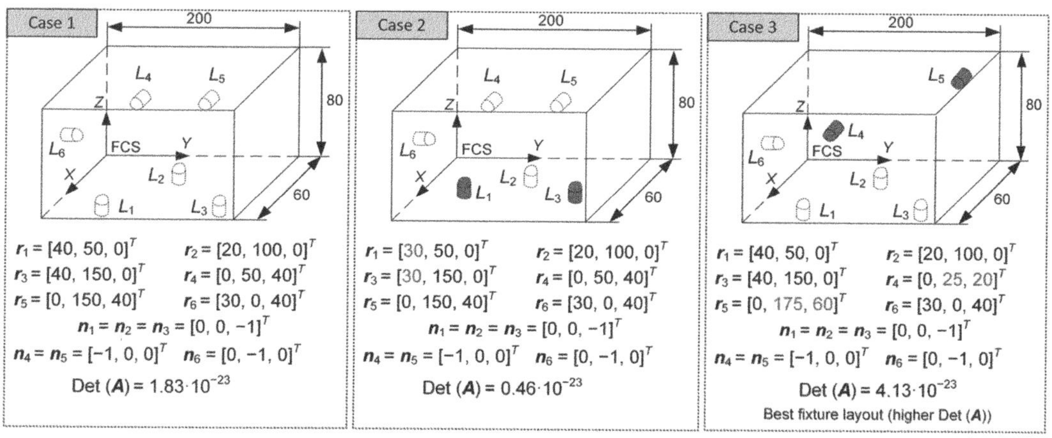

Figure 7.11 Example of comparing different fixture layouts by evaluating the accurateness matrix of localization.

used for this purpose. Here are some guidelines for designing the clamping system in machining operations (Carrlane Manufacturing Co. 2023; Joshi 2003):

7.2.3.1 Positioning Clamps

- Clamp placement. Clamps are sized to secure the workpiece against the locators without causing deformation and resisting any secondary forces. Thus, clamps should contact the work at its most rigid point to avoid bending or damaging the part. For instance, clamps are aligned directly over the supporting or locating elements of the workholder as illustrated in Figure 7.12.
- Worker's interference. Ensure that the clamps are placed to let the operator work easily and safely without making the loading/unloading difficult.
- Tool path consideration. Check for interference between the cutters and the clamps during both the machining and return cycles. Clamps should be placed on a robustly supported part of the workpiece, away from the loading/unloading and cutting tool paths.
- Clamp height. Most clamps are positioned on or near the top surface of the workpiece. Keep the overall height of the clamp relative to the workpiece to a minimum to avoid tool path interferences, as shown in Figure 7.13. Gooseneck-type clamps can be employed to achieve a lower profile when reduced clamp height is necessary.
- Contact area. Consider the size of the clamp-contact area to minimize interference with the cutter. Smaller clamping areas reduce the likelihood of interference and enhance clamping pressure on the workpiece. Choose a clamp size that is large enough to securely hold the workpiece but small enough to avoid obstruction.
- Surface damage. If a finished surface is used to place a clamp, the clamp should prevent damage to the surface by using proper terminals. Furthermore, the clamping pressure may be limited to ensure holding the part without damaging the surface.

7.2.3.2 Avoiding Workpiece Deformation

As stated earlier, the main goal of a workholding system is to locate and hold the part during the machining operation without causing distortion or deformation. Therefore, the clamping force should be adjusted to avoid distortion of the component, and supports must be added if the part is not rigid enough and is expected to be deformed.

An example is shown in Figure 7.14, where a compliant zone of the part is used for clamping, and auxiliary support is added before applying the clamp to avoid deformations. This practice is

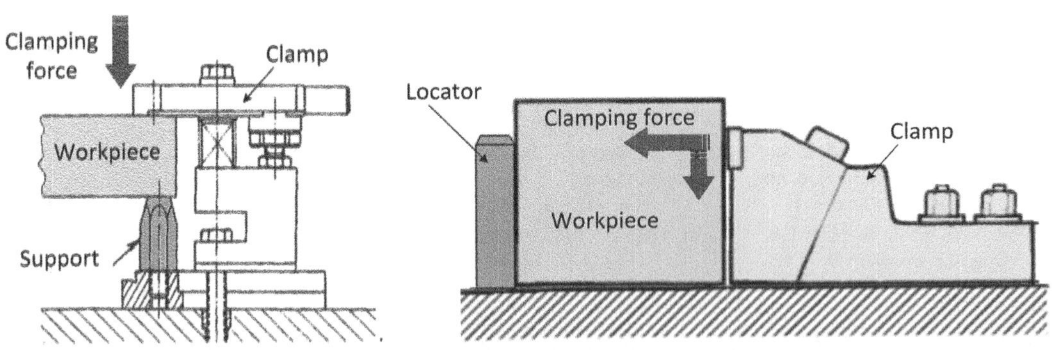

Figure 7.12 Clamps should always be positioned so the clamping force is directed into the supports or locators. *Source:* Adapted from Carrlane Manufacturing Co. (2023).

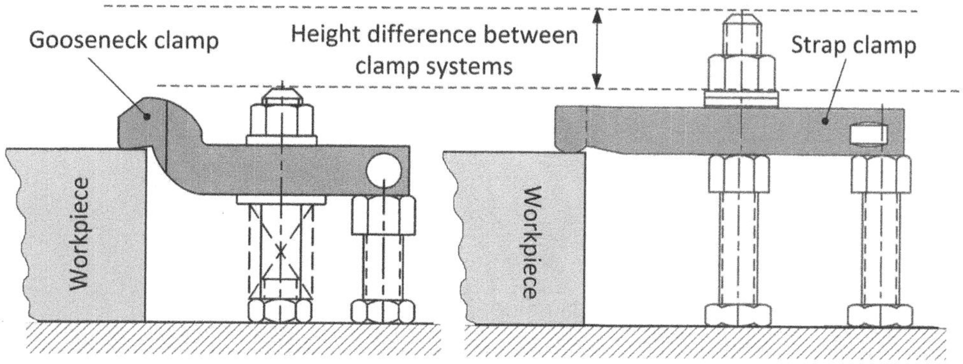

Figure 7.13 Using gooseneck clamps is one way to reduce the height of the clamps. *Source:* Adapted from Carrlane Manufacturing Co. (2023).

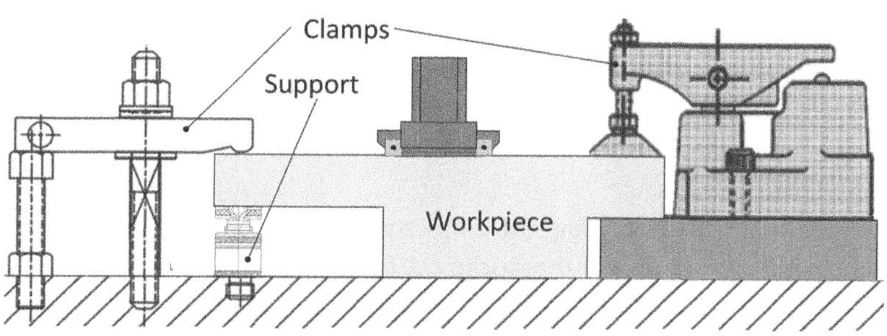

Figure 7.14 Use of supports to avoid workpiece deformation.

common in holding sheet parts, where an N-2-1 locating scheme is used, and $N-3$ is the number of auxiliary supports used to hold the part without deforming it due to clamping forces.

When the part is not rigid enough, the type of the workholding device may result in completely different holding results depending on the magnitude and direction of clamping forces. Figure 7.15a–c shows different workholding systems to securely position a thin-walled ring for boring its internal diameter. As can be seen, the performance of these workholding systems differs noticeably between them. The first workholding system is a four-jaw chuck that clamps the workpiece on its outer diameter at four points. This method is likely to cause deformation of the ring due to the direction and magnitude of the clamping force. The second workholding system is a face plate with four strap clamps where the clamping force is directed into the plate rather than against the spring force of the workpiece. Therefore, this clamping method eliminates the distortion that occurs with the previous workholding system. However, it could be the case that the part can only be clamped on its outer surface, and thus, this workholding system cannot be applied. Under these situations, the third example can be applied as an alternative method. This workholding system consists of a collet that completely encloses the part, and the clamping force generated is distributed along the whole perimeter of the part. Depending on the size of the part, either a collet or pie-shaped soft jaws can be employed for this arrangement.

7.2.3.3 Evaluation of Clamping Force Requirement

As shown earlier, workholding systems should securely position the part during machining operations. Therefore, cutting forces and the corresponding required clamping forces for every machining operation should be estimated, which can be indeed very challenging.

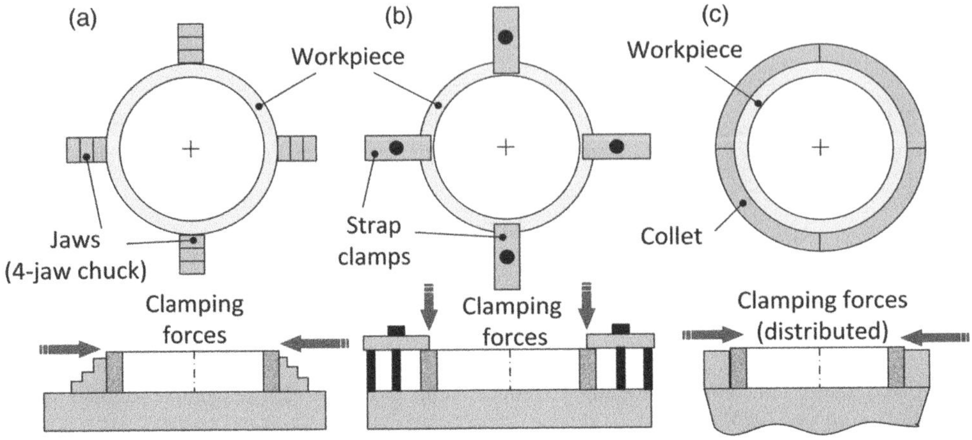

Figure 7.15 (a) Directing the clamping forces against an unsupported area will cause this cylindrical part to deform; (b) strap clamps eliminate deformation by directing the clamping forces into the plate under the part; (c) when the part can only be clamped on its outside surface, a collet or a pie-shaped chuck jaws can be used to hold the part and reduce deformation. *Source:* Adapted from Carrlane Manufacturing Co. (2023).

Cutting forces can be fairly estimated by analyzing the chip cross-section removed by the tool, the cutting speed, and the mechanical properties of the workpiece (refer to Chapter 10 for practical examples). Then, the cutting tool path is analyzed, and a diagram of cutting forces, locator reactions, and clamping forces is drawn and mathematically calculated. As a simple example, consider the machining operation shown in Figure 7.16, where face milling is conducted, and the workpiece is clamped on the machine tool table by strap clamps.

If it is assumed that the cutting force (F_c) is entirely horizontal and no workpiece locators are used, frictional forces alone are responsible for resisting the cutting forces. For this case, the required clamping force (F_f) can be evaluated as:

$$F_f = \frac{F_c}{\mu} \times S_f$$

where S_f is the safety factor, and μ is the friction coefficient, which depends on the type of materials and lubrication state. Table 7.1 shows common friction coefficients (dry and lubricated conditions) between different pairs of materials. Note that the friction coefficient depends on numerous factors (e.g., roughness, type of lubrication, thickness of layer lubrication, etc.), and the actual values may differ notably from those in Table 7.1.

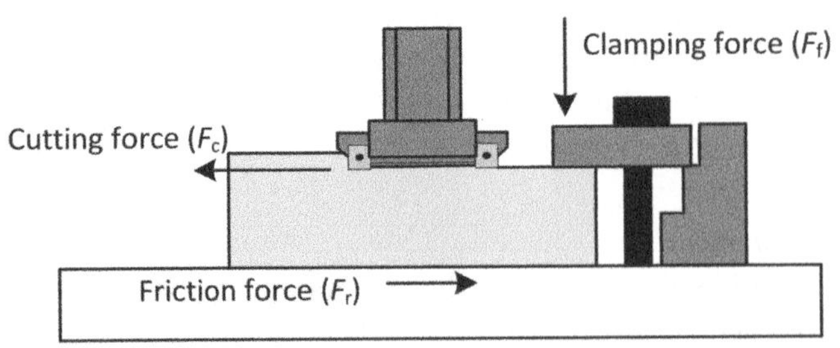

Figure 7.16 Simple example of clamping force calculation. Friction forces due to clamp force resist cutting forces.

Table 7.1 Friction coefficients for dry and lubricating contact between different materials.

Contact surfaces	Friction coefficient (dry)	Friction coefficient (lubricated)
Steel on steel	0.15	0.12
Steel on cast iron	0.19	0.10
Steel on aluminum	0.47	—
Steel on brass	0.15–0.23	0.10–0.16
Steel on bronze	0.15–0.19	0.10–0.15
Steel on graphite	0.1	0.1
Cast iron on cast iron	0.30	0.19

Source: Carrlane Manufacturing Co. (2023) and different databases.

Once the required clamping force is calculated, the metric selection for the strap clamps can be done by referring to engineering tables where the tensile force of the stud is given. The tensile force in the stud is usually 75% of the proof load, defined as the maximum tensile force that can be applied to a bolt without resulting in plastic deformation. For metric bolts, the strength is defined according to the property class (ISO 898-1: 2013 [International Standard Organization 2013]) with designations 4.6, 4.8, 5.8, 8.8, 9.8, 10.9, and 12.9. For instance, a property class of 4.6 refers to low or medium carbon steels with a metric range of M5-M36 and a minimum tensile strength of 400 MPa. For illustrative purposes, Table 7.2 shows the recommended torque, the clamping force, and the tensile force in the stud of different metric sizes, assuming a property class of 4.8. Note that for strap clamps, the tensile force in the stud is exerted on both sides of the strap (i.e., 2-to-1 clamping–force ratio). Thus, the clamping force is equal to the tensile force divided by two.

The calculations are usually more challenging for conventional workholding systems where workpiece locators and multidirectional forces should be taken into account. In those cases, the worst-case force scenario can be intuitively estimated and then represented as a free-body diagram to solve the acting forces in the system.

Table 7.2 Approximate clamping forces of different stud sizes in clamp straps with a 2-to-1 clamping-force ratio.

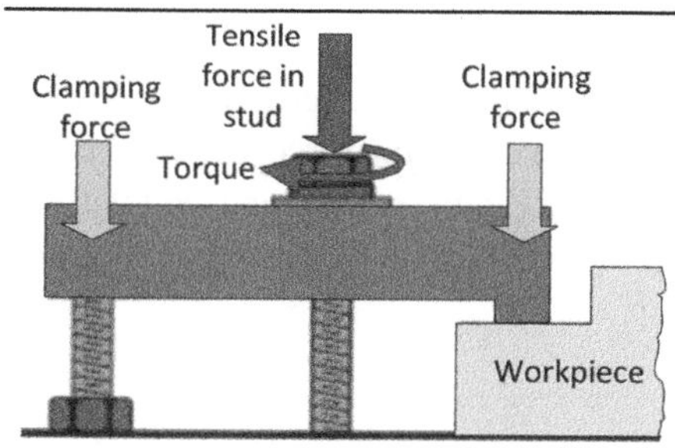

Metric stud size	Recommended torque (N·m)	Clamping force (N)	Tensile force in stud (N)
M6	5.4	2,200	4,400
M8	12.2	4,000	8,000
M12	51.5	10,200	20,400
M16	104.4	16,400	32,800
M22	301	33,800	67,600

Source: Adapted from Carrlane Manufacturing Co. (2023).

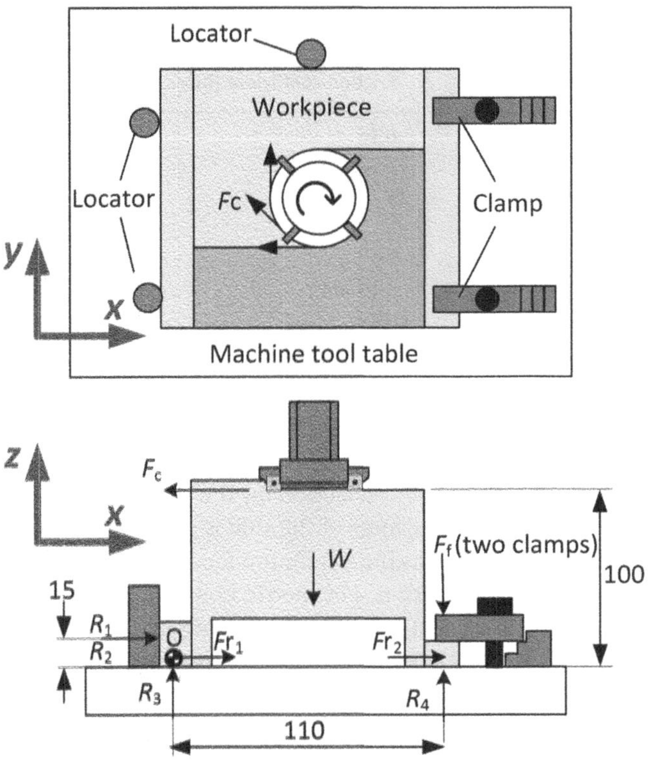

Figure 7.17 A more general clamping-force calculation, using a two-dimensional free-body diagram. *Source:* Adapted from Carrlane Manufacturing Co. (2023). Units in mm.

For illustrative purposes, consider the example shown in Figure 7.17 and assume that, based on a prior calculation, the cutting force F_c is 1,600 N. The weight of the workpiece is 250 N, and the friction coefficient between the workpiece and the machine tool table is assumed to be 0.19. According to the forces drawn in Figure 7.17, the unknown forces are the horizontal reaction force from locating pins (R_1 and R_2), the vertical reaction force from left side (R_3) and right side (R_4), the clamping force from the strap clamps (for each clamp, a force of F_f) and the friction force (F_{r1} and F_{r2}).

For a static condition, all forces in any direction and moments at any point must be equal to zero. Therefore, the following equations are given:

$$\Sigma F_x = 0 \rightarrow F_c = F_{r1} + F_{r2} + R_1 + R_2 \rightarrow F_c = \mu \times (R_3 + R_4) + R_1 + R_2$$

$$\Sigma F_z = 0 \rightarrow W + 2 \times F_f = R_3 + R_4$$

$$\Sigma M_O = 0 \rightarrow F_c \times 100 + 110 \times R_4 = 15 \times (R_1 + R_2) + 55 \times W + 110 \times 2 \times F_f$$

In this example, it initially appears to be statically indeterminate since there are only three equations and five variables. But analyzing the case for minimum clamping force, the workpiece would be barely touching the machine tool table and thus, R_4 can be assumed equal to 0. Furthermore, assuming $R_1 \approx R_2$, then the required clamping force is obtained by solving the following system of equations:

$$F_c = \mu \times R_3 + 2 \times R_1$$

$$W + 2 \times F_f = R_3$$

$$F_c \times 100 = 30 \times R_1 + 55 \times W + 220 \times F_f$$

where three equations and three variables are identified. Solving the equations, the results are $F_f = 574\,\text{N}$, $R_1 = R_2 = 667\,\text{N}$, $R_3 = 1{,}398\,\text{N}$. Therefore, a minimum clamping force per clamp of 574 N is required. Considering Table 7.2, a strap clamp with a metric stud size M6 and a clamping force of around 2,000 N would be more than enough for the application.

7.2.4 Basic Steps for Workholding/Fixturing Design

As a general guide, the following steps can be considered when designing a workholding/fixturing device:

- Analyze the type of workholding system required. Permanent fixtures, modular fixtures, and general-purpose workholders are used for different scenarios depending on the volume of production, accuracy, productivity, etc. A break-even analysis may be necessary to make the correct decision.
- Analyze the part. Identify the surfaces to be machined and locate surface candidates. For locating surface selection, the following order of preference is commonly used (Hoffman 2012):
  1) Holes for applying concentric and radial locators.
  2) Machined surfaces that form right angles to place locators and lock orthogonal movements.
  3) One machined surface and one unmachined surface form a right angle.
  4) Two unmachined surfaces that form a right angle.
- For fixturing layouts, apply the previous locating and clamping guidelines. The locating accuracy evaluation for 3-2-1 schemes may be used for selecting between different layouts.
- Estimate cutting forces and clamping forces requirements.
- Redesign the fixture according to force estimation. For instance, soft materials such as aluminum, magnesium, or polymers present lower cutting forces, and thus, less rigid fixtures are needed. If forces may deform some regions of the workpiece, supports may be required.

7.3 Types of Workholding Systems

A general classification of workholding systems is shown in Figure 7.18. While permanent fixtures are specifically designed for a large production volume with high requirements, general-purpose workholders are mainly used in shop-floor activities or very short runs with limited accuracy, and modular fixtures provide a flexible way of fixturing. Virtually every manufacturer finds suitable applications for each of these three options at various times, depending on its specific needs.

Next, some details about these three types of workholding systems are explained.

7.3.1 Permanent Fixturing

Workholding solutions designed for high-volume production typically take the form of permanent tools. These permanent jigs and fixtures are primarily tailored for a specific operation on a particular part and they are constructed using a combination of standard tooling components and custom-made parts.

The increased complexity of these permanent workholders offers advantages in terms of enhanced productivity, accuracy, and reduced reliance on operator decision-making, leading to a lower average cost per unit. Very often, the fixture elements are machined from a raw stock in the base structure of the fixture, which improves the accuracy of the fixture (i.e., increases workpiece

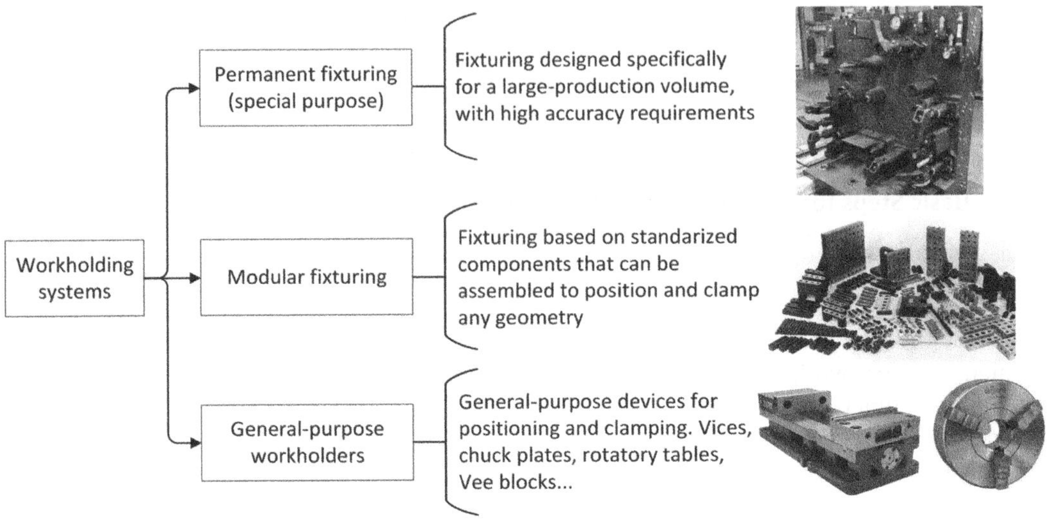

Figure 7.18 General classification of workholding systems.

location repeatability). In situations involving hydraulic or pneumatic fixtures, their inherent advantages can significantly boost productivity, further driving down per-unit costs despite the initially higher construction costs associated with these fixtures compared to other alternatives.

7.3.2 Modular Fixtures

Modular fixtures offer many of the benefits associated with permanent tools while employing only a temporary setup. The benefit of modular fixtures lies in their ability to harness the advantages of permanent tooling, such as reduced setup time, durability, productivity enhancements, and diminished reliance on operator decision-making, by utilizing a standardized set of components. Once a production run is completed, the fixture can be disassembled, allowing the components to be reused in constructing a different fixture. Later, the original fixture can be easily reconstructed using provided drawings, instructions, and photographic records. This reusability facilitates the creation of intricate, high-precision tools without the need for the permanent dedication of fixture components.

However, modular fixtures are used when low to medium cutting forces are applied due to stiffness limitations. Furthermore, dedicated fixtures are preferred to modular fixtures when large volumes of components are produced due to higher productivity and lower maintenance.

7.3.3 General-Purpose Workholders

In many cases, the part's shape and the required machining operations permit the use of versatile workholders such as vices, collets, or chucks. These workholding devices can adapt to various machines and a wide range of parts. Due to their nonspecific nature, they offer the flexibility to be employed repeatedly for different or small-scale production runs. The cost of these workholders is typically spread over several years and may not significantly impact job–cost calculations. However, because of their general-purpose design, they demand a higher level of operator vigilance and precision to maintain consistency and accuracy. Consequently, they are not the preferred choice for extended production runs.

7.3.4 Workholding System Selection

As shown previously, permanent fixturing, modular fixturing, and general-purpose workholding devices present different characteristics that make them suitable for specific production types. For instance, permanent fixtures are adequate for large production volumes. Longer production runs require fixtures to be in service longer, and production speeds are generally higher; thus, higher investment in an efficient fixture device is justified.

The main characteristics of these workholding solutions are qualitatively described in Table 7.3, which gives a general idea about which type of production may apply. As shown in the table, permanent solutions present higher investment costs than modular fixtures or general-purpose workholding devices. However, their accuracy and productivity are higher, which may result in lower production costs.

Besides this qualitative description, the selection of the proper fixturing solution may require a break-even analysis to select the most economical solution while ensuring part specifications. For this purpose, three cost terms are considered: initial fixturing cost (C_{ini}), setup cost (C_{setup}), and running cost (C_{run}).

The initial fixturing cost (C_{ini}) is estimated by calculating the design expenses and the cost of all fixture components including the cost of assembling. This cost is shared among all the expected parts to be produced under this workholding/fixturing device (N_{life}). Additional costs are related to the setup process (C_{setup}) for every batch operation such as inspection, calibration, cleaning, and greasing operations. This cost is shared among the parts expected to be produced in a batch (N_{batch}). Finally, the workholding/fixturing device may present a running cost (C_{run}) related to loading and unloading activities, power actuators, etc. Very often, the running cost includes the machining and nonproductive costs since the workholding selection may influence cutting tool trajectories, cutting parameters, etc. Therefore, the total fixturing cost per machined part can be formulated as follows:

$$C_{\text{fixturing}}(\$) = C_{\text{run}} + \frac{C_{\text{setup}}}{N_{\text{batch}}} + \frac{C_{\text{ini}}}{N_{\text{life}}}$$

The following example illustrates the fixturing selection among three tooling alternatives: (i) a modular fixture; (ii) a permanent fixture; and (iii) a permanent fixture actuated by hydraulic power. The costs of each term for each alternative and the break-even analysis to select the most adequate one are presented as follows.

- Running cost (C_{run}). The running costs vary among the alternatives due to changes in the machining cycle time and power consumption. For this example, it is assumed that a running cost of \$5 for the modular fixture, \$4 for the permanent fixture, and \$3 for the permanent hydraulic fixture.

Table 7.3 The three broad alternatives for workholding: permanent fixtures, modular fixtures and general-purpose devices. Relative comparison for cost, accuracy, and productivity.

	Permanent fixture	Modular fixture	General-purpose workholders
Cost	\$\$\$	\$\$	\$
Accuracy	+++	++	+
Productivity	+++	++	+
Typical application	Large-production volumes	short-to medium-production volumes	Short-production volumes, maintenance operations

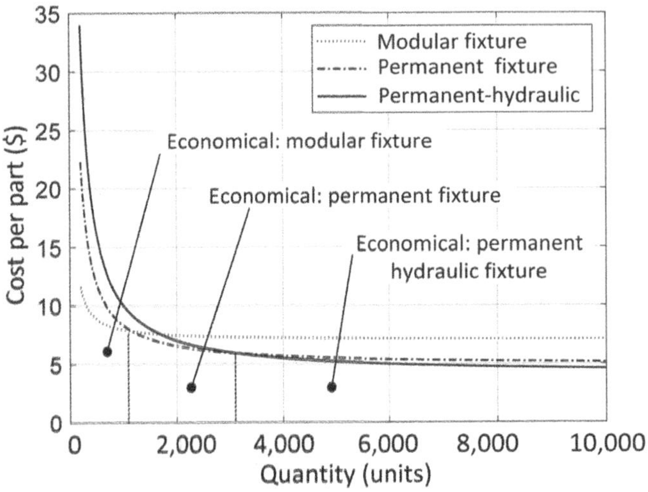

Figure 7.19 Break-even analysis for the example presented.

- Setup cost (C_{setup}). The permanent fixture is the quickest to set up, followed by the power-workholding fixture, which is slightly slower due to hydraulic connections, and finally, the modular fixture, which is the slowest due to the assembly required. For this example, it is assumed a setup cost of $400, $160, and $200 for the modular, permanent, and hydraulic fixtures, respectively.
- Lot size (N_{batch}). The lot size remains consistent across all three options. For this example, it is assumed that each batch contains 200 parts.
- Initial tooling cost (C_{ini}). The modular fixture is the most cost-effective because its components can be reused, followed by the permanent fixture and the hydraulic fixture, which is the most expensive option. For this example, the initial tooling costs are assumed to be $920, $3,500, and $6,000 for the modular, permanent, and hydraulic fixtures, respectively.
- Total quantity produced over the tooling's lifespan (N_{life}). The analysis is conducted considering up to 10,000 production volumes.

According to the preceding assumptions, the break-even analysis for the three fixturing alternatives is shown in Figure 7.19. Modular fixtures are unquestionably the most affordable option for a single run of 200 units. The permanent fixturing is ideal if 10 runs (2,000 parts) are anticipated. The hydraulic fixturing device would be the ideal option for orders of 3,000 parts or more. This example assumes that every noneconomic element is equal in all cases.

7.4 Permanent and Modular Fixtures

7.4.1 Basic Fixture Elements

A permanent and modular fixture are commonly composed of similar elements, and they can be mainly grouped in the following categories:

- Structural components. Clamping cubes, blocks, and bases are used as the body of the fixture system where the rest of the components are mounted. These elements are commonly made of nodular cast iron to impair properties such as vibration–damping, torsion–resistance, and high stability. The surfaces are precision ground for proper placement of support and clamping devices. Elements for positioning and quick change are commonly included, and they can be

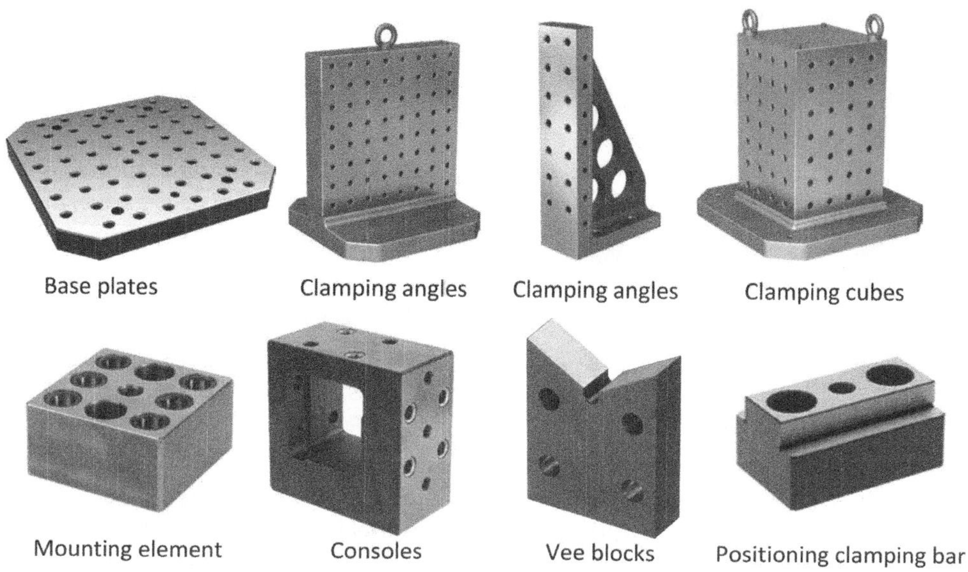

Figure 7.20 Basic structural elements for modular/permanent fixtures. *Source:* Halder (2024).

easily mounted on a machine pallet. Other structural elements can be used to locate the clamping and locating systems at the correct part surfaces. For instance, base plates (T-slot or hole threads), angle plates, consoles, position blocks, V-blocks, etc. (Figure 7.20).

- Rest pads. Rest pads are used to form a plane where the workpiece will lay. Rest pad material is hardened and ground for accurate positioning. When clamping rough surfaces such as casting and forging surfaces, serrated and hardened pads are recommended. In some cases, the rest of the pads are self-aligned, which means that they can be slightly tilted to accommodate nonhorizontal surfaces with respect to the base of the pad (Figure 7.21). Different types of rest pads are used to contact the workpiece and act as seats or stops (Figure 7.22).
- Locating pins. Locating pins are used to locate the workpiece on the fixture, and their function is similar to rest pads; they provide a surface contact to position the part and lock the corresponding part movements to provide location repeatability. Locating pins are cylinders

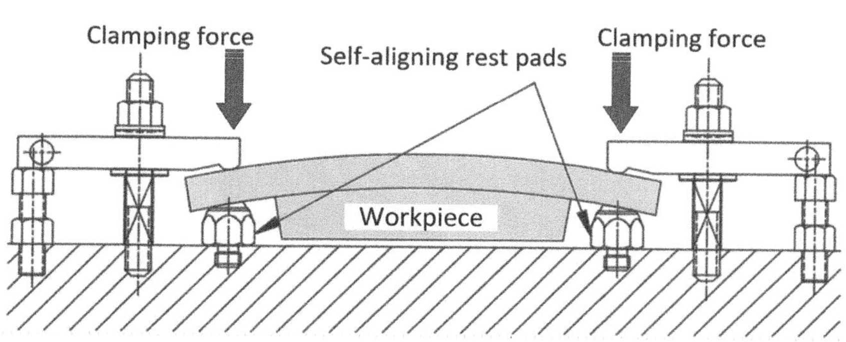

Figure 7.21 Example of self-aligning rest pads to support sloped surfaces. *Source:* Carrlane Manufacturing Co. (2023).

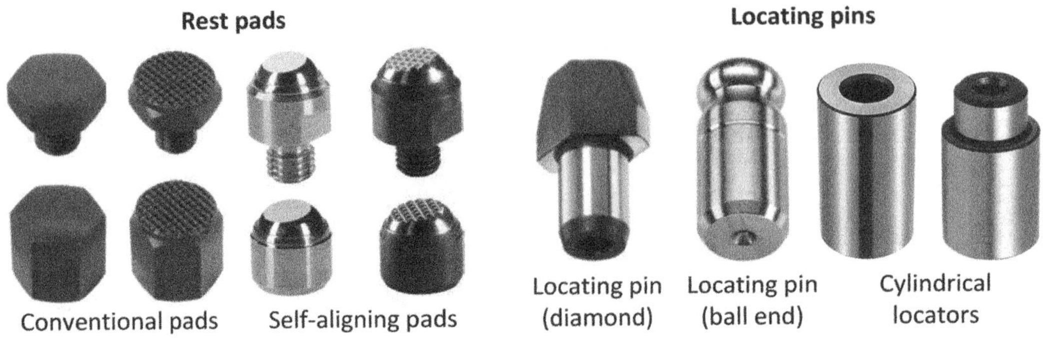

Figure 7.22 Different types of rest pads and locating pins. *Source:* Halder 2024 / Erwin Halder KG.

with dimensions and types of heads depending on the application (Figure 7.22). For instance, bullet or chamfered head pins provide ease of loading the workpiece, and long pins are used to lock eight potential movements of the part. Noncylindrical shapes such as diamond pins are used to lock only two potential movements, for instance, to locate a slot.

- Clamping devices. To hold the part in position during machining, clamping elements are applied. A large variety of clamping devices can be mounted on fixtures. Figure 7.23a, b shows some clamping devices commonly applied in fixturing. As shown in the figure, toggles must be applied with care since the clamping force is limited, although they provide a fast and easy clamping method. Other

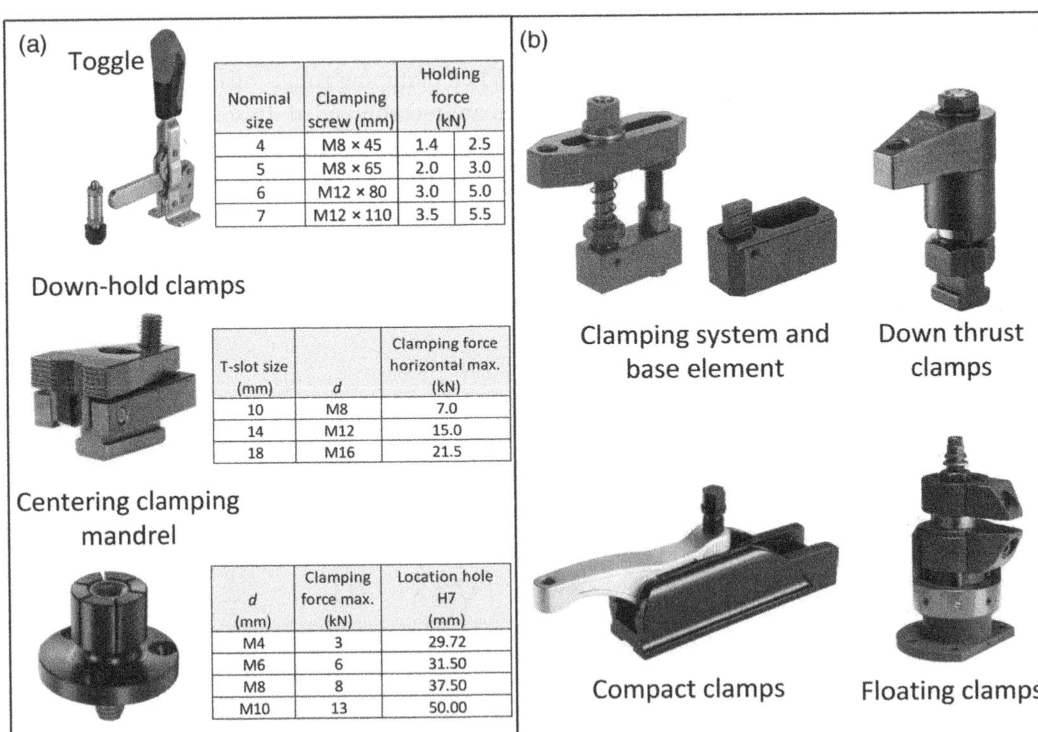

Nominal size	Clamping screw (mm)	Holding force (kN)	
4	M8 × 45	1.4	2.5
5	M8 × 65	2.0	3.0
6	M12 × 80	3.0	5.0
7	M12 × 110	3.5	5.5

T-slot size (mm)	*d*	Clamping force horizontal max. (kN)
10	M8	7.0
14	M12	15.0
18	M16	21.5

d (mm)	Clamping force max. (kN)	Location hole H7 (mm)
M4	3	29.72
M6	6	31.50
M8	8	37.50
M10	13	50.00

Figure 7.23 (a) Clamping forces for toggle, down-hold clamps, and centering clamping mandrels; (b) different types of clamps. *Source:* Halder 2024 / Erwin Halder KG.

mechanical clamps, such as down-hold clamps or centering clamping mandrels, present 3–10 times more clamping force. For automated fixtures, pneumatic and hydraulic systems are applied. The pneumatic systems are more economical and faster but provide low clamping forces (2–3 kN maximum), need more space to be implemented, and present low reliability due to the compressibility of air. Unlike pneumatic actuated clamps, hydraulic actuated clamps exert higher clamping forces (above 5 kN), and they are more compact and reliable. Although hydraulic systems can be applied for thin-section components, pneumatic fixtures may be preferred.

- Supports. Auxiliary devices such as support blocks, adjustable leveling supports, bedding supports or additional rest pads may be also added to stabilize the workpiece during cutting, avoiding deformations (Figure 7.24).

To illustrate the application of the preceding listed elements, Figure 7.25 presents two types of modular fixtures and the components required, according to a well-known manufacturer of fixture systems (Halder 2024).

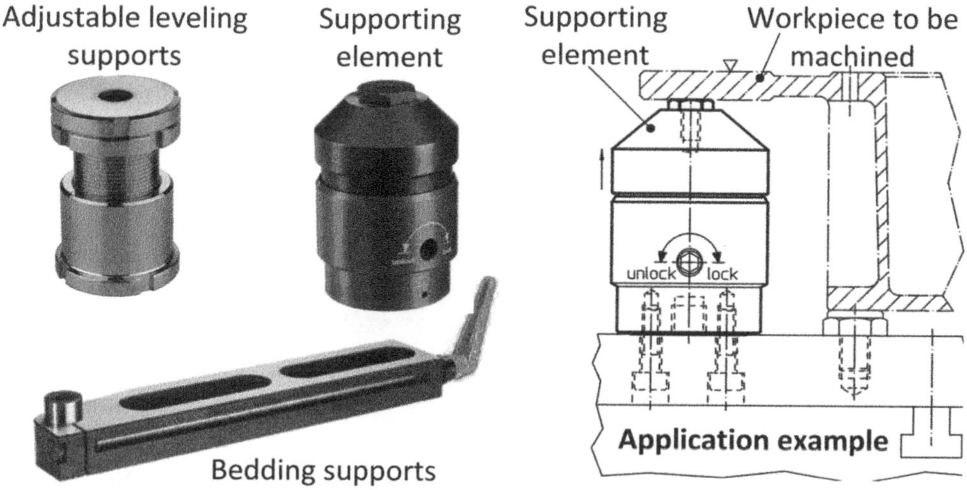

Figure 7.24 Different types of supporting elements and application example. *Source:* Halder 2024 / Erwin Halder KG.

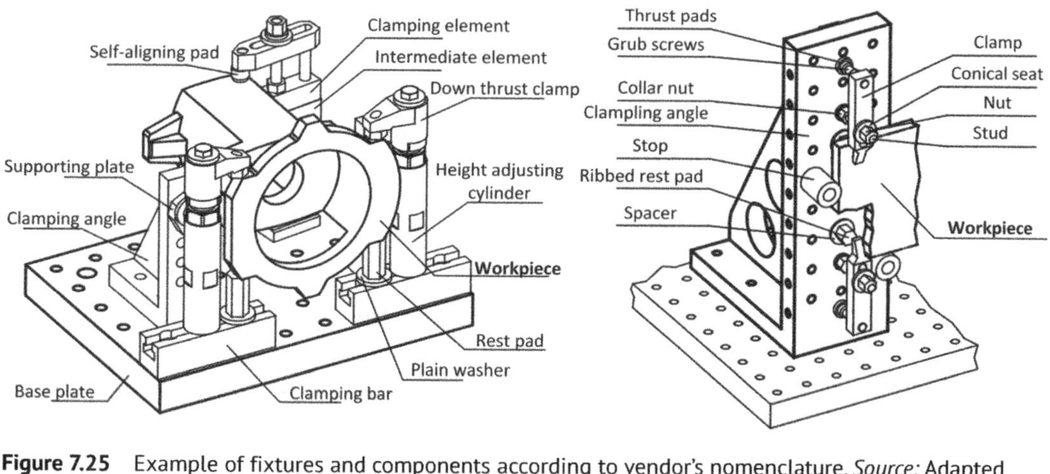

Figure 7.25 Example of fixtures and components according to vendor's nomenclature. *Source:* Adapted from Halder (2024).

Figure 7.26 Examples of modular fixtures. *Source:* Courtesy of Halder Inc.

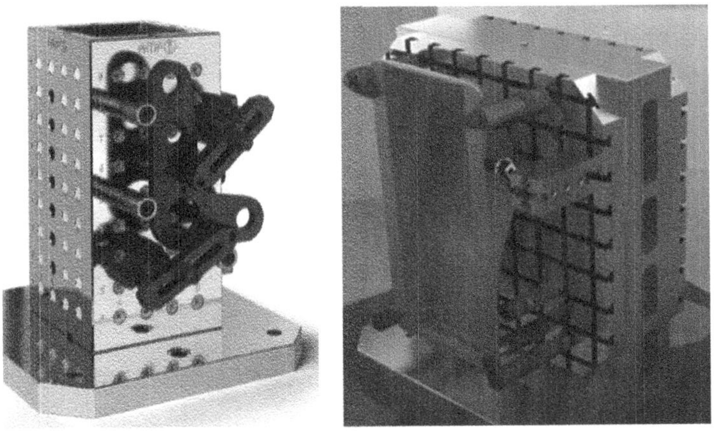

Figure 7.27 Examples of modular fixtures. *Source:* Courtesy of Andreas Maier Gmbh and Co (AMF) and Halder, Inc.

7.4.2 Fixturing Examples

Figures 7.26 and 7.27 show examples of modular fixtures for locating and clamping complex parts for machining operations. As it is shown, different types of basic elements such as T-slot base plates, clamping cubes, down thrust clamps, rest pads, locating pins, and manual clamping systems can be mounted to configure structures that are able to locate and clamp different geometries in a fast and efficient way.

7.5 General Purpose of Workholding Devices for Turning

Depending on the size, weight, diameter, shape, and location of the workpiece on the lathe, various general-purpose workholding systems are utilized in turning. They can be classified as chucks and plates, collets and mandrels, devices for holding between centers, and rests.

7.5.1 Chucks and Plates

Workpieces with a length-to-diameter (L/D) ratio lower than four may be clamped in a chuck quickly and rigidly without the need for additional support of the free end. There are different types of chucks commonly used in shop floor such as three-jaw self-centering chucks, four-jaw independent chucks, magnetic chucks, and face plates.

7.5.1.1 Three-Jaw Self-Centering Chucks

The three jaws installed in the three slots of the chuck can be simultaneously adjusted by rotating any of the three pinions using a chuck key (Figure 7.28a). This type of chuck is well suited for securing and rotating regularly shaped workpieces, such as round or hexagonal rods, along the lathe's axis. However, it cannot effectively hold workpieces with irregular shapes. The advantage here lies in the quick and straightforward clamping process, as all three jaws move simultaneously. As a drawback, its position repeatability is low compared to other general-purpose workholding systems.

7.5.1.2 Four-Jaw Independent Chucks

This chuck features four jaws, and each jaw can be adjusted independently by turning a screw with the assistance of a chuck key (Figure 7.28b). It allows for precise positioning of the jaws to accommodate irregularly shaped workpieces. However, it does require more time to align the workpiece with the lathe's axis. The chuck's face is marked with concentric circles to facilitate the rapid centering of the workpiece.

7.5.1.3 Magnetic Chucks

The magnetic chuck relies on the magnetic field generated by an electromagnet housed within it to hold the workpiece in place (Figure 7.29a). Magnets inside the chuck can be adjusted to secure or release the work. This type of chuck is only suitable for workpieces made of magnetic materials. It is especially useful for very small, thin, and lightweight workpieces that a standard chuck cannot effectively hold. However, its use is limited to machining under very low cutting forces.

7.5.1.4 Face Plates

A face plate is employed to secure large, heavy, and irregularly shaped workpieces that cannot be conveniently accommodated by other types of chucks (Figure 7.29b). It takes the form of a circular disc that is bored out and threaded to fit onto the lathe spindle's nose. The face plate is equipped with radial plain and T-slots, allowing workpieces to be secured using bolts and clamps.

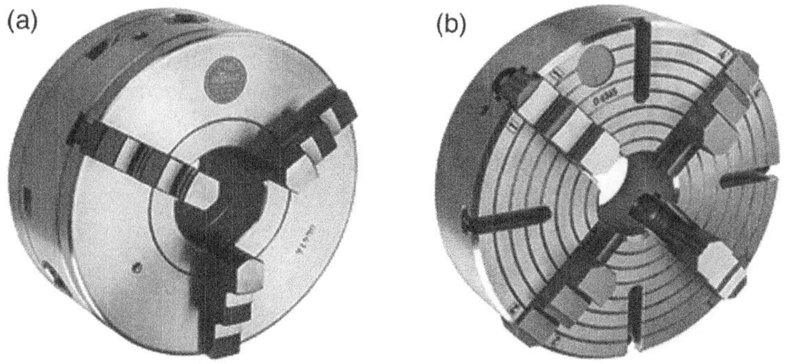

Figure 7.28 (a) Three-jaw chuck and (b) four-jaw independent chuck. *Source:* Courtesy of Röhm GmbH.

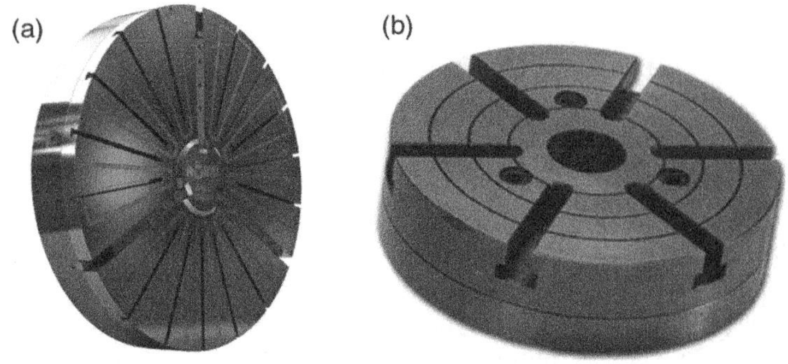

Figure 7.29 (a) Magnetic chuck and (b) face plate. *Source:* Figure 7.29a: Courtesy of Röhm GmbH; Figure 7.29b: Courtesy of Detroit Machine Tools (smithy.com).

7.5.2 Collets and Mandrels

For clamping small parts using the external surface, an interesting clamping option is the use of a collet and a collet chuck. The collet, a cylindrical bushing, is part of the collet chuck. The collet presents slots that have been carved lengthwise in spring steel (Figure 7.30) and serve to grip the work with high accuracy when the collet is inserted into the collet chuck. Although collets are used for holding small bars with different regular geometries (cylindrical, square, hexagonal, etc.), in turn, similar holding systems can be found for holding solid mills in milling.

For internal clamping, chucks with internal jaws can be used. However, when small internal surfaces are used for clamping, mandrels can be the best choice. This type of workholding device expands its cylindrical surface to compress and clamp the workpiece over a hole (Figure 7.31). For instance, tubes can be held using a mandrel mounted on the spindle lathe on one side and a live center at the opposite side of the lathe. Under this workholding scheme, the whole length of the workpiece can be machined.

7.5.3 Devices for Holding between Centers

A popular method of holding a workpiece in a lathe when accuracy and repeatability are required is between centers. Long workpieces (shafts and axles) with an L/D ratio above four are turned between centers. Figure 7.32 shows an example of a workpiece held between centers. The workholding system is based on centers, a carrier and a catch plate.

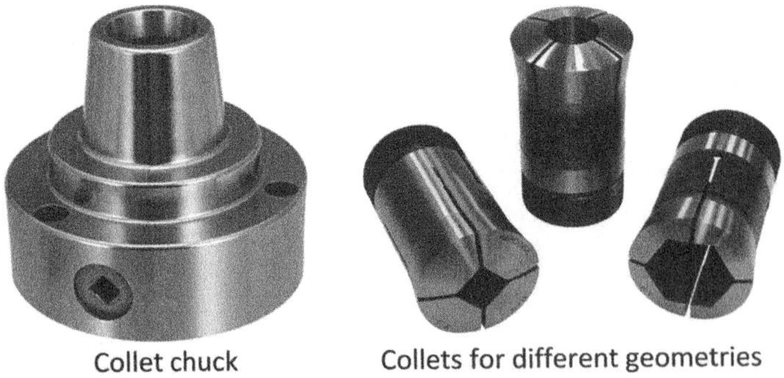

Figure 7.30 Collet and collet chuck for lathes. *Source:* Courtesy of Hardinge, Inc.

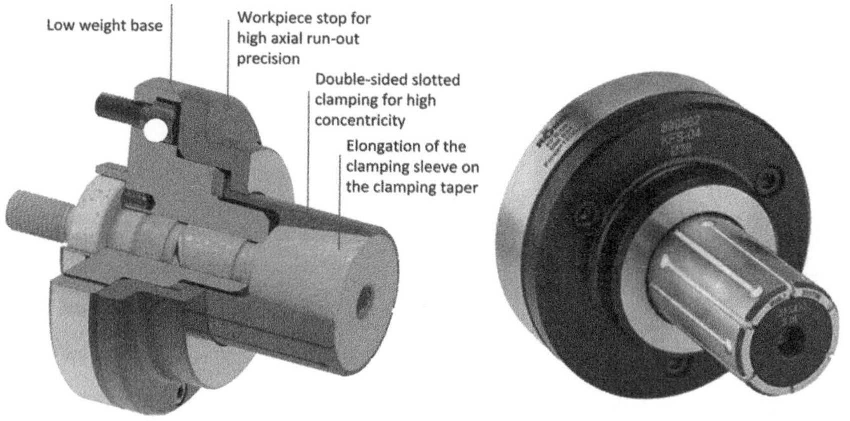

Figure 7.31 Mandrel for internal clamping in turning operations. *Source:* Courtesy of Röhm GmbH.

Figure 7.32 Holding a workpiece between centers.

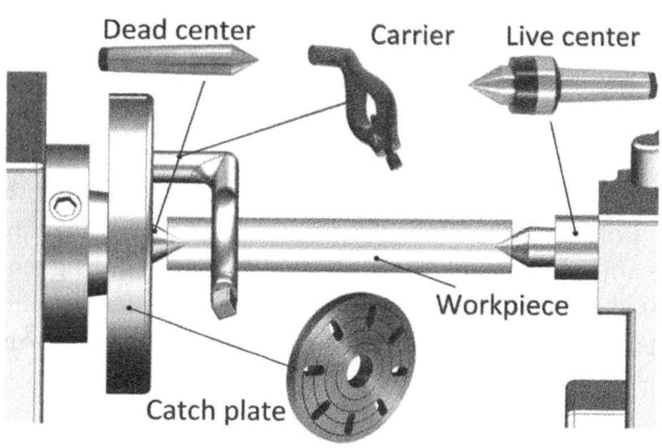

There are two types of centers: live centers and dead centers. Live centers have internal bearings that make the center rotate with the workpiece. This type of center is usually placed at the tailstock. Dead centers stay stationary while the conical surface of the center and the workpiece are rubbing. Dead centers may be placed at the tailstock or the headstock of the lathe. In the latter case, since the dead center rotates with the work, they may also be named as live centers. The catch plate is used to drive the workpiece during the cutting process. Catch plates are designed with U-slots or elliptical slots to receive the bent tail of the carrier. The carrier is used to transmit the driving force of the spindle to the work through the catch plates. The work is held inside the eye of the carrier and tightened by a screw.

7.5.4 Rests

A rest is a mechanical tool utilized to provide support for a lengthy and slender workpiece, whether it is being turned between centers or held by a chuck. Positioned at an intermediate point, its purpose is to prevent the workpiece from bending and to counteract vibrations induced by cutting forces. Two distinct types of rests are commonly used: the steady rest and the follower rest (Figure 7.33a, b).

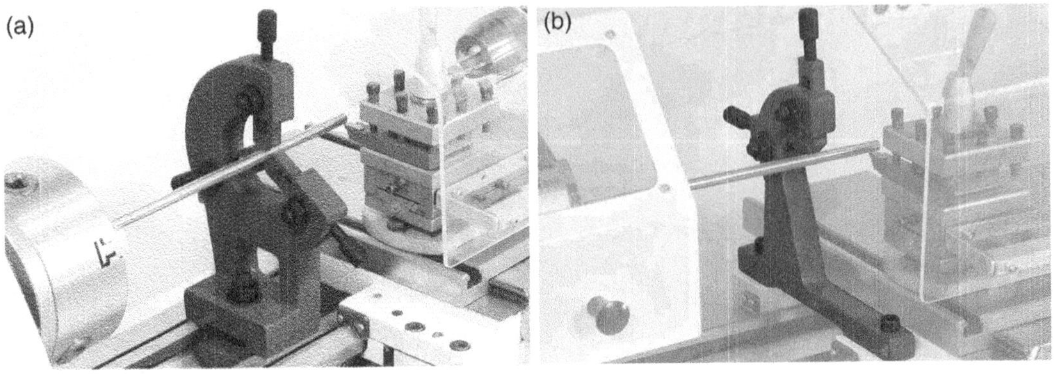

Figure 7.33 Common steady rest (a) and follower rest (b) for shop-floor turning operations. *Source:* Courtesy of Probois Machinoutils.

The steady rest can be adjusted to slide along the lathe bedways, securing itself at the desired location where the workpiece requires support. Comprising three jaws, it can be tailored to the workpiece's diameter. Machining operations are conducted within the span extending from the headstock to the point where the rest provides support.

Conversely, the follower rest presents two adjustable jaws designed to uphold the workpiece. It is affixed to the rear end of the carriage and, during machining, moves in tandem with the carriage itself. As a result, it closely follows the tool, offering continuous support to the workpiece throughout its entire length. Both steady and follower rests can also be automated and operated by hydraulic or pneumatic power (Figure 7.34).

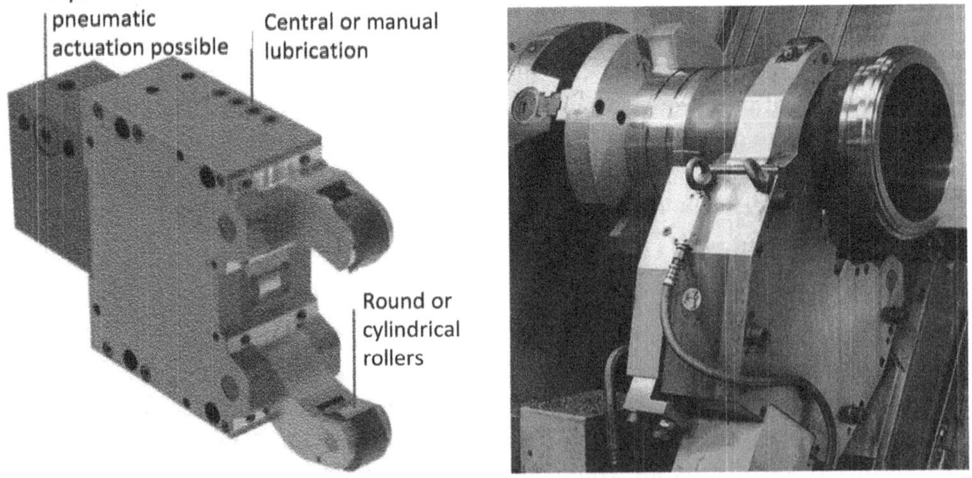

Figure 7.34 Automated steady rest. *Source:* Courtesy of Röhm GmbH.

7.6 General Purpose of Workholding Devices for Milling

7.6.1 Vices

Vices are the most employed devices for securing workpieces on the milling machine table, allowing for efficient loading and unloading. Vices come in three primary types:

a) Plain vice. The plain vice is directly affixed to the milling machine table and is the prevailing choice for plain milling operations that involve substantial material removal. Its compact design ensures that the workpiece remains close to the table, minimizing the risk of vibrations. Workpieces are clamped between the stationary and movable jaws, and specialized jaws may be used for securing irregularly shaped workpieces (Figure 7.35).
b) Swivel vices. Swivel vices are employed for milling angular surfaces in relation to straight ones without the need to remove the workpiece from the vice. They feature a circular base for a rotation movement and are fastened to the machine tool table using T-bolts or similar fasteners (Figure 7.36a).
c) Universal vices. Universal vices possess the ability to swivel in a horizontal plane similar to swivel vices and can also be tilted to assume various vertical positions for angled cuts. They facilitate the milling of surfaces at different inclinations without the necessity of removing the workpiece (Figure 7.36b).

7.6.2 Work Directly Mounted on Table

With the aid of slots, T-bolts, clamps, and nuts, heavy or oddly shaped work that cannot be supported by ordinary holding devices can be directly mounted and clamped on the machine tool table. There are numerous different types of clamping systems that can be directly mounted on the machine tool table. One of the most common types is based on step blocks, nuts, and

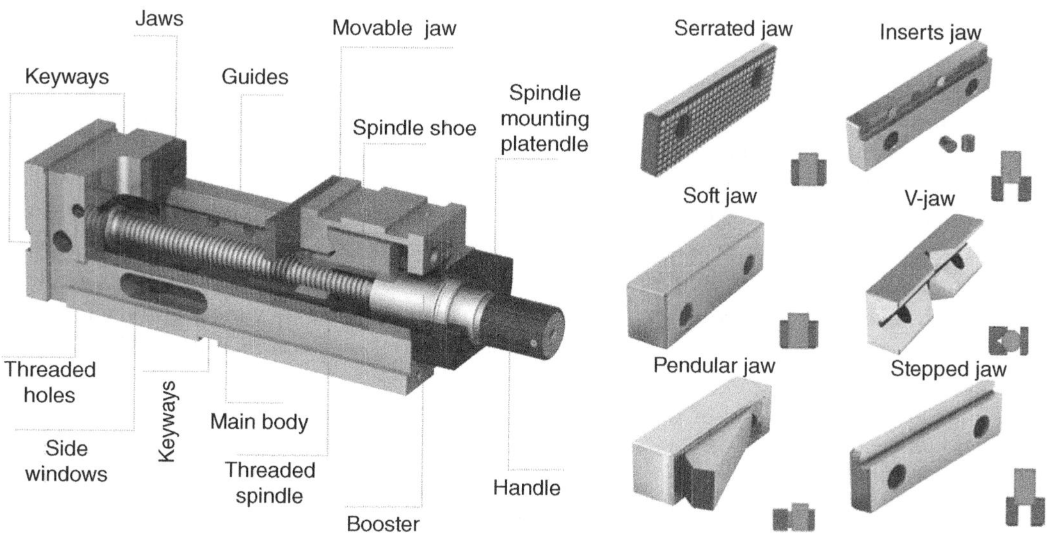

Figure 7.35 Plain vice and different types of jaw for vices. *Source:* Courtesy of Fresmak S.A.

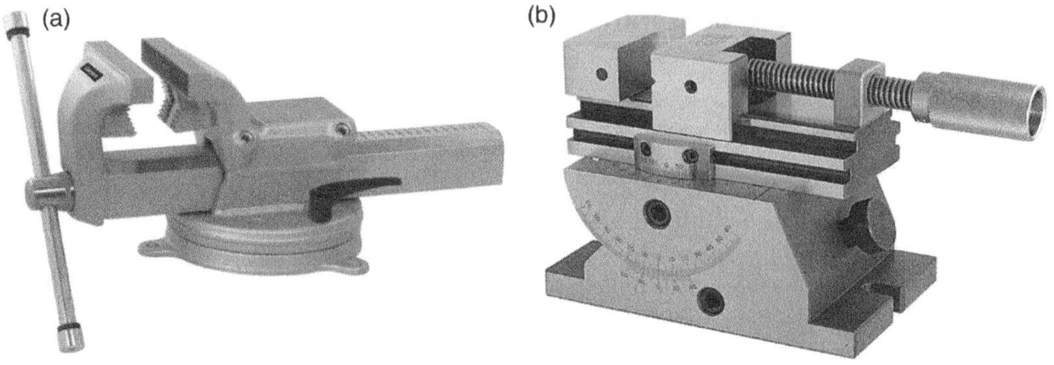

Figure 7.36 (a) Bench vice with swivel base and (b) universal vice for milling operations. *Source:* Courtesy of Hoffmann GmbH Quality Tools.

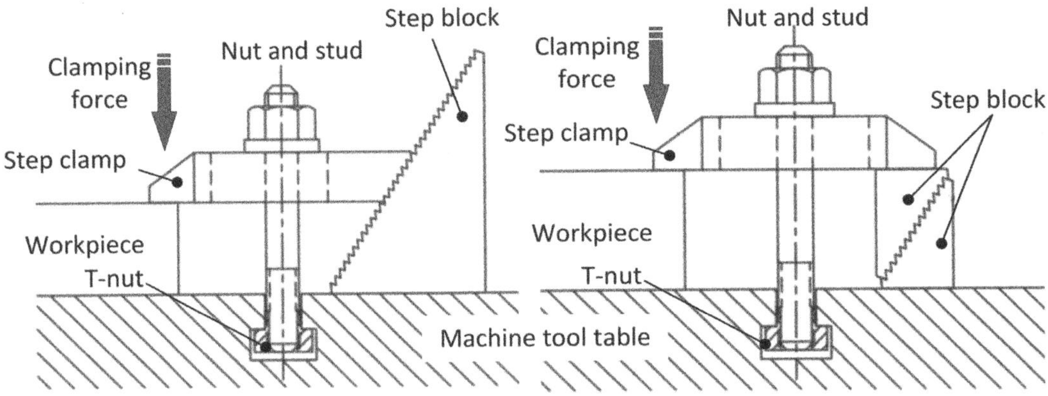

Figure 7.37 Different clamps and step blocks for workholding.

studs that can be easily installed using the T-slots in the machine tool table as shown in Figure 7.37. Figure 7.38 shows a common set of components used for clamping workpieces directly on the machine tool table.

Other popular clamping devices to be mounted on the machine tool table are clamping jaws. Heavy-duty clamping jaws present higher clamping force, height, and width. Low-height clamping jaws, also called bulles, can be used for clamping short parts. Similarly, mini bulles can be applied when only a small portion of the part can be used for clamping since they do not protrude above the workpiece. Figure 7.39 shows some examples of clamping jaws and their assembly with the workpiece.

Additionally, other devices used directly on the machine tool table are (Figure 7.40):

- Angle plates (Figure 7.40a). When work surfaces are to be milled at right angles to another face, angle plates are used to support the work.
- Vee-blocks (Figure 7.40b). Vee-blocks are used for holding cylindrical workpieces on the machine table in which keyways, slots and flats are to be machined.

Figure 7.38 Set of components for clamping a workpiece on machine tool table. *Source:* Courtesy of CarrLane Manufacturing Co.

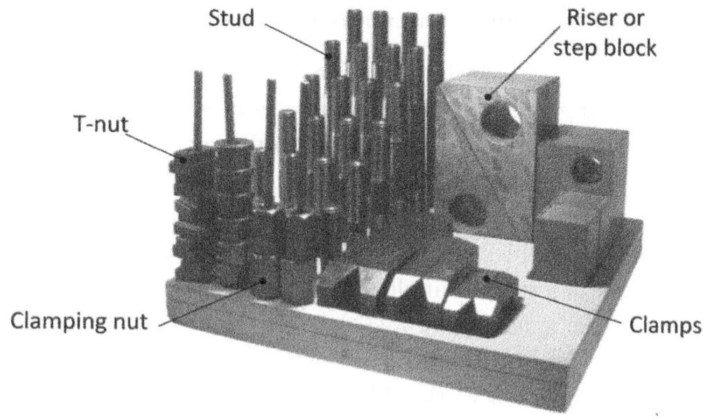

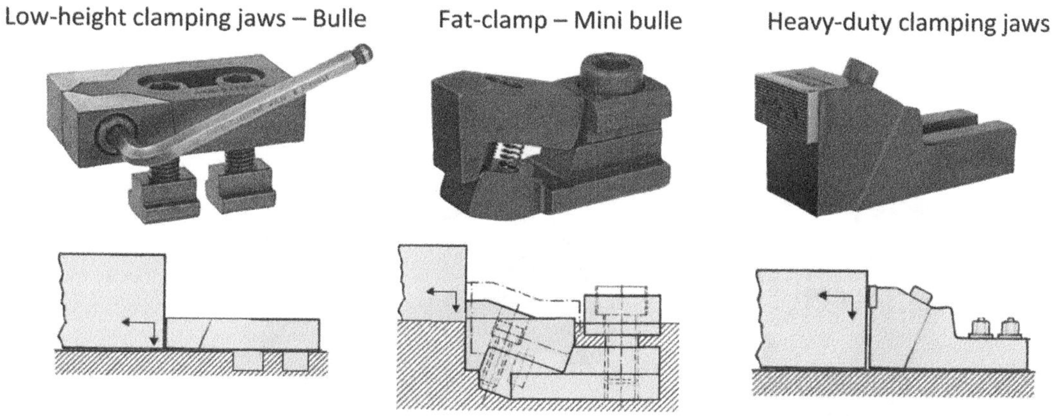

Figure 7.39 Different types of clamps for mounting on a machine tool table. *Source:* Courtesy of Hoffmann GmbH Quality Tools.

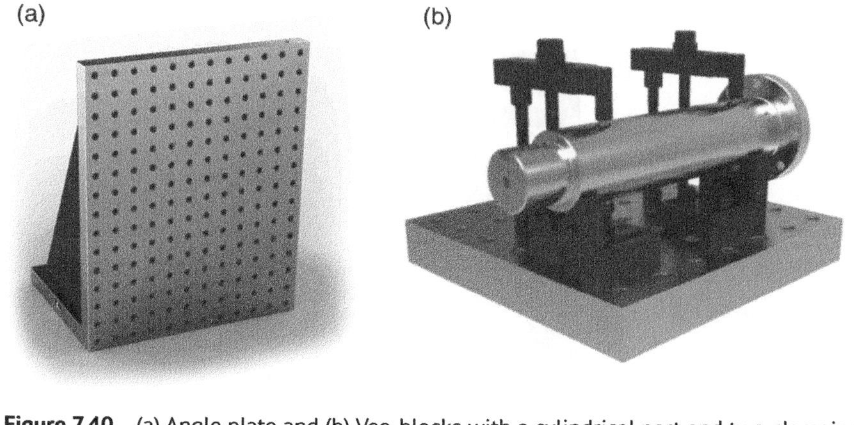

Figure 7.40 (a) Angle plate and (b) Vee-blocks with a cylindrical part and two clamping devices. *Source:* Courtesy of Advance Machine & Engineering Co.

7.6.3 Rotary Tables and Magnetic Plates

Rotary tables are circular, movable flat surface with a worm wheel mechanism to turn the table at a specific angle (Figure 7.41a). This system is commonly used in manual milling to machine circular slots or to provide a fourth axis to the machine tool. The workpiece is clamped in the rotary table using clamping devices such as step blocks, bulles, etc. The rotary table can be positioned vertically or horizontally according to the purpose of the machining operation.

Tilt milling tables are also an interesting fixture for milling operations since they can set up the workpiece at a specific angle. The table has machined T-slots along the length and can swing up to 45° on either side (Figure 7.41b).

For light milling operations of ferromagnetic workpieces that present difficulties for clamping using conventional workholding devices, magnetic clamping plates may be an option. However, the clamping force is limited, and it depends on the surface contact between the workpiece and the plate. Figure 7.42a shows a magnetic plate and Figure 7.42b shows a machining operation on a workpiece clamped using a magnetic plate.

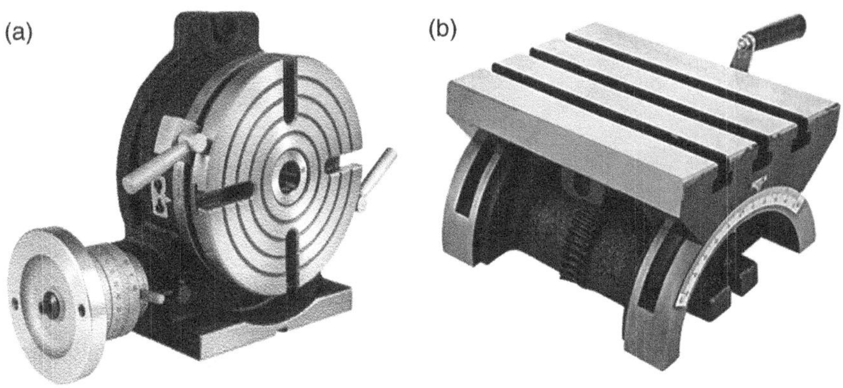

Figure 7.41 (a) Rotary table and (b) tilt milling table. *Source:* Courtesy of Vevor.

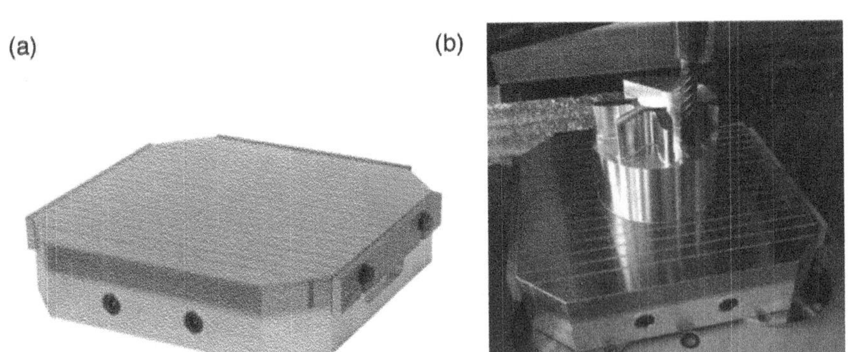

Figure 7.42 (a) Example of permanent magnetic clamping plate; (b) use of magnetic clamping plate in milling. *Source:* Courtesy of Hoffmann GmbH Quality Tools.

7.7 Workholding Drawing Representation

To create a common language for defining and locating symbols in fixture design, various initiatives for standardization can be found in the literature. Although not widely adopted, the French standard under the norm NF E04-013: 2015 (AFNOR 2015) proposes a symbology to communicate workpiece holding schemes in technical drawings that can be useful for process planners. The symbology used in this norm defines the position of locators and clamps and the part movements that are locked by each element. The basic symbol used is shown in Figure 7.43, where the following characteristics are identified: technology type, type of surface, function of technological element, and the type of contact on the surface.

The technology type refers to the type of position or clamping element. The following types are distinguished: fixed support, fixed centering, a system with clamping, a system with concentric clamping, irreversible fixed support, reversible support, and sliding positioning. Table 7.4 shows the symbols for each technology type.

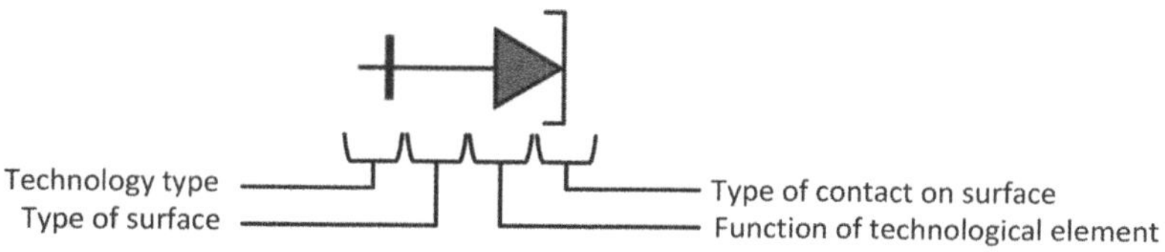

Figure 7.43 Basic symbol used to define positioning and clamping according to norm NF E04-013: 2015 (AFNOR 2015).

Table 7.4 Symbols for positioning and clamping according to each technology type.

	Symbol	
Type of technology	**Function: exact position**	**Function: clamping**
Fixed support	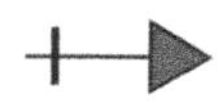	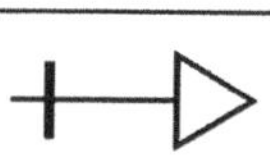
Fixed centering	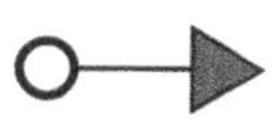	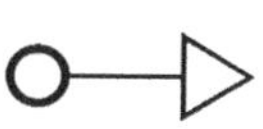
System with clamping	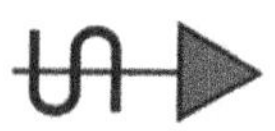	
System with concentric clamping	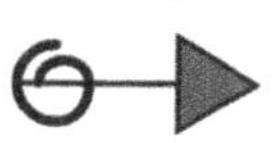	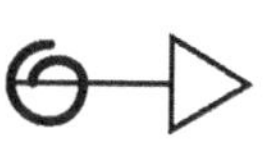
Irreversible fixed support	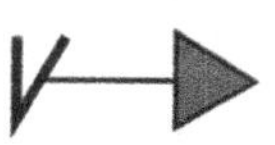	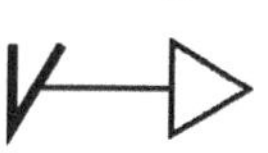
Reversible support		
Sliding positioning	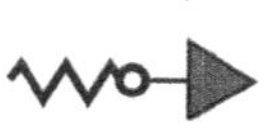	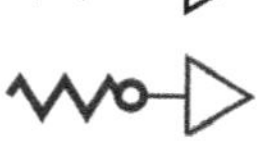

Table 7.5 Symbols for the type of surface.

Nature of surface	Symbol
Surface machined	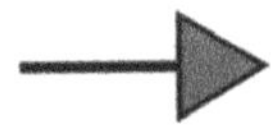
Surface raw	

The type of surface refers to the condition of the surface where the position/clamping element is placed. The following types are distinguished (Table 7.5): surface machined and surface raw.

The function of the technological element refers to the function of the position or clamping element. The main functions are exact positioning, which refers to locators or position elements that position the workpiece in the space, and clamping, where the elements exert a clamping force to hold and prevent the movement of the part due to machining forces. Additional functions are also identified such as datum origin, definition of axis, prepositioning, and part holding against vibrations and deformations. Table 7.6 shows the symbol that identifies each group of functions. Note that the representation of the symbol changes according to the drawing view, and projection views are defined where the element is in a top view.

The type of contact on the surface refers to the geometrical characteristics of the contact between the position/clamping element and the surface where it is placed. The following types are distinguished (Table 7.7): flat, curved, serrated, free, pan, swing bar, vee, fixed center, and turning center.

Additionally, Table 7.8 shows some physical realizations of these types of contacts in fixturing systems. Table 7.9 also shows a physical realization and associated symbols for concentric and radial locations.

To illustrate the use of the standard NF E04-013: 2015, Table 7.10 presents some common workholding systems and their corresponding drawing representation.

Table 7.6 Symbols for the function of the technological elements. Normal and projected views.

	Symbol			
Function	**Normal view**	**Projected view**		
• Exact positioning • Datum origin • Definition of axis	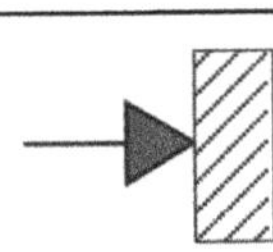	 Type: support	 Type: concentric	 Type: radial
• Clamping • Prepositioning • Opposition against vibrations and deformations	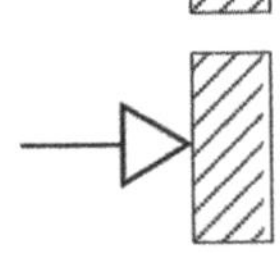			

Table 7.7 Symbols related to the type of contact.

Nature of contact	Symbol	Nature of contact	Symbol	Nature of contact	Symbol
Flat	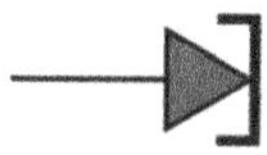	Fixed center		Free	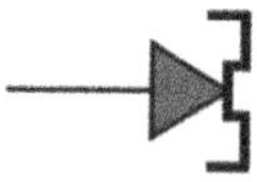
Serrated	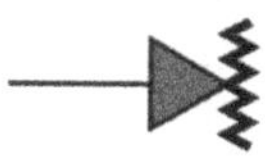	Turning center	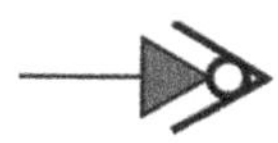	Pan	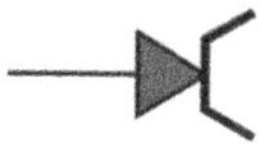
Curved	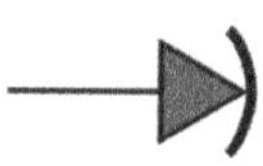	Swing bar	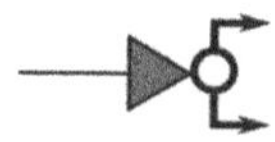	Vee	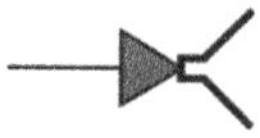

Table 7.8 Example of some physical elements in fixtures and their associated type of contact symbol.

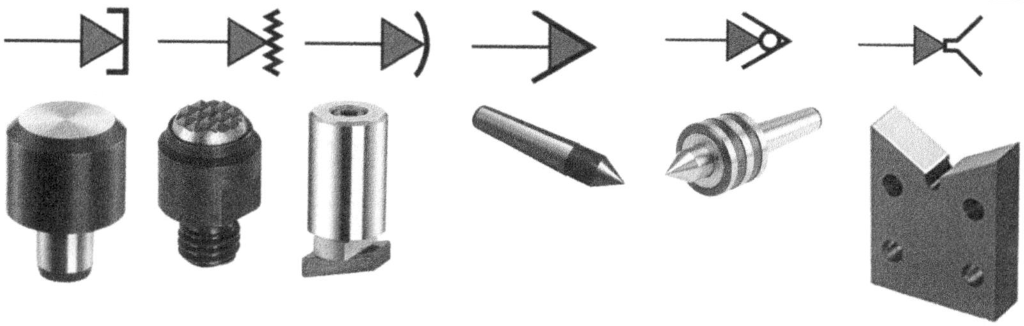

Flat rest pad	Serrated rest pad	Cylindrical locator (stop pin)	Dead center	Live center	Vee-block

Table 7.9 Concentric and radial location. Symbols and examples of physical realization.

	Symbol	Symbol (projected view)	Example
Concentric location	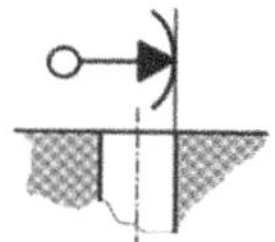		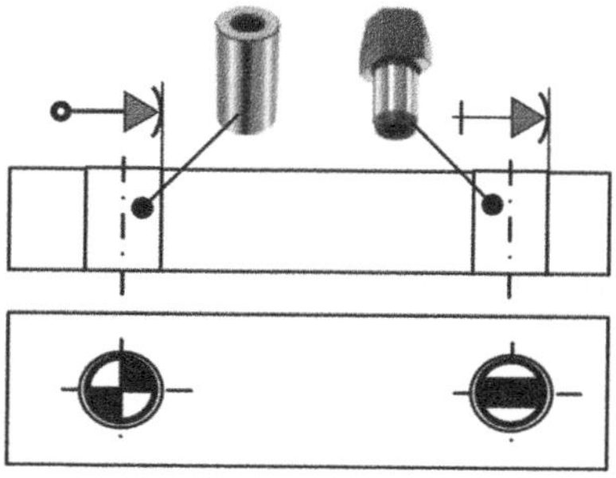
Radial location	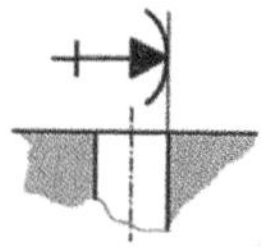		

Table 7.10 Examples of different workholding systems and their corresponding drawing representation according to the standard NF E04-013: 2015.

Physical realization of workholding system	**Workholding drawing representation and part movements locked**
Example 1: Three-jaw chuck with axial positioning on workpiece face	
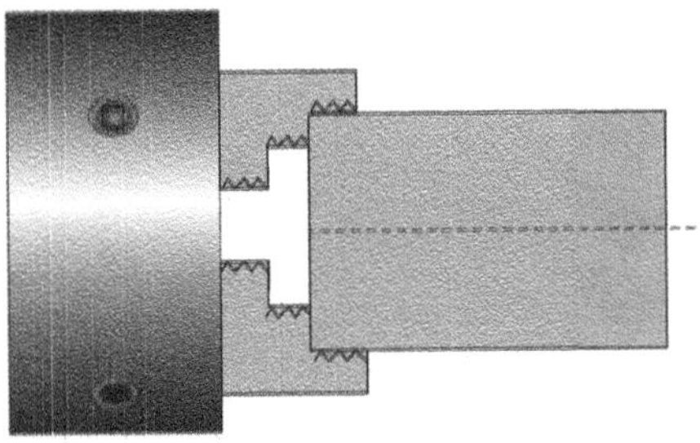 Three-jaw chuck with axial support. Raw surfaces.	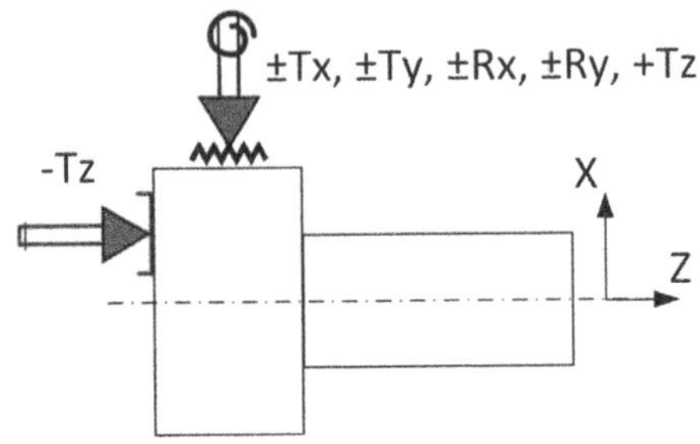
Example 2: Between centers, carrier and steady rest	
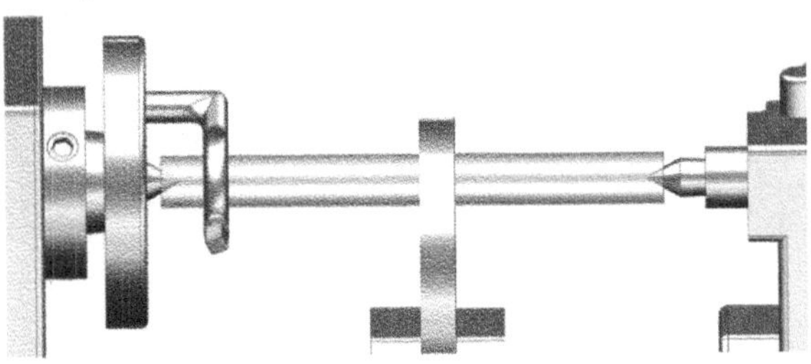 Catch plate, live center, dead center, carrier, and steady rest. Raw surfaces	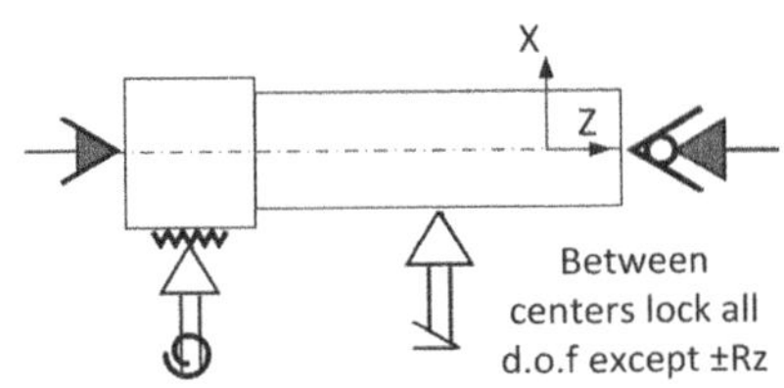
Example 3: 3-2-1 locating scheme	
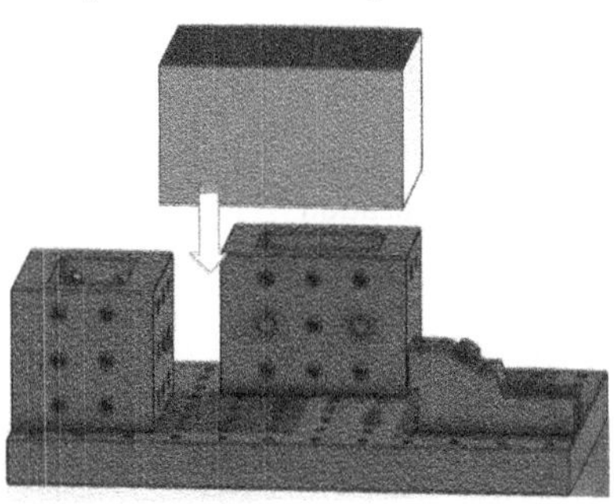 Six flat rest pads on tooling blocks, a heavy-duty clamping jaw, and all items mounted on a base plate. Datum and clamping surfaces are previously machined	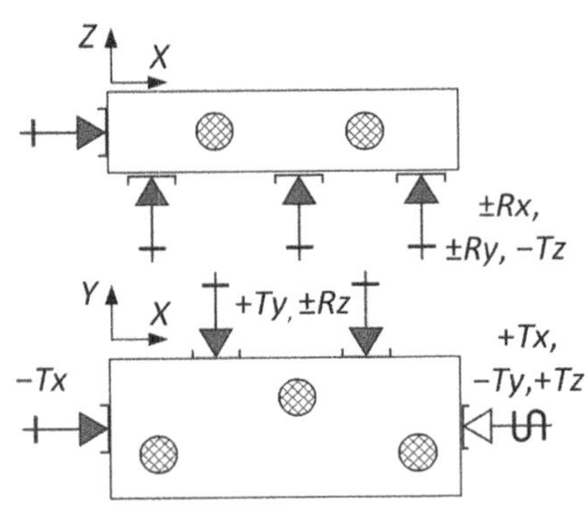

Table 7.10 (Continued)

Physical realization of workholding system	Workholding drawing representation and part movements locked
Example 4: Concentric and radial locating scheme	
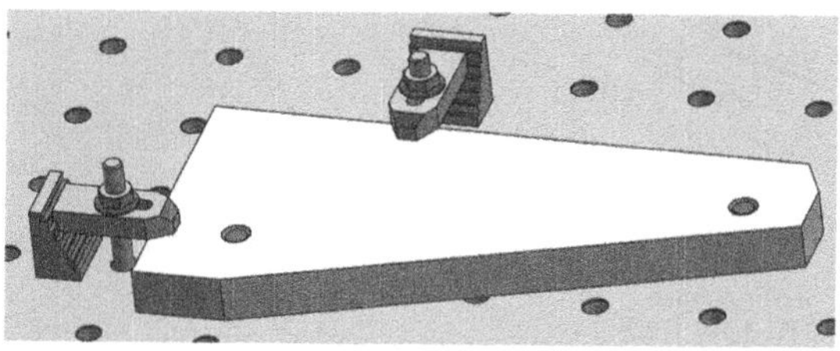 Cylindrical locator, diamond pin, strap clamp, step block and bolt, and modular base	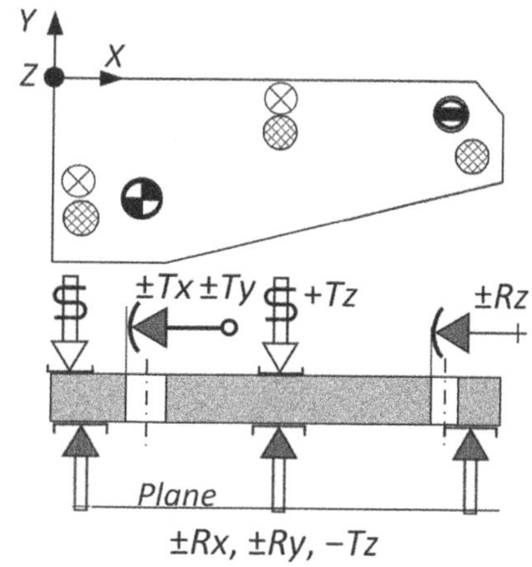

Questions and Exercises

1 Explain what a "3-2-1 fixture scheme" means. Use the concept of degrees of freedom to answer the question.

2 Which are the part movements locked in a prismatic part under a 3-2-1 fixturing scheme?

3 When designing a fixture device, should the main machining force be directed toward the clamps, locators, or supports? Explain the answer.

4 Write two possible workholding devices that grip the workpiece by its interior surface.

5 Two fixturing options are being evaluated to manufacture a part. The first fixturing option is a manual workholding system with an initial setup time of 45 minutes per batch, and the complete purchase of the system costs $650. The second fixturing option is a semi-automated system where the initial setup time is 2 hours per batch and the purchase of the system is around $2,000. The second fixturing option reduces the time of loading, clamping, and unloading the part. The total time per part under the semi-automated system is estimated to be 10 seconds while the manual system requires 75 seconds per part. Calculate the break-even process between both fixturing options. Consider a labor rate of $50/hour and a batch size of 500 units.

6 Complete the name of the elements of the fixture shown in Figure 7.44.

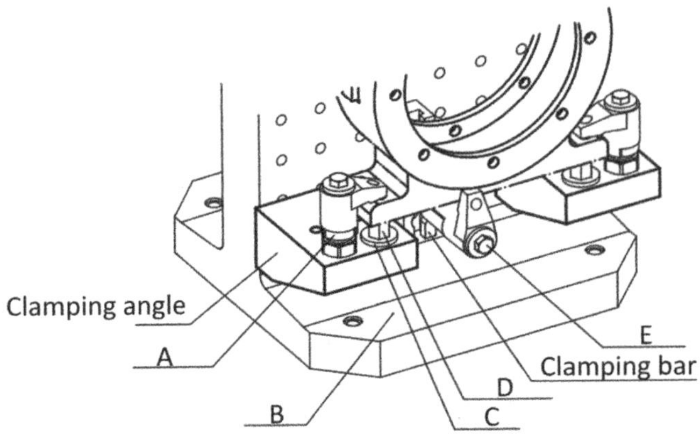

Figure 7.44 Exercise 6. Complete the name of the components. *Source:* Courtesy of Halder, Inc.

7 Complete the name of the elements of the fixture shown in Figure 7.45.

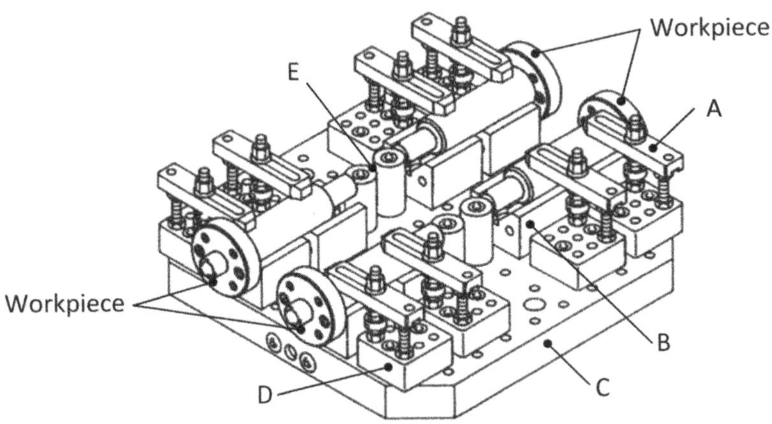

Figure 7.45 Exercise 7. Complete the name of the components. *Source:* Courtesy of CarrLane Manufacturing Co.

8 Complete the name of the elements of the fixture shown in Figure 7.46.

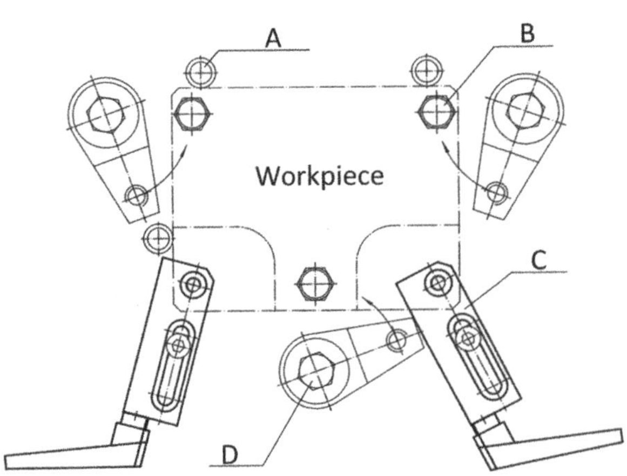

Figure 7.46 Exercise 8. Complete the name of the components. *Source:* Courtesy of Halder, Inc.

9 Write the name of the workholding devices/elements that physically realize the following representations defined by the standard NF E04-013: 2015.

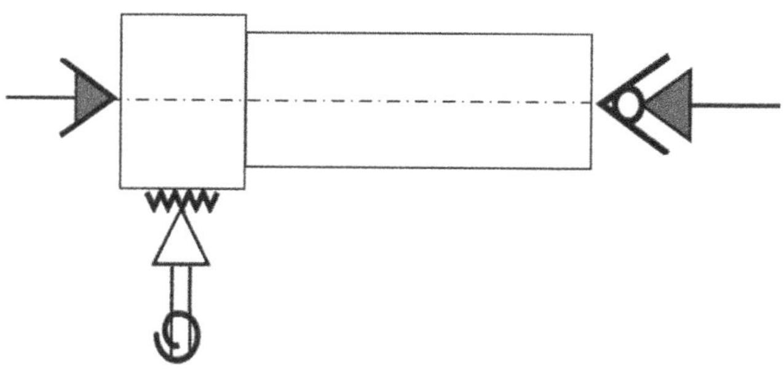

(a)

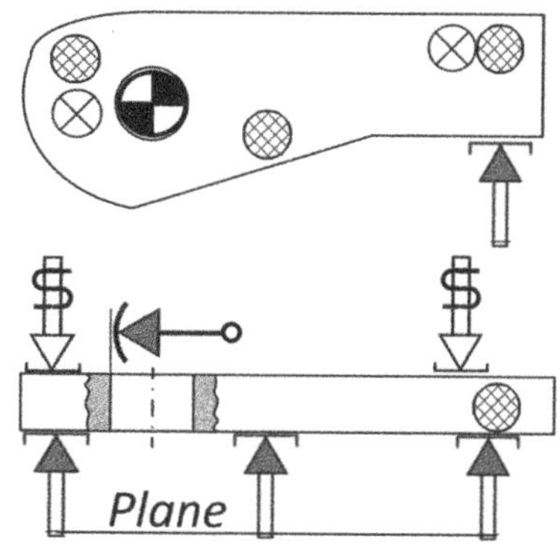

(b)

Write the name of the following devices:

(a)

(b)

(c)

(d)

(e)

(f)

(g)

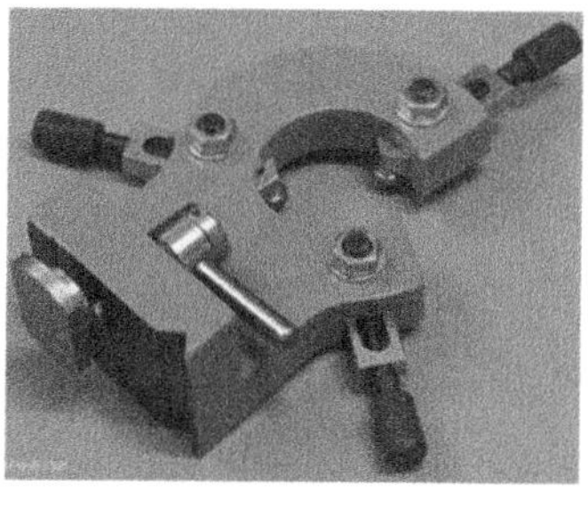

(h)

(i)

References

AFNOR. (2015). *NF E04-013: 2015. Geometrical Product Specification (GPS) – Symbol of Workpiece Holding*. AFNOR.

Carrlane Manufacturing Co. (2023). Jig & fixture design – installation & use information. https://www.carrlane.com (accessed May 2023).

Halder, K.G. (2024). *Standard Parts Catalogue*, vol. 2024. Ascon.

Hoffman, E. (2012). *Jig and Fixture Design*. Cengage Learning.

International Standard Organization. (2013). ISO 898-1:2013: mechanical properties of fasteners made of carbon steel and alloy steel – part 1: bolts, screws and studs with specified property classes – coarse thread and fine pitch thread.

Joshi, P.H. (2003). *Jigs and Fixtures: Design Manual*. McGraw-Hill Professional Publishing.

Nee, A.Y.C., and Tao, Z.J. (2004). *An Advanced Treatise on Fixture Design and Planning*. World Scientific.

Wang, M.Y. (2002). Characterizations of localization accuracy of fixtures. *IEEE Transactions on Robotics and Automation* 18 (6): 976–981.

8

Geometry Analysis and Machining Operations Sequence

8.1 Introduction

After the part drawing has been appropriately examined from a geometric point of view (described in Chapter 2) and primary manufacturing processes have been selected and planned (described in Chapters 3–6), machining process plans related to secondary manufacturing processes can then be defined. Process planning in machining requires the execution of the tasks listed in Figure 8.1, which ends with the definition of an adequate process plan subjected to economical and technical constraints (Zamanillo 1993).

This chapter covers the first four steps involved in machining process planning: (i) geometrical analysis of the part; (ii) identification of surfaces, machining volumes and operations; (iii) determination of machining processes and operations sequence; and (iv) selection of locating surfaces and workholding devices. A practical example is provided at the end of the chapter to illustrate the methodology proposed to conduct these four machining planning tasks.

The rest of the steps to complete the machining process-planning methodology are covered in this book in the subsequent chapters. Chapter 9 presents the selection of machine tools and cutting tools, and the selection of proper cutting parameters are explained in detail in Chapter 10. Chapter 11 shows the geometric validation of process plans through tolerance charts. Chapter 12 presents computer numerical control (CNC) programming fundamentals, and finally, Chapter 14 shows the details for machining time and cost estimations.

8.2 Definition of Terms

The following terms are used throughout this chapter:

- Job. A job in machining is the collection of all the operations that must be carried out on a single machine tool. A job may be a machining center, a CNC lathe, or any other machine tool.
- Subjob or setup. A subjob is the term used to describe a set of operations that are conducted under the same fixturing. A machine's fixturing can be changed for a second set of operations and both subjobs can be produced under the same machine, that is, the same job.
- Machining operation. The term machining operation refers to a simple task carried out with a single tool over the course of multiple cuts or passes. An operation, for instance, is the cylindrical

Manufacturing Process Planning: A Practical Approach for Mechanical Engineering, First Edition.
José V. Abellán-Nebot, Carlos Vila Pastor, and Héctor R. Siller.

Companion website: www.wiley.com/go/Manufacturing_Process_Planning_1e

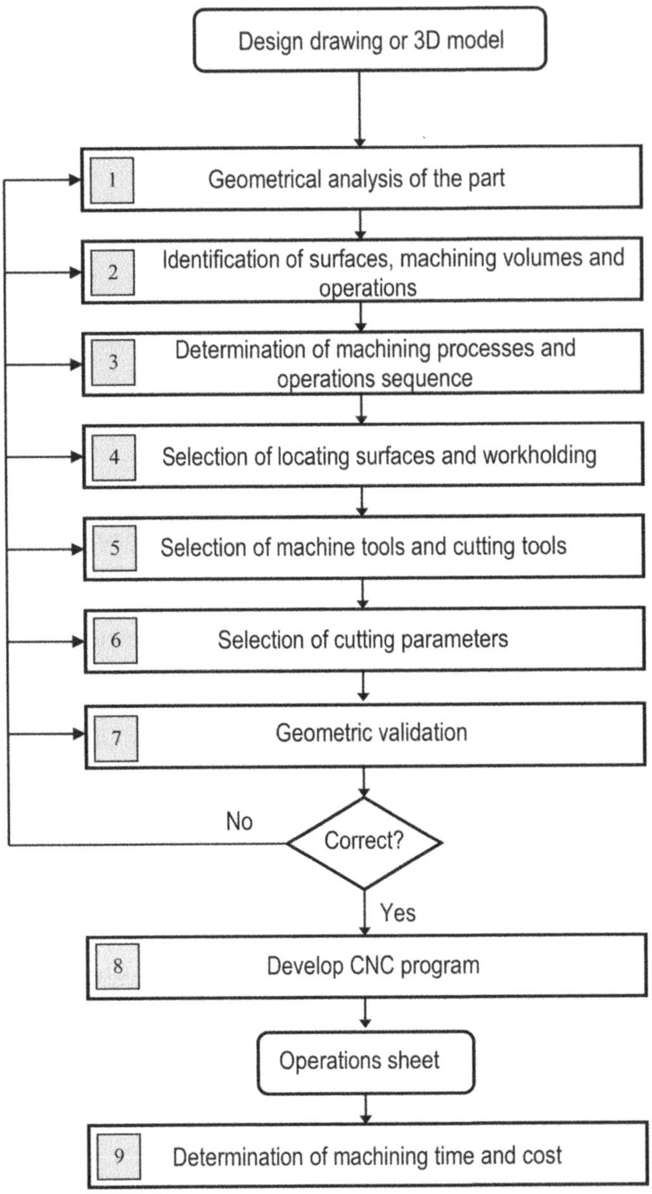

Figure 8.1 Process planning methodology for machining operations.

turning of a workpiece on a lathe. Figure 8.2 illustrates the concepts of operations, subjobs, and jobs in machining processes.

- Tool approach direction (TAD). The TAD of a machining feature refers to the tool direction for machining the feature without being obstructed by any part of the workpiece. A machining feature can have more than one TAD, and those features with the same TAD are candidates to be produced under the same setup/subjob. For machining a prismatic part, six possible TAD cases can be found: $+X$, $-X$, $+Y$, $-Y$, $+Z$, $-Z$. In the case of a turning part, four possible cases are available: $+X$, $-X$, $+Z$, $-Z$. Figure 8.3a, b shows two examples and their corresponding TADs for each feature to be machined.

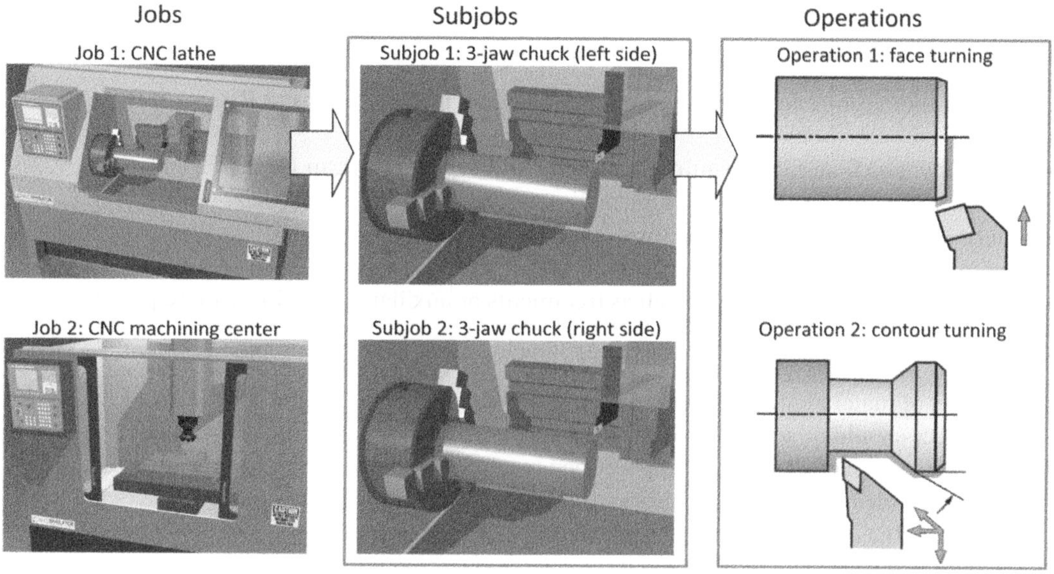

Figure 8.2 Definition of job, subjob, and operation.

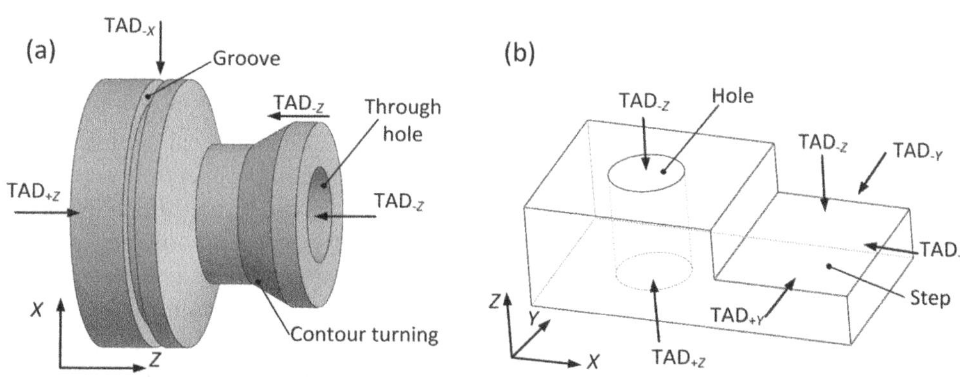

Figure 8.3 Tool approach directions (TADs) for some machined features in a revolution part (a) and prismatic part (b).

- Datum surface. A surface of a workpiece that is used as a datum (reference) for controlling the position and orientation of another surface produced by a manufacturing process.
- Locating surface. A surface of a workpiece that is used to locate the workpiece relative to a machine tool. This surface is also named as setup datum or locating datum. A datum surface may or may not be a locating surface.

8.3 Geometry Analysis

The first step in machining process planning is to analyze the part design to evaluate the manufacturability of the part and the fulfillment of all specifications. Note that the part may come from a primary manufacturing process such as casting or bulk forming processes, and some specifications may have already been reached at this stage (hence, there is no need for machining). The analysis of the geometry should consider the following actions:

- Check if dimensions and tolerances can be uniquely interpretable and if some information is missing.
- Determine which surfaces require machining and which ones can be kept from the previous manufacturing process (casting, preform dimensions, etc.).
- Determine if there are available processes to reach dimensions and tolerances defined on drawings.
- Analyze if additional processes, such as treatments or auxiliary operations, are required.

To elaborate a correct geometry analysis, the following methodology is proposed (Zamanillo 1993):

1) Analysis of material and initial workpiece geometry. Dimensions and tolerances of the initial workpiece should be analyzed to check if some specifications are already obtained.
2) Numbering surfaces. To distinguish between machined and raw surfaces that may be not necessarily machined, machined surfaces are numbered as 1, 2, 3, etc., while raw surfaces are numbered with the prefix "*R*" (from "raw surface") as *R*1, *R*2, *R*3, etc.
3) Analysis of surface specifications:
   - Dimensional specifications. Dimensions of size such as pin or hole diameters, width of a slot, etc., are intrinsic specifications. When geometrical product specifications (GPS) are not applied, ± dimensional specifications are commonly used to relate one geometrical feature with another one, and ambiguity issues may arise (refer to Chapter 2). These dimensional specifications are, in fact, extrinsic specifications that would have been specified using geometrical tolerances such as profile or position tolerances if GPS had been applied. It is important to distinguish between intrinsic and extrinsic dimensional/geometrical specifications since the extrinsic ones will have a significant influence on the machining operation sequence.
   - Geometrical tolerances. Two types of geometrical tolerances should be identified: intrinsic specifications and extrinsic specifications. Intrinsic specifications depend only on the geometrical feature itself, and thus, there is no datum surface required. These are form specifications such as straightness, flatness, roundness, cylindricity, etc. On the contrary, extrinsic specifications require a datum surface. Some of these geometric specifications are parallelism, perpendicularity, run-out, coaxiality, etc. Extrinsic specifications should be clearly identified at this stage since they have a direct impact on setup definitions and precedence constraints.
   - Surface roughness specifications. Surface roughness values should be identified to define if finishing and high-quality operations are required.
4) Analysis of other specifications. Other specifications should be also identified such as heat treatments, chemical treatments, and special surface roughness requirements.

8.4 Machining Volumes

8.4.1 Determination of Machining Volumes, Features, and Tool Approach Directions (TADs)

Material volumes to be removed can be identified by knowing the workpiece dimensions previous to machining and the dimensions of the final machined part. The total material to be removed can be divided into subvolumes, which are those volumes that are potentially eliminated with different operations. These subvolumes are related to machining features that can be identified in the final part facilitating the subsequent link to machining operations. For each machining subvolume, all possible TADs are also identified.

Common machining features that are found in prismatic (milling subvolumes) and revolution parts (turning subvolumes) are defined as follows (Ma et al. 2020; Mokhtar and Xu 2011):

- Turning:
  - One-face feature:
    - Face
    - Cylinder
    - Through hole
    - Taper (internal and external)
    - Chamfer and fillets (internal and external)
  - Two-face feature:
    - Taper step, arc step, rectangular step (internal and external)
    - Blind hole
    - Groove (internal and external)
  - Three-face feature:
    - Groove (internal and external)
  - More than three-face feature:
    - Profile (internal and external)
    - Threading (internal and external)
- Milling:
  - One-face feature:
    - Planar face
    - Through hole
    - Chamfer and fillets
  - Two-face feature:
    - Step
    - Boss
    - Blind hole
  - Three-face feature:
    - Slot
  - More than three-face feature:
    - Open pocket
    - Pocket
    - Profile
    - Threading (internal and external)

Figures 8.4 and 8.5 show some examples of these machining features and associated subvolumes for both milling and turning features.

8.4.2 Linking Machining Features and Potential Machining Operations

The possible machining operations that could be applied to obtain each machining feature/subvolume are listed. At this point, only operations compatibility with the volume to be removed are taken into account and tolerances or surface roughness specifications are not considered. These potential operations will generate different process plans to be evaluated. Table 8.1 shows common machining features and potential machining operations applied. Some examples of machining operations in milling and turning are also illustrated in Figure 8.6.

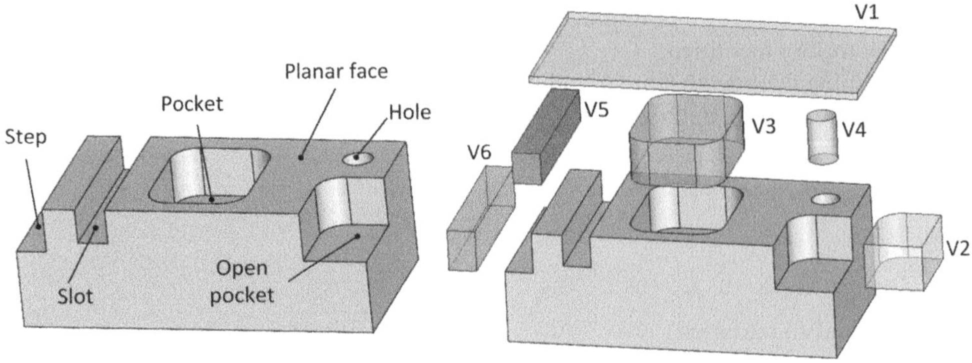

Figure 8.4 Example of machining features and machining volumes in milling operations.

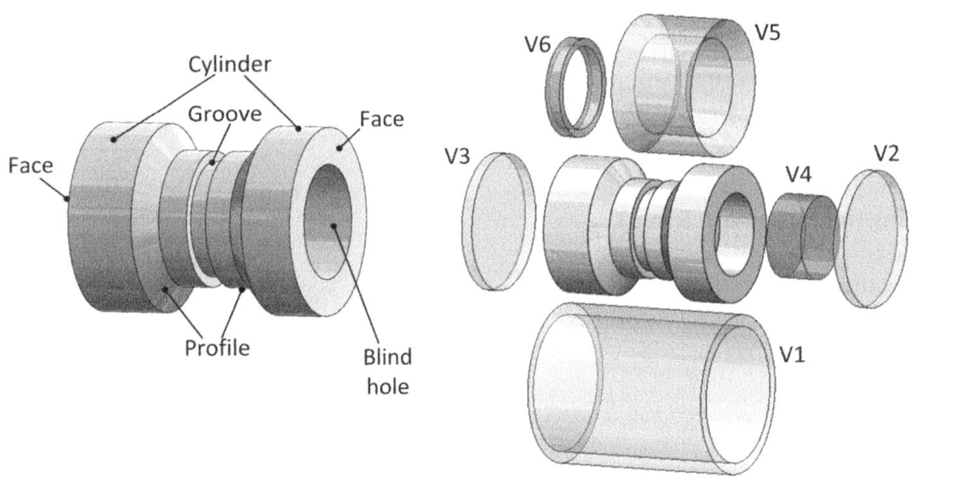

Figure 8.5 Example of machining features and machining volumes in turning operations.

Common turning and milling operations are shown in Figure 8.6 for different turning and milling features.

It should be noted that machining operations may be also classified as roughing, semi-finishing, and finishing operations. When a large volume of material should be removed, a first roughing operation using high depth of cuts and feed rates increases the material removal rate reducing machining costs. This operation keeps some material on the workpiece to be removed through a semi-finishing or finishing operation to reach surface roughness and dimensional tolerances. Thus, the machining volume to be removed and the final quality of the part should be taken into account to determine if different roughing and finishing operations are needed. As a general view, the following guidelines are given:

- Roughing. Operation that gets close to the final shape of the part, removing the maximum quantity of material at high depth of cuts and feed rates. In some situations, after roughing operations other operations may be not needed if geometric specifications are already fulfilled. Common values obtained in this type of machining operations are:

  Surface roughness $12.5\,\mu m \geq R_a \geq 5\,\mu m$
  Dimensional tolerances $12 \leq IT \leq 16$

Table 8.1 Common machining features and corresponding machining operations.

Machining features: turning	Potential machining operations
Face	Face turning
Cylinder	Straight turning
Rectangular step (external)	Straight and face turning
Rectangular step (internal)	Boring (internal straight and face turning)
Taper and arc step (external)	Straight and face turning, contour turning
Taper and arc step (internal)	Boring (internal straight and face turning, internal contour turning)
Through holes, blind holes	Drilling, boring, reaming
Taper, chamfers, fillets, profiles (external)	Contour turning
Taper, chamfers, fillets, profiles (internal)	Boring
Groove	Grooving
Threads	Threading
Machining features: milling	**Potential machining operations**
Planar face	Face milling, plain milling
Through hole, blind hole	Drilling, boring, reaming, end milling
Step, boss, slot, pocket, open pocket	Shoulder milling, end milling, slot milling
Profile	Profile milling
Threads	Threading, tapping
Chamfer, fillets	Angle milling, end milling

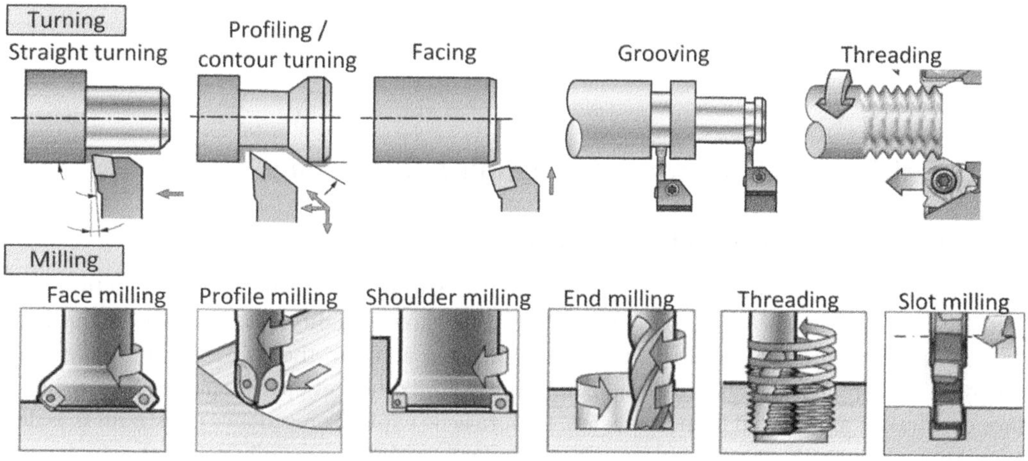

Figure 8.6 Examples of milling and turning operations.

- Semi-finishing. Operation that removes machining allowances from roughing operations and keeps a constant allowance for a later finishing operation. These operations are commonly applied in mold-making processes. Common values obtained in this type of machining operations are:

Surface roughness	$3.2\,\mu m \geq R_a \geq 2.5\,\mu m$
Dimensional tolerances	$9 \leq IT \leq 11$

Table 8.2 Overview of dimensional and geometrical capabilities of machining processes.

		Dimensional tolerances		Geometrical tolerances (mm)			
Process	Surface roughness (μm)	(mm)	IT	Cylindricity	Flatness/ straightness	Parallelism/ perpendicularity	Concentricity
Drilling	0.4–6	±0.1	13–9	0.05–0.1	0.04	0.1–0.2	0.1
Turning	0.1–12	±0.02	13–7	0.01	0.01	0.01–0.02	0.01
Milling	0.5–20	±0.05	13–8	0.02	0.02	0.02	—
Boring	0.1–2	±0.02	9–6	0.005–0.01	0.005	0.005–0.01	0.005–0.01
Broaching	0.4–6	±0.02	9–6	0.01	0.01	0.01–0.02	—
Grinding	0.05-2	±0.005	6–4	0.004	0.004	0.002–0.005	0.002–0.005
Honing	0.025–0.5	±0.002	5–2	0.002	0.002	0.002	0.002
Lapping	0.01–0.4	±0.002	3–01	0.001	0.001	0.002	0.002

Source: Adapted from Scallan (2003), Chapter 4 and Halevi and Weill (1995), Chapter 6.

- Finishing. Operation that generates the final surface with higher accuracy than previous operations. Common values obtained in this type of machining operations are:

  Surface roughness $3.2\,\mu m \geq R_a \geq 0.4\,\mu m$
  Dimensional tolerances $6 \leq IT \leq 8$

Note that if higher dimensional or surface roughness quality is needed, additional machining processes should be added after conventional machining operations. These operations are grinding, honing, lapping, or polishing and they can achieve values of surface roughness from 0.25 μm to 0.01 μm and dimensional tolerances from IT6 to IT01.

Table 8.2 shows a global view of machining process capabilities to facilitate the machining process selection.

8.5 Definition and Sequence of Machining Operations

8.5.1 Grouping Features According to Tool Approach Direction (TAD)

The main goal of determining the machining sequence is to minimize the number of setups and tool changes, which means lower costs and higher dimensional quality. The first step in sequencing machining operations is grouping the operations according to their TADs to minimize the number of setups required (Liu and Wang 2007). To sort the different operations and setups, precedence constraints are studied in the next step. If there is no explicit precedence constraint to order the operations/setups, those with bigger volumes to be machined are conducted first, which usually means higher cutting forces and heat that may result in deformation issues.

8.5.2 Precedence Constraints

Precedence constraints may be divided into hard constraints or soft constraints (Faheem et al. 1998; Mokhtar and Xu 2011). Hard constraints are precedence relationships that must be followed or an infeasible process plan will result. Soft constraints describe manufacturing preferences that are not necessary to be followed but which would increase the quality of the plan in terms of costs or dimensional quality.

Some precedence constraints that are classified as hard constraints are related to manufacturing interactions. Manufacturing interactions can be fixture–feature, tool–feature, and tool–fixture interaction. Some typical examples of manufacturing interactions are presented below:

- Fixture–feature interaction. Fixture–feature interactions occur when machining a feature may destroy other features that are needed for proper clamping in a subsequent machining operation. Figure 8.7a shows an example of fixture–feature interaction where the slot should be machined before the boss to facilitate fixturing when machining the slot. A similar interaction is shown in Figure 8.7b where the angle cut of the feature prevents holding the part in a subsequent operation.
- Tool–feature interaction. This interaction occurs when machining one feature destroys a condition required by the tool used in a subsequent machining operation. Figure 8.7c shows an example of tool–feature interaction where the drilling operation cannot be conducted correctly due to the chamfer previously machined.
- Tool–fixture interaction. This interaction occurs when the fixture and the tool used in the operation may collide during cutting (Figure 8.7d). In some cases, a proper sequence of operations may facilitate or avoid this type of interaction.

Additionally, to manufacturing interactions, other hard constraints are listed in the following:

- Roughing/finishing operations. Surfaces that are machined using roughing and finishing operations, obviously require roughing operations to be conducted first.
- Drilling/threading. Threaded holes require first drilling operations to conduct later threading or tapping operations.
- Drilling/boring/reaming. To increase the quality of a hole, drilling operations (IT13–IT9) are conducted first, and then boring operations or reaming operations (IT9–IT6) may be applied.

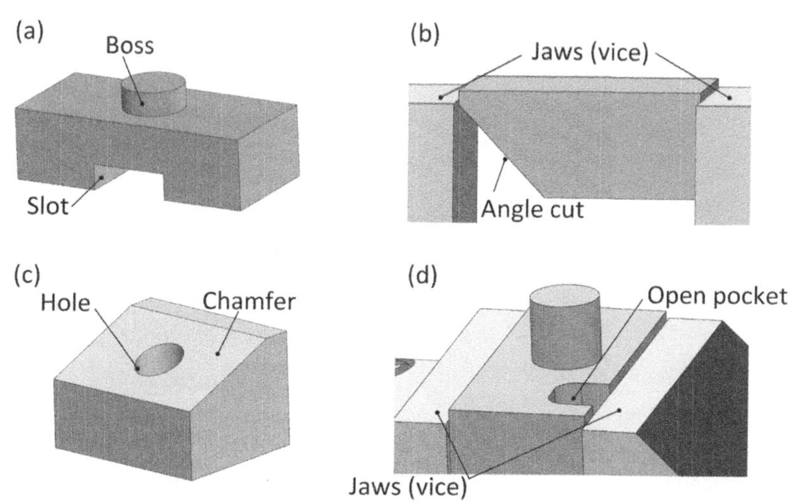

Figure 8.7 Different types of interactions: (a) and (b) fixture–feature interaction; (c) tool–feature interaction; and (d) tool–fixture interaction.

On the contrary, good manufacturing practices provide guidelines for defining machining sequencing that may result in soft precedence constraints. Some typical examples of these recommended precedence constraints classified as soft constraints are shown below (Faheem et al. 1998; Mokhtar and Xu 2011):

- Drilling/counterboring/countersinking. Counterboring and countersinking operations are conducted after drilling operations. If threading or reaming operations are also needed, they are conducted after counterboring/countersinking operations.
- Holes in thin machined walls. When drilling a hole in a thin wall that is produced after a milling/slot operation, drilling is conducted before milling to avoid deformation of the thin wall (Figure 8.8a).
- Transition features (e.g., chamfer, fillets). Machining a transition feature is done after machining the main features that interact with it (Figure 8.8b).
- Interacting features:
  - Nesting/nested features. When one feature is nested into another one, the top feature (nesting feature) is machined before the bottom feature (nested feature). Under this strategy, air-cutting is minimized, and chip removal is facilitated (Figure 8.8c). This is the case when drilling and milling should be sequenced (Figure 8.8d). Additionally, under this strategy burr formation is minimized.
  - Side approach cutting. When two features are interacting and they are not nested/nesting features, the feature with easier tool access (side access instead of plunging) is machined first. In Figure 8.8e, both open pockets with side tool access are machined before the third pocket.
  - Interacting without side access. If two features are interacting and they are not nested/nesting features and there is no side access, the feature with greater depth is machined first (Figure 8.8f).

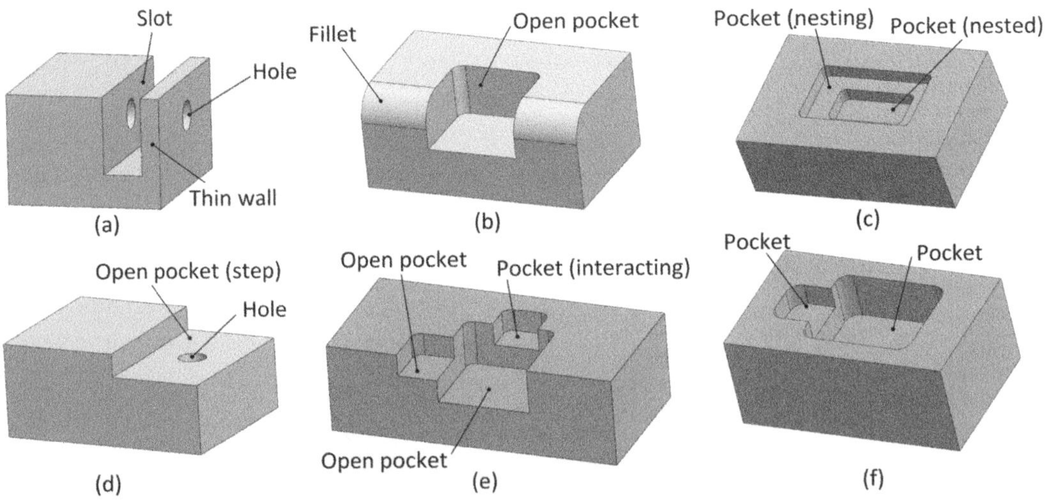

Figure 8.8 Examples of recommended precedence constraints: (a) hole before slot machining to avoid deformation in thin walls; (b) fillets after main feature; (c) nesting features before nested features; (d) milling before drilling; (e) side accessed features first; and (f) features no nested with no side access, greater depth features are machined first.

8.5.3 Feature–Datum Interactions. Critical Extrinsic Relationships

As shown earlier, precedence constraints are extracted from manufacturing interactions such as fixture–feature, tool–feature, and tool–fixture interactions. Additionally to these interactions, feature–datum interactions are also relevant for defining a proper machining process sequence.

When analyzing a feature–datum interaction where two features present a dimensional or geometrical tolerance relation, three main options can be considered (Bansal et al. 2008): machine both features in the same setup; use one feature as the setup datum and machine the other; or use an intermediate setup datum to machine the two features in different setups.

Employing the same setup ensures that the differences in inaccuracy between various surfaces are solely caused by the limits of the machine-tool (such as the straightness of the axis, thermal distortions, etc.). Any error due to the workholding system (e.g., a calibration error of the fixture coordinate system) is added to all machined surfaces and, thus, the error between them is canceled. Thus, machining both features under the same setup presents the highest possible quality in a machining process. Feature–datum interactions where a tight tolerance is needed will require this setup. These cases are called critical extrinsic relationships.

An example is shown in Figure 8.9 to illustrate this critical extrinsic relationship. According to the drawing, surfaces B and C must be perpendicular with a tolerance of Ø0.02 mm, which is a tight machining tolerance. Both surfaces should be machined within the same subjob to ensure high accuracy. If the perpendicular tolerance were Ø 0.2 mm, then, there is no need to machine both surfaces at the same time, and thus, any sequence would be feasible. Additionally, since surfaces B, C, and D have the same TAD, all three are machined in the same subjob.

If machining both features under the same setup is not possible, the second-best alternative in terms of machining accuracy is using one feature as setup datum (the one indicated in the datum reference frame or the larger feature among them), which should have been previously machined, and machine the other one (Liu and Huang 2003). A common example of this situation is machining a hole where a position tolerance applies. In this tolerance type, the hole should be positioned with respect to some references that define the datum surfaces. These datum surfaces should be machined in a previous subjob to be used later as locating surfaces when machining the holes. If a tight tolerance is required and machining both features in the same setup is not possible, this option should be applied and a hard constraint is imposed.

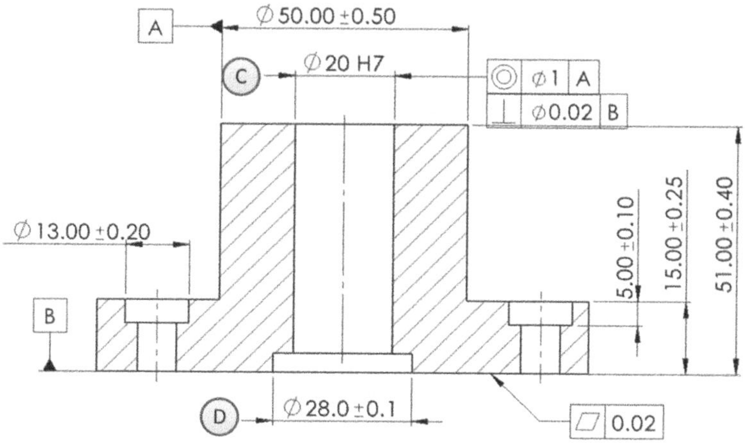

Figure 8.9 Example of a drawing of a mechanical part.

Table 8.3 Estimated values of factors contributing to inaccuracy in machining.

Workholding inaccuracies			
Turning	**Estimated error (mm)**	**Milling**	**Estimated error (mm)**
Three-jaw chuck (hard jaws)	0.1	Three-jaw chuck	0.05–0.1
Three-jaw chuck (soft jaws)	0.05	Plain vice	0.1
Four-jaw independent chuck	0.05	3-2-1 Scheme	0.04
Chuck collet	0.005	Concentric and radial locator	0.08
Between centers	0.04	Mounting with pressure on surface/support	0–0.05
Between plate and center	0.08		
Machining inaccuracies			
Cutting tool related	**Estimated error (mm)**	**Machine tool related**	**Estimated error (mm)**
Tool wear	0.02–0.05	Spindle run-out	0.005
Tool holder repeatability	0.0025–0.025	Machine lead screw	0.005
Tool setting	0.0025–0.025	Machine slide-tracking	0.005

Source: Adapted from Halevi and Weill (1995).

Finally, if the datum–feature interaction does not require a tight tolerance, a third option where both features are machined under different setups indifferently may be used. Under this scheme, a tolerance stack-up arises increasing manufacturing variability.

To establish if the dimensional and geometrical relationships between surfaces can be obtained in multiple subjobs or if the same subjob should be applied to ensure specifications, common machining errors and setup errors from different workholding systems are provided in Table 8.3. More details about setup and machining errors can be found in (Halevi and Weill 1995; Murty 2005). Specifications lower than the workholding accuracy means that those surfaces should be machined in the same subjob to avoid workholding setup errors (i.e., they should be considered as critical extrinsic relationships). It should be highlighted that multiple setups for carrying out different machining operations would result in an unfavorable accumulation of tolerances. Chapter 11 will show how to estimate the tolerance stackup of these setup errors to validate the process plan according to the design tolerances.

8.5.4 Machining Sequence Diagram

Once the geometry of the part has been analyzed, the machining operations have been grouped according to their TADs and their precedence constraints have been identified, a diagram to gather all relevant information is drawn as follows:

- Step 1. Add all surfaces/operations identified. The first step in drawing the diagram is to introduce all surfaces/operations as columns. Since these columns refer to the machining operations

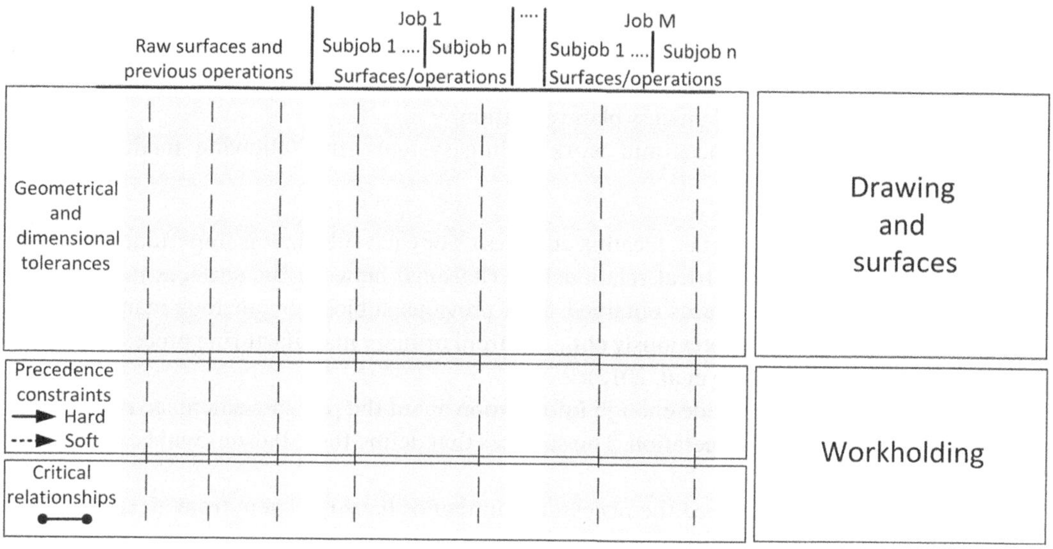

Figure 8.10 Machining sequence diagram template.

that produce the surfaces, those surfaces produced with the same machining operation are introduced in the same column. These surfaces are grouped according to their TADs.

- Step 2. Include extrinsic relationships. In this diagram, all extrinsic relationships between surfaces/operations are included. For instance, positional tolerances, perpendicularity and parallelism tolerances, dimensional tolerances between features, etc.
- Step 3. Add hard and soft precedence constraints. At the bottom of the diagram, each precedence constraint identified is graphically indicated using an arrow that indicates the order between operations. Soft constraints are added as dashed arrows to indicate that this precedence is not mandatory. For the case of hard precedence constraints, solid lines are used.
- Step 4. Identify critical extrinsic relationships. Those surfaces with extrinsic tolerances that are too tight may require to be machined under the same setup. Highlighting this type of relationship may help to define properly the machining sequence. The identified critical extrinsic relationships are illustrated in the diagram by a black line below the technological precedence.

Once the diagram is built, all the information presented is analyzed and the columns are exchanged to comply with the precedence constraints, the TADs, and the critical extrinsic relationships. For instance, in the case of hard precedence constraints, all arrows should be pointing to the right to illustrate that all precedence constraints are correct. Then, the diagram is used to identify jobs, subjobs, and the final machining operations sequence. As a general rule, the number of jobs and subjobs should always be minimized and the higher machining volumes should be machined first. Figure 8.10 overviews the parts of the diagram to be elaborated for a correct machining sequence analysis.

8.6 Definition of Locating Surfaces and Workholding

Once the jobs, subjobs, and machining operations sequence has been defined, it is necessary to determine the locating surfaces and the corresponding workholding devices for each subjob. Since not all surfaces may be available for locating, a tolerance stackup may arise. This is a critical issue since the process plan candidate may result in an unfeasible plan due to an excessive tolerance accumulation. In that case, a step backward should be taken to analyze other process plan candidates that minimize this tolerance stackup issue. Note that Chapter 11 presents in detail tolerance stackup calculations and process plan validation.

To define the locating surfaces and workholding systems, the following methodology is presented (Zamanillo 1993):

1) Determine the setup datum (i.e., locating surfaces). For each subjob, it is important to identify the dimensional and geometrical relationships (liaisons) between the surfaces machined in the current subjob and surfaces obtained from previous subjobs or previous manufacturing processes (i.e., raw surfaces previously obtained from primary manufacturing processes). Three cases may be found (Romero et al. 2012):
   a) The existing liaisons provide enough information about the part movements to be locked to conduct the machining operation. The surfaces that define these liaisons will be considered as locating surfaces.
   b) The existing liaisons exceed the necessary number of liaisons. Then, from all the surfaces that define the liaisons, the ones that are more critical and can guarantee the isostatic location of the workpiece are defined as locating surfaces. If there is no possible way to ensure all liaisons due to tolerance stackup issues, the subjob is divided into different subjobs.
   c) The existing liaisons between the current subjob and previous ones are not enough to define the locating surfaces. In that case, it will be analyzed both previous and subsequent subjobs to establish new liaisons between the current subjob and the previous ones that could define the appropriate datum surfaces. If there are no new liaisons to define the locating surfaces, convenient surfaces to use conventional workholding systems are selected.
2) Determine the clamping surfaces and the fixturing method. Clamping surfaces are usually defined according to the locating surfaces by following the guidelines shown in Chapter 7. Additional supports may be added to avoid deformation during machining. The fixture device or fixturing elements used to locate and clamp the part are defined according to the tooling available. The accuracy of different fixturing methods (Table 8.3) should be considered to ensure the geometrical and dimensional tolerances of the diagram.
3) Represent graphically the fixturing method. Symbols according to international norms (French Standard NF E04-013: 2015 is commonly used [AFNOR 2015]) are used to represent the locating and clamping method.

8.7 Practical Example

The proposed methodology is applied to the case study shown in Figure 8.11. The part is assumed to be cast using a sand-casting process with a machining allowance of 3 mm and a casting variability of ±0.5 mm. After casting, machining operations are required to reach the design tolerances from the technical drawing. The part is made of aluminum AISI SC64D (SAE 326), the general tolerance is ±0.5 mm and the general surface roughness is 1.6 μm to all machined features and 12.5 μm to casting features. It is assumed that the external surfaces of the part that do not require tight tolerances are kept from casting.

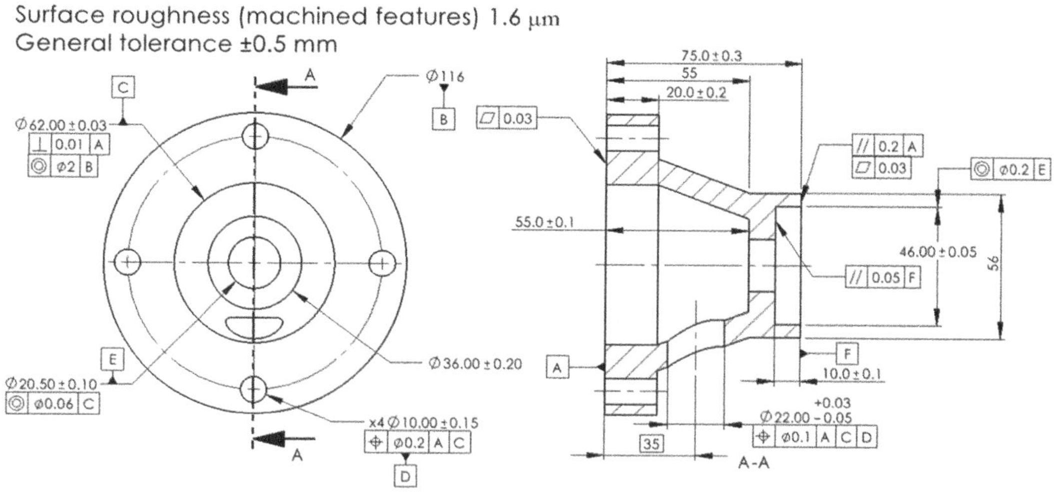

Figure 8.11 Practical example. Part drawing.

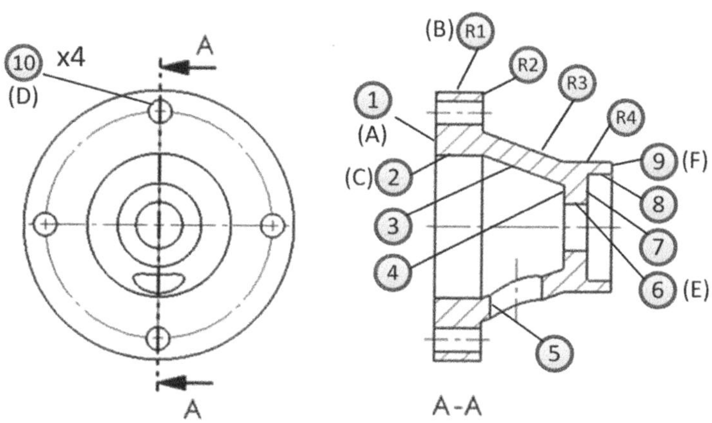

Figure 8.12 Numbering of surfaces. The datum letter used in the drawing is shown in parentheses.

For this case study, the proposed methodology is applied and it is discussed in the following.

a) Geometry analysis

The initial workpiece is a preform from a casting process with 3 mm machining allowance. For this case study, it is considered that surfaces from R1 to R4 (Figure 8.12) are directly obtained from casting and do not require machining.

All surfaces are numerated except those that have been previously numerated as raw surfaces. Figure 8.12 shows the numbering of the surfaces for this case study indicating in parenthesis the datum letter used in the drawing.

All dimensional and geometrical specifications are reviewed and included in Table 8.4.

Table 8.4 Analysis of surface specifications.

		Specification					
		Intrinsic			Extrinsic		
N° and type of surface[a]		**Dimension (mm)**	**Geometrical (mm)**	**Surface roughness (μm)**	**Between machined surfaces (mm)**	**In relation to raw surfaces (mm)**	**Geometrical (mm)**
R1 (B)	C	Ø116±0.5	—	$R_a = 12.5$	—	—	—
R2	P	—	—	$R_a = 12.5$	—	20±0.2 (R2–1)	—
R3	C	—	—	$R_a = 12.5$	—	55±0.5 (R3–1)	—
R4	C	Ø56±0.5	—	$R_a = 12.5$	—	—	—
1 (A)	P	—	⏥ | 0.03	$R_a = 1.6$	75±0.3 (1–9) 55±0.1 (1–4)	—	—
2 (C)	C	Ø62±0.03	—	$R_a = 1.6$	—	—	◎ | Ø 2 | B ⊥ | 0.01 | A
3	C	Ø36±0.2	—	$R_a = 1.6$	—	—	—
4	P	—	—	$R_a = 1.6$	—	—	—
5	C	$Ø22^{+0.03}_{-0.05}$	—	$R_a = 1.6$	—	—	⌖ | Ø 0.1 | A | C | D
6 (E)	C	Ø20.5±0.1	—	$R_a = 1.6$	—	—	◎ | Ø 0.06 | C
7	P	—	—	$R_a = 1.6$	10±0.1 (7–9)	—	∥ | 0.05 | F
8	C	Ø46±0.05	—	$R_a = 1.6$	—	—	◎ | Ø 0.2 | E
9 (F)	P	—	⏥ | 0.03	$R_a = 1.6$	—	—	∥ | 0.2 | A
10 (D)	C	Ø10±0.15	—	$R_a = 1.6$	—	—	⌖ | Ø 0.2 | A | C

C, cylindrical surface; H, helicoidal; P, planar surface.
Other specifications: none.
Observations: All surfaces with a surface roughness of 1.6 μm require finishing operations.
[a] The datum letter used in the drawing is shown in parentheses.

b) Machining volumes, operations and TADs

After the analysis of part geometry, the machining volumes shown in Figure 8.13 can be identified. As it can be seen, surfaces 2, 3, and 4 or surfaces 7 and 8 can be considered as one surface since they are generated after one machining operation. For each identified machining volume, its possible TADs are studied. The defined subvolumes and TADs are included in Table 8.5.

Since the part is a revolution part, most of the features are machined in a lathe. After turning operations, the part is finished in a machining center to conduct additional operations that cannot be conducted in a lathe.

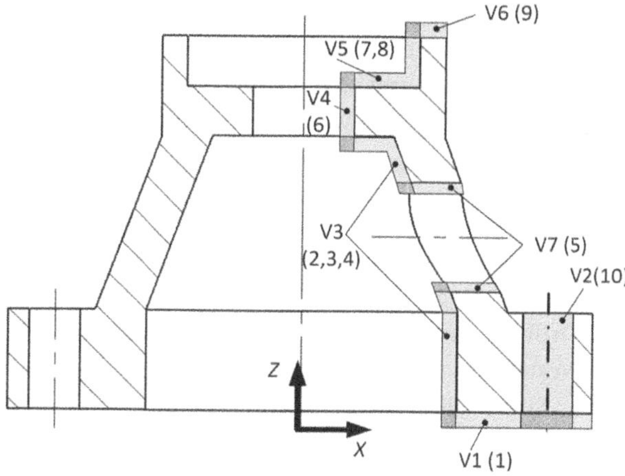

Figure 8.13 Machining volumes and surfaces generated.

Table 8.5 Machining operations and TADs for identified machining volumes

# surface	# volume	Volume/feature type	# operation	Machining operation	TAD
1	V1	Face	1f	Face turning	$+X$, $-X$
2–3–4	V3	Profile	2–3–4f	Boring	$+Z$
5	V7	Through hole	5d + 5b	Drilling, boring/reaming[a]	$-X$ (not lathe)
6	V4	Through hole	4f	Boring	$+Z$, $-Z$
7–8	V5	Internal rectangular step	7–8f	Boring	$-Z$
9	V6	Face	9f	Face turning	$+X$, $-X$
10	V2	Through hole	10d	Drilling	$+Z$, $-Z$ (not lathe)

[a] To reach dimensional and geometrical specifications, a boring or reaming operation after drilling is needed.

Table 8.5 shows the machining operations and the corresponding TADs to machine all surfaces/volumes. Note that the machining allowance is 3 mm and, in this example, it is considered that there is no need to add roughing and finishing operations, only finishing operations to reach 1.6 μm is enough. Thus, the operations are numbered by the surface produced but with the letter "f" to indicate finishing operations. If roughing and finishing should be distinguished, "r" and "f" letters would be used. In the case of surface 5, both drilling and boring operations are needed to reach the geometrical specifications ($\varnothing 22^{+0.05}_{-0.03}$ cannot be reached only with drilling operations), then, 5d and 5b operations are included for drilling and boring, respectively.

c) Sequence of machining operations

Analyzing the machining operations and the different TADs, the following groups of operations can be identified:

- Group 1: surfaces 7 and 8, and surface 9. TAD $+X$, $-X$ and $-Z$.
- Group 2: surface 1 and surfaces 2–4. TAD $+X$, $-X$ and $+Z$.
- Group 3: surface 5. TAD $-X$ (not lathe).
- Group 4: surface 10. TAD $+Z$ and $-Z$ (not lathe).

Surface 6 can be included in group 1 or 2 since its TAD is $+Z$ and $-Z$.

As a result, two different jobs are needed, a CNC lathe for groups 1 and 2, and a machining center for groups 3 and 4, and a total of four different setups or subjobs.

Furthermore, the following hard and soft precedence constraints are identified:

- Hard constraints:
  - Surface 5 requires drilling and boring/reaming operation to reach dimensional and geometrical tolerances. Boring or reaming operations should be conducted after drilling.
- Soft constraints:
  - Drilling versus milling: surface 1 before surface 10.
  - Nesting prior to nested features:
    - Surface 1 before surfaces 2–4.
    - Surface 9 before surfaces 7 and 8.

Analyzing the feature–datum interactions, the following cases are identified:

- Case 1. Both features are machined in the same setup due to too tight a tolerance relationship (critical extrinsic relationships). In this case, a critical extrinsic relationship is highlighted in the diagram:
  - Perpendicularity between surfaces 1 and 2 with a tolerance value of 0.01 mm.
  - Parallelism between surfaces 7 and 9 with a tolerance value of 0.05 mm.
  - Coaxiality between surfaces 2 and 6 with a tolerance value of 0.06 mm.
- Case 2. One feature (usually the one indicated in the datum reference frame) is machined first and used as locating datum for the second feature:
  - Position tolerance between surface 5 and the datums 1, 2, and 10 (primary, secondary, and tertiary) with a tolerance value of 0.1 mm. Surface 5 should be machined after surfaces 1, 2, and 10, and a hard precedence constraint is drawn.
  - The dimensional tolerance of 55 ± 0.1 between surfaces 1 and 4 may require attention. Surface 1 should be machined first or both features should be machined in the same setup. Since surfaces 1 and 4 are grouped in the same setup, no constraint is added.
- Case 3. Both features are machined in different setups. In this case, no precedence constraints are required:
  - Position tolerance between surface 10 and the datums surface 1 and 2 (primary, secondary) with a tolerance value of 0.2 mm.
  - Parallelism between surfaces 9 and 1 with a tolerance value of 0.2 mm.
  - Coaxiality between surfaces 8 and 6 with a tolerance value of 0.2 mm.
  - Dimensional tolerances that present a high tolerance value: 20 ± 0.2, 55 ± 0.5, and 75 ± 0.3.

At this point, the machining sequence diagram is drawn including all previous information, as shown in Figure 8.14. Note that in the first columns, the raw surfaces are indicated (R1, R2, R3, and R4) together with any other surface that may be of interest, for instance, surface R9, which refers to surface 9 prior to machining, that is, in its raw state.

Given the machining sequence diagram, the columns are exchanged to analyze possible alternatives ensuring that hard precedence constraints are not violated and critical relationships are kept in the same setup. For this case study, the order of the jobs is defined by the hard constraints. In relation to the subjobs order, at the machining center job the order is also defined by

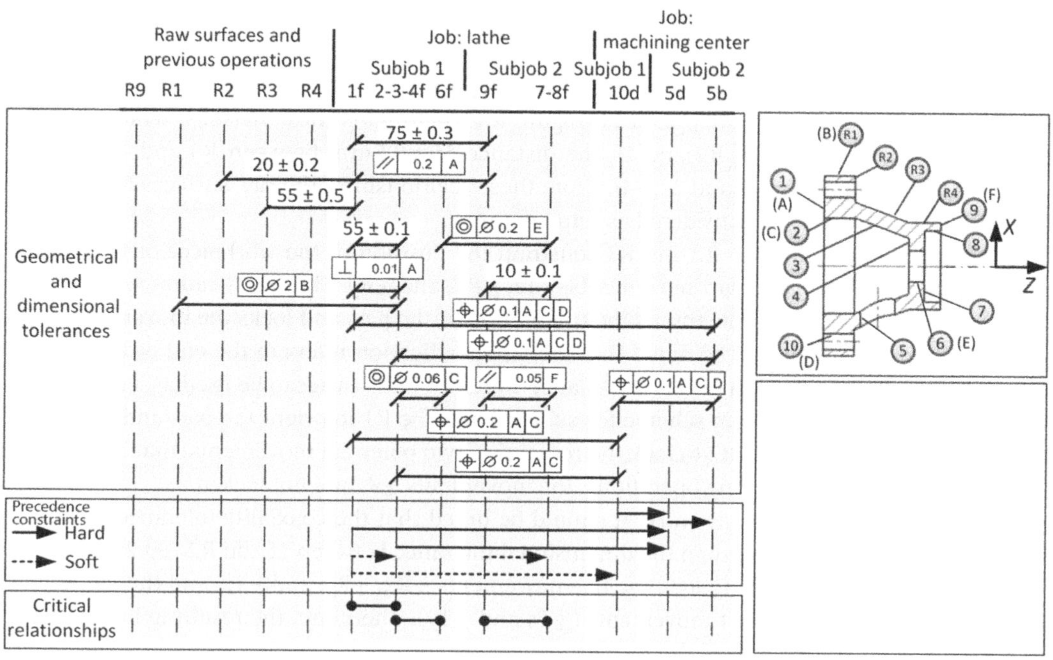

Figure 8.14 Machining sequence diagram of the case study.

the hard constraints. However, at the lathe job, it is possible to exchange the order of both subjobs without violating any hard constraints. Thus, two alternatives in the sequence of subjobs are found. However, it must be remembered that, as a general rule, if there are no constraints, jobs and subjobs with higher material removal are recommended to be conducted first. Furthermore, looking at the diagram it can be seen that a parallelism between surface 9 and surface 1 of 0.2 mm should be ensured, and surface 1 (datum A in the reference frame) is larger than surface 9 and it would be preferrable to be machined first. As a conclusion, the diagram shown in Figure 8.14 gives the correct machining sequence. Any other exchange of columns to define an alternative machining sequence is not possible without violating precedence constraints or critical relationships.

Thus, the sequence of operations is as follows (refer to Figure 8.14):

- Job 1. Lathe
  - Subjob 1: machining operations 1f, 2–3–4f and 6f.
  - Subjob 2: machining operations 9f, 7–8f.
- Job 2. Machining center
  - Subjob 1: machining operation 10f.
  - Subjob 2: machining operation 5d and 5b.

Note that, if the coaxiality of surface 6, Ø0.06 mm, were higher than the expected setup error (e.g., higher than 0.05–0.1 mm), then surface 6 would not be considered critical surface with surface 2, and it would be preferred to be machined in subjob 2 since the boring tool would require less overhang.

d) Locating datum and workholding

Subjob 1 in lathe

The first step is to determine the setup datum. For this purpose, the liaisons involved in this subjob and the precedence surfaces are analyzed. Figure 8.15 highlights these liaisons, which are: the distance 20±0.2 mm between R2 and 1f, the distance 55±0.5 mm between R3 and 1f, and the coaxiality Ø2 mm between R1 and 2–3–4f. From these relationships, the raw surfaces R1, R2, and R3 should be analyzed to be considered as setup datums.

The specifications related to R2 and R3 constrain the position of the workpiece on the *Z*-axis, and thus, they lock the same part movements, −Tz, ±Rx, and ±Ry. The specification related to R1 is a coaxiality tolerance and constraints the orientation of the part and locks the movements ±Tx, ±Ty, ±Rx, and ±Ry. It should be noted that the R3 specification refers to the end of the conical shape, and the R2 specification refers to a plane, therefore R2 is easier to be used as a datum surface. Hence, the possible datum scheme for subjob 1 may be R1 to orient the part and lock eight possible part movements and R2 to lock the remaining two potential movements, that is, the ±Tz. Note that since these are turning operations, the movements ±Rz are not locked.

Although this alternative may work, it should be noted that the coaxiality tolerance specification, Ø2 mm, is less restrictive than the dimensional tolerance between 1f and R2, ±0.2 mm. Thus, a better approach is to use R2 to lock the five movements ±Rx, ±Ry, and −Tz, and the R1 surface to lock the remaining four part movements (±Tx and ±Ty). Therefore, the resulting locating surfaces would be R2 and R1 in this order.

Next, the clamping surfaces and the fixturing device are selected. To facilitate the clamping process, instead of using R2 as the datum surface, the use of surface R9 (surface 9 before machining) may be an option. However, although this datum surface facilitates the clamping process from the right side of the part, the distance controlled during machining would be the distance between surface R9 and surface 1, not the distance between R2 and 1. Therefore, a stackup tolerance arises and this phenomenon should be taken into account if this option is selected. The stackup is shown Figure 8.16 where the resulting distance between surface 1 and surface R2 is obtained as the result of dimension R9-1f minus dimension R9-R2, since the machining operations are being referenced

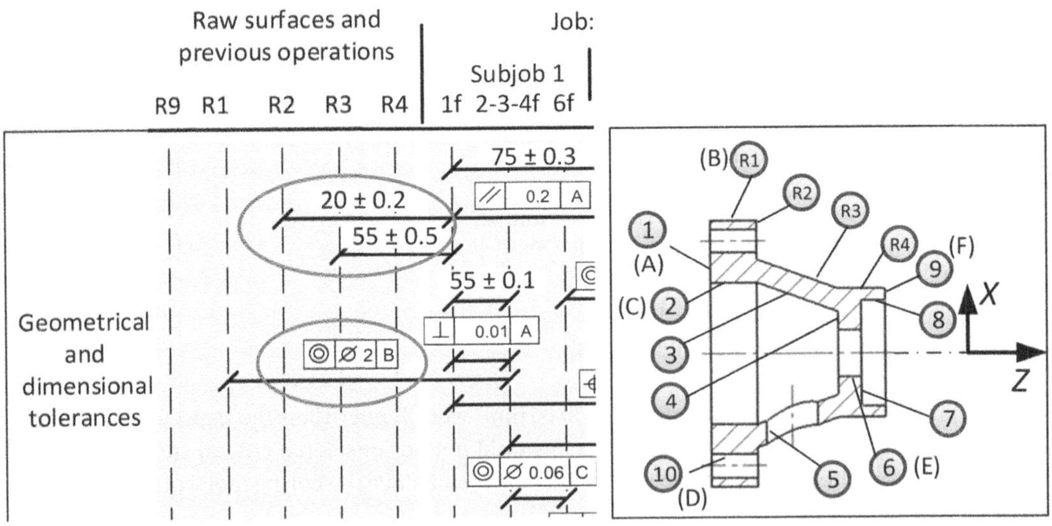

Figure 8.15 Liaisons between currently analyzed subjob and previous subjobs/raw surfaces.

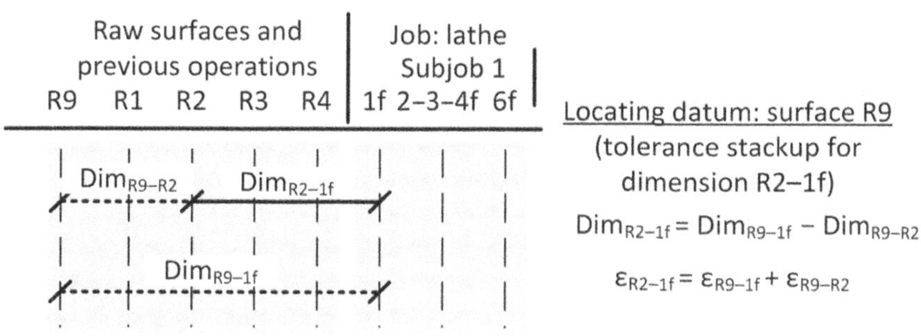

Figure 8.16 Tolerance stackup if surface R9 is used as datum surface. Design specification R2-1f is 20 ± 0.2 mm, but tolerance stackup results in the summation of machining/fixturing error and casting error (±0.56 mm).

from surface R9. Assuming as manufacturing errors a machining error of ±0.01 mm, a fixturing error of ±0.05 mm (a three-jaw chuck with soft jaws), and a casting tolerance of ±0.5 mm, the tolerance stackup for R2-1f is ±0.56 mm whereas the admissible design tolerance is ±0.2 mm. Therefore, the possible use of surface R9 as the locating surface is discarded and the final locating surfaces are R2 and R1. It can be seen that if the datum surface used is R2, the distance R2-1f is directly the machining error plus the fixturing error, that is, ±0.06 mm, lower than the design tolerance ±0.2 mm.

Note that the specification related to the distance between R3 and 1f is obtained through a tolerance stackup, since the machining distance is R2-1f, and the distance R2-R3 comes from casting. However, since the tolerance in this specification is high (±0.8 mm), the resulting stackup is within design specifications.

The workholding device to be used must lock five movements on R2 (±Rx, ±Ry, and −Tz), so a planar locating surface is required. Additionally, four additional movements should be locked on R1 (±Tx and ±Ty). A possible workholding system is a three-jaw chuck, using the jaws to support the R2 plane and closing the jaws to lock the R1 movement. The clamping pressure will lock the remaining three potential part movements (Tz and ±Rz). The chuck may have hard or soft jaws since part tolerances (±0.2 and ±0.5 mm) are high. In this case, hard jaws are finally selected.

As a final step, the fixturing scheme is graphically represented (Figure 8.17) and the diagram is updated by indicating the setup datum for each subjob and the number of part movements constrained (Figure 8.18).

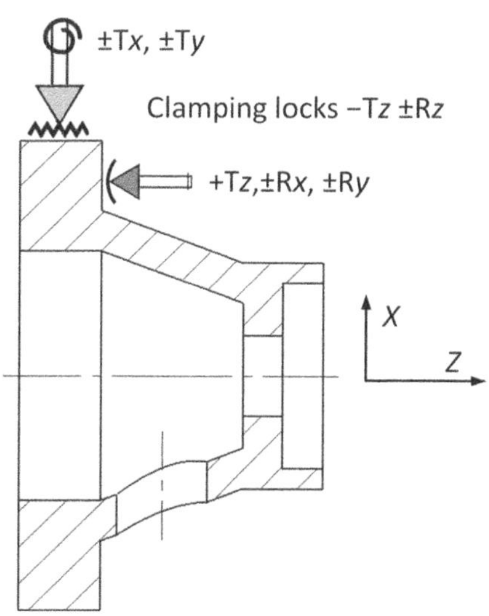

Figure 8.17 Graphical representation of the location and clamping method in subjob 1, job 1. Note that holes (surface 10) are not included since these holes are not obtained in the casting process.

Subjob 2 in lathe

In this case, the following relationships will give relevant information about the setup datum definition:

- Distance between 1f and 9f is 75 ± 0.3 mm.
- Parallelism between 1f and 9f is 0.2 mm.
- Coaxiality between 6f and 7–8f is Ø0.2 mm.

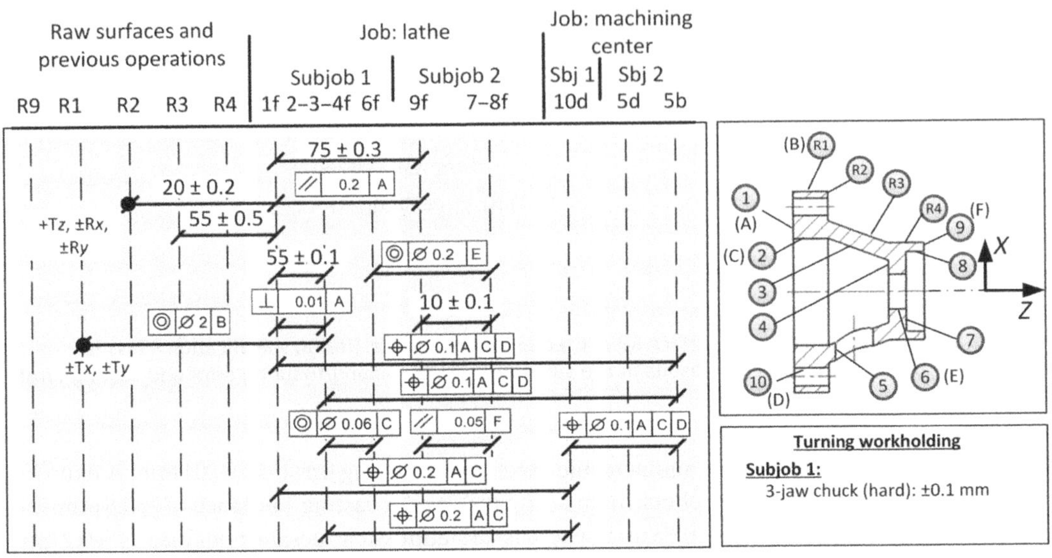

Figure 8.18 Diagram updated with the setup datum in subjob 1, job 1.

Therefore, there are liaisons that compromise the same part movements, and the most restrictive should be selected. The most restrictive tolerances are the coaxiality and parallelism, both 0.2 mm, but coaxiality is more restrictive since involves orientation and position whereas parallelism only involves orientation. However, surface 6 is difficult if not impossible to be used as locating surface. At this point, the following alternatives are identified:

- Alternative 1. Use as locating datum surface 1, which locks ±Rx, ±Ry, and −Tz, and surface R1 to lock ±Tx and ±Ty.
- Alternative 2. Use as locating datum surface 2, which locks ±Tx, ±Ty, ±Rx, and ±Ry, and surface 1 to lock −Tz.
- Alternative 3. Use as locating datum surface R1, which locks ±Tx, ±Ty, ±Rx, and ±Ry, and surface 1 to lock −Tz. This is similar to alternative 2 but using surface R1 instead of surface 2.

Alternative 1 ensures the parallelism tolerance since surface 9 is machined using surface 1 as the datum. However, the coaxiality will depend on a tolerance stackup where a rough cast surface is involved since R1 is locking the movements ±Tx and ±Ty. Then, this alternative is discarded due to the impossibility of ensuring coaxiality.

The alternative 2 orients the part through surface 2 (locks ±Rx, ±Ry, ±Tx, and ±Ty) which has been previously produced using the same subjob as for machining surface 6. Thus, the coaxiality between 2 and 6 is high (only machining errors may influence). Since surfaces 2 and 6 present a high coaxiality, machining surface 8 using surface 2 as the locating surface will also ensure a high coaxiality between surfaces 6 and 8. On the contrary, the parallelism of surface 1 is also ensured since surfaces 1 and 2 are also machined in the same previous subjob and thus, the possible orientation error between both surfaces may be only caused by machining errors, not by the fixturing error. Therefore, in both parallelism and coaxiality tolerances there is a tolerance stackup, but since the tolerances are not too tight (0.2 mm), it can be expected that both tolerances will be reached. Finally, alternative 3 may be more appropriate to keep the tooling similar to subjob 1, but it is discarded by the same reason than alternative 1, that is, the use of a rough cast surface as an intermediate surface to ensure coaxiality. Figure 8.19 shows the tolerance stackups identified in the diagram for alternatives 2 and 3.

Therefore, alternative 2 is chosen. It should be noted that the distance tolerance between 1f and 9f is ensured since surface 1 locks the movement −Tz. Figure 8.20 shows the machining sequence diagram adding the setup datums and the part movements constrained.

A possible physical realization of this fixturing scheme is based on an internal gripping device that will lock eight-part movements by acting on surface 2. Considering that surface 2 has a dimension of Ø62 mm, a three-jaw chuck using internal jaws can be used. Additionally, since surface 1 locks the *Z*-axis movement, a locator or the internal jaw itself placed on surface 1 can be used for this purpose. The accuracy of the three-jaw chuck depends on the type of jaws used (Table 8.3). In this case, since there is a coaxiality and parallelism tolerance of 0.2 mm and a tolerance stackup arises, soft jaws are selected to increase the location accuracy. The graphical representation of the fixturing scheme applied is illustrated in Figure 8.21.

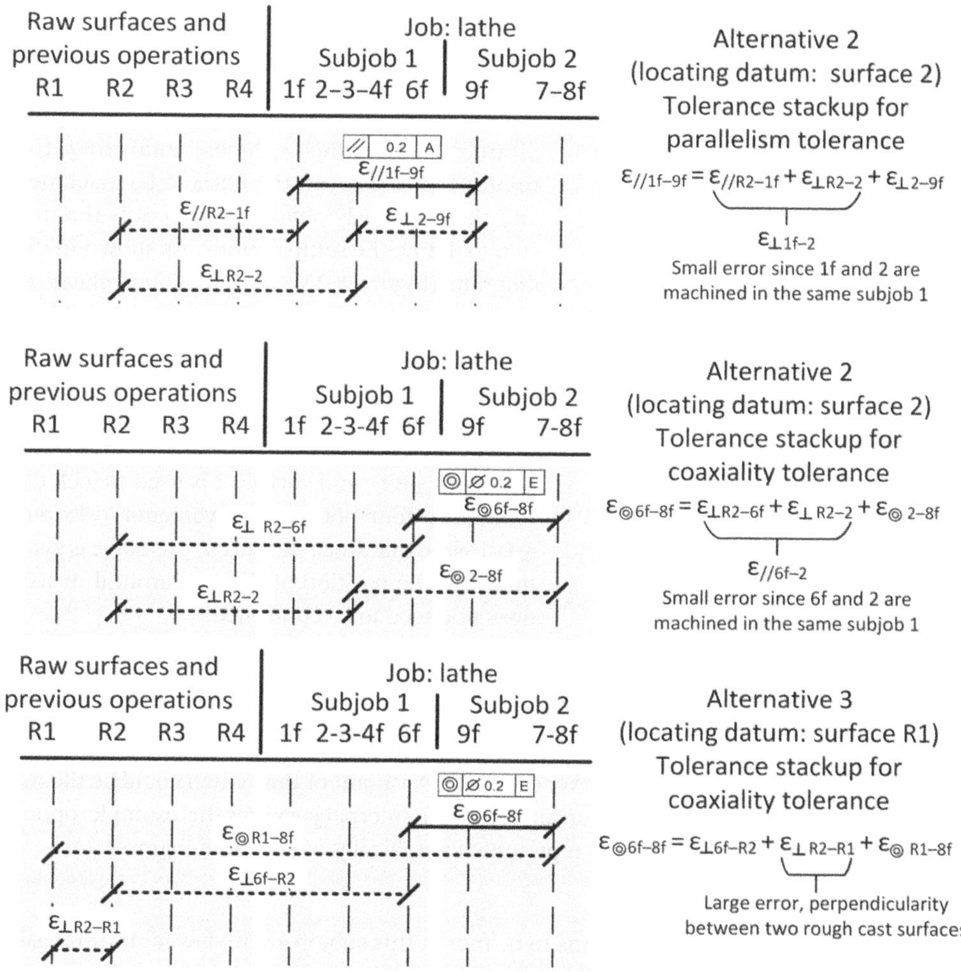

Figure 8.19 Tolerance stackups for different alternatives of locating datum surfaces in subjob 2, job 1.

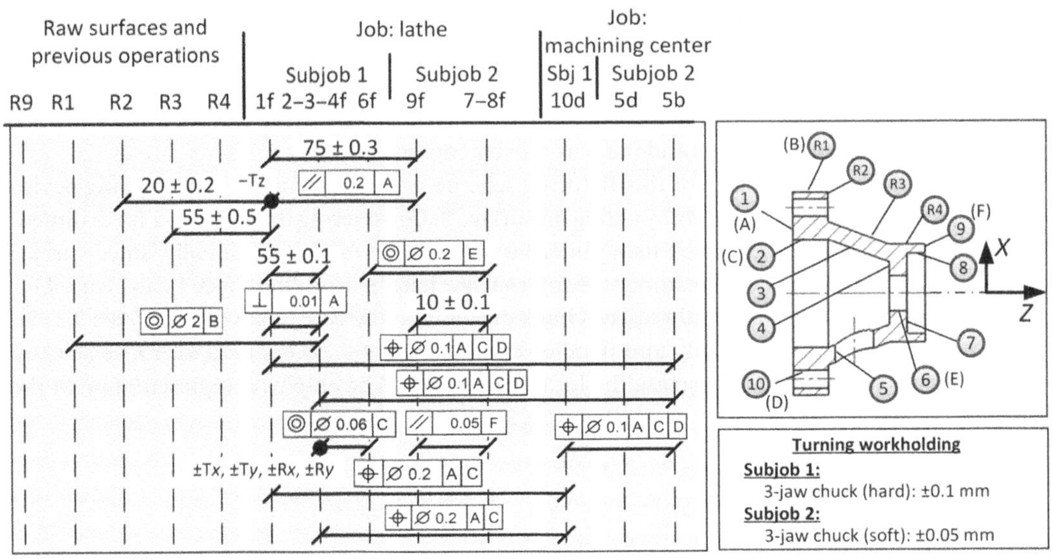

Figure 8.20 Updated diagram including the datum surface for subjob 2, job 1.

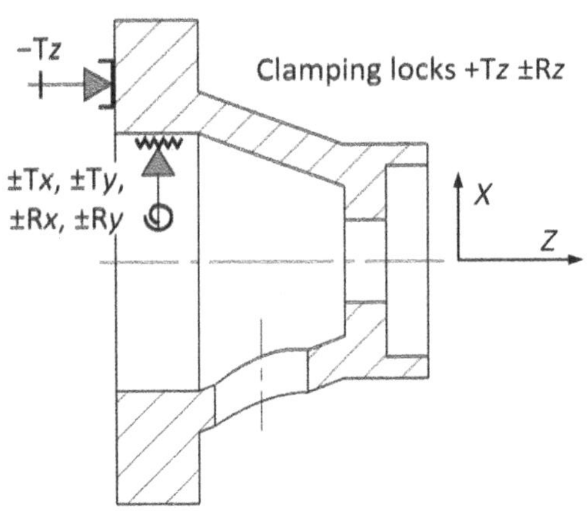

Figure 8.21 Graphical representation of the location and clamping scheme in subjob 2, job 1.

Subjob 1 in machining center

Similar to previous subjobs, analyzing the relationships between the holes to be machined and the previously obtained surfaces is the first step to define the setup datums. As shown in the diagram (Figure 8.22), the holes are related to surfaces obtained in operations 1f and 2-3-4f by a position tolerance. Since surface 1 is the primary datum in the position tolerance and surface 2 is the secondary one, surface 1 is used as a plane–plane contact to lock five-part movements (−Tz, ±Rx, and ±Ry), and surface 2 is used to lock the remaining four potential movements (±Tx and ±Ty). Note that since the part at this stage is symmetrical, the position of the part around its axis does not need to be controlled.

To constrain the explained part movements, a planar surface (e.g., a plate) can be used in surface 1 and a concentric pin in surface 2. To clamp the part, two options can be easily devised. As the first option, a strap clamp can be used to exert a clamping force over surface R2. As a second option, instead of using a basic concentric pin, a mandril can be used for clamping the part exerting force over surface 2. A variant of this option could be the use of a three-jaw chuck for clamping the part over surface 2 with internal jaws. For this example, option 1 is selected and the final fixturing scheme is represented graphically, as shown in Figure 8.23.

Subjob 2 in machining center

Similar to previous subjobs, the specifications to be met in this subjob are reviewed. In this case, this subjob machines surface 5 (drilling and boring) and requires the positioning of this feature with respect to references A, C, and D, which means surfaces 1, 2, and 10. As the position

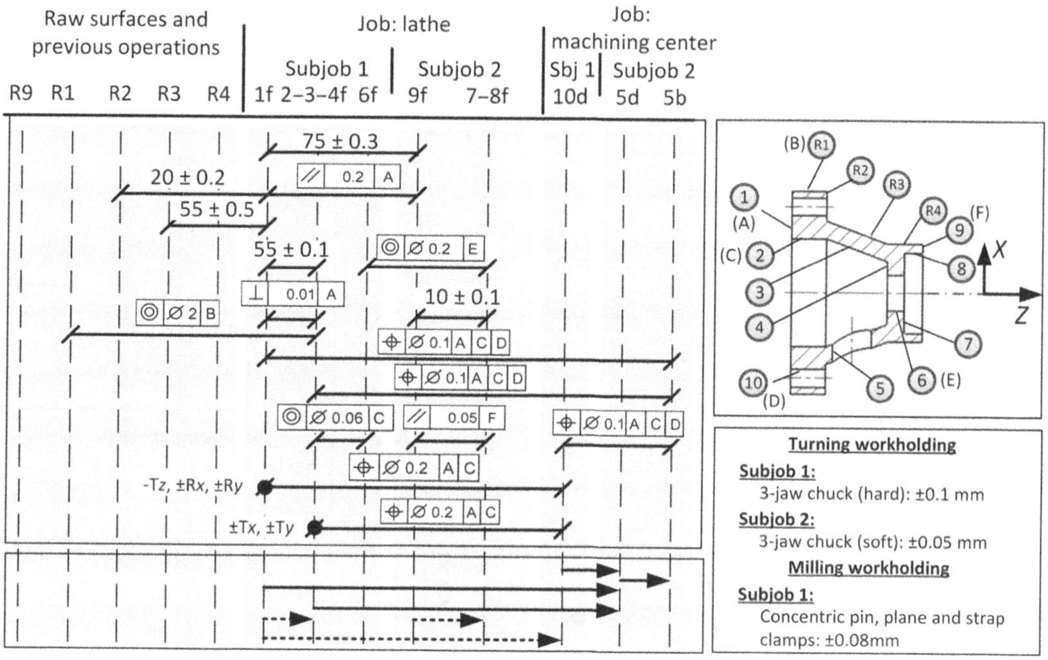

Figure 8.22 Updated diagram, including the datum surfaces for subjob 1, job 2.

tolerance includes a primary, a secondary, and a tertiary datum surface, the locating scheme to hold the part must keep this datum scheme to avoid a stackup tolerance. All three surfaces are accessible during machining, and thus, the following scheme is proposed: surface 1, a plane-to-plane contact, locks five-part movements (primary datum); surface 2 locks the position of the part with respect to the axis of this cylindrical geometry and locks four movements (secondary datum); and finally, the holes of surface 10 lock the two remaining part movements related to part rotation (tertiary datum). Figure 8.24 shows the diagram of the machining sequence with the datum surfaces selected.

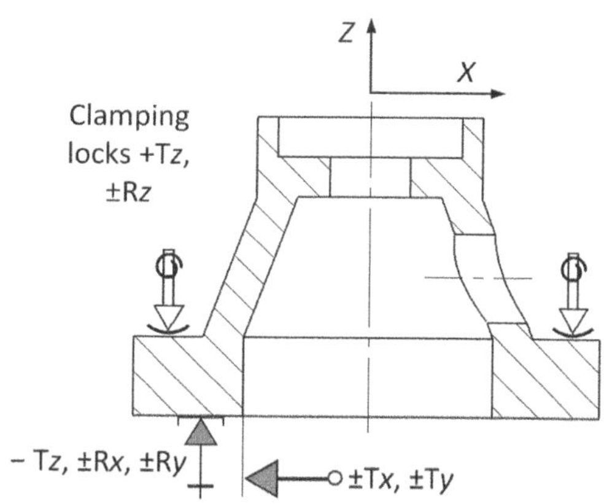

Figure 8.23 Graphical representation of subjob 1, job 2.

To hold the part using as primary datum surface 1, surface R2 could be the first option to be used as a clamping surface. Since R2 is accessible, this surface will be used for locating strap clamps or similar clamps and exerting a clamping force toward the primary datum surface. Other options similar to previous subjob 1 can be studied, but R2 may be the most straightforward one. Therefore, the workholding system used in this subjob is similar to the previous one. A plane to lock five movements and a concentric pin to locate the part on the axis of surface 2. Additionally, a diamond pin is used on hole 10 to lock the rotation of the part around its axis. Finally, two strap clamps or similar devices are used to hold the part. Figure 8.25 shows graphically the proposed workholding system.

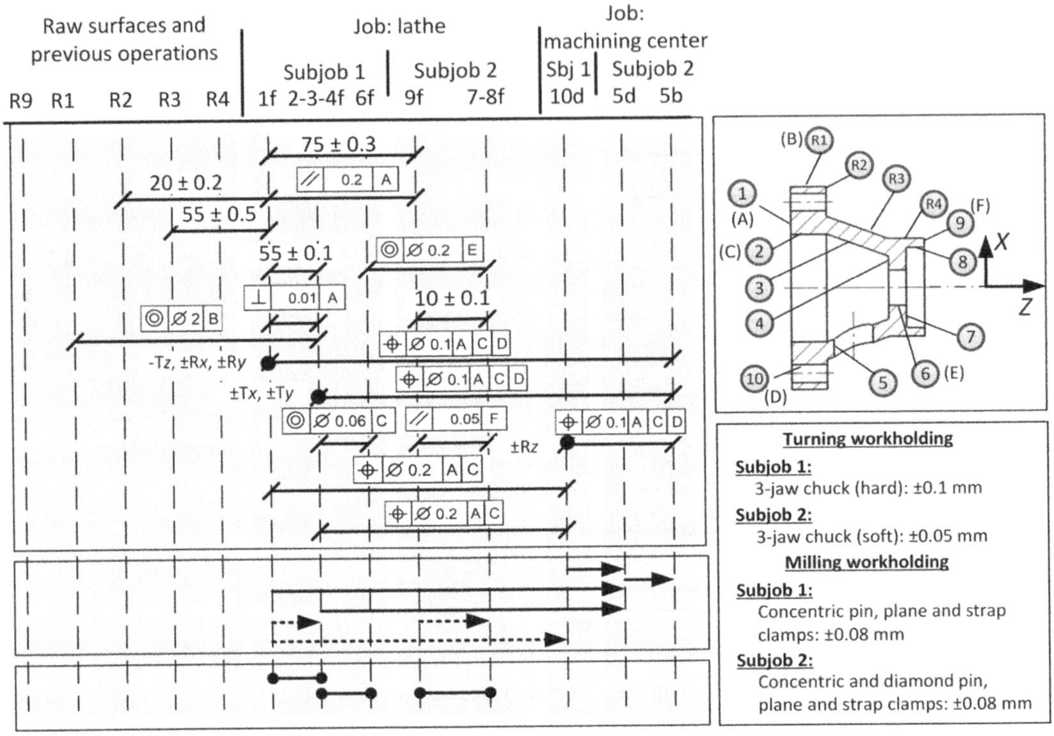

Figure 8.24 Updated diagram, including the datum surfaces for subjob 2, job 2.

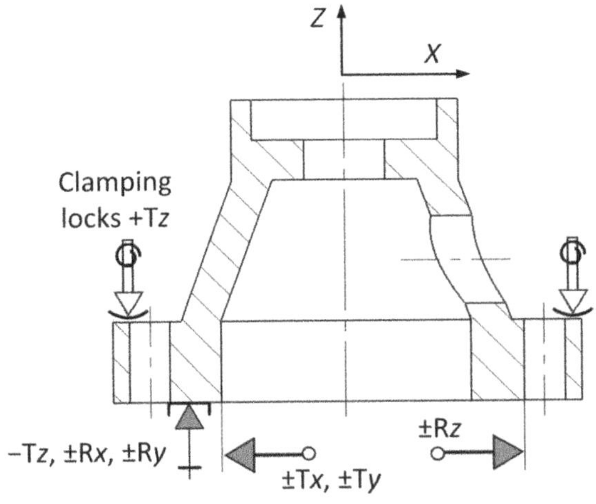

Figure 8.25 Graphical representation of subjob 2, job 2.

Questions and Exercises

1 Explain the meaning of the terms "job" and "subjob."

2 In terms of locating accuracy, sort from highest to lowest accurate the following workholding solutions: chuck collect, four-jaw chuck, three-jaw chuck, between centers with chuck and a carrier.

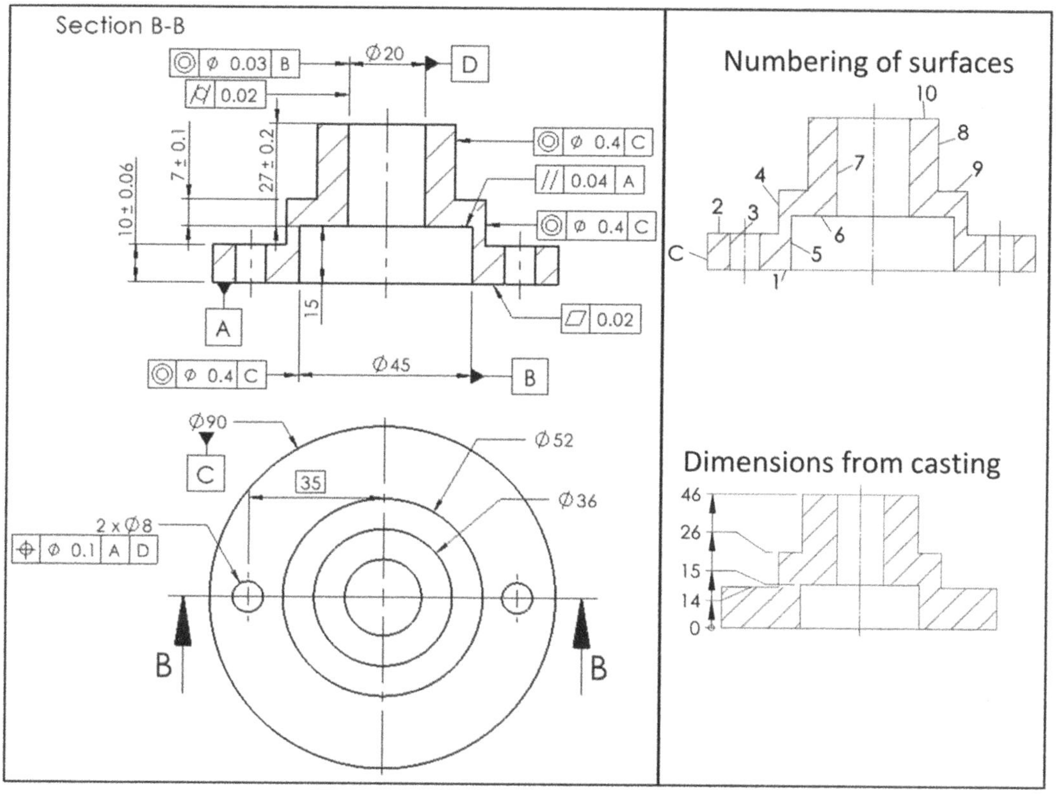

Figure 8.26 Exercise 3 to carry out the machining process planning. Units in mm.

3 The part shown in Figure 8.26 is manufactured by casting and machining processes. The casting process gives 2 mm machining allowance to all surfaces except for surface C, and then, all surfaces are machined (except C; C is kept as a raw surface). To define the machining operation sequence (assume only finishing machining operations), follow the proposed methodology for machining process planning in this chapter and define:

- Analysis of surface specifications. Table with the intrinsic and extrinsic specifications of the part. Consider surface roughness of 3 μm for all surfaces except for surface C and holes, both with 12 μm.
- Machining volumes, TADs, and machining operations.
- Diagram with:
  - Sequence of jobs.
  - Sequence of subjobs.
  - Datum surfaces.
  - Precedence constraints and critical feature relationships.
- Graphical and physical description of the workholding system applied for each subjob using the French Standard NF E04-013 (AFNOR 2015).

4 The part shown in Figure 8.27 is manufactured from a bar of Ø70 mm. Define a proper machining operation sequence following the methodology presented in this chapter. Identify the locating surfaces and the potential workholding system for each subjob.

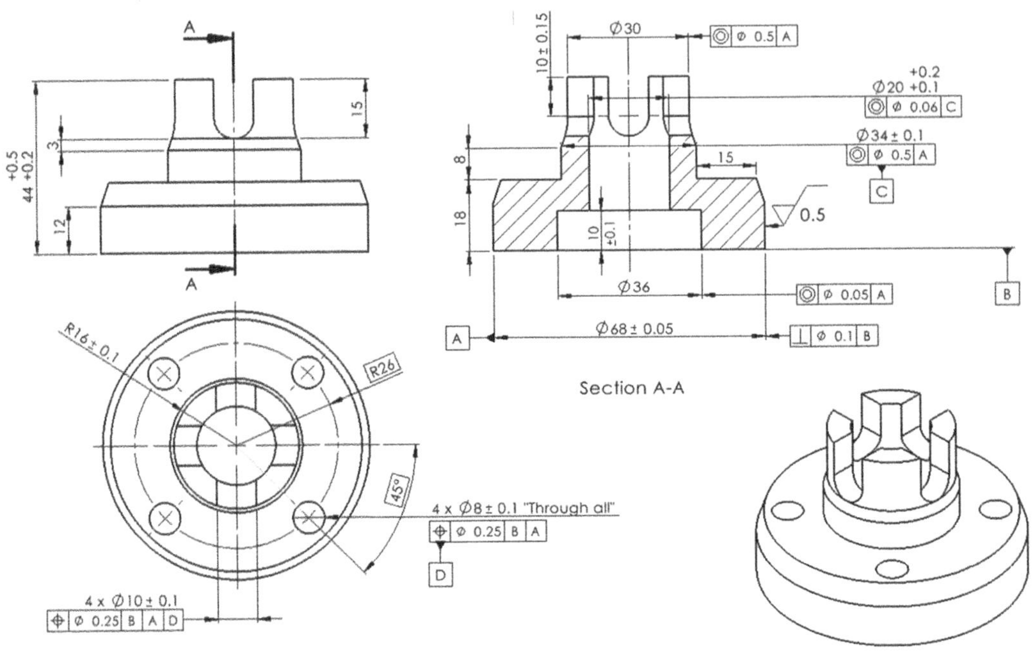

Figure 8.27 Exercise 4 to carry out the machining process planning. Units in mm. Assume general tolerances ISO 2768-mH and general surface roughness class N8 according to ISO 1302.

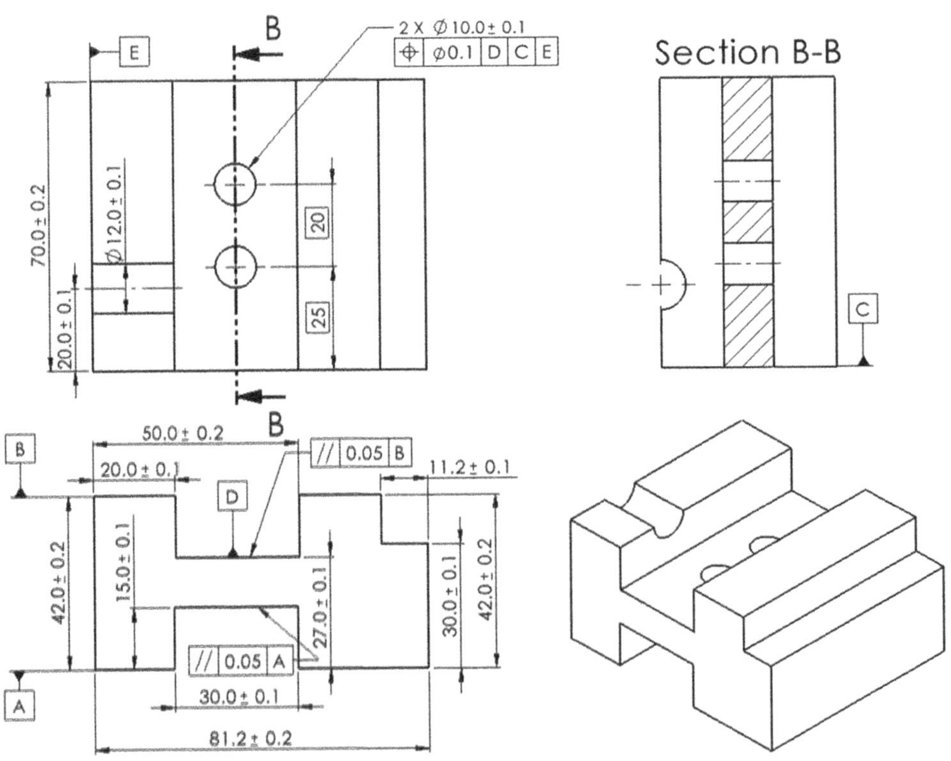

Figure 8.28 Exercise 5 to carry out the machining process planning. Units in mm. Assume a general surface roughness of 3 µm.

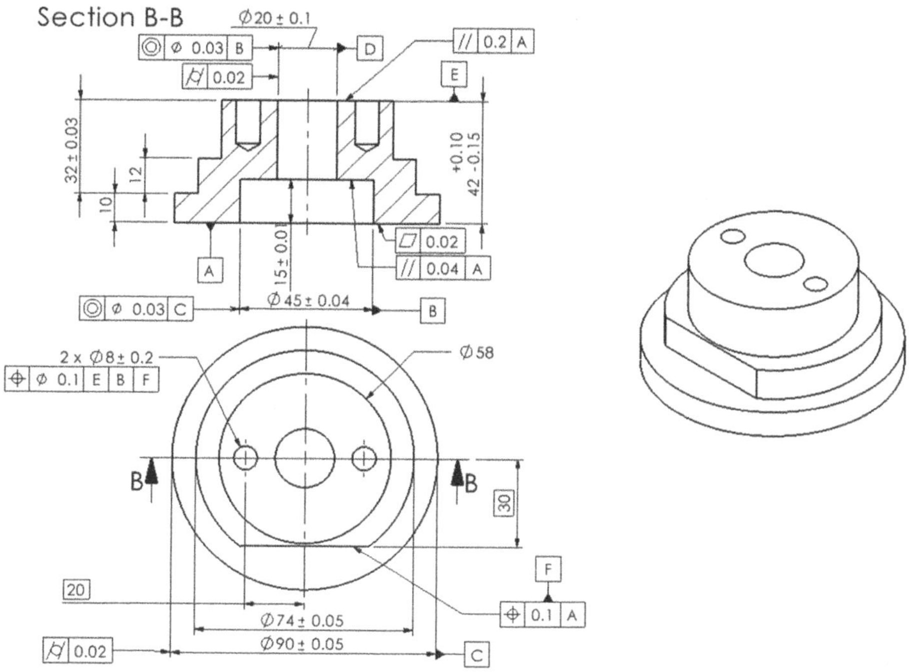

Figure 8.29 Exercise 6 to carry out the machining process planning. Units in mm. Assume a general surface roughness of 3 μm.

5 The part shown in Figure 8.28 is manufactured from a prismatic workpiece previously squared to the dimensions 81.2 × 70 × 46 mm. Define a proper machining operation sequence following the methodology presented in this chapter. Identify the locating surfaces and the potential workholding system for each subjob.

6 The part shown in Figure 8.29 is manufactured from a bar of Ø100 mm and a length 50 mm. Define a proper machining operation sequence following the methodology presented in this chapter. Identify the locating surfaces and the potential workholding system for each subjob.

References

AFNOR. (2015). *NF E04-013: 2015. Geometrical Product Specification (GPS) – Symbol of Workpiece Holding*. AFNOR.

Bansal, S., Nagarajan, S., and Reddy, N.V. (2008). An integrated fixture planning system for minimum tolerances. *The International Journal of Advanced Manufacturing Technology* 38: 501–513.

Faheem, W., Hayes, C. C., Castaño, J. F., and Gaines, D. M. (1998), What is a manufacturing interaction? *Proceedings of the ASME 1998 Design Engineering Technical Conferences. Volume 4: 3rd Design for Manufacturing Conference*. Atlanta, Georgia, USA (September 13–16).

Halevi, G., and Weill, R. (1995). *Principles of Process Planning: a Logical Approach*. Springer Science & Business Media.

Liu, Q., and Huang, S.H. (2003). Rigorous application of tolerance analysis in setup planning. *The International Journal of Advanced Manufacturing Technology* 21 (3): 196–207.

Liu, Z., and Wang, L. (2007). Sequencing of interacting prismatic machining features for process planning. *Computers in Industry* 58 (4): 295–303.

Ma, H., Zhou, X., Liu, W. et al. (2020). A customizable process planning approach for rotational parts based on multi-level machining features and ontology. *The International Journal of Advanced Manufacturing Technology* 108: 647–669.

Mokhtar, A., and Xu, X. (2011). Machining precedence of 2½D interacting features in a feature-based data model. *Journal of Intelligent Manufacturing* 22: 145–161.

Murty, R.L. (2005). *Precision Engineering in Manufacturing*. New Age International.

Romero, F., Bruscas, G.M., and Serrano, J. (2012). A new methodological approach for the machining process planning. In: *Key Engineering Materials*, vol. 502, 13–18. Trans Tech Publications Ltd.

Scallan, P. (2003). *Process Planning: the Design/Manufacture Interface*. Butterworth-Heinemann.

Zamanillo, J. (1993). Planificación de procesos. *Universidad Politécnica de Valencia (Spanish).*

9

Machine Tool/Cutting Tool Selection

9.1 Introduction

Machine tool and cutting tool selection are important activities that process planners must conduct when machining processes are planned. Factors such as productivity, machine tool accuracy, power requirements, workpiece material, and part geometry should be considered at this stage. This chapter details some of these factors and how to conduct an efficient machine tool and cutting tool selection. For this purpose, the chapter briefly reviews the cutting tool materials and the most common cutting tools employed in milling, turning, and drilling. The International Organization for Standardization (ISO) standard that describes the characteristics of tool holders and cutting inserts is also presented, and the influence of cutting tool parameters on cutting performance, such as lead angles, rake angles, nose radius, etc., is analyzed. The chapter also provides practical examples to reinforce the theoretical explanations related to machine tool and cutting tool selection.

9.2 Machine Tool Selection

9.2.1 Factors in Machine Tool Selection

Machine tool selection is conducted considering several factors that can be categorized as either technical or operational factors. In terms of technical considerations, the primary focus lies in ensuring that the chosen machine tool possesses the capacity to produce the desired part in accordance with the specified requirements. The key aspects to take into account are as follows:

- Physical dimensions and robustness. The machine tool should be able to accommodate the workpiece's dimensions and, at the same time, must be robust enough to support the weight of the workpiece and the dynamic load during cutting.
- Power requirements. The machine tool should be able to provide the required cutting power to conduct all machining operations efficiently. For this purpose, calculations must be performed to estimate the power requirements for critical operations given workpiece/tool materials and proper ranges for cutting parameters (i.e., feed rates, cutting speed, and depth of cut). Chapter 10 provides the technical background to conduct a proper power consumption calculation in machining operations.

Manufacturing Process Planning: A Practical Approach for Mechanical Engineering, First Edition.
José V. Abellán-Nebot, Carlos Vila Pastor, and Héctor R. Siller.

Companion website: www.wiley.com/go/Manufacturing_Process_Planning_1e

- Machine tool capability. The machine tool should be able to machine the parts within the prescribed dimensional/geometric tolerance limits and surface roughness requirements. In other words, the manufacturing variability should be lower than these tolerance limits, otherwise, additional machining or finishing operations will be required, increasing the total manufacturing cost.

Unlike technical factors that are centered on ensuring that the chosen machine tool can meet the required part specifications, operational factors are more focused on considering an efficient utilization of available resources to fulfill the master production schedule. The key aspects of operational considerations encompass (Scallan 2003):

- Batch size. Similar to how each manufacturing process has an economic batch quantity (EBQ) that must be reached for it to be economically viable, specific machinery also has its own batch size requirements. A common approach involves comparing different machines for a given batch size using break-even analysis to determine the most cost-effective choice.
- Capacity. Equally significant as the EBQ is the production rate achievable by the machines under evaluation. Every machine has the capacity to produce a certain output per unit of time. Consequently, it is crucial to assign parts to machines with output capabilities that align with the requirements of the master production schedule for a given period.
- Availability. This refers to whether the required machine is currently in use or not, in other words, whether it is available or unavailable.

While all these factors are equally important, the primary process planner's concern is typically the batch size analysis. According to the batch size, the selection of the optimal machine tool can be done for the most economical production or for the maximum production.

To select the machinery for the most economical production, the cost of using a machine tool for machining a part can be mathematically derived as follows:

$$C = \frac{C_{\mathrm{T}}}{n \times L} + \frac{C_{\mathrm{L}}}{n} + C_{\mathrm{u}}$$

where C_{T} is the cost incurred only once in the manufacturing of the part (for example, process planning cost, computer numerical control (CNC) programming cost, fixture design cost, etc.); C_{L} is the cost incurred every batch is manufactured (for instance, machine tool and fixtures setups); L is the number of batches to be manufactured; n is the number of parts per batch; and C_{u} is the unitary manufacturing cost per part (e.g., labor cost, tooling cost, etc.).

On the contrary, to select the machinery for maximum production, the goal is to minimize the manufacturing time of the parts. Thus, the equation to be evaluated is the manufacturing time that is described as:

$$T = \frac{T_{\mathrm{T}}}{n \times L} + \frac{T_{\mathrm{L}}}{n} + T_{\mathrm{u}}$$

where the variables refer to the time required for each of the items expressed above in the manufacturing cost function.

9.2.2 Machine Tool Selection Method

The procedure to select the most adequate machine tool is based on the following steps:

- Discard the machine tools that do not meet the technical factors.
- From the remaining machine tools, choose the one that minimizes the cost/time according to the desired production approach (maximum production or economical production). If batch size is not known or is still under consideration, evaluate the break-even for all machine tools.

9.2.3 Practical Example

Consider a CNC machine tool A with a setup cost of \$500 per batch and an operation cost of \$3 per part, and a CNC machine tool B with a setup cost of \$1,500 and an operation cost of \$1. Select the machine tool for the most economical production as a function of the number of parts per batch.

Under the assumption that the C_T cost is the same for any machine tool (process planning cost, CNC programming, etc.), the cost expression has a fixed cost (C_L/n) and a variable cost (C_u, which is the machining cost per part), a break-even analysis as in the following can be conducted to evaluate the quantity at which the same cost is given for each machine tool:

$$C_{L_A} + C_{uA} \times n = C_{L_B} + C_{uB} \times n \rightarrow n = \frac{C_{L_B} - C_{L_A}}{C_{uA} - C_{uB}} = \frac{1{,}500 - 500}{3-1} = 500 \text{ parts}$$

Thus, for a production with more than 500 parts per batch, the machine tool B will be used. If the quantity is lower than 500, machine tool A will be used. Figure 9.1 shows graphically the break-even analysis.

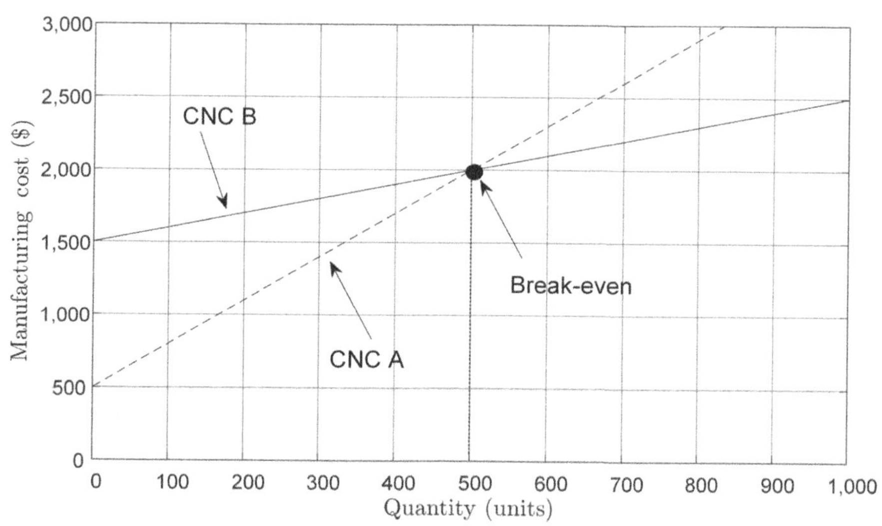

Figure 9.1 Example of break-even point between two CNC machines with different fixed and operating costs.

9.3 Cutting Tool Selection

9.3.1 Factors in Cutting Tool Selection

In machining, using a single tool to complete all cutting processes is quite uncommon. As a result, several tools are frequently needed for machining a part. For instance, different tools are used for roughing and finishing. Additionally, specific tools are needed for tasks such as thread cutting, drilling, and slot machining. Therefore, it is crucial to choose the appropriate tool for the specific task that is needed.

Numerous factors influence the tool selection decision but the most important can be summarized as follows:

- Workpiece material. The type of material to be machined will influence on the selection of cutting tool geometry and cutting tool material. Some aspects to consider on tool material and tool geometry selection are briefly explained in the following.
  - Influence on cutting tool materials. The main cutting tool materials are high-speed steels (HSS), carbides, cermets, ceramics, cubic boron nitrides (cBN), and polycrystalline diamond (PCD). PCD is a costly material that cannot be used in machining ferrous alloys nor titanium or Ni–Co alloys due to their chemical affinity. Despite their high cost, for nonferreous and nonmetallic materials, the use of PCD tools can produce excellent surface finish and exceptionally long tool life. cBN is another costly material that can increase productivity (up to 100 times better than carbides or ceramics in terms of longer tool life and/or higher removal rates) in many difficult metal cutting operations such as machining hardened steels and heat-resistant materials. Cermets and ceramics present high wear resistance and work well at high cutting temperatures, but they are not tough enough to be used in noncontinuous operations such as milling or interrupted turning operations. HSS and carbides are more generally used; HSS for low cutting speeds and light or medium-duty applications whereas carbides are more appropriate for higher cutting speeds and heavy-duty applications. Carbides are more expensive but they can work 2.5 times faster than their equivalent HSS tools. Carbide inserts are usually coated (titanium nitride [TiN], titanium carbonitride [TiCN], etc.), which increases tool life, cutting speeds, and feeds allowable by 15–25%. The ISO 513: 2012 "Classification and application of hard cutting materials for metal removal with defined cutting edges – Designation of the main groups and groups of application," defines six main application groups for hard cutting materials which correspond to six classes of workpiece materials, classified by letters and colors. The six classes are steels, code P (blue color); stainless steel, code M (yellow color); cast iron, code K (red color); nonferrous metals, code N (green color); hard materials (hardened steels), code H (gray color); and exotic materials (titanium and heat resistance alloys), code S (brown color). For each application group, different subgroups are defined by using the same code and a number (e.g., P10, P20, ...). The higher the number within each application group, the lower is the material's wear resistance but the higher the toughness. To properly select the application group and subgroup, besides workpiece material, it is considered the type of operation (roughing, medium, or finishing) and the stability of the cutting operation (good, average, or difficult). In general, the subgroup is selected for high wear resistance in finishing and stable cutting operations, and for more demanding operations in terms of forces (roughing) or interrupted cutting, the subgroup is selected for higher toughness but with less wear resistance. Figure 9.2 summarizes the types of cutting tool materials, their properties, and applications. Table 9.1 shows the application groups and subgroups according to ISO 513

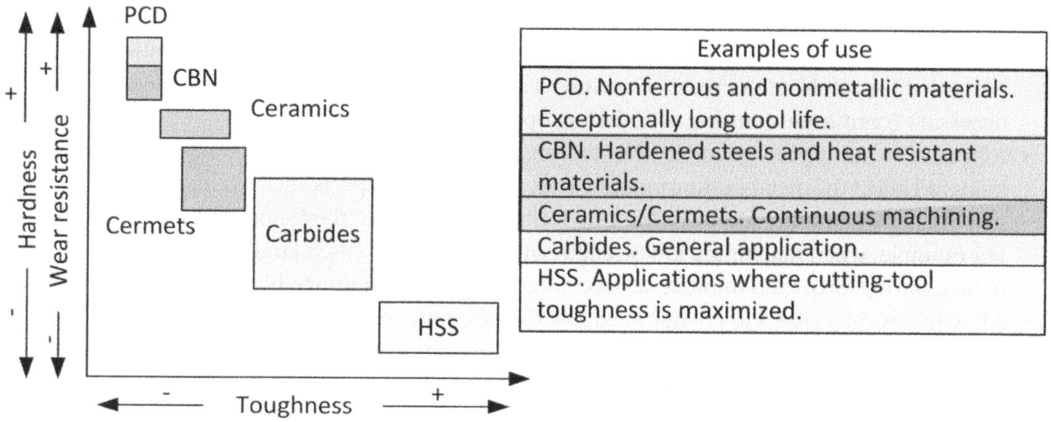

Figure 9.2 Overview of cutting tool materials, properties, and applications. *Source:* Adapted from Mitsubishi Materials (2023).

Table 9.1 Application group and subgroups of hard cutting tool materials, according to ISO 513.

Main application group			**Subgroup**	
Code letter	**Code color**	**Example**	**Subgroup code**	**Characteristics**
P	Blue	Steels: all kinds of steels except hardened and stainless steels	P01	Higher wear resistance
			P10	...
			...	...
			P50	Higher toughness
M	Yellow	Stainless steels	M01	Higher wear resistance
			M10	...
			...	...
			M40	Higher toughness
K	Red	Cast irons	K01	Higher wear resistance
			K10	...
			...	...
			K40	Higher toughness
N	Green	Nonferrous metals: aluminum, magnesium, and other nonferrous metals.	N01	Higher wear resistance
			N10	...
			...	...
			N30	Higher toughness
S	Brown	Heat-resistant alloys: nickel and cobalt alloys, titanium alloys	S01	Higher wear resistance
			S10	...
			...	...
			S30	Higher toughness
H	Gray	Hard materials: hardened steels, hardened cast irons	H01	Higher wear resistance
			H10	...
			...	...
			H30	Higher toughness

that are commonly found in catalogs for cutting tool selection. For instance, grade GC4305 from Sandvik Coromant manufacturer is recommended for P05 and P10 applications; thus, it is applied for steels where wear resistance wants to be maximized and high toughness is not necessary (continuous cutting or finishing operations).

- Influence on cutting tool geometry. The tool angles, rake angles, cutting-edge angles, and tool nose radius all affect the metal-cutting process. They influence on aspects such as chip generation, thermal dissipation, cutting forces, or built-up formation, and thus, they should be properly selected. For example, when machining low-strength and ductile materials, a large rake angle is applied to reduce cutting forces and improve surface roughness. On the contrary, high-strength materials are cut with low rake angles to provide adequate tool strength and better heat dissipation.

- Part geometry. The geometry of the part has a significant impact on the choice of cutting tools. For instance, geometric features such as T-slots required the specific use of T-slot cutting tools, pockets with corner radii at the bottom require specific bull nose end mills, or areas with low accessibility may require specific cutting tools to avoid collisions. The compatibility of cutting tool geometry and part geometry should always be studied in detail.
- Capability. Cutting tools should be selected to ensure that dimension/geometrical tolerances together with roughness specifications are met. For instance, holes with dimensional quality IT8-12 can be produced with conventional drills but for holes with dimensional quality IT6-7, reamers or boring tools must be used.

9.3.2 Cutting Tools and Selection Recommendations

9.3.2.1 Cutting Tools in Turning

The cutting tool selection in turning depends on the type of operation to be conducted, as shown in Figure 9.3. The cutting tools for external cutting differ notably from similar operations that are conducted internally, and grooving, threading, or cutting-off operations require specific tools.

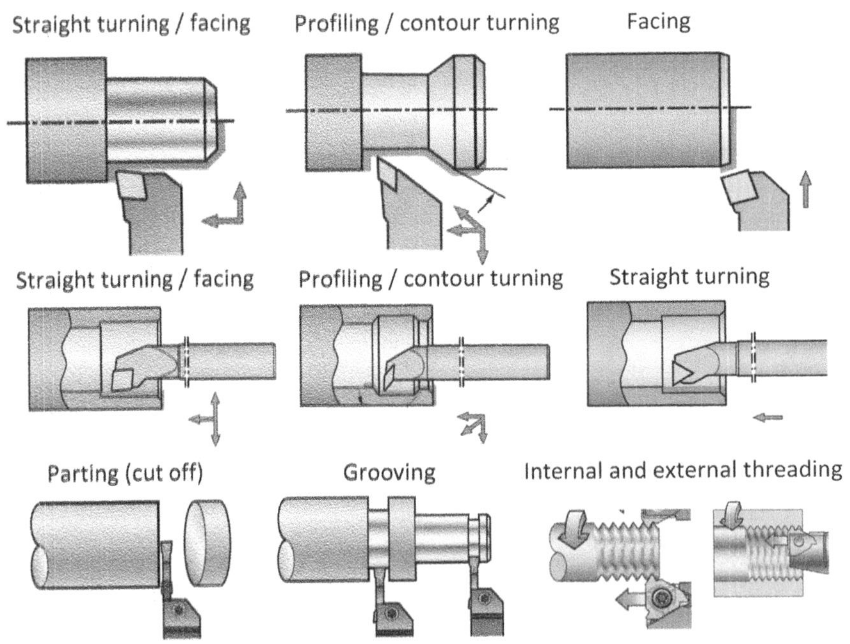

Figure 9.3 External and internal turning operations. *Source:* Courtesy of Sandvik Coromant.

Although HSS tools can be found, the most common cutting tools applied in turning are those based on cutting inserts, usually carbide inserts. To hold the turning insert, a turning tool holder must be used which should be selected in accordance with the cutting insert as they both consist of one working tool. The main parameters that define cutting inserts and turning holders are shown in the following together with some practical recommendations.

9.3.2.2 Turning Holders and Cutting Inserts

The ISO standard "ISO 5608: 2012. Turning and copying tool holders and cartridges for indexable inserts – Designation" describes the shape and size of turning holders using a specific code based on numbers and letters. An example of the use of this ISO standard is given in Figure 9.4 where a tool holder code PCLNL2525M12 with each parameter description is presented. The meaning of each parameter in this ISO standard is briefly described in the following.

- Parameter 1. Insert clamping method. This parameter is commonly selected according to the expected cutting forces of the operation and its intermittent nature. Top clamping is the sturdiest and screw clamping is the weakest. Some clamping methods are shown in Figure 9.5.
- Parameter 2. Insert shape. Many insert shapes are available as shown in Figure 9.6. Note that the insert geometry is selected based on the type of operation and the accessibility of the cutting tool required (see Figures 9.7 and 9.8). Additionally, it should be taken into account that large point angle inserts present higher edge strength but tend to produce higher vibrations.

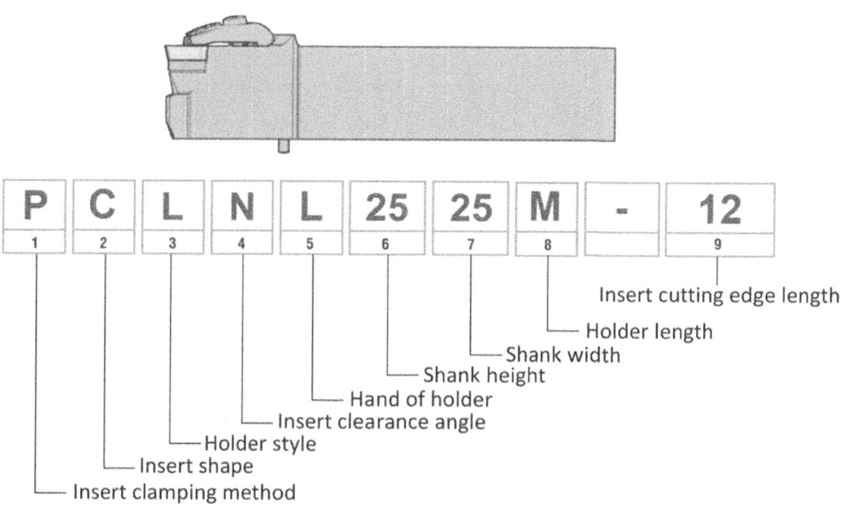

Figure 9.4 Turning holders. Example of ISO nomenclature. *Source:* Seco Tools AB (2023a).

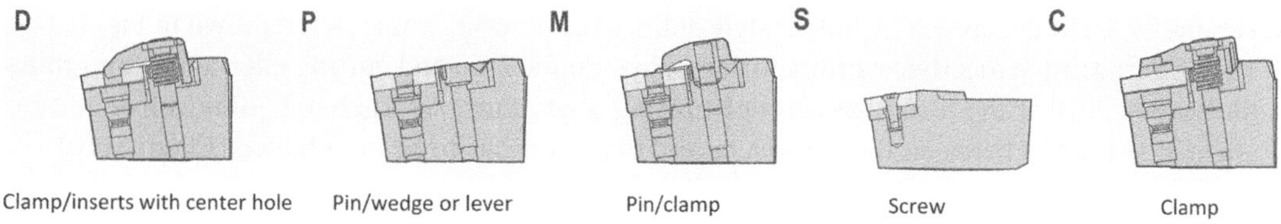

Figure 9.5 Clamping method of cutting inserts. *Source:* Seco Tools AB (2023a).

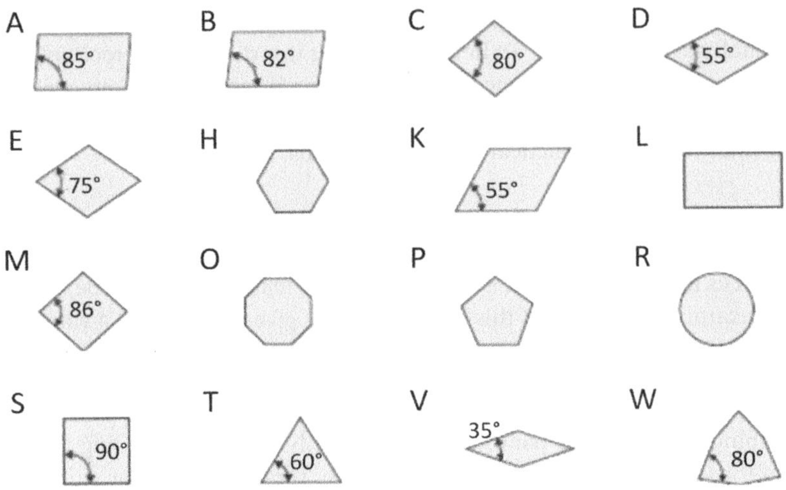

Figure 9.6 Insert shape according to insert geometry code. *Source:* Seco Tools AB (2023a).

Insert shape ++ = Recommended + = Alternative	Longitudinal turning	Profiling	Facing	Longitudinal turning	Profiling	Facing
Rhombic 80°	++		+	+		++
Rhombic 55°	+	++	+	+	++	+
Round	+	+	+	+		+
Square	+		++	+		
Triangular	+	+	+	++		+
Trigon 80°	+		+	+		+
Rhombic 35°		+			+	

Figure 9.7 Recommended cutting insert geometry for external and internal turning operations. *Source:* Courtesy of Sandvik Coromant (2017).

- Parameter 3. Holder style. The holder style defines the entering angle (K) as shown in Figure 9.9. The entering angle together with the insert shape defines the end cutting-edge angle, and thus, the accessibility of the insert for contour turning operations. As a general guide, a free cutting angle of at least 2° between the workpiece and the insert has to be maintained (Figure 9.10).

Large entering angles (small lead angles) produce cutting forces that are aimed at the chuck, which reduces the likelihood of vibrations. With small entering angles the radial forces increase, resulting in a clear impact on vibration tendencies and workpiece deflection. However, small

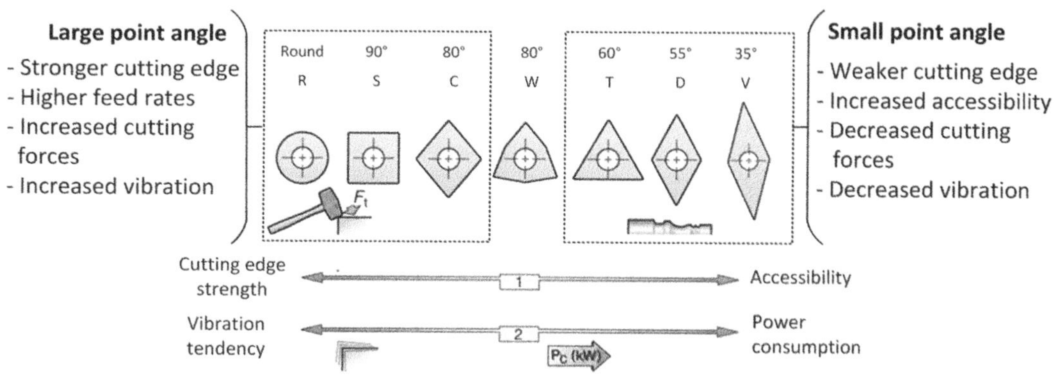

Figure 9.8 Influence of insert geometry with respect to toughness, vibrations, accessibility, and power consumption. *Source:* Courtesy of Sandvik Coromant (2017).

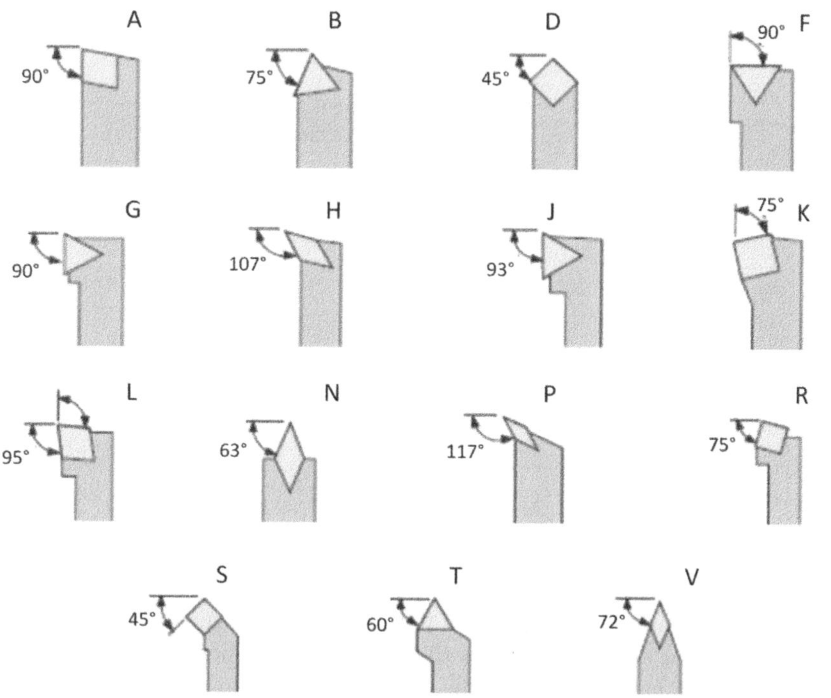

Figure 9.9 Holder style. *Source:* Seco Tools AB (2023a).

Figure 9.10 Relationship between entering angle and end cutting-edge angle, which determines the accessibility for contour turning operations.

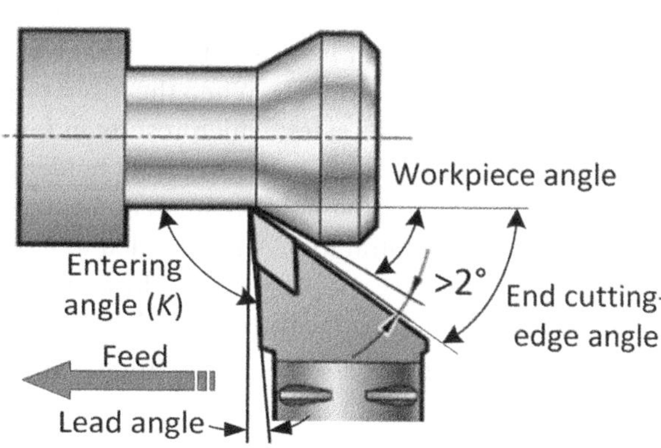

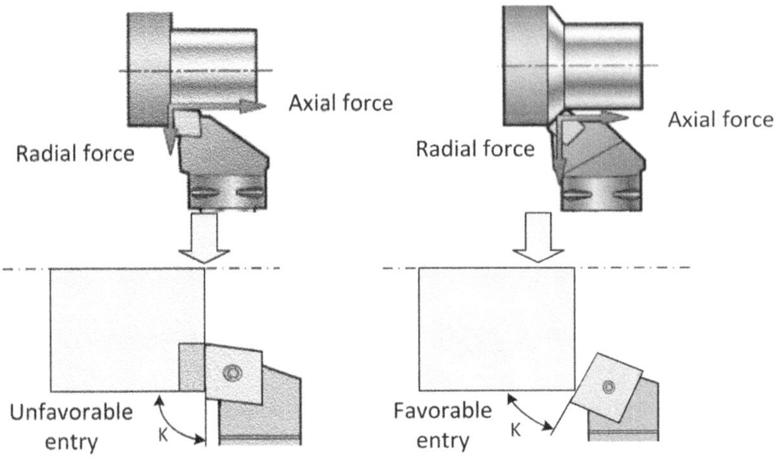

Figure 9.11 Initial contact of the insert with the workpiece according to the entering angle (K) and resulting axial and radial forces.

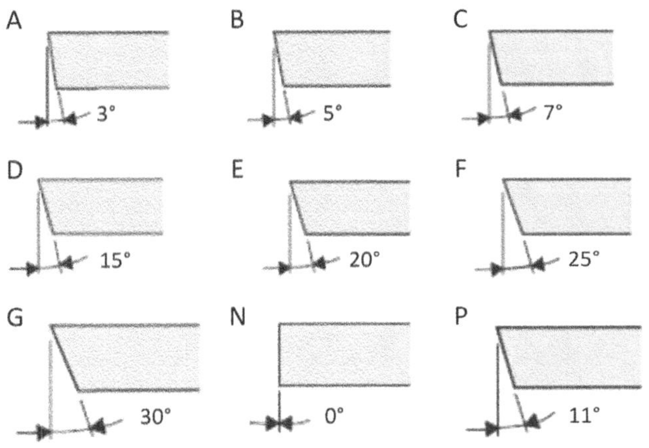

Figure 9.12 Insert clearance angle. *Source:* Seco Tools AB (2023a).

entering angles provide the cutting-edge with a larger corner and can add cutting-edge strength by distributing machining pressure along a greater length of the cutting edge (Figure 9.11). Note that the preceding definition of the lead angle is based on the ISO standard whereas the American National Standards Institute (ANSI) standard defines the lead angle similar to the entering angle.

- Parameter 4. Insert clearance angle. Different clearances are defined in Figure 9.12. To increase edge strength, the clearance should be as low as possible. However, low clearance angles may produce rubbing and an increase of tool wear. As a general rule, machining ductile materials require larger clearance angles. When the insert clearance angle is 0, the insert is reversible and can be mounted in two sides, which results in lower cutting tool costs. However, this kind of inserts should be only used in negative tool holder configurations to avoid rubbing.
- Parameter 5. Hand of the holder. "R" denotes right-handed tool, "L" denotes left-handed tool, and "N" denotes neutral tool. A left-handed tool can remove material while moving rightward (top view keeping rake surface visible). The name is derived from its analogy with the human left hand as the left-hand thumb direction indicates the tool feed direction. Figure 9.13 illustrates the left- and right-handed tools and the neutral tools that can move in both directions.

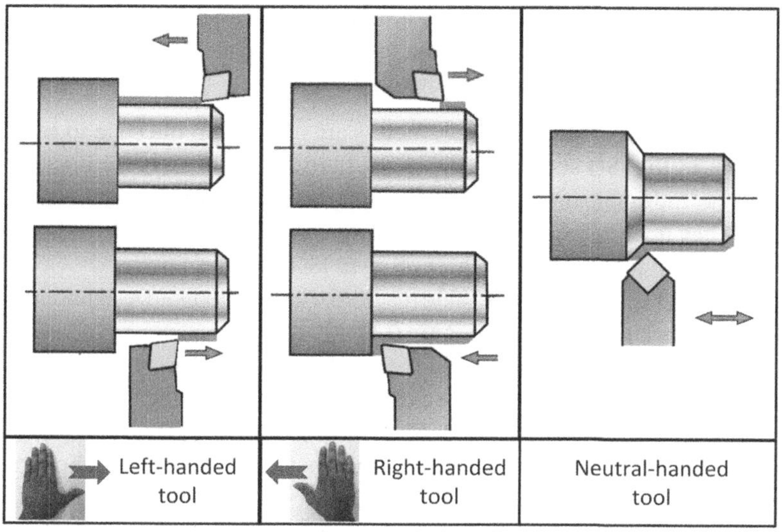

Figure 9.13 Hand of tool. Left-handed, right-handed, and neutral-handed tools.

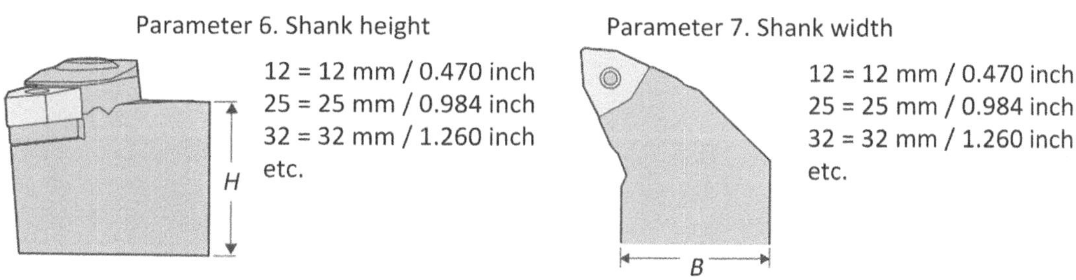

Figure 9.14 Height of shank and width of shank. *Source:* Seco Tools AB (2023a).

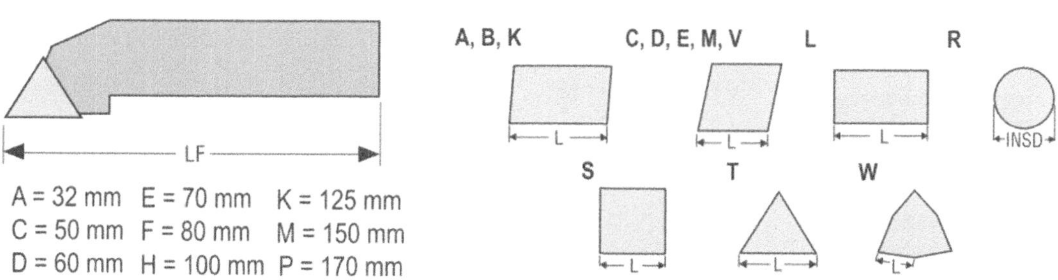

Figure 9.15 Length of holder, LF, and length of insert cutting edge, L. *Source:* Seco Tools AB (2023a).

- Parameters 6 and 7. Shank height and shank width are defined according to the holder size of the machine tool (Figure 9.14).
- Parameters 8 and 9. These parameters define the length of the holder and the length of the cutting insert (Figure 9.15). The holder length, LF, is defined according to the machine tool and the operation itself. For external turning operations, it is recommended to use the largest tool holder shank for maximum stability. For internal turning operations, the tool holder used is a

boring bar, and to minimize vibrations and deflections, a tool holder with $L/D<4$ should be used. The length of the insert cutting edge, L, is defined according to the type of insert geometry and it depends on the required or desired depth of cut during machining.

For internal turning (boring bars), three additional parameters are added into the code to refer to the bar dimensions as shown in the ISO standard "ISO 5609-1:2012. Tool holders for internal turning with cylindrical shank for indexable inserts." These parameters are tool holder type, shank diameter, and tool length, and they refer to parameters 1, 2, and 3, respectively. Figure 9.16 shows a general ISO designation for internal tool holder bars according to ISO 5609.

According to the ISO standard "ISO 1832: 2017. Indexable inserts for cutting tools – Designation," the cutting inserts follow the nomenclature shown in Figure 9.17.

From the selection of the turning holder, some characteristics of the cutting insert are already defined such as insert shape, insert clearance, and size (cutting-edge length). Other parameters such as insert tolerance, thickness, nose radius, etc., are defined during the selection of the cutting insert. Figures 9.18–9.20 show the meaning of the insert parameters not defined yet by the tool holder according to ISO 1832.

In the selection of the cutting insert, nose radius is a critical property. Higher nose radius leads to more robust inserts and can be applied for heavy feed rates and large depths of cut. A smaller nose radius is ideal for small cutting depths and low vibrations although it presents a weaker cutting edge. Table 9.2 shows common feed rate ranges according to nose radius values.

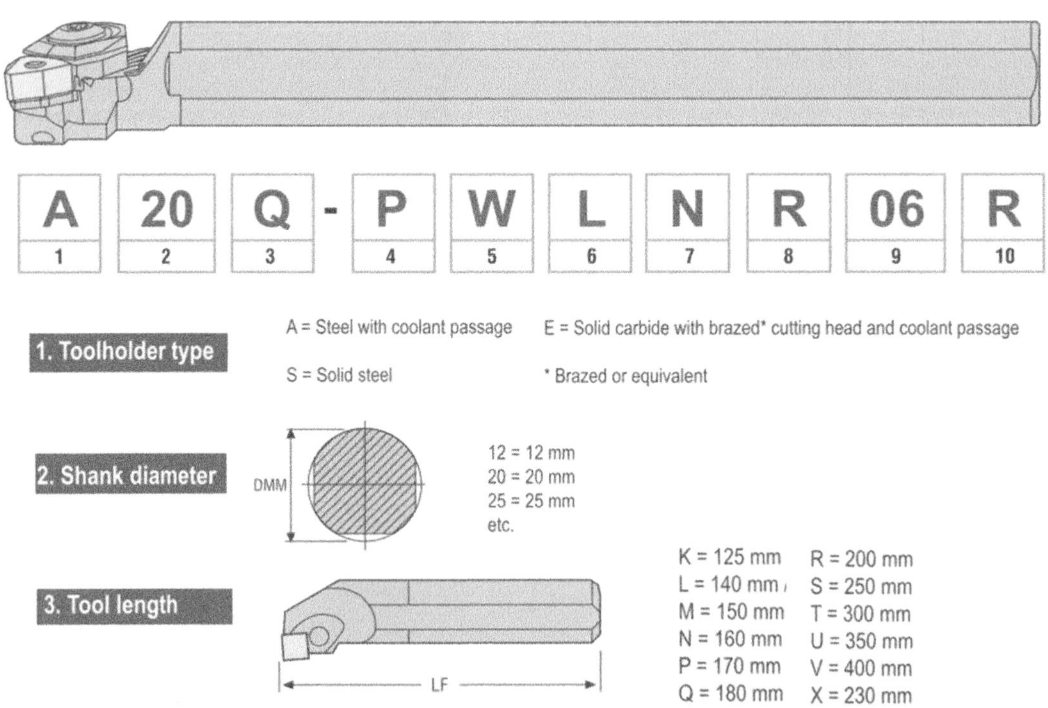

Figure 9.16 Designation for internal tool holder bars. Parameters 1–3 are specific. The rest of the parameters are similar to external tool holders. Parameter 4: insert clamping; parameter 5: insert shape; parameter 6: holder style; parameter 7: insert clearance angle; parameter 8: hand of tool; parameter 9: insert size; and parameter 10: vendor designation. *Source:* Seco Tools AB (2023a).

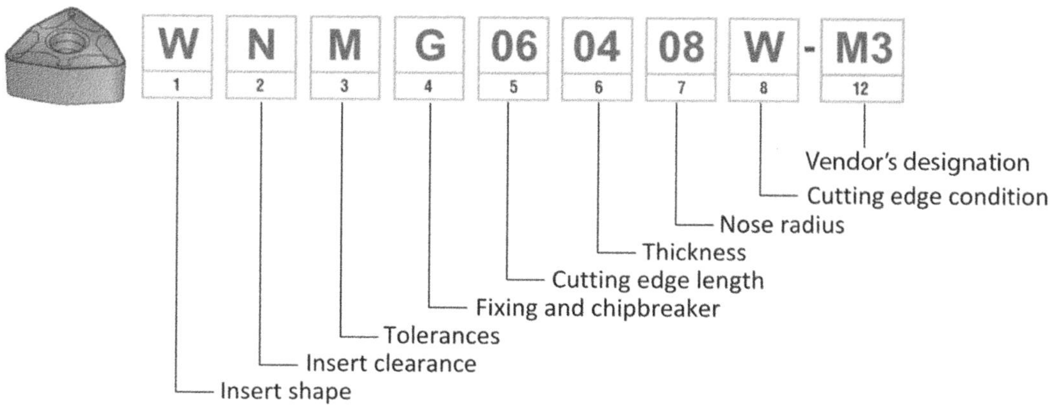

Figure 9.17 Cutting inserts nomenclature according to ISO 1832. *Source:* Adapted from Seco Tools AB (2023a).

3. Tolerances

Tolerance class	Tolerances: mm/inch			
	S mm	IC mm	S inch	IC inch
A	0.025	0.025	0.001	0.001
C	0.025	0.025	0.001	0.001
E	0.025	0.025	0.001	0.001
F	0.025	0.013	0.001	0.0005
G	0.130	0.025	0.002	0.001
H	0.025	0.013	0.001	0.0005
J	0.025	0.050	0.001	0.002
K	0.025	0.050	0.001	0.002
M	0.130	0.050	0.002	0.002

Figure 9.18 ISO 1832. Nomenclature of cutting inserts I. *Source:* Seco Tools AB (2023a).

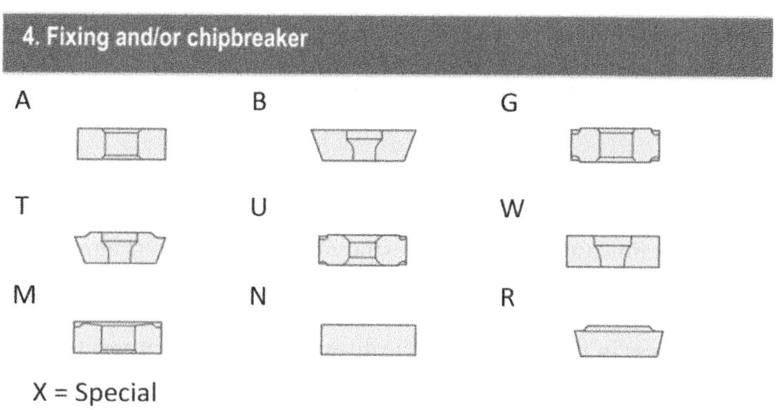

Figure 9.19 ISO 1832. Nomenclature of cutting inserts II. *Source:* Seco Tools AB (2023a).

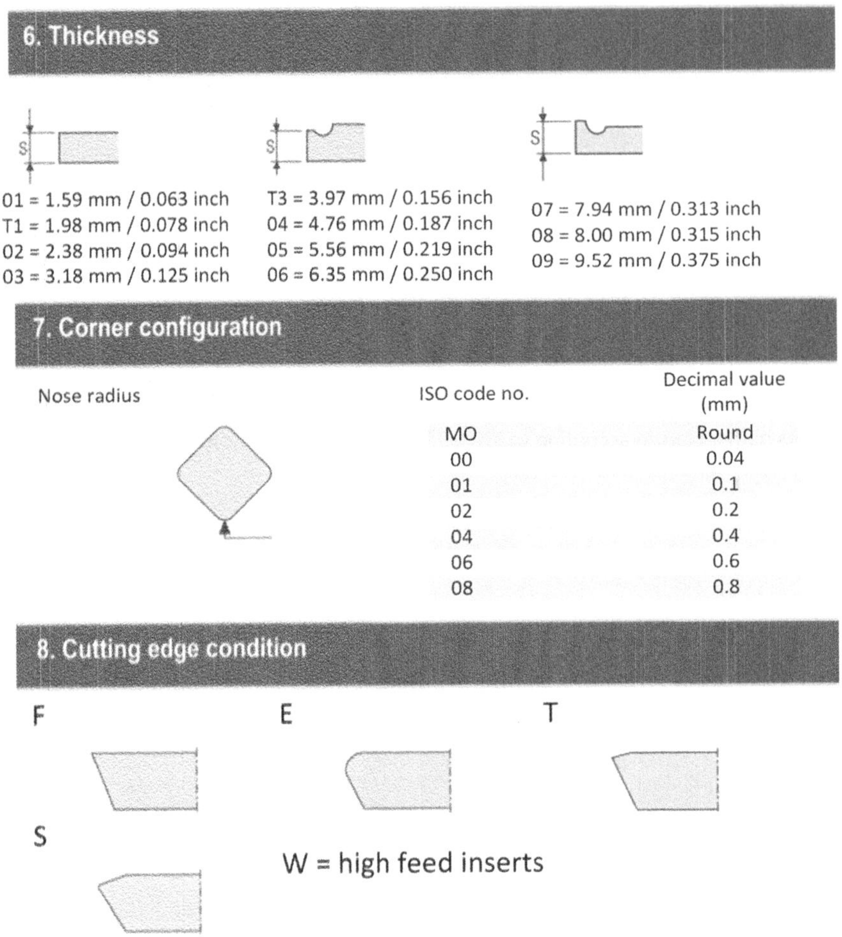

Figure 9.20 ISO 1832. Nomenclature of cutting inserts III. *Source:* Seco Tools AB (2023a).

Table 9.2 Typical feed rates according to cutting tool nose radius for turning operations.

Nose radius (mm)	**0.4**	**0.8**	**1.2**	**1.6**	**2.4**
Feed rate (mm/rev)	0.12–0.25	0.25–0.5	0.36–0.7	0.5–1	0.7–1.6

For finishing operations, special inserts named "wiper" inserts are also available. The nose on a wiper insert is a combination of radii instead of a single radius which makes the nose flatter and more adequate for high-quality surface finish although increases the vibration tendency (Figure 9.21).

9.3.2.3 Cutting Tools in Milling

Milling is a versatile machining process in part because of the diversity of milling cutters that are available. HSS are used to make certain milling cutters, while others use carbide tips and many have replaceable or indexable inserts made of cBN, cermets, ceramics, PCD, and more commonly, coated carbides.

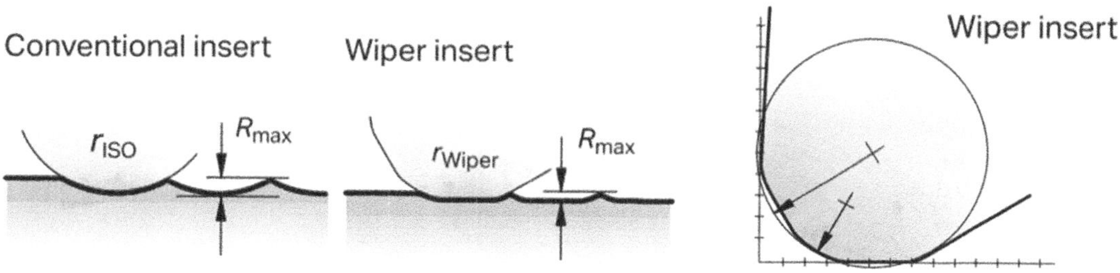

Figure 9.21 The geometry of conventional inserts versus wiper inserts. *Source:* Courtesy of Sandvik Coromant (2017).

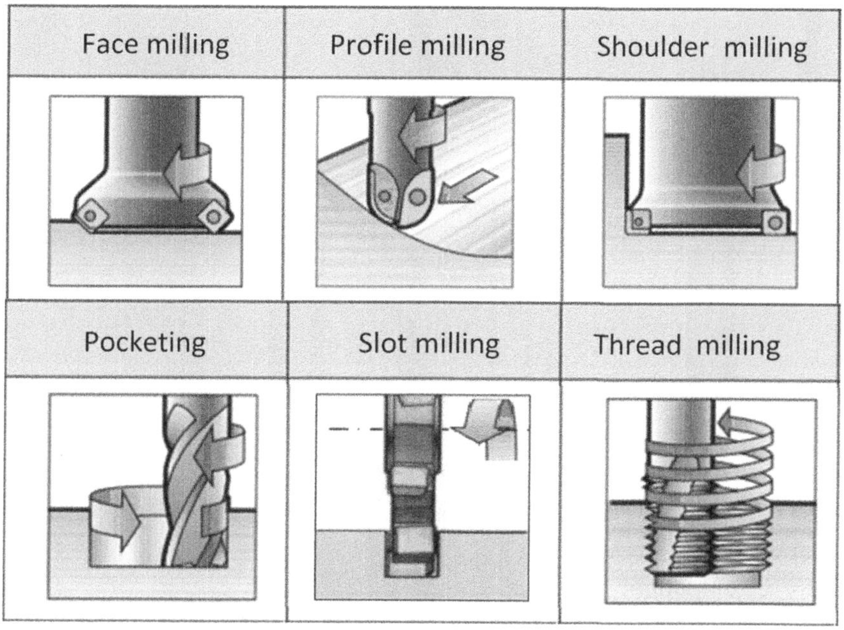

Figure 9.22 Common milling operations. *Source:* Courtesy of Sandvik Coromant.

Cutting tool selection depends on the type of operation to be conducted, as shown in Figures 9.22 and 9.23.

Once the type of cutting tool is chosen, the specific characteristics of the cutting tool must be defined. At this step, the following good practices can be followed:

- Tool size. As a general rule, productivity should be the main goal in machining, and thus, bigger tools with higher removal rates are recommended for any operation. Workpiece dimensions and power limitations may limit the final cutting tool dimension.
- Entering angle. Cutter entering angles also affect the machining performance. A 45° entering tool reduces chip thickness by about 30% as compared to a 90° entering angle. Since smaller entering angles result in smaller chip thickness, cutting tool vendors recommend higher feed rates when smaller entering angles are applied to maintain a proper chip load. High feed cutters usually present very small entering angles (e.g., 10–15°) which lead to very fast feed rates although the maximum depth of cut is highly limited by the entering angle itself. Additionally, increasing the entering angle reduces the axial component of the cutting force as the radial

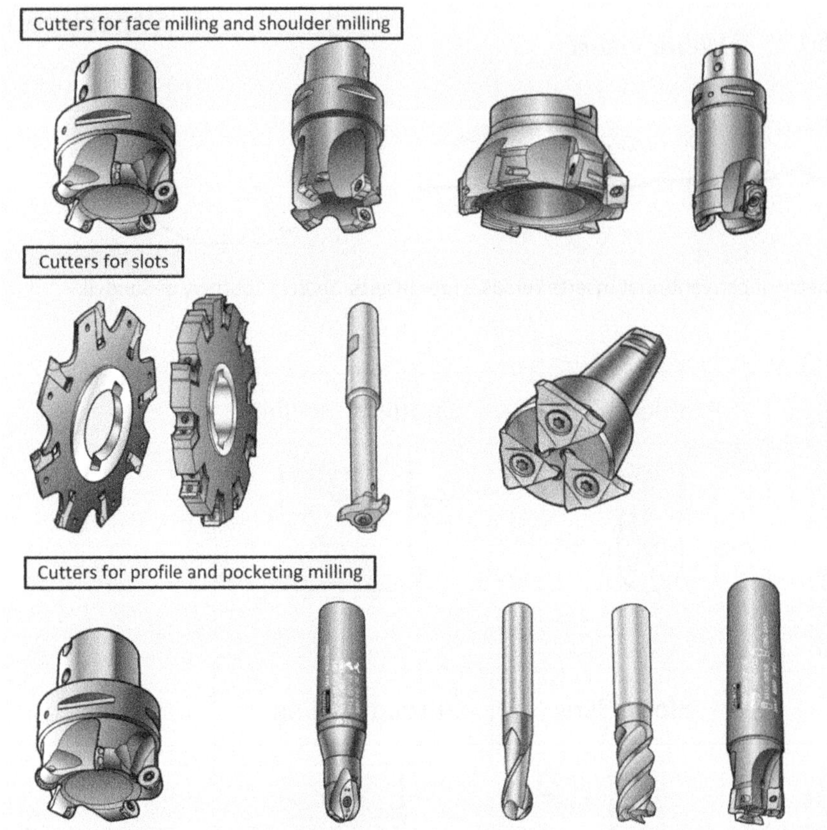

Figure 9.23 Common cutting tools for different milling operations. *Source:* Courtesy of Sandvik Coromant.

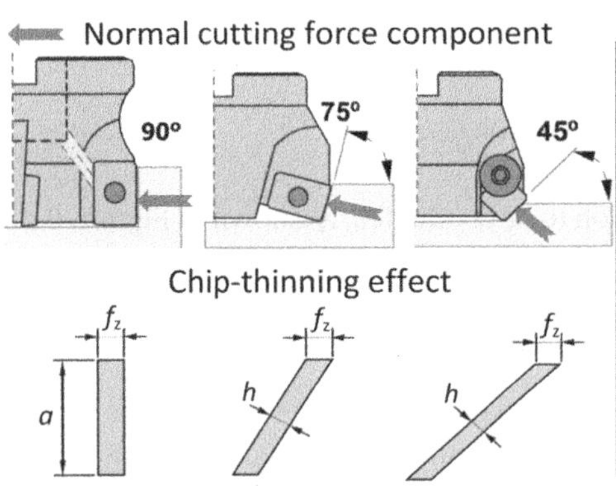

Entering angle	Radial force	Axial force
90°	100%	0%
88°	99%	1%
75°	93%	7%
45°	50%	50%
21°	13%	87%
15°	7%	93%

Figure 9.24 Entering angle and its relationship with normal cutting force direction and chip-thinning phenomena.

component increases (Figure 9.24). This can cause deflection of the machined surface when the workpiece has a thin cross-section. As recommended, use 90° only for operations that require a final 90° geometry, 45° for heavy machining such as face milling or roughing operations, and 60–75° for general and frontal milling operations.

- Cutter geometry. Insert cutting edges can be positioned in positive, neutral, or negative rakes with respect to both radial and axial planes. There are three standard combinations available (see Table 9.3):
  i) Double negative (negative radial and axial rake angles). When horsepower and stiffness are sufficient, double negative geometry is the conventional starting point for rough milling cast iron, steels, and hard materials. The double negative insert design offers the strongest cutting edge available and can withstand greater cutting pressures as well as higher chip loads. Additionally, double-negative cutters call for increased rigidity in the machine, workpiece, and fixture. The employment of reversible inserts is the natural result of this configuration which reduces cutting tool costs.
  ii) Double positive (positive radial and axial rake angles). The power consumption is lower and facilitates chip extraction. However, the tool impact is higher and it should not be used in machining hard materials. Due to this limitation, it works best with nonferrous and soft materials.
  iii) Positive/negative (negative radial rake and positive axial rake angles). Some of the benefits of both double negative and double positive are combined in this setup. It is commonly used in steels and cast irons.
- Cutter density. Different cutter densities can be found in tool holders and very often three categories are distinguished: coarse pitch, normal pitch, and close pitch (Figure 9.25). Chips must be able to correctly form and pass through the cut due to the cutter density. Insufficient chip room can result in chips clogging the gullet, harming the workpiece, and damaging the cutting edge. For soft materials that create continuous chips, in large cuts with a long insert

Table 9.3 Double negative, double positive, and positive/negative geometry.

	Double positive	Positive/negative	Double negative
Graphical representation	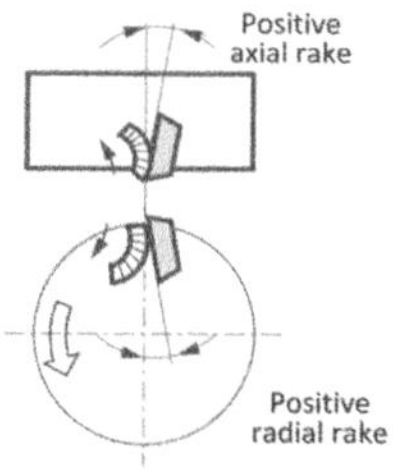	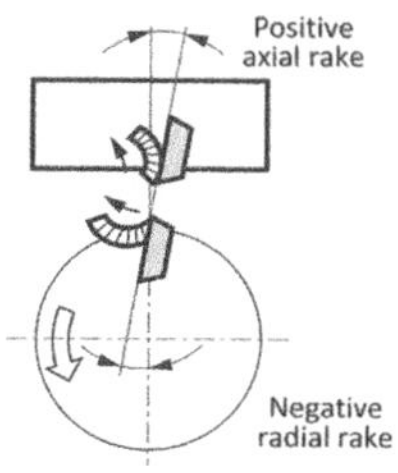	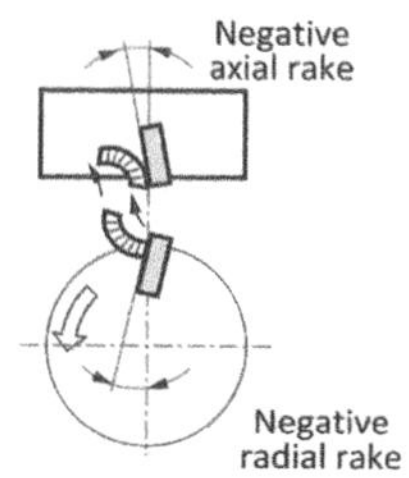
Advantages	Smooth cutting, good chip removal, good surface roughness	Good chip removal, favorable cutting forces	High cutting-edge strength, high productivity, use of both sides of the insert
Disadvantages	Low cutting-edge strength, unfavorable entry contact, only single-sided inserts	Only single-sided inserts can be used	Large cutting forces, chip obstruction
Applications	Machining aluminum alloys and other soft materials that require sharpness	Machining steels, cast iron, and stainless steels, with suitable strength and good evacuation	For hard materials such as cast iron, hardened steels, or irregular surfaces such as castings

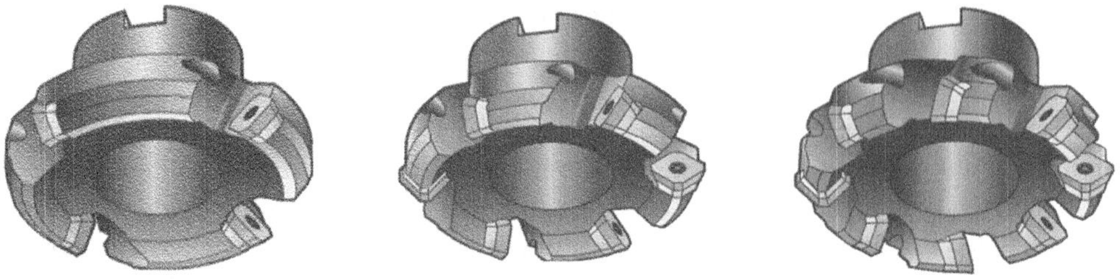

Figure 9.25 Different types of cutter density in face cutters. *Source:* Courtesy of Sandvik Coromant (2017).

engagement, and when stability and power are constrained (small machines), coarse pitch tools are advised since they provide more chip gullet area. On the contrary, close-pitch (high cutter density) tools are better for shallow or narrow cuts where light chip loads are taken. Smaller chips need less gullet space, enabling higher cutter densities. As a general recommendation, a close pitch is preferred for best productivity as long as chip flow is guaranteed.

- Rigidity. Use the most rigid cutter possible, that is, tools with larger diameters and shorter lengths.

A special remark should be made about milling aluminum alloys and other soft materials. The main problem with these alloys is that the chip tends to clog the cutter. To overcome this problem, it is required the use of a cutter that has few teeth (coarse pitch) and larger positive rake angles than cutters for milling steels. Sharp cutting edges are employed to generate good surface roughness and enough clearance is required to prevent heeling. For instance, clearances around 10° are recommended with axial rake angles between 15° and 45°.

HSS mills and indexable carbide mills are the most common type of tools applied in milling. For the case of indexable carbide mills, the milling holder and the type of carbide inserts should be defined at the same time. The main characteristics of holders and carbide inserts for milling are defined in the following.

9.3.2.4 Milling Holders and Cutting Inserts

There is no ISO standard for milling tool holders and thus, tool holders' nomenclature depends on cutting tool vendors although they usually present similar information as those from turning tools. For instance, Figure 9.26 shows the specific designation system used by a tool vendor for milling holders' designation.

The cutting inserts used in milling follow the standard ISO 1832: 2017. To select the proper cutting insert, the main characteristics of cutting insert geometry defined in Figures 9.27–9.30 can be followed.

9.3.2.5 Cutting Tools in Drilling

A general overview of cutting tools in drilling according to the drill diameter and the length-to-diameter ratio is shown in Figure 9.31. The most common tools for drilling are solid drills such as twist drills or solid carbide drills. When closer hole tolerance is required, solid carbides are the first choice, especially for smaller diameters and short to relatively deep holes. Exchangeable tip drills are the first choice for medium-diameter holes as an economical solution. For medium to large diameter holes and medium tolerance demands, indexable insert drills are the first option. In this case, blind holes require a flat bottom. Trepanning drills are used for larger through-hole diameters and where machine power is limited. Gun drills are applied for very large length-to-diameter ratios, for instance, for drilling cooling channels in injection molds.

Milling cutters

In milling, Seco uses product specifi designation systems, there is no ISO system available for cutters. See example below.

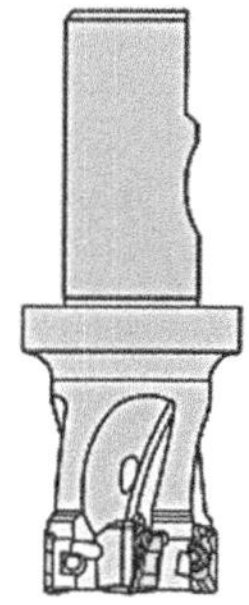

Code Key for Turbo milling cutter 217/220.69

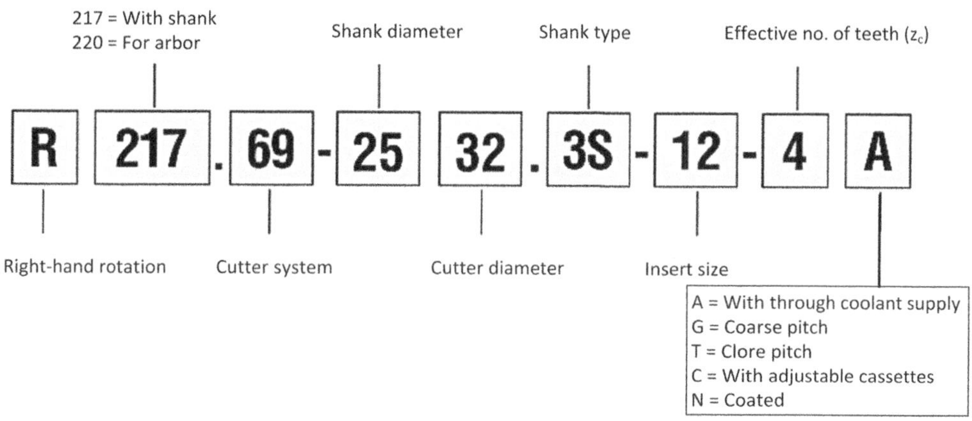

Figure 9.26 Nomenclature used by Seco Tools AB for milling tool holders. *Source:* Seco Tools AB (2023b).

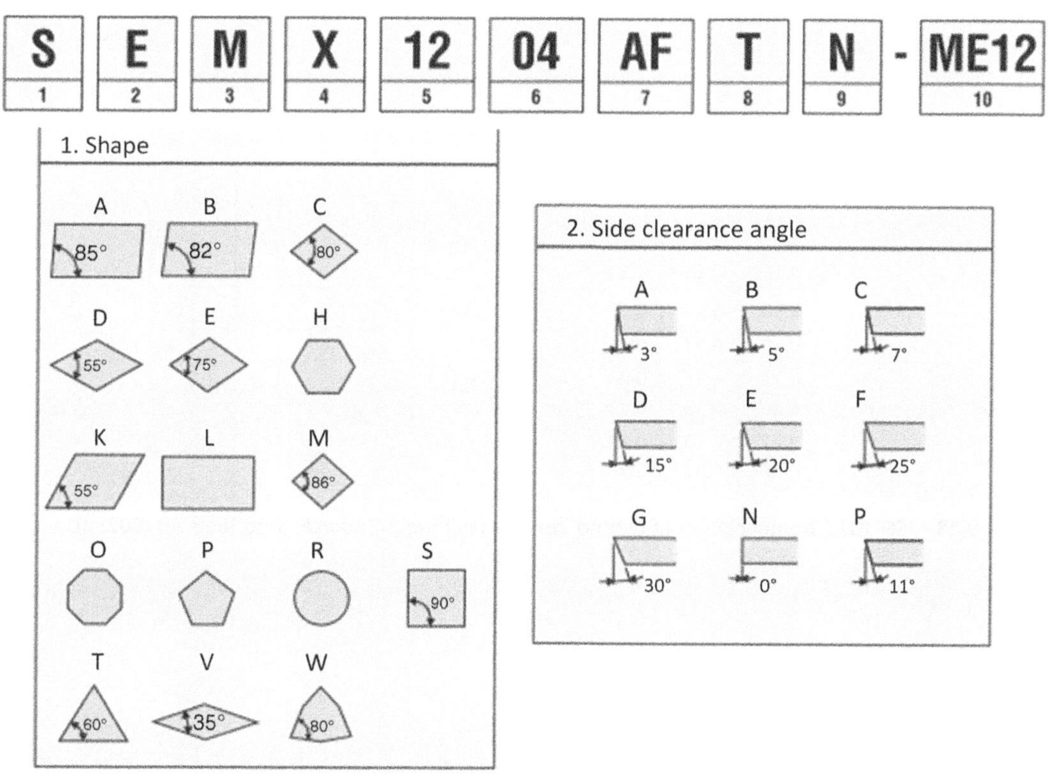

Figure 9.27 ISO 1832. Nomenclature of cutting inserts in milling (I). *Source:* Seco Tools AB (2023b).

S	E	M	X	12	04	AF	T	N	-	ME12
1	2	3	4	5	6	7	8	9		10

3. Tolerances

Tolerance class	Tolerance +/− mm			For *d*, dimension mm									
	m	*s*	*d*	3.175	4.76	6.35	9.525	12.7	15.875	19.05	25.4	31.75	38.1
A	0.005	0.025	0.025	•	•	•	•	•	•	•	•	•	•
E	0.025	0.025	0.025	•	•	•	•	•	•	•	•	•	•
F	0.005	-0.025	0.013	•	•	•	•	•	•	•	•	•	•
G	0.025	0.13	0.025	•	•	•	•	•	•	•	•	•	•
H	0.013	0.025	0.013	•	•	•	•	•	•	•	•	•	•
J	0.005	0.025	0.05	•	•	•	•						
	0.005	0.025	0.08					•					
	0.005	0.025	0.10						•	•			
	0.005	0.025	0.13								•		
	0.005	0.025	0.15									•	•
K	0.013	0.025	0.05	•	•	•	•						
	0.013	0.025	0.08					•					
	0.013	0.025	0.10						•	•			
	0.013	0.025	0.13								•		
	0.013	0.025	0.15									•	•
M	0.08	0.13	0.05	•	•	•	•						
	0.13	0.13	0.08					•					
	0.15	0.13	0.10						•	•			
	0.18	0.13	0.13								•		
	0.20	0.13	0.15									•	•
U	0.13	0.13	0.08	•	•	•	•						
	0.20	0.13	0.13					•					
	0.27	0.13	0.18						•	•			
	0.38	0.13	0.25								•	•	•

Figure 9.28 ISO 1832. Nomenclature of cutting inserts in milling (II). *Source:* Seco Tools AB (2023b).

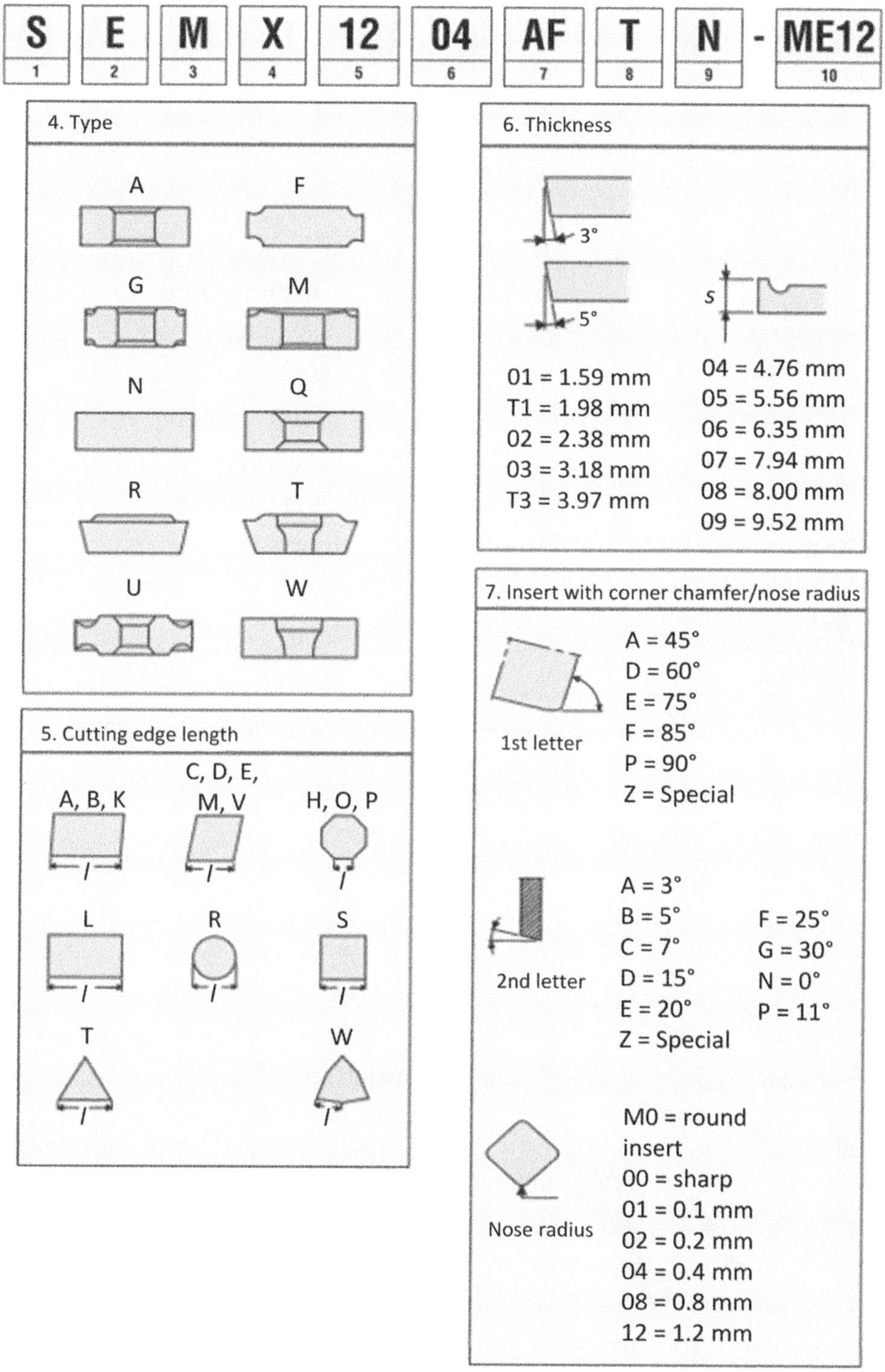

Figure 9.29 ISO 1832. Nomenclature of cutting inserts in milling (III). *Source:* Seco Tools AB (2023b).

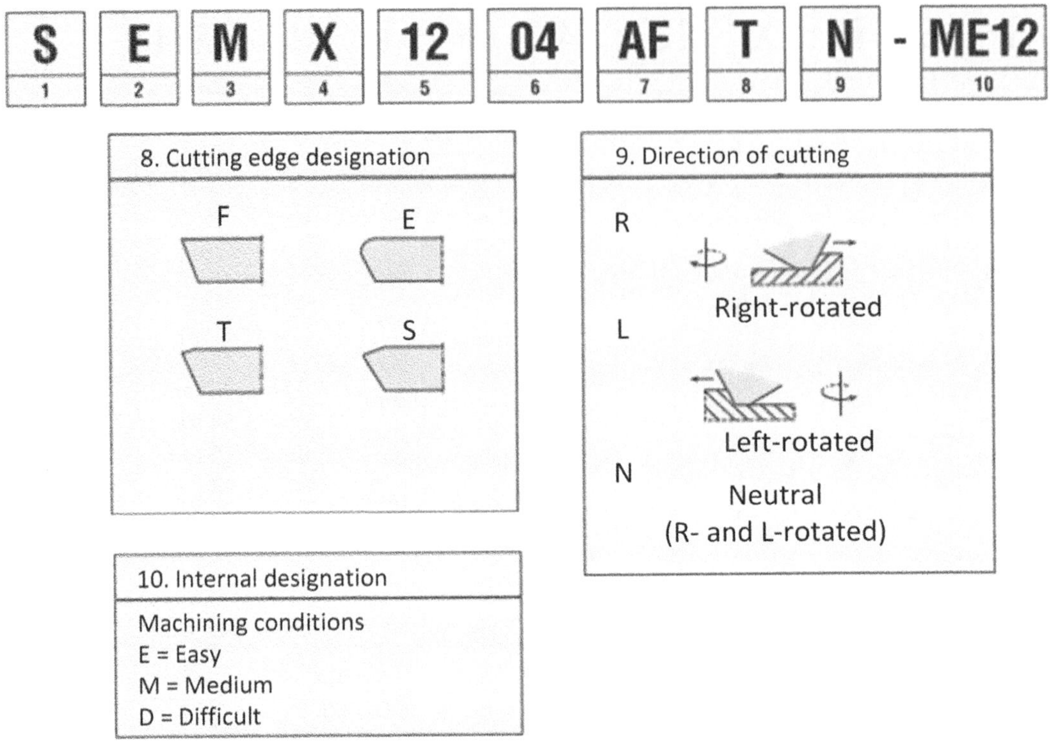

Figure 9.30 ISO 1832. Nomenclature of cutting inserts in milling (IV). *Source:* Seco Tools AB (2023b).

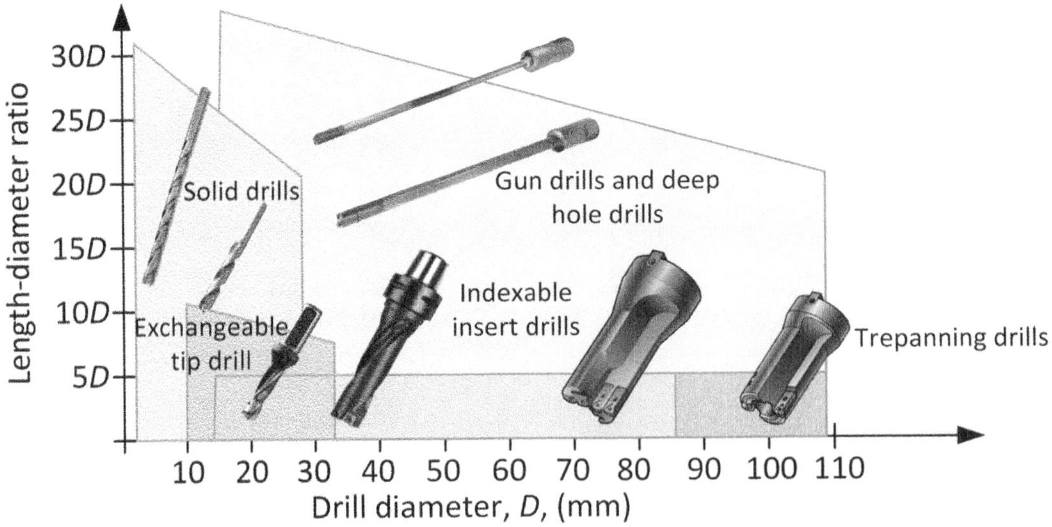

Figure 9.31 General overview of drilling tools and their application according to length-diameter ratio. *Source:* Adapted from Sandvik Coromant (2017).

From the preceding types of drills, twist drills are the most common tools employed. Figure 9.32 shows the main characteristics of a twist drill. For a correct selection of the drill geometry, the following recommendations are given:

- Drill angle (point angle). To drill soft materials (magnesium, aluminum), drill angles are usually 80–110°. For steels and cast irons the drill angle is 118°, whereas for hard materials such as tool steels or stainless steels the drill angle is usually 118–135°.
- Helix angle. Small helix angles are used for drilling hardened materials (Inconel, titanium, etc.) whereas large angles are applied for soft materials (aluminum, copper, etc.).
- Flute length. Flute length is determined by the cutting depth; however, the length must be as short as possible due to tool life issues.
- Internal coolant. Internal coolant is usually recommended to facilitate the chip extraction. This is a key issue in holes deeper than three times the tool diameter, where internal coolant is highly recommended.

Drilling accuracy ranges from IT12 to IT8, and the type of drill, the type of drill holder, and cutting conditions have a clear influence on the resulting hole quality. For instance, a drill clamped using a collet chuck has a good run-out accuracy (usually lower than 5–10 μm), but a standard drill chuck may present a lower run-out accuracy (usually lower than 50 μm).

When drilling with HSS drills, the irregularity of the surface and the unbalanced contact of drilling edges may deflect the entrance of the drill, producing a nonperpendicular hole. If perpendicularity of the hole, position accuracy, or run-out is an issue, a spot drill can be firstly used to guide and limit the drill deflection. Center drills can be also used to guide the drilling operation but they are mainly applied to create a hole to act as a guide/support for a live and dead center in turning operations. Note that if carbide drills are used, spot drilling is not required due to its rigidity compared with HSS drills. Spot drill diameters are chosen to be 50–70% of the ultimate drill diameter.

For drilling holes larger than 8–10 mm in diameter, a sequence of drilling operations is recommended. For this purpose, the rule 1/2-1/3 is applied, which means that a previous hole between 1/2 and 1/3 of the desired drill must be machined. For instance, to drill a hole of Ø36 mm, it is required a drilling operation of Ø15 mm ($1/2\times36 = 18$ mm; $1/3\times36 = 12$ mm), a drilling operation of Ø7 mm ($1/2\times15 = 7.5$ mm; $1/3\times15 = 5$ mm) and a spot drilling, obviously in the inverse order.

For large holes, indexable insert drills are usually the most common tool employed. However, when limited machine power is available, the following methods are recommended:

- Use a trepanning tool, only for through-hole applications.
- Enlarge the hole with a boring tool.
- Use helical interpolation with milling tools.

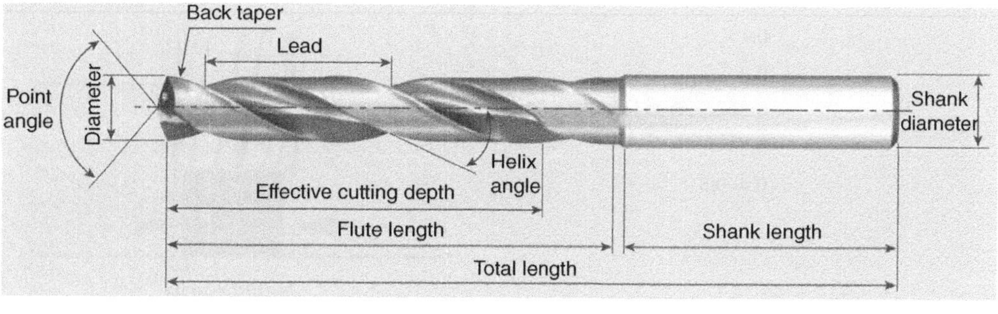

Figure 9.32 Main geometrical characteristics of a twist drill. *Source:* Courtesy of TaeguTec (2023a).

Drilling tools are a better option from a productivity standpoint than milling. However, milling is recommended when chip breakage is an issue or when torque and power should be limited since milling a hole requires less of both.

For machining the entrance of a hole, counterboring, and countersinking operations are conducted (Figure 9.33). They are always conducted before threading or reaming operations.

9.3.2.6 Cutting Tools in Reaming and Boring

Reaming and boring operations require premachined holes that are finished with special tools named reamers and boring tools to increase dimensional/geometrical tolerances and surface quality. For reaming operations, reamers are designed to work with very small depth of cut values, which means that the previous hole should be machined with a slightly smaller diameter than the final reamed hole. Table 9.4 shows the depth of cut recommended for reaming operations according to the final hole diameter. For example, to obtain a final hole with specifications 18H7, a previous hole of 17.75 mm should be drilled, and later, a reamer 18H7 should be used.

Reamers are usually employed to produce holes with IT7-8 since holes drilled usually present a quality of IT9-12. Although adjustable reamers can be used, reamers are usually applied for small holes. When machining medium to large holes with high quality, boring operations are more commonly applied. Boring operations can reach up to IT6, and tools such as single-edge or two-edge tools can be used (Figure 9.34).

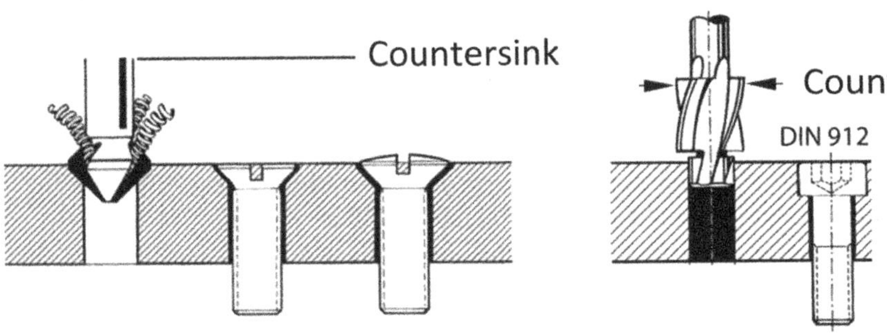

Figure 9.33 Counterboring and countersinking operations. *Source:* Courtesy of Hoffmann GmbH Quality Tools.

Table 9.4 Depth of cut for reaming operations.

Hole diameter (mm)	Material to remove (mm)
1–5	0.15
5–10	0.2
10–20	0.25
20–30	0.3
>30	0.4–0.5

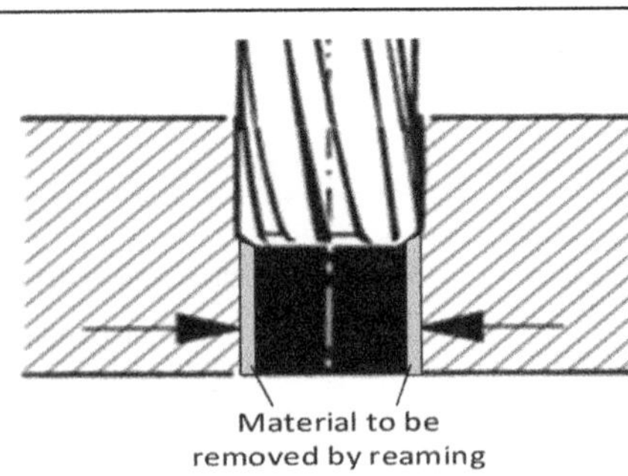

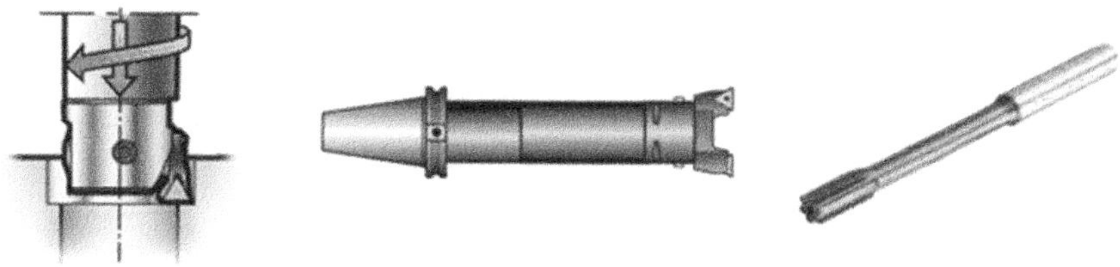
Single-edge boring tool Two-edge boring tool Reamer

Figure 9.34 Boring and reamer tools. *Source:* Courtesy of Sandvik Coromant.

9.3.2.7 Cutting Tools in Threading

For threading holes, different machining operations can be conducted: tapping, thread turning, and thread milling (Figure 9.35). In tapping, a tool named tap is used and its selection depends on the thread metric required. When threading counterbored holes, threading is conducted last (Figure 9.36).

The diameter of the previous hole required before tapping is defined as

$$D_{\mathrm{b}} = d_{\mathrm{ext}} - p$$

where d_{ext} is the nominal diameter and p is the pitch of the threaded hole. For instance, to obtain a threaded hole with the specification $\mathrm{M}10 \times 1.50$, a previous hole of $10 - 1.50 = 8.50\,\mathrm{mm}$ diameter is required.

For internal and external threading operations in turning and milling operations, special cutter inserts are used. Special care should be paid depending on the type of thread (right- or left-hand thread) and the type of hand tool (right or left), as it can be seen in Figures 9.37 and 9.38.

For external operations, the diameter required previous to the threading operation is defined as

$$D_{\mathrm{p}} = d_{\mathrm{ext}} - 0.1 \times \mathrm{p}$$

For instance, threading a bar with the specification $\mathrm{M}8 \times 1.25$ requires an initial bar diameter of $8 - 1.25 \times 0.1 = 7.875\,\mathrm{mm}$.

Thread turning Tapping Thread milling

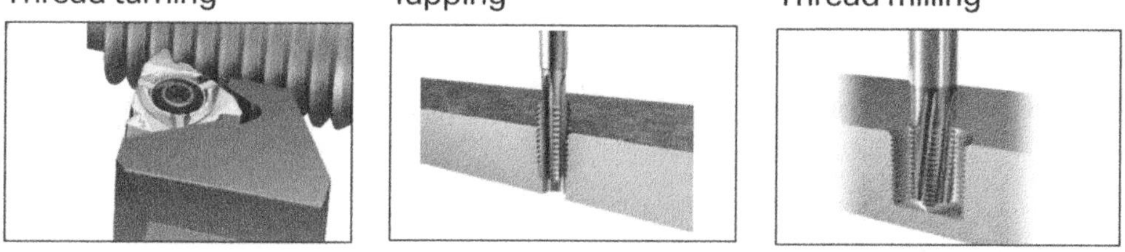

Figure 9.35 Methods for threading. *Source:* Courtesy of Sandvik Coromant.

Figure 9.36 Sequence for threading a counterbored hole: drilling, counterboring, tapping. *Source:* Courtesy of Hoffmann GmbH Quality Tools.

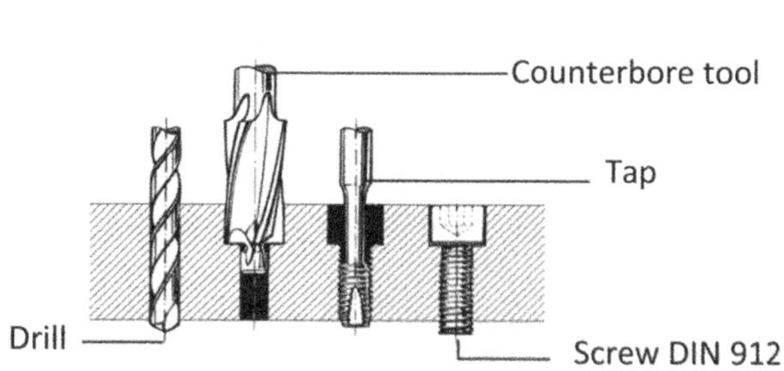

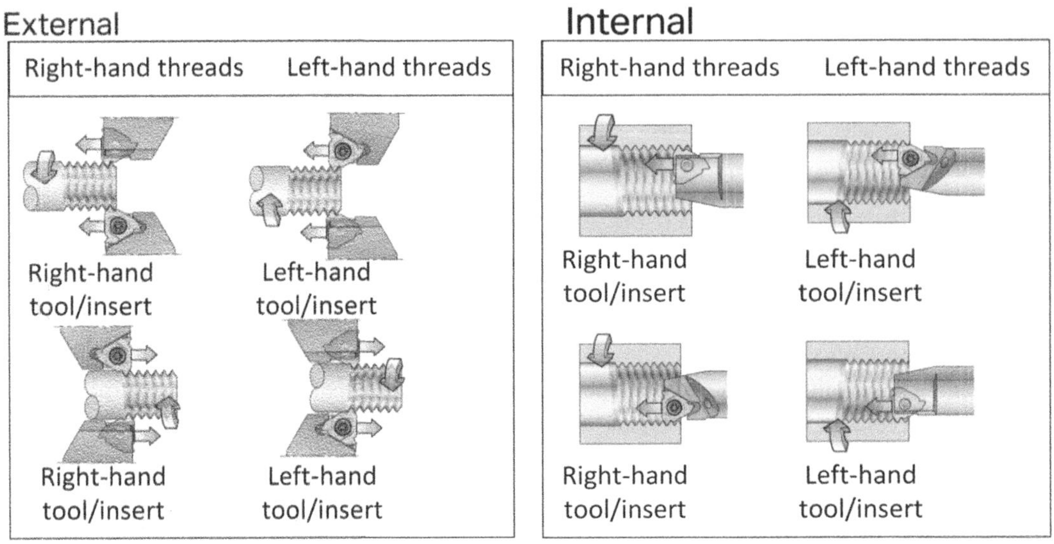

Figure 9.37 External and internal thread turning. *Source:* Courtesy of Sandvik Coromant.

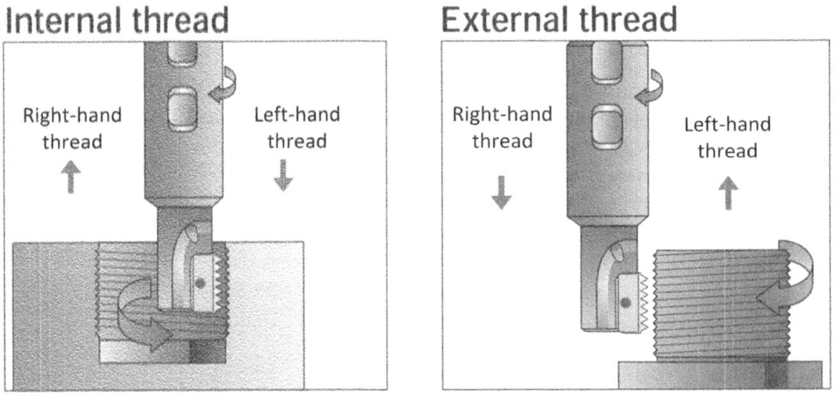

Figure 9.38 External and internal thread milling. *Source:* Courtesy of TaeguTec LTD (2023b).

9.3.3 Practical Example

In this section, cutting tool selection is conducted for the example studied in Chapter 8. To remind the example, Figure 9.39 and Table 9.5 show the engineering drawing and the resulting machining sequence of the example, respectively. After determining the machining sequence, the next step is the selection of machine tools and cutting tools. For this practical example, it is assumed that the machine tools are predefined (i.e., only one CNC lathe and one CNC machining center is available), and the cutting tools available are assumed to be those shown in Section 9.4.

The cutting tool selection is mainly done by analyzing the operation to be conducted, accessibility issues, and the compatibility with the workpiece material, aluminum AISI SC64D

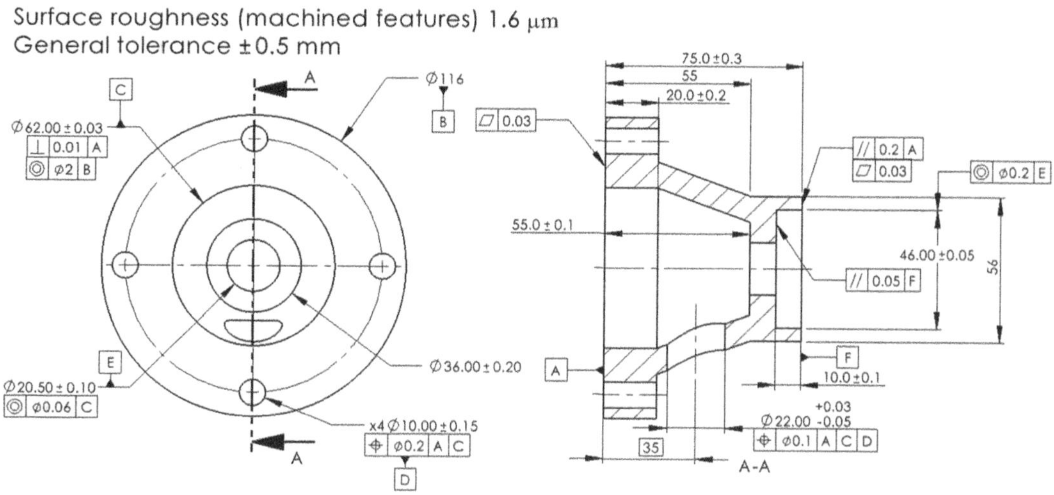

Figure 9.39 Part drawing of example in Chapter 8.

Table 9.5 Machining sequence for the example in Chapter 8.

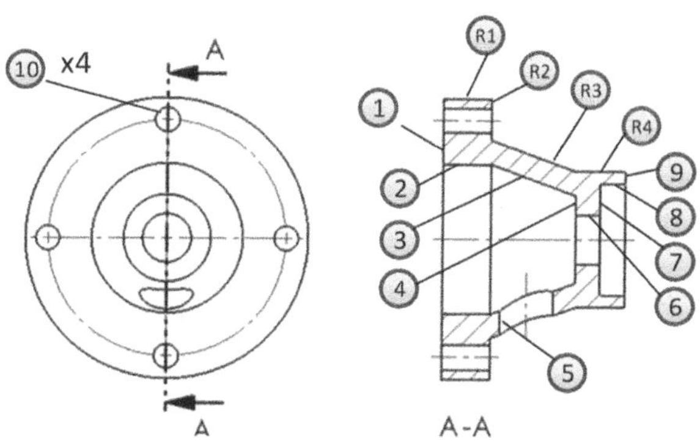

- Job 1 (Lathe)
  - Subjob 1:
    - Face turning (1f)
    - Internal turning (2–3–4f)
    - Internal turning (6f)
  - Subjob 2:
    - Face turning (9f)
    - Internal face turning (7–8f)
- Job 2 (Machining center)
  - Subjob 1:
    - Drilling (10d)
  - Subjob 2:
    - Drilling (5d)
    - Boring (5b)

(SAE 326), which corresponds to the N code according to ISO 513. The selected cutting tools and a brief explanation of their choice are shown below.

- Job 1 (Lathe)
  - Subjob 1:
    - Face turning (1f)
      - Tool T11 D3. Tool holder PCLNL 2020K12 Insert CNMG 120408 Iscar IC20. To machine surface 1, a left-handed tool is required. From the straight turning/face turning tools, two left options are available: tool T2 (SSSCL 1212F09) and tool T11 (PCLNL 2020K12). However, tool T2 is not indicated for machining N code materials, and thus, tool T11 is selected.
    - Internal turning (2-3-4f)
      - Tool T13 D31. Tool holder S12M-STFCL-11 and inserts TCMT 110202 Garant HU7310. According to the tools available and the application material (code N), two boring tools can be used, T13 and T4. Note that T4 is a boring bar of Ø25 mm, and surface 4 ends with a diameter of 20.5 mm. Thus, only T13 can be employed.
    - Internal turning (6f)
      - Tool T13 D31. Tool holder S12M-STFCL-11 and inserts TCMT 110202 Garant HU7310. Similar to the previous operation, only T13 can be used since surface 6 is Ø20.5 mm.
  - Subjob 2:
    - Face turning (9f)
      - Tool T11 D3. Tool holder PCLNL 2020K12 Insert CNMG 120408 Iscar IC20. The same reasoning as subjob 1 is conducted.
    - Internal turning (7-8f)
      - Tool T13 D31. Tool holder S12M-STFCL-11 and carbide inserts TCMT 110202 Garant HU7310. The same reasoning as subjob 1 is conducted.
- Job 2 (Machining center)
  - Subjob 1:
    - Drilling (10d)
      - Tool T132. Drill holder with a spot drill TR-BR-2F-6-HSS. This is the only spot drill available.
      - Tool T140. Drill holder with twist drill TR-BR-2F-10-HSS. Drill selected since is the only one compatible with the diameter required, 10 mm.
  - Subjob 2:
    - Drilling (5d)
      - Tool T143. Drill holder PP-IN-2P-21 with carbide inserts WOEX 040304 Garant HB7310. Drill selected to machine a hole to Ø21 mm and keep 1 mm thickness for a later boring operation.
    - Boring (5b)
      - Tool T150. Fine boring head PP-BO-1P-(18-25) with carbide inserts TOGX 06 T104-14 Komet BK8430. Boring tool compatible with the final diameter that is being machined.

9.4 Example Data: Shop-Floor Tool Tables

Tables 9.6–9.9 show examples of tool tables for a CNC lathe and a CNC machining center.

Table 9.6 Example of cutting tools available for a CNC lathe (external tools).

Operation	Holder (ISO)/ vendor's reference	Insert (ISO)/vendor's reference – Brand and grade	Application group (ISO)	Tool position
Grooving (cut-off)	XLCT N2603 J04	PTNT04 -Canela TN30-	P	T12 D60
Grooving (cut-off)	G2042.32N-4T50SX-P	SX-4E400R6-CF5 -Walter WSM33S-	P M K S	[a]
Grooving (cut-off)	QD-NN2G36-21A	QD-LG-0300-0501-CO -Sandvik 1125-	P M	[a]
Straight turning Face turning (left hand)	PCLNL 2020K12	CNMG 120408 -Iscar IC20-	N S	T11 D3
Straight turning Face turning (left hand)	SSSCL 1212F09	SCMT 09T3 -Sandvik MF 1125-	P M S	T2 D5
Straight turning Face turning (right hand)	SSSCR 1212F09	SCMT 09T3 -Sandvik MF 1125-	P M S	[a]
Straight turning Face turning (right hand)	SSSCR 2525M12	SCMT 1204 -Tungaloy AH725-	P M	[a]
Contour turning (left hand)	SVLCL 2020K13	VCMT 130304 - Iscar IC830-	P20-P40 M25-M35 S15-S30	T9 D06
Contour turning (left hand)	SVHCL 2020K16	VCGX 160404 K10 -STD U4010-	N	[b]D12
Contour turning (left hand)	SDJCL 1010E07	CCGW 060202 -Garant BU7220-	H	[a]
Contour turning (right hand)	SVLCR 2020K13	VCMT 130304 - Iscar IC830-	P20-P40 M25-M35 S15-S30	T3 D07
Contour turning (right hand)	SVHCR 2020K16	VCGX 160404 K10 -STD U4010-	N	[b]D11
Contour turning (right hand)	SDJCR 1010E07	CCGW 060202 -Garant BU7220-	H	[a]
Contour turning (neutral)	SVVCN 2020K16	VCGX 160404 K10 -STD U4010-	N	[b]D10
Contour turning (neutral)	SRDCN 1212F06	RC.T 0602 -Garant HU7310-	N	[a]
External threading (left handed)	SEL1212H16	16ER AG60 -Hard Tech HM3020-	M20-M30 P15-P30	[a]
External threading (right handed)	SER1010H11	11ER AG60 -Hard Tech HM3020-	M20-M30 P15-P30 P15-P30	[a]

[a] Without assigned tool position and corrector value.
[b] Without assigned tool position.

Table 9.7 Example of cutting tools available for a CNC lathe (internal tools and drills).

Operation	Holder (ISO)/ vendor's reference	Insert (ISO)/vendor's reference – Brand and grade	Application group (ISO)	Tool position
Boring	S12M-STFCL-11	TCMT 110202 - APT Carbide M250-	P35 M25 S15	[a]
		TCMT 110202 - Garant HU7310-	N	T13 D31
Boring	E12M-STLDR-07	TDAT 07010 -Garant HB7035-	P K	T1 D1
Boring	S25T-STLDLS4	CDCT S4010 -Garant HU7310-	N	T4 D4
Drilling	Drill holder	HSS drill, 135°, Ø6 mm, $Lc = 0$ mm, $L = 70$ mm	P30-P40 M S	[a]
Drilling	Drill holder	HSS drill, 135°, Ø8 mm, $Lc = 75$ mm = 100 mm	P30-P40 M S	[a]
Drilling	Drill holder	HSS + TiAlN, 118°, Ø14 mm, $Lc = 51$ mm, $L = 70$ mm	P M	[a]
Centering	Drill holder	Center drill, HSS, 60°, Ø5 mm, $L = 35$ mm	P N	[a]
			P15-P30	
Internal threading (left handed)	SNL0013M16	16IR AG60 -Haas HU30T-	M K P	[a]
Internal threading (right handed)	SNR0010K11	11IR AG60 -Hard Tech HM3020-	M20-M30 P15-P30	[a]

L, overhang length; Lc, flute length.
[a] Without assigned tool position and corrector value.

Table 9.8 Example of tool table for a CNC machining center (indexable tools).

Operation	Holder tool designation	Insert (ISO) – Brand and grade	Application group (ISO)	Dimensions (in mm)	Tool
Face milling	PP-PL-6P-80	SEKW 1204 AFTN -Garant HB7515-	P M K	Ø80, $\kappa = 45°$, $z = 6$	1
		SEHT 1204 -Garant HB7810-	N		
Face milling	PP-PL-4P-50	SEKT 1204 AFTN -Garant HB7635-	M	Ø50, $\kappa = 45°$, $z = 4$	3
		SEHT 1204 -Garant HB7810-	N		
Face milling	PP-PL-3P-40	SEKW 1204 AFTN -Garant HB7515-	P M K	Ø40, $\kappa = 45°$, $z = 3$	[a]
		SEHT 1204 -Garant HB7810-	N		
Shoulder milling	PP-ES-3P-25	ADHX 110305 ER -Fette LC240T-	P M K	Ø25, $\kappa = 90°$, $z = 3$	11
Shoulder milling	PP-ES-4P-32	ADHX 110305 ER -Fette LC630T-	P	Ø32, $\kappa = 90°$, $z = 4$	12
Profile milling	PP-CO-2P-16	RDHX 0501 MOFN -Garant HB7810-	N	Ø16, $z = 2$	17
Profile milling	PP-CO-2P-25	RDHW 0802 MO -Fette LC610T-	K N	Ø25, $z = 2$	22
Profile milling	PP-CO-4P-50	RDHW 1003 MO -Fette LC630T-	P	Ø50, $z = 4$	[a]
Drilling	PP-IN-2P-20	WCMT 040208 -Saratools APM25T-	M	Ø20, $z = 2$, $L = 80$	31
Drilling	PP-IN-2P-25	WCMT 050308 -Saratools APM25T-	M	Ø25, $z = 2$, $L = 85$	32
Drilling	PP-IN-2P-21	WOEX 040304 -Garant HB7640-	P M S	Ø27, $z = 2$, $L = 75$	[a]
		WOEX 040304 -Garant HB7310-	N		143
Fine boring head	PP-BO-1P-(18-25)	TOGX 1403 06 T102 -Komet BK60-	P M K	Ø18–Ø25 (adjustable), $z = 1$, $L = 63$	[a]
		TOGX 06 T104-14 -Komet BK8430-	P M K N		150

Holder/solid tool designation AA-BB-CC-DD-EE-FF:

- AA. Type of tool, PP: Indexable cutter, FR: Solid mill, TR: Drill.
- BB. Type of application, PL: Face cutter, ES: Shoulder milling, CO: Profiling, BR: Twist drill, IN: Insert drill, BO: Boring. FR-AC: Flat endmill, FR-CO: Ball endmill, AV: Countersink, CB: Counterboring, MA: Tap.
- CC. Number of inserts/teeth. 6P: 6 inserts. 2F: 2 teeth.
- DD. Diameter.
- EE. Tool material. HSS: High-speed steel, SC: Solid carbide.
- FF: Coating if applicable. TiCN: Titanium Carbo-Nitride, TiAlN: Titanium Aluminum Nitride, AlCrN: Aluminum Chrome Nitride, TiN: Titanium Nitride.

L, overhang length; z, number of teeth.

[a] Without assigned tool position and corrector value.

Table 9.9 Example of tool table for a CNC machining center (solid tools).

Operation	Solid tool designation	Insert (ISO)	Application group (ISO)				Dimensions (units in mm)	Tool
End milling	FR-AC-2F-06-HSS	—	P		K	N	Ø6, $\kappa = 90°$, $z = 2$, $Lc = 12$, $L = 35$	75
End milling	FR-AC-3F-08-HSS-TiAlN	—	P	M	N	K	Ø8, $\kappa = 90°$, $z = 3$, $Lc = 15$, $L = 45$	77
End milling	FR-AC-2F-10-HSS	—	P		N	K	Ø10, $\kappa = 90°$, $z = 2$, $Lc = 22$, $L = 55$	79
End milling	FR-AC-3F-12-SC-TiAlN	—	P	M	N	K	Ø12, $\kappa = 90°$, $z = 3$, $Lc = 25$, $L = 60$	82
Profile milling	FR-CO-2F-06-HSS	—	P		N	K	Ø6, $z = 2$, $Lc = 12$, $L = 60$	116
Profile milling	FR-CO-2F-10-HSS-TiAlN	—	P	M	N	K	Ø10, $z = 2$, $Lc = 20$, $L = 60$	121
Profile milling	FR-CO-3F-12-HSS-TiAlN	—	P	M	N	K	Ø12, $z = 3$, $Lc = 24$, $L = 65$	122
Profile milling (engraving)	FR-CO-2F-2-HSS-TiAlN	—	P	M	N	K	Ø2, $z = 2$, $Lc = 4$, $L = 25$	[a]
Spot drill	TR-BR-2F-6-HSS	—	P		N		Ø6, 145°, $L = 40$	132
Center drill	TR-BR-2F-3-HSS	—	P		N		Ø3, 60°, $L = 35$	133
Drilling	TR-BR-2F-06-HSS	—	P				Ø6, 135°, $Lc = 40$, $L = 70$	136
Drilling	TR-BR-2F-8-HSS-TiAlN	—	P		K	N	Ø8, 118°, $Lc = 50$, $L = 85$	138
Drilling	TR-BR-2F-10-HSS	—	P		N		Ø10, 118°, $Lc = 60$, $L = 90$	140
Drilling	TR-BR-2F-4-SC-TiCN	—	P				Ø4, 118°, $Lc = 25$, $L = 45$	142
Countersinking	TR-AV-3F-8-HSS	—	P		N		Ø8, 120°, $z = 3$, $L = 30$	182
Countersinking	TR-AV-3F-6-HSS-TiAlN	—	P		N	M	Ø6, 90°, $z = 3$, $L = 30$	185
Counterboring	TR-CB-3F-06-HSS-TiAlN	—	P		N	M	Ø4, $z = 3$, $L = 30$	202
Counterboring	TR-CB-3F-08-HSS-TiAlN	—	P		N	M	Ø8, $z = 3$, $L = 30$	204
Threading (tap)	TR-MA-3F-04-HSS	—	P				Ø4, $L = 30$	222
Threading (tap)	TR-MA-3F-08-HSS-TiN	—	P				Ø8, $L = 40$	225

Holder/solid tool designation AA-BB-CC-DD-EE-FF:

- AA. Type of tool. PP: Indexable cutter, FR: Solid mill, TR: Drill.
- BB. Type of application. PL: Face cutter, ES: Shoulder milling, CO: Profiling, BR: Twist drill, IN: Insert drill, BO: Boring. FR-AC: Flat endmill, FR-CO: Ball endmill, AV: Countersink, CB: Counterboring, MA: Tap.
- CC. Number of inserts/teeth. 6P: 6 inserts. 2F: 2 teeth.
- DD. Diameter.
- EE. Tool material. HSS: High-speed steel, SC: Solid carbide.
- FF. Coating if applicable. TiCN: Titanium Carbo-Nitride, TiAlN: Titanium Aluminum Nitride, AlCrN: Aluminum Chrome Nitride, TiN: Titanium Nitride.

L, overhang length; Lc, flute length; z, number of teeth.
[a] Without assigned tool position and corrector value.

Questions and Exercises

1 Select the most appropriate cutting tool material for the following cases: (a) machining a mold made of AISI D3 steel (high alloy steel), (b) machining operation with interrupted cuts where cutting tool toughness should be maximized, and (c) nonferrous operations where tool's life should be maximized.

2 Identify the main characteristics of the cutting inserts with the following ISO codes: (a) VCMT 130304, (b) CNMG 120408, and (c) TCMT 110202.

3 For a boring operation, select the most appropriate tool from Table 9.10.

Table 9.10 Tools available for Exercise 3.

Tool	Holder (ISO)	Insert (ISO)
T1 D5	SVLCL 2020K13	VCMT 130304
T3 D10	SSSCR 2525M12	SCMT 1204
T6 D20	S12M-STFCL-11	TCMT 110202
T11 D03	PCLNL 2020K12	CNMG 120408

4 Explain the main benefits and drawbacks of wiper inserts with respect to conventional ones.

5 Consider a part where a M12×1.75 thread hole must be machined. Explain the machining sequence required to produce the part.

6 Consider a cylindrical bar in which one of its sides is threaded with the specification M20×1.5. Explain the machining sequence required to produce the part.

7 For the 3D part shown in Figure 9.40, write the operations sheet for machining the part indicating the machining operation (e.g., face milling, drilling, etc.) and the cutting tool used. Assume that the raw material is a prismatic part made of low-carbon steel with dimensions

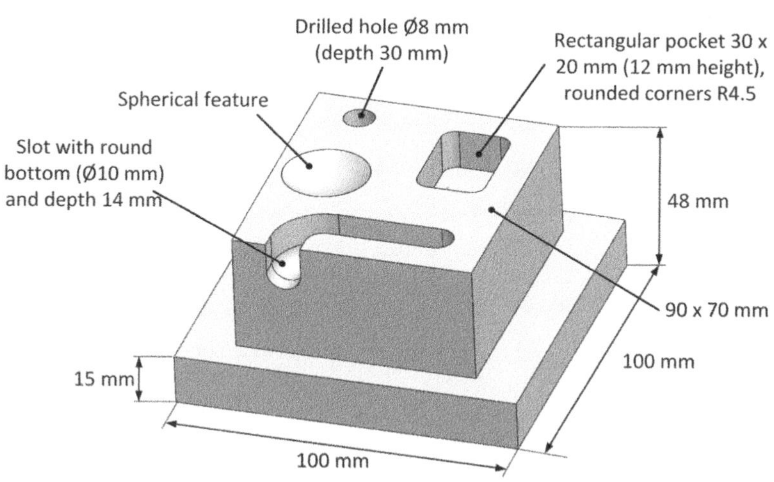

Figure 9.40 3D part to be machined. Exercise 7.

100 × 100 × 50 mm, and only one setup is required. For the cutting tools, select the most adequate from the tables in Section 9.4.

8 For the engineering drawing shown in Figure 9.41, write the operations sheet for machining the part indicating the machining operation (e.g., face turning, contour turning, etc.) and the cutting tool used. Assume that the raw material is a stainless-steel bar of Ø62 mm, only one setup is required, and the final operation is a cut-off operation. For the cutting tools, select the most adequate from the tables in Section 9.4.

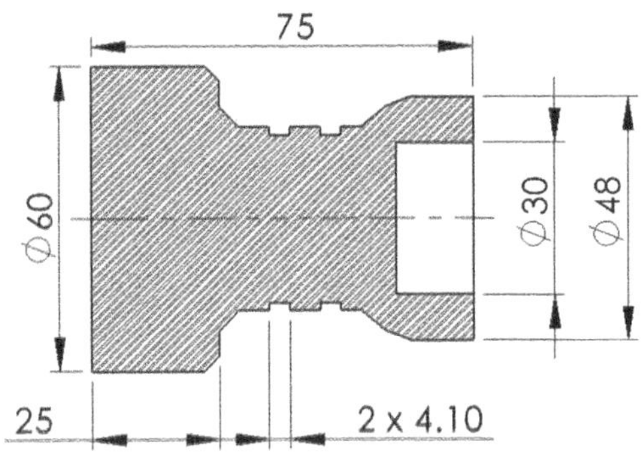

Figure 9.41 Engineering drawing for Exercise 8. Units in mm.

References

Mitsubishi Materials. (2023). Technical information: cutting tool materials. http://www.mitsubishicarbide.net (accessed May 2023).

Sandvik Coromant. (2017). *Training Handbook*. Metal Cutting Technology http://www.sandvik.coromant.com.

Scallan, P. (2003). *Process Planning: The Design/Manufacture Interface*. Butterworth-Heinemann.

Seco Tools AB. (2023a). Catalog and technical guide. Turning. http://www.secotools.com (accessed May 2023).

Seco Tools AB. (2023b). Catalog and technical guide. Milling. http://www.secotools.com (accessed May 2023).

TaeguTec LTD. (2023a). Technical guide – holemaking. https://www.taegutec.com (accessed May 2023).

TaeguTec LTD. (2023b). Technical guide – thread making. https://www.taegutec.com (accessed May 2023).

10

Process Parameter Selection

10.1 Introduction

In machining operations, process parameters should be carefully selected for achieving high-quality, efficient, and cost-effective machining processes. Machining costs due to cutting tool costs and power consumption are directly related to cutting parameter optimization. At the same time, cutting parameters have a direct impact on surface roughness and dimensional accuracy on machined parts, so process parameter selection may be not a straightforward operation.

In this chapter, it is explained how to select the cutting conditions for machining processes such as turning, milling, or drilling. Cutting conditions usually refer to cutting speed, feed rate, and depth of cut, and their selection mainly depends on the employed cutting tool and workpiece material, the limitations of the machine tool, and the dimensional and geometrical specifications of the part. For instance, surface roughness may limit feed rates, and geometrical errors due to cutting tool deflection may limit the depth of cut and feed rates.

This chapter shows a stepwise procedure to determine the cutting conditions for machining operations and it also describes some specific guidelines and good practices. A brief explanation about which type of cutting fluids should be employed is also provided since they are also considered part of cutting conditions.

10.2 Methodology

10.2.1 Stepwise Procedure

Typically, the cutting parameters are determined and optimized by adopting a stepwise procedure. The first group of parameters to be set up is the cutting speed (V_c) and the feed rate (f_n) or feed per tooth (f_z). These two parameters are defined according to the type of cutting tool and workpiece material. Cutting tool vendors provide tables and cutting data for the correct selection of their values. For the reader's convenience, Appendix 2 shows some generic tables for cutting speed and feed rate selection for different types of tools, workpiece materials, and operations.

Although ranges of cutting speed and feed rate are defined according to the tool vendor's recommendations, the optimal parameters from the recommended ranges can be obtained in terms of costs or productivity. If cutting is conducted at high-speed ranges, tool life is determined by

Manufacturing Process Planning: A Practical Approach for Mechanical Engineering, First Edition.
José V. Abellán-Nebot, Carlos Vila Pastor, and Héctor R. Siller.

Companion website: www.wiley.com/go/Manufacturing_Process_Planning_1e

the temperature-dependent wear processes involving deformation and diffusion. Under these conditions, the relationship between cutting speed, V_c, and tool life, T, can be expressed using Taylor's equation (Stephenson and Agapiou 2016, Chapter 9):

$$V_c \times T^n = C$$

where C and n are constants for the material and conditions used. Typical values of n are 0.1–0.17 for HSS tools, 0.2–0.25 for uncoated WC tools, 0.3 for TiC or TiN-coated WC tools, and 0.4 for Al_2O_3-coated WC tools. This empirical equation shows that when cutting speed V_c increases, the feasible tool life T decreases and vice versa. A greater value of V_c will decrease the machining time and hence lower the cost of machining time per workpiece, but it will also shorten tool life, increasing the cost of the tool per workpiece. At a specific cutting speed V_{cc}, when these two expenses move in opposite directions, their combined sum per workpiece crosses a minimum. The position of the cost-optimized cutting speed V_{cc} is shown in Figure 10.1, along with how the various cost components change as a function of cutting speed. Any deviation from this cost-optimized cutting speed V_{cc} will increase the unit cost of production as either the machining time or the tool costs will rise. A similar reasoning can be conducted considering the time per piece required for machining. A higher cutting speed increases production so that the time required per piece is reduced, but the wear rate increases so the tool replacement time increases. It can be seen that V_{cp} is the cutting speed that maximizes the production since the time required for machining a part is minimal.

Furthermore, for roughing operations a higher value of feed rate within the recommended range is used to increase productivity since tolerance requirements apply to finishing operations. In case

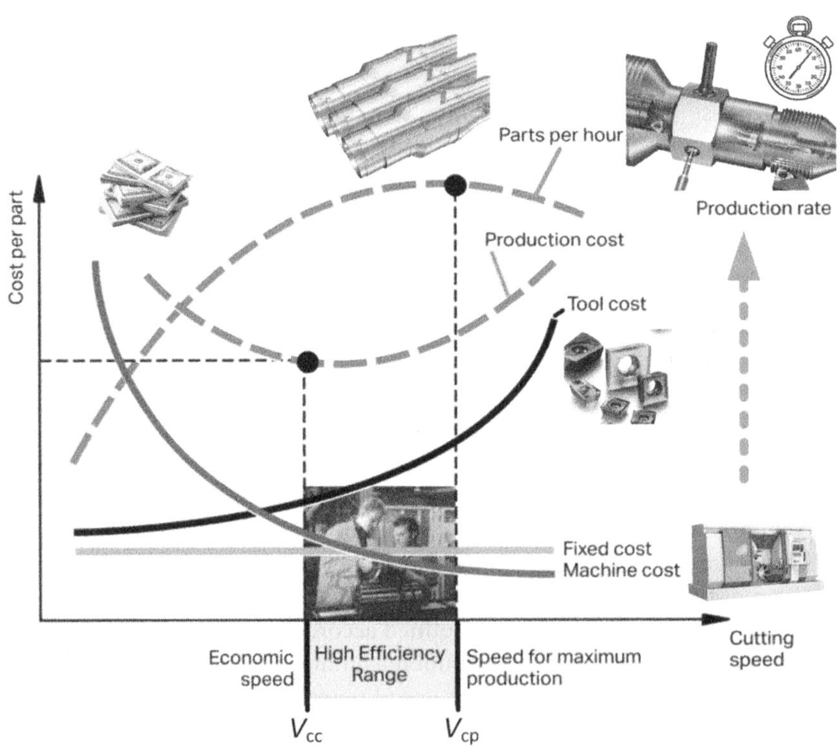

Figure 10.1 Production cost and parts per hour as a function of cutting speed. Cutting speeds for minimum production cost (V_{cc}) and maximum production rate (V_{cp}). *Source:* Sandvik Coromant (2017).

of finishing operations, the feed rate is constrained by the required surface roughness or geometrical quality and therefore will be fixed by these requirements.

A second group of parameters define the depth of cut, which is defined by two parameters in milling operations: the radial depth of cut (a_e) and the axial depth of cut (a_p). These parameters define the material removal rate, and therefore, the power requirement to conduct the operation. For the sake of simplicity, only the axial depth of cut is considered since the radial depth of cut is usually defined by the operation itself, for example, face milling is usually conducted under an a_e of 70–80% cutting diameter, or slot operations are conducted with a_e equal to tool diameter. The axial depth of cut is then limited by the stock allowance of the workpiece. For roughing operations, the depth of cut should initially be chosen to be as large as possible, to reduce the number of cuts required. For finishing operations, it can be considered a fixed value that corresponds to the remaining material from the last roughing operation.

The stepwise procedure for defining the cutting parameters is defined as follows:

1) **Select the cutting speed** according to the vendor's recommendations and the desired production requirement, that is, maximum production or minimum cost as shown in Figure 10.1. The cutting speed for each production strategy is defined as (DeVries 1991, Chapter 7):

   a) Minimum production cost. $V_{cc} = V_r \times \left(\frac{n}{1-n} \times \frac{C_o \times T_r}{C_o \times t_c + C_t} \right)^n$

   b) Maximum production rate. $V_{cp} = V_r \times \left(\frac{n}{1-n} \times \frac{T_r}{t_c} \right)^n$

   where T_r denotes the tool life (minute) for the reference cutting speed V_r (m/min); n is Taylor's tool-life exponent; t_c is the tool changing time (minute); C_t is the tool cost ($), and C_o is the total machine and operator's rate ($/min). In relation to the tool cost, regrindable and disposable tools can be distinguished. Depending on the case, the tool cost is defined as:

   - Disposable inserts. $C_t = \frac{P_t}{n_e}$, where P_t is the cost of the insert and n_e is the number of cutting edges per insert.
   - Regrindable tools. $C_t = \frac{P_t}{n_e} + T_g \times C_g$, where P_t is the cost of the tool, n_e is the number of times the tool can be ground, T_g is the time required for grinding the tool (minute), and C_g is the grinder's rate ($/min).

   In industry, the term economic tool life (T_e) refers to the tool life's time when the minimum cost approach is applied, and the term optimum tool life (T_p) when applying the maximum production rate strategy. The values of 20 and 15 minutes for T_e and T_p, respectively (Grzesik 2008), are usually accepted. Using Taylor's expression, the cutting tool speed is straightforwardly obtained for each scenario.

   To illustrate the evaluation of the cutting speed for maximum production or minimum cost, consider a turning operation with the following information: tool cost $12, 2 cutting edges per insert, tool change time 1 minute, $84/hour machine/labor rate ($1.4/min), Taylor's coefficient 0.25, and tool's life is 20 minutes at 150 m/min. Therefore, we have the following:

   a) Minimum production cost. $V_{cc} = 150 \times \left(\frac{0.25}{1-0.25} \times \frac{1.4 \times 20}{1.4 \times 1 + 6} \right)^{0.25} = 159\,\text{m/min}$

   b) Maximum production rate. $V_{cp} = 150 \times \left(\frac{0.25}{1-0.25} \times \frac{20}{1} \right)^{0.25} = 303\,\text{m/min}$

Table 10.1 Table to correct cutting speed for different tool-life values.

Tool life (minutes)	10	15	20	25	30	45	60
Correction factor	1.11	1.0	0.93	0.88	0.84	0.75	0.70

Source: Sandvik Coromant (2017).

Table 10.2 Table to correct cutting speed for variations of material hardness with respect to the nominal one.

			Correction factor for variations of material hardness								
			Reduced hardness				Increased hardness				
ISO	Group	HB HRC	−60	−40	−20	0	+20	+40	+60	+80	+100
P	Low-alloy steels	HB 180	1.44	1.25	1.11	1.0	0.91	0.84	0.77	0.72	0.67
M	Austenitic stainless steels	HB 180	1.42	1.24	1.11	1.0	0.91	0.84	0.78	0.73	0.68
K	Gray cast iron	HB 220	1.21	1.13	1.06	1.0	0.95	0.90	0.86	0.82	0.79
	Ductile cast iron	HB 250	1.33	1.21	1.09	1.0	0.91	0.84	0.75	0.70	0.65
N	Aluminum alloys	HB 75	—	—	1.05	1.0	0.95	—	—	—	—
S	Heat-resistance alloys	HB 350	—	—	1.12	1.0	0.89	—	—	—	—
H	Hardened cast iron	HRC 60	—	—	1.07	1.0	0.97	—	—	—	—

HB, hardness Brinell; HRC, hardness Rockwell C.
Source: Sandvik Coromant (2017).

Very often, the cutting speed is straightforwardly selected from the vendor's recommendations. The tables provided by tool manufacturers usually define the cutting speed for a tool life of 15 minutes of continuous machining operation. The cutting speed is selected as a function of the type of workpiece material and its hardness. For cutting the same material with different hardness, a corrector factor is applied to proper adjust the cutting speed. Similarly, to adjust the cutting tool life for other values such as 20 minutes, 25 minutes, etc., instead of applying Taylor's equation, a coefficient can be applied to tune the cutting speed. Tables 10.1 and 10.2 show the corrector factors for both tool life and material hardness according to Sandvik Coromant vendor (Sandvik Coromant 2017). Note that Table 10.2 shows some correction factors on specific materials within the workpiece group material defined by the letter code (e.g., P, N, etc.) according to ISO 513.

As an example of application, consider as cutting conditions from the catalog a cutting speed of 225 m/min for 15 minutes of tool life when machining low-alloy steel with HB 180. If the workpiece to be machined is a similar low-alloy steel with hardness HB 200 and the desired tool life is 20 minutes, the correctors to be applied are 0.93 for tool life and 0.91 due to hardness change. Therefore, the cutting speed required is $225 \times 0.93 \times 0.91 = 190.4\,\text{m/min}$.

2) **Select the feed rate** according to the vendor's recommendations. For rough operations, a feed rate to maximize material removal rates is the most economical approach. For finishing operations, constraints on surface roughness will limit the feed rate. It should be noted that too low feed rates may make the cutting process unfeasible and, instead of shearing the material, a

rubbing effect may be produced, which leads to worse surface roughness and larger tool wear rates. In both roughing and finishing operations, feed rates should be chosen according to tool nose radius and chip-breaking diagrams of the cutting inserts used.

3) **Select the depth of cut** to minimize the number of passes. This increases productivity and reduces the cutting tool wear in comparison with a higher number of passes. The main constraint here is the power consumption and cutting forces that should be compared with the machine tool and tool holder capabilities, and the maximum depth of cut limited by the size of the insert.

This stepwise procedure requires checking, at each step, that the constraints that apply to each cutting parameter are taken into consideration. In case of doubt or for a straightforward parameter selection in machining tests, middle cutting speeds, lower feed rates, and lighter depth of cuts within the vendor's recommended ranges are considered the general practice on the shop floor.

The next subsection shows the constraints that should be considered during the selection of cutting parameters.

10.2.2 General Constraints

10.2.2.1 Constraints on Cutting Speed

- Limitations due to the maximum spindle speed of the machine tool. The machine tool presents a maximum spindle speed that cannot be exceeded.
- Limitations due to maximum spindle speed allowed by the tool holder/workholding device. Tool holders in milling/drilling operations and workholding devices in turning operations may present a maximum spindle speed that must be not exceeded.

As a result from previous constraints, the allowable cutting speed will be limited in the form of:

$$V_{\mathrm{c}} \leq V_{\mathrm{c\text{-}max}}$$

Note that these are constraints related to maximum cutting speed. Although initially there is no limit to minimum cutting speed, the vendor's recommendations provide a range since too low cutting speeds produce built-up edge, vibration, and low tool-life values. Thus, cutting speed constraints are defined as follows:

$$V_{\mathrm{c\text{-}min}} \leq V_{\mathrm{c}} \leq V_{\mathrm{c\text{-}max}}$$

10.2.2.2 Constraints on Feed Rate

- Limitations due to nose radius (r_{e}). The following expressions are commonly applied to limit the maximum feed rate in turning operations:

  Roughing operations: $f_{\mathrm{n}} \leq 0.5 \times r_{\mathrm{e}}$

  Finishing operations: $f_{\mathrm{n}} \leq 0.15 \times \sqrt{r_{\mathrm{e}} \times R_{\mathrm{a}}}$ where R_{a} is the desired surface roughness.

  For finishing operations, the theoretical surface roughness value can be estimated as (Stephenson and Agapiou 2016, Chapter 10):

$$R_{\mathrm{a}}(\mu\mathrm{m}) = \frac{f_n^2}{32 \times r_{\mathrm{e}}} \times 1{,}000;\ R_z(\mu\mathrm{m}) = \frac{f_n^2}{8 \times r_{\mathrm{e}}} \times 1{,}000$$

However, it should be noted that the actual surface finish after machining is usually around 1.5–3 times higher than the theoretical ones. Table 10.3 shows the relationships among nose radius, feed rate, and surface roughness requirements.

Table 10.3 Feed rate according to nose radius and surface roughness requirement.

Surface roughness (μm)		Nose radius (mm)				
		0.4	0.8	1.2	1.6	2.4
R_a	R_z	Feed rate (mm/rev)				
0.6	1.6	0.07	0.1	0.12	0.14	0.17
1.6	4	0.11	0.15	0.19	0.22	0.26
3.2	10	0.17	0.24	0.29	0.34	0.42
6.3	16	0.22	0.30	0.37	0.43	0.53
8	25	0.27	0.38	0.47	0.54	0.66
32	100	—	—	—	1.08	1.32

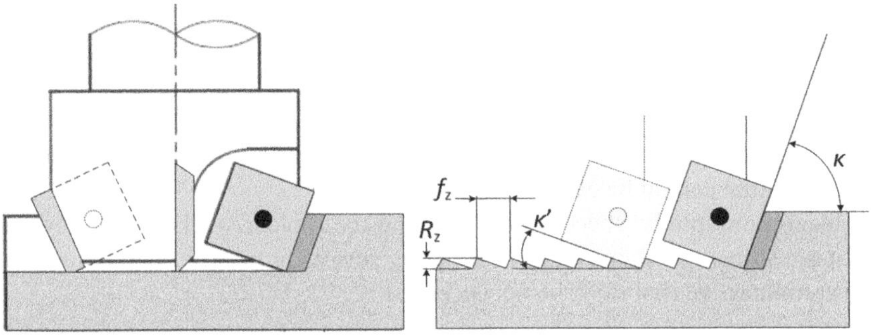

Figure 10.2 Frontal milling and surface roughness

For milling operations, similar equations apply but f_z replaces f_n values. However, the final roughness in milling is usually higher due to run-out errors when mounting the cutting inserts in the cutter. When the nose radius is negligible, insert angles define the theoretical surface roughness as follows (DeVries 1991, Chapter 5):

$$R_z = \frac{f_z}{\cot(\kappa) + \cot(\kappa')}; \; R_a = \frac{f_z}{4 \times [\cot(\kappa) + \cot(\kappa')]}$$

where κ' is the face cutting-edge angle and κ is the principal cutting-edge angle, that is, the entering angle (see Figure 10.2).

Similarly, in peripheral milling (Figure 10.3), the surface roughness is estimated as

$$R_z = \frac{f_z^2}{8 \times \frac{D}{2}}; \; R_a = \frac{f_z^2}{32 \times \frac{D}{2}}$$

The estimations for peripheral milling are usually modified to include the well-known effect of conventional and climb milling. Some authors (DeVries 1991, Chapter 5) correct them by including an additional term as shown in the following:

$$R_z = \frac{f_z^2}{8 \times \left(\frac{D}{2} + \frac{f_z \times z}{\pi}\right)}; \; R_a = \frac{f_z^2}{32 \times \left(\frac{D}{2} + \frac{f_z \times z}{\pi}\right)} \quad \text{Climb milling}$$

Figure 10.3 Peripheral milling and surface roughness.

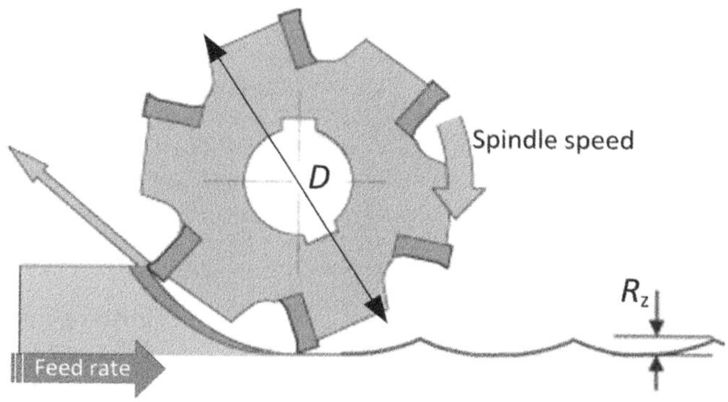

$$R_z = \frac{f_z^2}{8\times\left(\frac{D}{2}-\frac{f_z\times z}{\pi}\right)};\ R_a = \frac{f_z^2}{32\times\left(\frac{D}{2}-\frac{f_z\times z}{\pi}\right)} \quad \text{Conventional milling}$$

where z is the number of teeth in the cutter. As it can be seen by the above expression, climb milling presents lower surface roughness values than conventional milling.

- Limitations due to chip-breaking diagram. Both feed rate and depth of cut should be properly selected to make the chip-breaking features of the insert work. Diagrams are provided by insert vendors to show ranges of operations where the chip breakers work correctly facilitating the chip removal (Figures 10.4 and 10.5).

As a result of previous constraints, the allowable feed rate will be limited by a minimum and maximum value in the form of

$$f_{\text{n-min}} \leq f_{\text{n}} \leq f_{\text{n-max}}$$

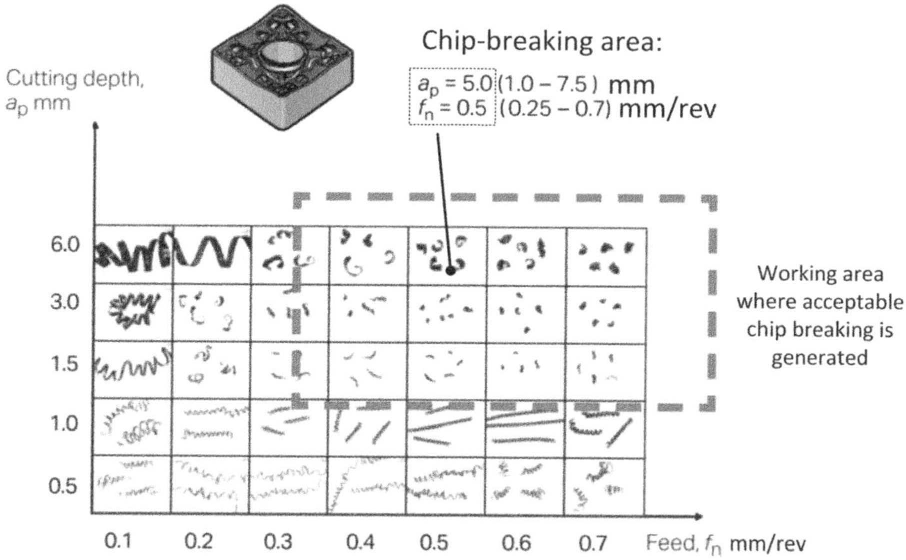

Figure 10.4 Chip-breaking diagram for an insert designed for roughing operations, and type of chips generated under different conditions. *Source:* Courtesy of Sandvik Coromant (2017).

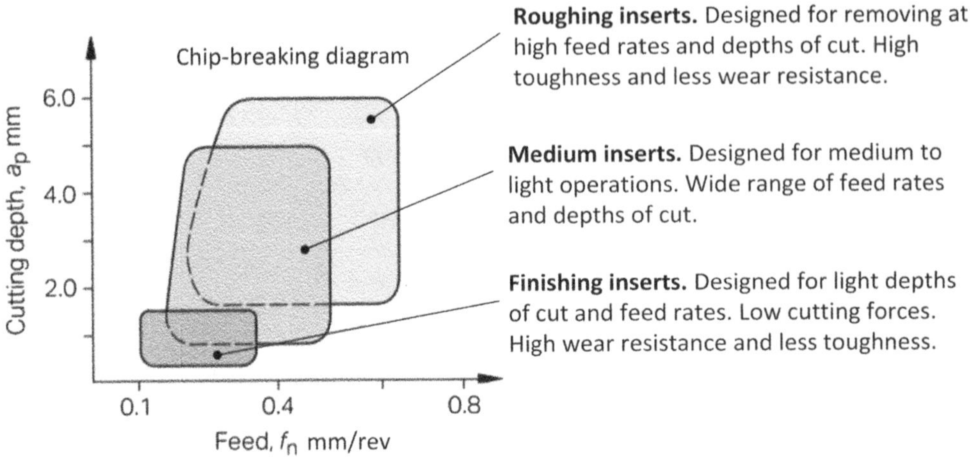

Figure 10.5 Chip breakers, as part of the insert geometry, are designed to work at different feed/depth of cut areas. *Source:* Courtesy of Sandvik Coromant (2017).

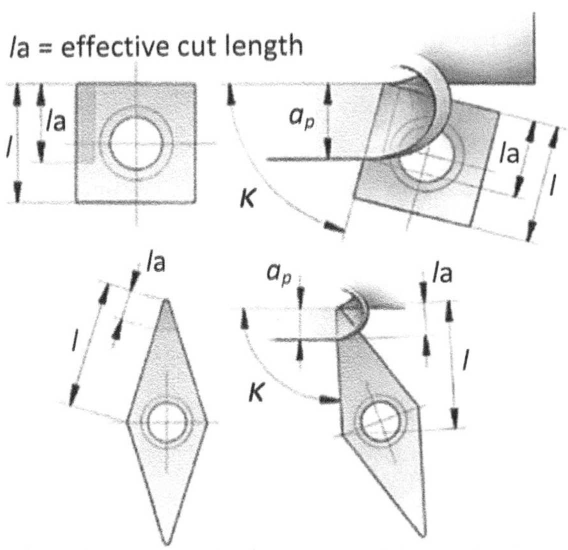

Insert shape		*la* (*l* = insert length)
S		2/3 × *l*
C		2/3 × *l*
T		1/2 × *l*
D		1/4 × *l*
K		1/4 × *l*
R		0.4 × *d*

Figure 10.6 Effective cutting-edge length. *Source:* Courtesy of Sandvik Coromant (2017).

10.2.2.3 Constraints on Depth of Cut

- Limitations due to cutting-edge length. The allowable cutting edge in a cutting tool is indicated by the vendor. For indexable cutting inserts, the allowable cutting edge is around 2/3–1/4 the cutting edge according to the geometry of the insert. Figure 10.6 shows some effective cut lengths according of to different insert shapes.
- Limitations due to chip-breaking diagram. As shown previously, the depth of cut together with the feed rate should be selected to ensure that chip breakers work correctly providing a good chip release.

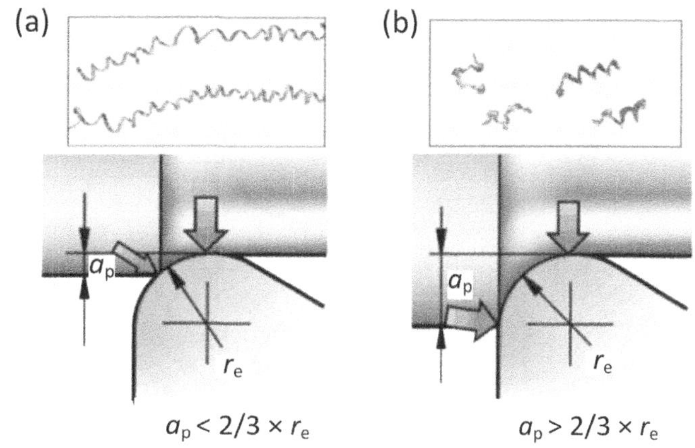

Figure 10.7 Chip formation issues. (a) Problematic long chips due to low depth of cut. (b) Correct chip formation. *Source:* Courtesy of Sandvik Coromant (2017).

- Limitations due to nose radius. A depth of cut too small may produce vibrations, excessive heat, and difficulties in shearing the workpiece material. As a rule of thumb, the depth of cut should be greater than or equal to 2/3 of the nose radius. Figure 10.7a, b illustrates the problem of chip formation for low-depth of cuts.

As a result from previous constraints, the allowable depth of cut will be limited by a minimum and maximum value in the form of

$$a_{\text{p-min}} \le a_{\text{p}} \le a_{\text{p-max}}$$

10.2.2.4 Constraints on Cutting Forces and Power Consumption

- Maximum cutting forces due to the following features:
  - Tool deformation (Figure 10.8). Deflection produces higher tool wear rates due to runout effects and quality errors in finishing operations. In general, it is recommended to keep tool deflection below 0.025 mm, and much less for small tools. Cutting tool deflection can be estimated as (Goodno and Gere 2020) follows:

$$\delta = \frac{F_{\text{c}} \times L^3}{3 \times E \times I}$$

    where δ is the deflection value (mm); F_{c} is the cutting force (N); L is the overhang length of the cutting tool (mm); E is the Young's modulus of the cutting tool shank ($E = 210{,}000\ \text{N/mm}^2$ for steel); and I is the inertial moment of the shank cross-section (mm^4). Some common I values are: $I = \pi \times D^4/64$ (round section); $I = 1/12 \times B \times H^3$ (rectangular section).
  - Workpiece deformation. Similar to tool deflection, the workpiece may deform due to the cutting forces. Milling thin-walled parts or turning slender parts may require limiting cutting forces to avoid excessive deflection and geometrical quality issues.
  - Tool breakage. If cutting forces are too high, a breakage may occur. Keeping the depth of cuts and feed rates within the vendor's recommendations according to insert geometry will avoid premature failure. Appropriate reduced feed rates when entering to cut also avoid potential early breakages.
- Maximum machine tool power consumption. This constraint should be considered in roughing operations where the power demand is higher. This limitation may keep the cutting speed, feed rate, and depth of cut to be lower than the initial ones. To calculate machine tool power consumption, machine tool efficiency is usually considered in the range of 0.5–0.8.

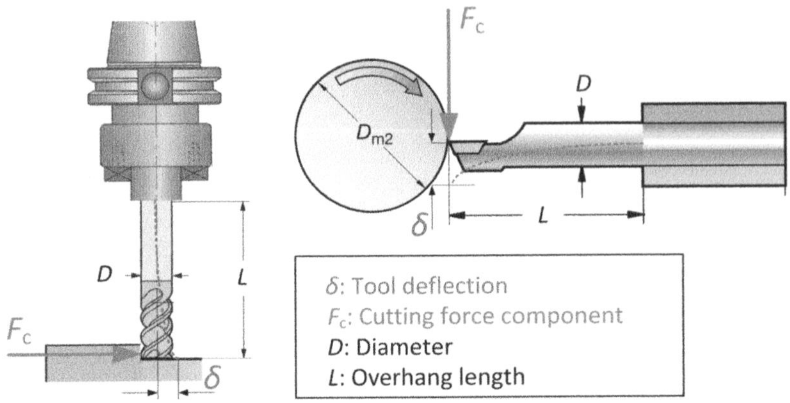

Figure 10.8 Cutting tool deflection in milling and turning.

10.2.3 Cutting Force and Power Calculation

The estimation of cutting forces and power consumption is necessary to conduct efficient process planning in machining. For instance, maximizing the material removal rate in roughing operations requires the estimation of the cutting force and power consumption to avoid tool breakage or machine tool failures.

Most of cutting tool vendors provide tables, equations, and online calculators for estimating cutting forces and power requirements. Tables 10.4 and 10.5 show the expressions that can be applied for this purpose in turning, milling, and drilling operations. Cutting data such as k_{c1} and m_c coefficients can be found in cutting tool vendors' catalogs and they are also reported in Appendix 2, A2.2 Specific Cutting Force.

Besides previous expressions, it should be noted that tool wear and machining efficiency can have an important impact on the final machine tool power consumption. For instance, common machine tool efficiency values are 0.5–0.8, and cutting forces and power consumption may increase up to 10–30% due to tool wear.

To illustrate the calculus of cutting forces and power requirements, consider as an example, a face milling operation conducted with a face cutter of Ø200 mm, 12 inserts, an axial depth of cut 4 mm, and an entering angle 60°. The inserts present a square geometry with a rake angle of −1°. The cutting speed is 150 m/min, the feed rate is 0.25 mm/tooth, the workpiece material is AISI 1020 steel (nonalloyed steel) and the width of the workpiece is 140 mm. For this operation, the cutting force and cutting power can be obtained as follows.

First, the operation is graphically represented to identify the main cutting parameters, as shown in Figure 10.9. Then, according to Appendix 2, the material AISI 1020 corresponds to the ISO group P (nonalloyed steels, with carbon content around 0.2%), and thus, the specific cutting force k_{c1} and m_c coefficient are as follows:

$$k_{c1} = 1{,}500\ \text{N/mm}^2,\ m_c = 0.25$$

The average chip thickness is found as follows:

$$h_m = \sin\kappa \times \frac{360}{\pi} \times \frac{f_z \times a_e}{\varphi \times D} = \sin 60 \times \frac{360}{\pi} \times \frac{0.25 \times 140}{88.85 \times 200} = 0.195\,\text{mm}$$

Table 10.4 Cutting forces in machining.

Turning		
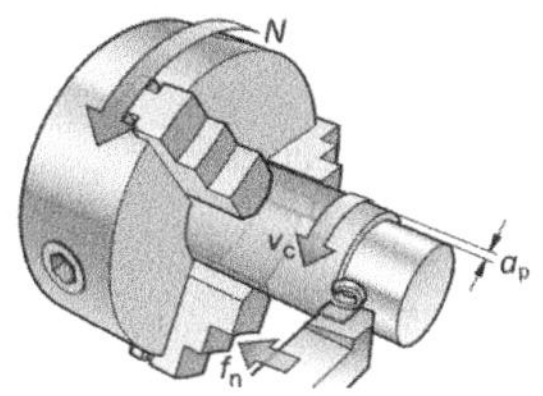	Specific cutting forces, N/mm^2 $k_c = k_{c1} \times \left(\frac{1}{t}\right)^{m_c} \times \left(1 - \frac{\gamma_o}{100}\right)$ Cutting forces, N $F_c = k_c \times a_p \times f_n$	Chip thickness, mm $t = f_n \times \sin\kappa$
Milling		
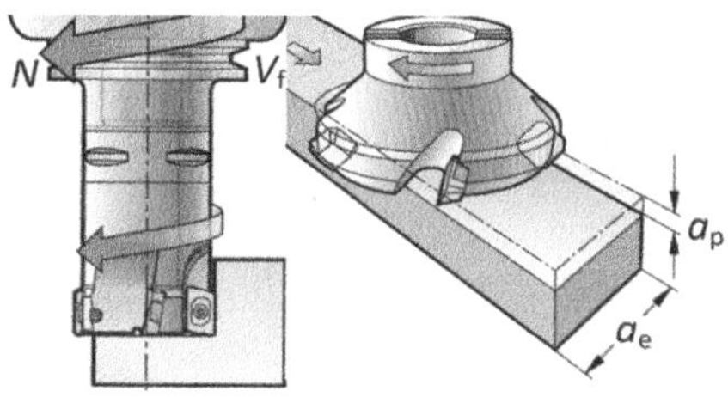	Specific cutting forces, N/mm^2 $k_c = k_{c1} \times \left(\frac{1}{h_m}\right)^{m_c} \times \left(1 - \frac{\gamma_o}{100}\right)$ Cutting forces, N $F_c = z_c \times k_c \times \frac{a_p}{\sin\kappa} \times h_m$	Average chip thickness, mm $h_m = \sin\kappa \times \frac{360}{\pi} \times \frac{f_z \times a_e}{\varphi \times D}$ Number of teeth cutting $z_c = z \times \frac{\varphi}{360}$ φ : engagement angle
Drilling		
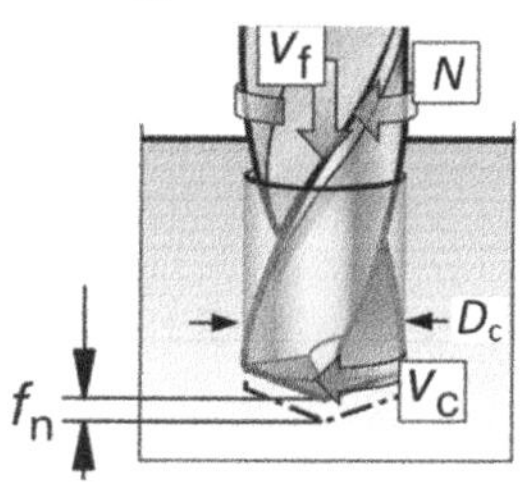	Specific cutting forces, N/mm^2 $k_c = k_{c1} \times \left(\frac{1}{t}\right)^{m_c} \times \left(1 - \frac{\gamma_o}{100}\right)$ Cutting forces, N $F_c = k_c \times a_p \times f_n$	Chip thickness, mm $t = f_z \times \sin\kappa$ Feed force, N $F_f \approx 0.5 \times k_c \times \frac{Dc}{2} \times f_n \times \sin\kappa$

Source: Adapted from Sandvik Coromant (2017).

since the face cutter is centered over the workpiece and the resulting engagement angle is calculated as follows:

$$\varphi = 2 \times \varphi_1 = 2 \times \arcsin\left(\frac{140}{2 \times 100}\right) = 88.85^{\mathrm{o}}$$

Therefore, the specific cutting force and the cutting force are calculated as follows:

$$k_c = k_{c1} \times \left(\frac{1}{h_m}\right)^{m_c} \times \left(1 - \frac{\gamma_o}{100}\right)$$

$$k_c = 1{,}500 \times \left(\frac{1}{0.195}\right)^{0.25} \times \left(1 - \frac{(-1)}{100}\right) = 2{,}280 \text{ N/mm}^2$$

$$F_c = z_c \times k_c \times \frac{a_p}{\sin\kappa} \times h_m = 2.96 \times 2{,}280 \times \frac{4}{\sin 60} \times 0.195 = 6{,}094 \text{ N}$$

Table 10.5 Cutting power requirement in machining.

Turning

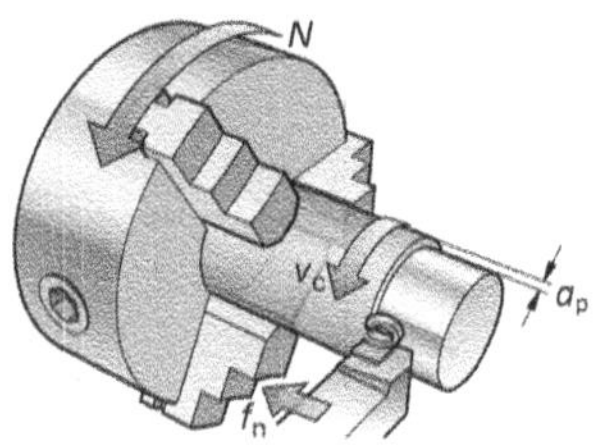

Metal removal rate, cm^3/min

$Q = V_c \times a_p \times f_n$

Cutting power, kW

$$P_c = \frac{V_c \times a_p \times f_n \times k_c}{60 \times 10^3}$$

Cutting torque, N·m

Depends on tool position (d) with respect to the spindle axis

$M_c = F_c \times d$

Milling

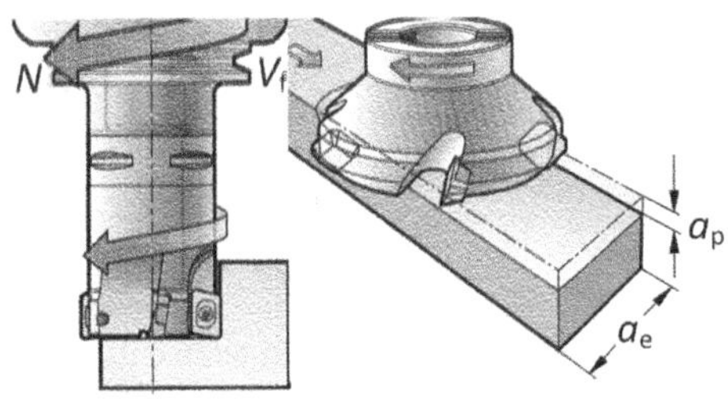

Metal removal rate, cm^3/min

$$Q = \frac{a_p \times a_e \times V_f}{1{,}000}$$

Cutting power, kW

$$P_c = \frac{a_e \times a_p \times V_f \times k_c}{60 \times 10^3}$$

Cutting torque, N·m

$$M_c = \frac{P_c \times 30 \times 10^3}{\pi \times N}$$

Feed rate, mm/min

$V_f = N \times f_n$

Drilling

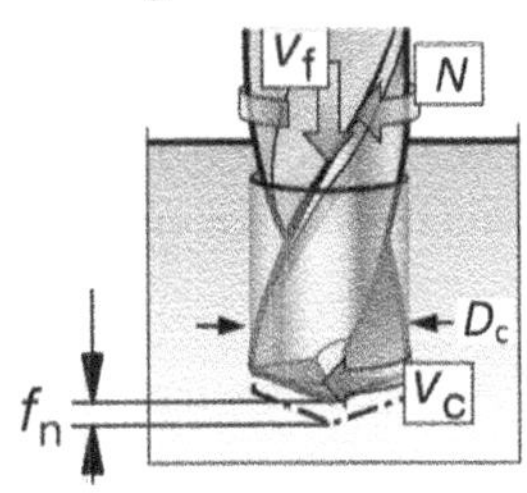

Metal removal rate, cm^3/min

$$Q = \frac{V_c \times Dc \times f_n}{4}$$

Cutting power, kW

$$P_c = \frac{a_e \times a_p \times V_f \times k_c}{60 \times 10^3}$$

Cutting torque, N·m

$$M_c = \frac{P_c \times 30 \times 10^3}{\pi \times N}$$

Source: Adapted from Sandvik Coromant (2017).

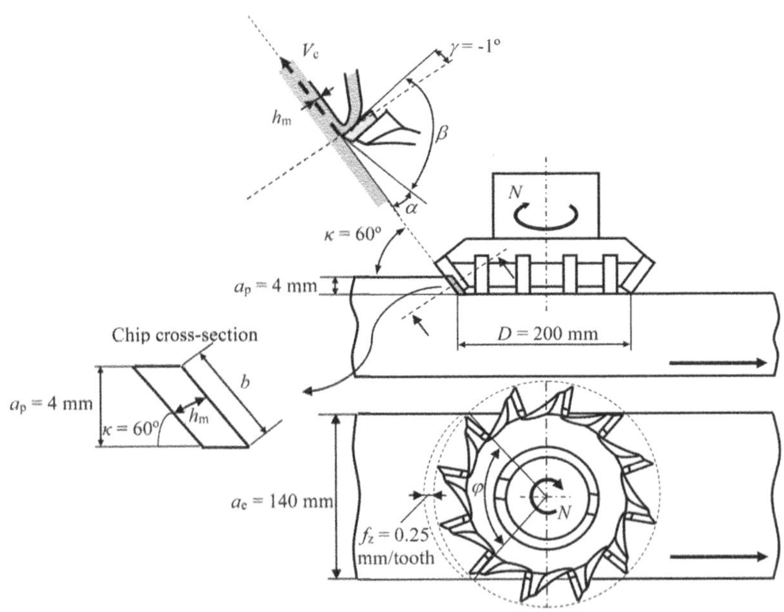

Figure 10.9 Drawing to illustrate the milling operation for cutting forces and power estimation.
Source: Abellán-Nebot (2024).

where z_c is the number of teeth due to simultaneous cutting, which is calculated as

$$z_c = \frac{z}{360} \times \varphi = \frac{12}{360} \times 88.85 = 2.96$$

Finally, the feed rate and the cutting power are expressed as follows:

$$V_f = f_n \times N = z \times f_z \times \left(\frac{V_c \times 1{,}000}{\pi \times D} \right)$$

$$V_f = 12 \times 0.25 \times \left(\frac{150 \times 1{,}000}{\pi \times 200} \right) = 716.2 \text{ mm/min}$$

$$P_c = \frac{a_p \times a_e \times V_f \times k_c}{60 \times 10^6} = \frac{4 \times 140 \times 716.2 \times 2{,}280}{60 \times 10^6} = 15.24 \text{ kW}$$

The preceding example shows how to calculate cutting forces and power consumption in machining operations. These values should be then analyzed in terms of machine tool capabilities, potential cutting tool deflections, etc. Readers may refer to Abellán-Nebot et al. (2024) for additional examples related to cutting forces and power consumption in milling, turning, and drilling.

10.3 Specific Guidelines for Turning

In turning operations, the following aspects should be considered:

- Minimum chip thickness is 5–20% of the cutting-edge radius. Below that level, chips may not form properly, and the cutter will "plow" across the workpiece causing plastic deformation and considerable heat. As a general recommendation, keep the feed rate higher than 0.025 mm/rev to ensure a minimum chip thickness.
- Workpiece/tool holder deflections due to cutting forces. The deflection of the workpiece will depend on the way it is held and the cutting force value. Table 10.6 shows some common workholding systems and estimated workpiece/tool holder deflections.
- Required clamping torque. The maximum torque generated by the cutting force should be supported by the clamping system during cutting. The cutting torque is calculated as follows:

  $$M_c = F_c \times d$$

  where F_c is the cutting force calculated as explained previously and d is the distance between the cross-section cut and the spindle axis. The clamping torque that the fixturing system can generate should be higher than this cutting torque. A safety factor is usually added, and a value of 2 or 2.5 is used for new or old workholding devices, respectively. Fixture vendors usually provide technical information about maximum clamping torques. Figure 10.10 shows an example of the maximum clamping torque for three-jaw chucks from Röhm GmbH.

10.4 Specific Guidelines in Milling

In milling operations, the following aspects should be considered:

- Depth of cut recommendations:
  - Slot milling.
    - Flat mills. The maximum axial depth of cut is the tool diameter.
    - Ball mills. The maximum axial depth of cut is half the tool diameter.

Table 10.6 Deflections in turning with different workholding systems.

Workholding system	Example	Equivalent schematic system	Deflection
Three-jaw chuck	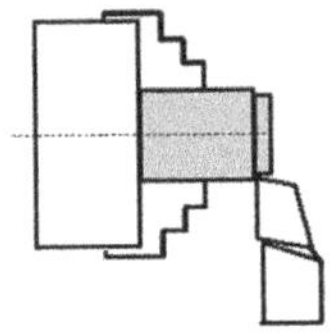	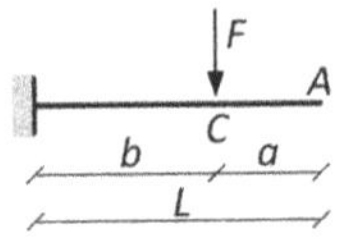	$y_C = \frac{F\times b^3}{3\times E\times I}$ $y_A = \frac{F\times b^2}{6\times E\times I}\times\left(2b+3a\right)$
Three-jaw chuck and center	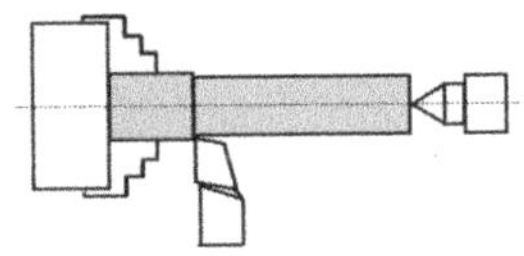	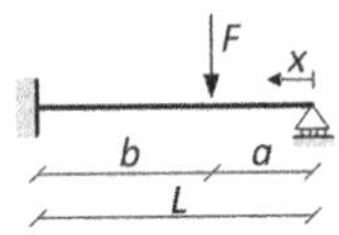	$y_{\max} = \frac{F\times b^2\times a}{6\times E\times I}\times\sqrt{\frac{a}{2L+a}}$ for $x = L\times\sqrt{\frac{a}{2L+a}}$
Between centers and carrier	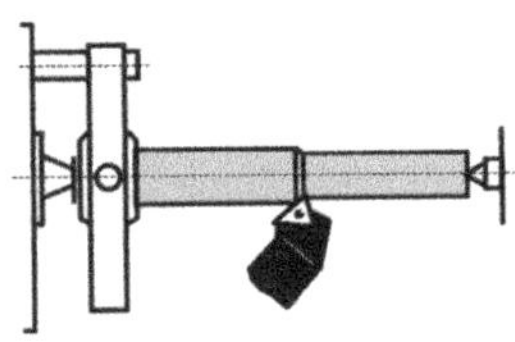	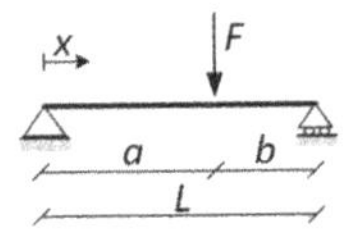	$y_{\max} = \frac{F\times b}{9\times E\times I\times L\times\sqrt{3}}\times\left(L^2-b^2\right)^{3/2}$ for $x = \sqrt{\frac{L^2-b^2}{3}}$
Cutting tool holder	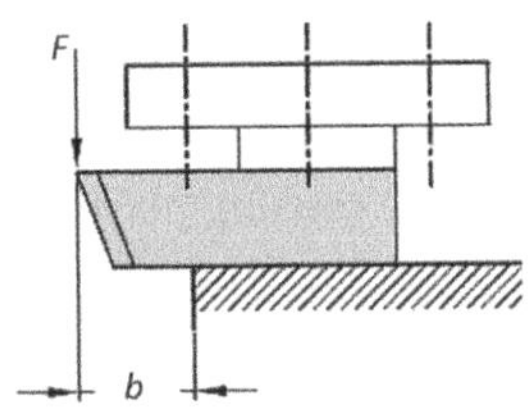	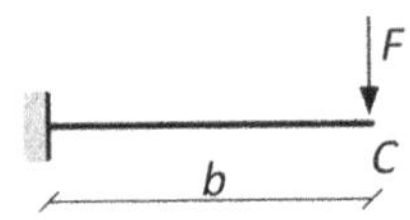	$y_C = \frac{F\times b^3}{3\times E\times I}$

a and *b*, dimensions related to cutting force position (mm); *E*, Young's modulus of the workpiece (MPa); *F*, cutting force (N); *I*, inertial moment of the workpiece (mm^4); *L*, length of the workpiece (mm).
Values: $I = \pi\times D^2/64$ for round cross-section, $I = \pi\times[(D_e)^4 - (D_i)^4]/64$ for tubular cross-section, $I = 1/12\times B\times H^3$ for rectangular cross-section.

Size	Inch	Maximum speed min^{-1}	Maximum torque Nm	Maximum total clamping force kN
125	5	6,000	40	23
160	6 1/4	5,400	120	73
200	8	4,600	155	114
250	10	4,200	190	185
315	12 1/2	3,300	210	240
400	15 3/4	2,200	260	260
500	20	1,900	320	290
630	25	1,100	350	320

Figure 10.10 Example of three-jaw chuck specifications. *Source:* Röhm (2024).

In both cases, the maximum radial depth of cut is the tool diameter.
- Square shoulder milling. The maximum axial depth of cut is 1.5 times the tool diameter when the radial depth of cut is lower than 50% of the tool diameter.
- Profile milling. In roughing operations, the radial depth of cut should be less than 0.7 times the tool diameter, and the axial depth of cut lower than 0.3–0.5 times the tool diameter. For finishing operations, although it depends on surface roughness, it is recommended axial depth of cuts of 0.025–0.05 times the tool diameter and radial depths of cut of 0.1 times the tool diameter.

- Feed rate correction factors due to chip thinning in side milling. For side milling or low working engagement with cutter positioned off center, a feed rate corrector should be applied. This is done to keep chip thickness similar to cutting operations with large engagement. As an example, consider an end milling operation with a cutter diameter of 40 mm, radial depth of cut 2 mm, and feed rate of 0.11 mm/tooth. The cutter diameter – radial depth of cut ratio is 40/2 = 20. According to Figure 10.11, the correction factor is 2.8, and thus, the correct feed rate to generate a proper chip thickness is $0.11 \times 2.8 = 0.31$ mm/tooth.
- Feed rate correction factors due to chip thinning in profile milling. The correction factor of the feed rate depends on the axial depth of cut according to Table 10.7. Note that profile milling employs ball end mills or round inserts with indexable tools.
- Surface roughness in profile milling. In profile milling, the radial depth of cut defines the resulting surface roughness (Figure 10.12). As a general practice, a sequence of roughing, semi-finishing, and finishing operations is conducted. In the finishing operation, the theoretical surface roughness value can be estimated as follows:

$$R_z = R - \sqrt{R^2 - \left(\frac{a_e}{2}\right)^2} \text{ and } R_a \approx 0.25 \times R_z$$

where R is the tool radius. From this equation, the radial depth of cut to achieve a specific theoretical surface roughness value is calculated as follows:

$$a_e = 2\sqrt{R^2 - (R - R_z)^2}$$

- Vibrations. Vibrations are always present during machining although they do not suppose a critical problem. However, in some operations, the frequency of vibrations may be close to the

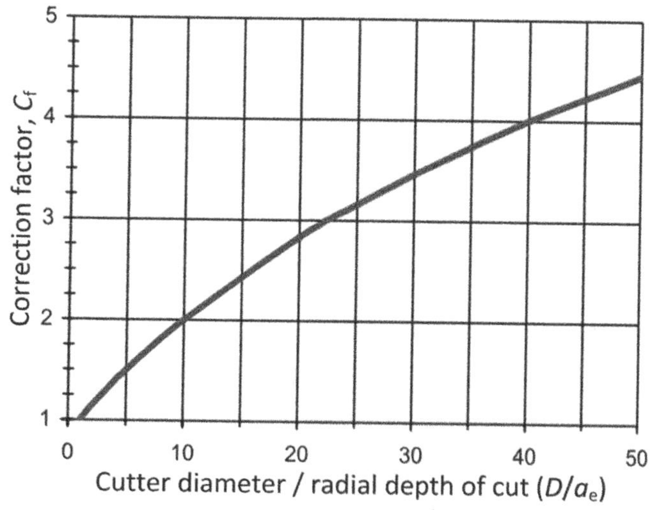

Figure 10.11 Correction factor for side milling.

Table 10.7 Correction factor of feed rate for profile milling according to axial depth of cut.

	Tool diameter (mm)							
Depth of cut (mm)	**6**	**8**	**10**	**12**	**16**	**20**	**25**	**32**
0.1	3.6	4.0	4.4	5.0	5.6	6.1	7.1	7.9
0.25	2.6	2.8	3.1	3.6	4.0	4.4	5.0	5.6
0.40	2.1	2.3	2.6	2.9	3.3	3.6	4.1	4.6
0.50	1.8	2.0	2.2	2.6	2.8	3.1	3.6	4.0
0.65	1.7	1.8	2.0	2.3	2.6	2.8	3.2	3.6
1.30	1.2	1.4	1.5	1.7	1.8	2.0	2.3	2.6
2.00	1.1	1.2	1.2	1.4	1.5	1.7	1.9	2.1
2.50	—	1.1	1.1	1.2	1.4	1.5	1.7	1.8
3.00	—	—	1.1	1.2	1.3	1.3	1.5	1.7
4.00	—	—	—	1.1	1.2	1.3	1.4	1.5
4.50	—	—	—	—	1.1	1.2	1.3	1.4

Source: Adapted from Dapra Milling Solutions (2023).

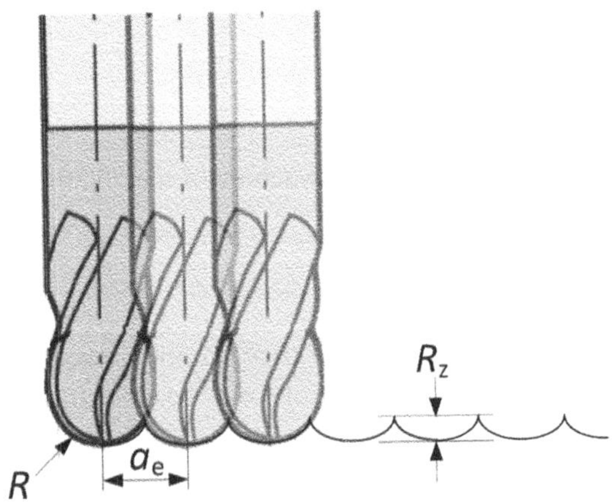

Figure 10.12 Theoretical roughness (R_z) in profile milling with ball end mills.

natural frequencies of machine tools or workpiece–fixture assemblies producing a severe problem in the resulting part quality. This phenomenon, called chatter, may be avoided by tuning the cutting speed and the depth of the cut. Other practices are to reduce cutting forces by reducing number of teeth, increase rake angle, and decrease tool overhang.

10.5 Specific Guidelines in Drilling

In drilling operations, the following aspects should be considered:

- Buckling. An excessive thrust on a drill may bend the tool leading to buckling and producing an unsafe condition for anyone in the immediate area. The critical thrust force that should not be exceeded is the critical buckling Euler force, defined as (Stephenson and Agapiou 2016, Chapter 4) follows:

$$F_{max} = \frac{k \times \pi^2 \times E \times I}{L^2}$$

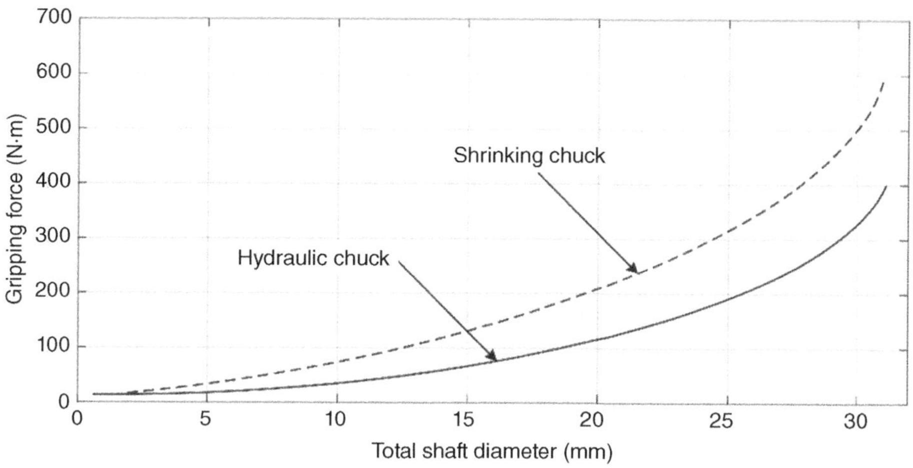

Figure 10.13 Gripping force for shrinking and hydraulic chucks. Elaborated from technical data. *Source:* TaeguTec LTD (2023) and Albrecht Präzision GmbH (2023).

where k defines the drill entry condition; k is 0.2–0.25 for drill-free point and k is 1–2 if a starting bushing or pilot hole is used. To avoid the buckling effect, long drills may require either a bushing or pilot hole. Note that buckling may be a severe issue when long HSS drills are used. On the contrary, carbide drills are tools three times stiffer than HSS drills of identical design ($E = 650$ GPa instead of $E = 220$ GPa), which make them more appropriate to avoid buckling when thrust forces are high.

- Limitations due to cutting torque on drills. The cutting torque can be estimated by knowing the cutting force and the distance of the geometric central point on the chip area from the drill axis. The drill holder has a limit cutting torque so the drilling operation should ensure that this maximum admissible cutting torque is not exceeded. Drill chucks present higher clamping torque than collet chucks, and thus, drill chucks are preferred when drilling higher diameters with larger cutting forces. Although it depends on the size and type of the drill holder, common maximum torque values are 10–20 N·m for drill chucks with a clamping range <Ø15 mm. For larger drills and more advanced chucks such as hydraulic or shrinking chucks, the admissible cutting torque can increase notably (Figure 10.13).

10.6 Use of Cutting Fluids

One of the primary methods for maintaining and often improving the performance of cutting tools involves the utilization of cutting fluids. These cutting fluids serve two essential functions during machining. First, they serve to cool both the workpiece and the cutting tool, effectively lowering the operating temperature of the tools. This cooling process aids in preserving the hardness of the tools and prevents the potential distortion of the workpiece due to excessive heat. Second, cutting fluids function as lubricants between the tool and the workpiece, reducing friction and contributing to the tool's resistance to wear. This lubrication also helps prevent the accumulation of chips on the cutting edge of the tool, particularly when machining materials like aluminum and its alloys. Additionally, cutting fluids offers supplementary benefits such as enhancing the surface finish and facilitating the removal of chips.

Practically all cutting fluids currently in use can be categorized into one of the following groups: neat cutting oils (often referred to as straight oils), soluble oils, synthetic oils, and semi-synthetic oils. Neat oils, which do not emulsify, are typically employed in an undiluted form for machining. In contrast, soluble oils, the most cost-effective among cutting fluids, are commonly diluted before use at typical concentrations of 5–10% oil in water to form an emulsion. Synthetic oils, known for their superior cooling properties, are composed of alkaline compounds and additives and are also generally employed in diluted form. Semi-synthetic oils combine characteristics from both synthetic and soluble oils, offering a balance between the two in terms of cost and heat dissipation properties.

The selection of cutting fluids is carried out according to the type of machining process, workpiece material, and cutting tool material. The following recommendations can be considered when choosing the proper cutting fluid in machining (Çakīr et al. 2007; Oberg et al. 2004):

- Cutting fluids and machining processes.
  - Heavy machining processes (e.g., broaching, tapping, reaming, and gear cutting) generally require neat cutting oils. These processes are conducted at low cutting speeds with high pressure in the cutting area which require more of lubrication than cooling to reduce friction at the tool and chip contacts to avoid built-up edge formation.
  - In turning, milling, and grinding machining processes, water-based cutting fluids are more suitable due to their cooling capabilities. At higher cutting speeds, temperatures are higher between the tool and the workpiece which may produce welds, so the cooling effect of the cutting fluid is mandatory.
  - In drilling operations, coolant supply can be internal or external. Internal coolant supply is preferred in long-chipping materials and when drilling deeper holes (higher than three times tool diameter) whereas the external coolant supply can be used when chip formation is good and when hole depth is shallow.
- Cutting fluids and workpiece material (carbide tools applied).
  - Gray cast iron is commonly machined dry since graphite content acts as a natural lubricant. A soluble oil or low-viscosity mineral oil may be used to prevent excessive metal dust.
  - Copper alloys are chemically attacked by sulfurs and chlorides, so cutting oils without these additives are required. Similar to cast iron, they lubricate well by themselves so very little extra lubrication is needed. If needed, soluble oil mixed to 1 part oil and 20–25 parts water may be satisfactory.
  - Aluminum alloys may be machined dry, although cutting fluids with high ability to act as coolant, for example, soluble oils, are preferred. Furthermore, aluminum alloys may require cutting fluids for a good chip formation and evacuation to avoid smearing on the insert edges and to improve surface finish.
  - Magnesium alloys require neat cutting oils of low viscosity. Water-based fluids should not be used because of the fire danger when magnesium chips react with water forming hydrogen gas.
  - Heat-resistant and difficult-to-cut steel alloys put significant thermal and mechanical strains on tools producing extremely high abrasive wear. Cutting fluids are mandatory in these applications using specific formulations with extreme pressure additives. Water-based cutting fluids are preferred because of the high temperature reached at the cutting area.
  - Low-carbon and hardened steels may be machined dry- or air-cooled assisted, especially in milling operations with carbide inserts to minimize thermal shock issues which could chip, crack, and eventually break the insert. To improve surface roughness, soluble oils are also typically applied.

- Stainless steels. Present lower heat conductivity and the cutting fluids usually applied here are water-soluble oils, usually in proportions of one part oil to 12–20 parts water and sulfochlorinated cutting oils.
- Cutting fluids and cutting tool material.
  - Ceramic cutting tools are generally used in finish machining operations with the use of sprayed air to facilitate chip removal. Oil-based cutting oil can be also employed but water-soluble cutting oil should be avoided.
  - Cubic boron nitrate (CBN) and polycrystalline diamonds (PCD) cutting tools can resist high-temperature machining conditions, and thus, cutting fluids are typically not required. They are typically employed in finish machining operations to provide high dimensional accuracy and great surface finish quality.

10.7 Practical Example

In this section, the methodology for process parameter selection is applied to the example discussed in Chapters 8 and 9. To remind the example, the engineering drawing and the machining sequence are shown in Figure 10.14 and Table 10.8, respectively.

For this purpose, it is assumed the following machine tool and workholding limitations:

- Computer numerical control (CNC) lathe:
  - Maximum power consumption: 10 kW.
  - Maximum spindle speed: 4,000 rpm.
  - Machine tool efficiency: 89%.
  - Three-jaw chuck clamping limitation: 40 N·m.
  - Three-jaw chuck spindle speed limitation: 5,000 rpm.

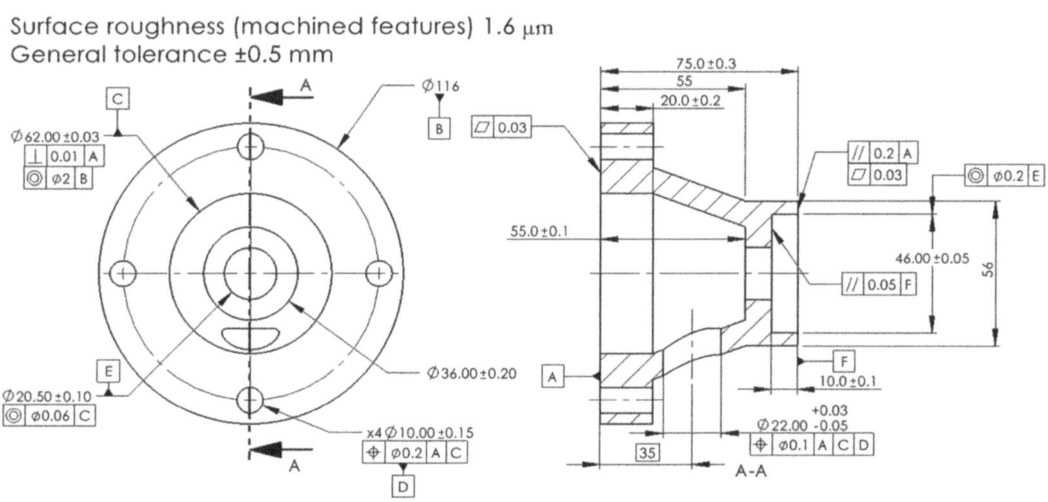

Figure 10.14 Part drawing of example in Chapters 8 and 9.

Table 10.8 Machining sequence and cutting tools of the example in Chapters 8 and 9.

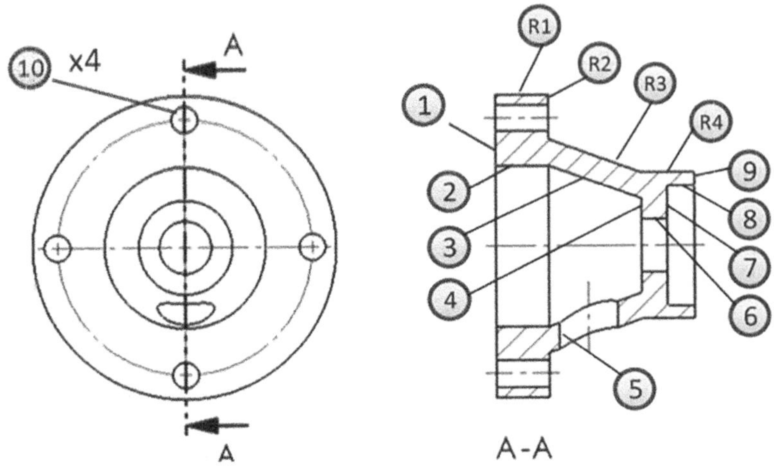

- Job 1 (lathe)
  - Subjob 1:
    - Face turning (1f).
      T11 D3. Tool holder PCLNL 2020K12, insert CNMG 120408 Iscar IC20.
    - Internal turning (2-3-4f).
      T13 D31. Tool holder S12M-STFCL-11, inserts TCMT 110202 Garant HU7310.
    - Internal turning (6f).
      T13 D31. Tool holder S12M-STFCL-11, inserts TCMT 110202 Garant HU7310.
  - Subjob 2:
    - Face turning (9f).
      T11 D3. Tool holder PCLNL 2020K12, insert CNMG 120408 Iscar IC20.
    - Internal face turning (7–8f).
      T13 D31. Tool holder S12M-STFCL-11, inserts TCMT110202 Garant HU7310.
- Job 2 (Machining center)
  - Subjob 1:
    - Drilling (10d).
      T132. Drill holder with spot drill TR-BR-2F-6-HSS.
      T140. Drill holder with twist drill TR-BR-2F-10-HSS.
  - Subjob 2:
    - Drilling (5d).
      T143. Drill holder PP-IN-2F-21 with carbide inserts WOEX 040304 Garant HB7310.
    - Boring (5b).
      T150. Fine boring head PP-BO-1F-(18-25) with carbide inserts TOGX 06 T104-14 Komet BK8430.

- CNC vertical machining center:
  - Maximum power consumption: 20 kW.
  - Maximum spindle speed: 15,000 rpm.
  - Machine tool efficiency: 85%.
  - Drill holder clamping limitation: 10 N·m.

To illustrate the methodology of setting the process parameters, operations 1f (face turning) and 10d (drilling) are detailed in the following. Note that to select the cutting parameters for the rest of machining operations the same procedure can be followed.

a) Face turning, operation 1f

- Tool T11 D3. Tool holder PCLNL 2020K12, insert CNMG 120408 Iscar IC20.
- The workpiece material is aluminum AISI SC64D, which corresponds to the N group according to ISO 513. Cutting data according to vendor's recommendations for straight and face turning operations assuming an insert grade with less wear resistance than toughness are (Appendix 2):

$$V_c = 770\,\text{m/min};\ f_n = 0.15 - 0.8\,\text{mm/rev}$$

The machining allowance is 3 ± 0.5 mm from the casting process. Only a finishing operation to remove 3 mm is the option considered in Chapter 8. The cutting speed selected is 770 m/min. The feed rate depends on the surface roughness specification. Since R_a should be 1.6 μm or lower, the theoretical maximum feed is expressed as

$$f_n \le 0.15 \times \sqrt{r_e \times R_a}$$
$$f_n \le 0.15 \times \sqrt{0.8 \times 1.6} = 0.17\ \text{mm/rev}$$

since r_e is 0.8 mm according to the insert ISO code. However, Table 10.3 indicates for these cutting conditions an actual feed rate of 0.15 mm/rev; thus, a feed rate of 0.15 mm/rev is selected, which is within the recommended vendor's range of [0.15 – 0.8] mm/rev.

- Spindle speed (N) and feed rate (V_f). Note that the machining operation is a face turning, so the spindle speed should vary to keep the same cutting speed (V_c) during turning. Thus, the maximum spindle speed and feed rate are given at the minimum workpiece diameter.

$$N = \frac{Vc \times 1{,}000}{\pi \times Dc} = \frac{770 \times 1{,}000}{\pi \times (62)} = 3{,}953\ \text{rpm};\ V_f = N \times f_n = 593\ \text{mm/min}$$

- Depth of cut (a_p).
  The maximum depth of cut is predefined from the machining allowance, 3 mm, and casting tolerance, ± 0.5 mm. Then,

$$a_p = 3.5\ \text{mm}$$

The maximum depth of cut due to the limitation of the length of the cutting edge is

$$a_p < \frac{2}{3} \times \text{L} = 8\ \text{mm}$$

where L is the length of cut, 12 mm according to the tool insert code. Then, the tool insert can be used for machining 3.5 mm depth. However, the feasibility of this operation depends on the machine tool and workholding constraints related to power consumption and clamping torque limitations. Therefore, the cutting force and power consumption should be estimated as follows.

First, a drawing with the main parameters of the operation studied is elaborated, as shown in Figure 10.15.

Then, the specific cutting force is estimated as

$$k_c = k_{c1} \times \left(\frac{1}{t}\right)^{m_c} = 600 \times \left(\frac{1}{0.149}\right)^{0.25} = 965\ \text{N/mm}^2$$

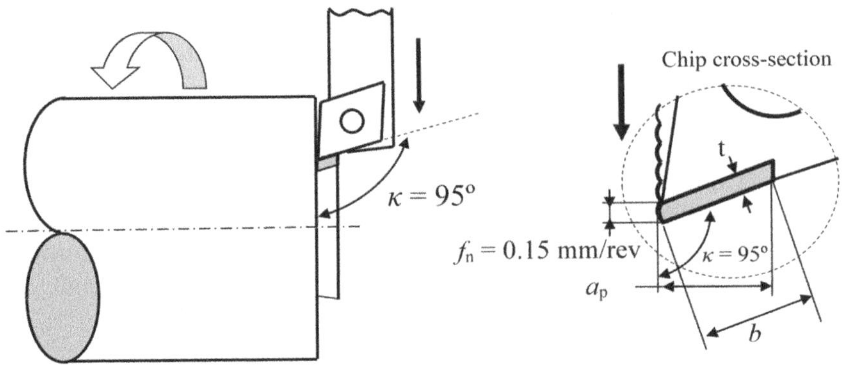

Figure 10.15 Face turning operation analyzed. Main parameters for cutting force calculation.

where the workpiece material is aluminum AISI SC64D which means k_{c1} and m_c are 600 N/mm^2 and 0.25 (ISO N and "cast, nonaging" subgroup), respectively (Appendix 2), and the chip thickness is calculated as

$$t = f_n \times \text{sen}(180 - \kappa) = 0.15 \times \text{sen}85 = 0.149 \text{ mm}$$

Next, the cutting force is estimated as

$$F_c = k_c \times a_p \times f_n = 965 \times 3.5 \times 0.15 = 507 \text{ N}$$

Thus, the cutting power is:

$$P_c = \frac{V_c \times a_p \times f_n \times k_c}{60 \times 10^3} = \frac{770 \times 3.5 \times 0.15 \times 965}{60 \times 10^3} = 6.5 \text{ kW}$$

and considering the machining efficiency, the power consumption is calculated as

$$P_{cons} = \frac{P_c}{\eta} = \frac{6.5}{0.89} = 7.3 \text{ kW}$$

The lathe machine has a power consumption limitation of 10 kW, so the operation is feasible. On the contrary, the maximum cutting torque is given at the outer diameter, thus

$$M_c = F_c \times \frac{D_c}{2} = 507 \times \frac{116}{2 \times 1000} = 29.4 \text{ N} \cdot \text{m}$$

Considering that the maximum torque due to the clamping system is 40 N·m, the operation is feasible.

b) Drilling, operation 10d

- Tool T132. Drill holder with spot drill TR-BR-2F-6-HSS and Tool T140: Drill holder with twist drill TR-BR-2F-10-HSS uncoated.
- Cutting data according to vendor's recommendations. The workpiece material is aluminum AISI SC64D with a Si content of 5.5–6.5%, therefore, the following parameters are defined (see Appendix 2):
  - Spot drilling:

$$V_c = 40 - 60 \frac{\text{m}}{\text{min}}; f_n = 0.29 \text{ mm/rev} (\text{curve } 16):$$

It is chosen to be $V_c = 50\,\text{m/min}$.

- Drilling:

$$V_c = 30 - 60\,\text{m/min};\ f_n = 0.315\,\text{mm/rev}(\text{curve } 7)$$

It is chosen to be $V_c = 30\,\text{m/min}$.

- Spindle speed (N) and feed rate (V_f).
  - Spot drilling:

$$N = \frac{Vc \times 1{,}000}{\times Dc} = \frac{50 \times 1{,}000}{\times 6} = 2{,}653\ \text{rpm};\ V_f = N \times f_n = 769\ \text{mm/min}$$

  - Drilling:

$$N = \frac{Vc \times 1{,}000}{\pi \times \text{Dc}} = \frac{30 \times 1{,}000}{\pi \times 10} = 955\ \text{rpm};\ V_f = N \times f_n = 301\ \text{mm/min}$$

- Depth of cut.

The depth of cut is predefined by the size of the hole to be machined. However, it is necessary to ensure that there is no problem with the machine tool's power consumption, the buckle effect of the drill, and the maximum permissible cutting torque applied to the drill holder.

The study for the spot drilling operation is skipped since is not a critical operation. For the drilling operation, a drawing to illustrate the main cutting parameters is shown in Figure 10.16. For this operation, the specific cutting force is found as follows:

$$k_c = k_{c1} \times \left(\frac{1}{t}\right)^{m_c} = 600 \times \left(\frac{1}{0.135}\right)^{0.25} = 990\ \text{N/mm}^2$$

since the workpiece material aluminum AISI SC64D has $k_{c1} = 600\text{N/mm}^2$ and $m_c = 0.25$, and the chip thickness is

$$t = f_z \times \text{sen}\ \kappa = \frac{0.315}{2} \times \text{sen}\left(\frac{118}{2}\right) = 0.135\ \text{mm}$$

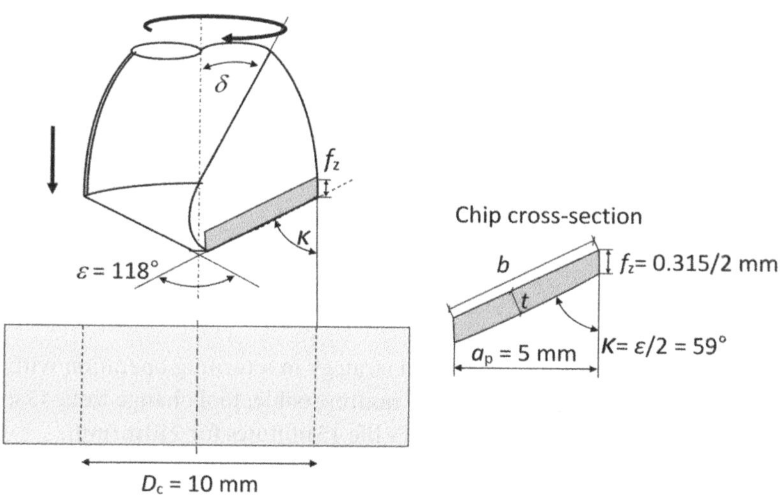

Figure 10.16 Drilling operation analyzed. Main parameters for cutting force calculation.

Assuming negligible the chisel edge, the cutting force per tooth is

$$F_{cz} = k_c \times a_p \times f_z = 990 \times 5 \times \frac{0.315}{2} = 780 \text{ N}$$

Note that as the drill has two teeth, the total cutting force is

$$F_c = 2 \times F_{cz} = k_c \times a_p \times f_n = 1{,}560 \text{ N}$$

Thus, the cutting power is

$$P_c = \frac{V_c \times D_c \times f_n \times k_c}{240 \times 10^3} = \frac{30 \times 10 \times 0.315 \times 990}{240 \times 10^3} = 0.39 \text{ kW}$$

and considering the machining efficiency, the power consumption is

$$P_{cons} = \frac{P_c}{\eta} = \frac{0.39}{0.89} = 0.44 \text{ kW}$$

which is lower than the machining center limitation, 20 kW.

The thrust force is estimated as

$$F_f = 0.5 \times k_c \times \frac{D_c}{2} \times f_n \times \sin\kappa$$

$$F_f = 0.5 \times 990 \times 5 \times 0.315 \times \sin\left(\frac{118}{2}\right) = 668 \text{N}$$

The critical thrust force to prevent the buckling effect is

$$F_{max} = \frac{k \times \pi^2 \times E \times I}{L^2}$$

$$F_{max} = \frac{0.22 \times \pi^2 \times 220 \times 10^3 \times \pi \times (10)^4 / 64}{70^2} = 47.8 \times 10^3 \text{N}$$

where k is 0.2–0.25, E = 220 GPa, L = 70 mm (overhang of the drill where the total length of the drill is 90 mm), $I = \pi \cdot D^4/64$. Thus, there is no buckling effect in this drilling operation.

Finally, the cutting torque applied to the drill holder is

$$M_c = \frac{P_c \times 30 \times 10^3}{\pi \times N} = \frac{0.39 \times 30 \times 10^3}{\pi \times 955} = 3.90 \text{ N} \cdot \text{m}$$

which is lower than the drill holder's torque limitation of 10 N·m. As a conclusion, the drilling operation is feasible.

Questions and Exercises

1 Calculate the cutting speed for the maximum production strategy in a turning operation with the following characteristics: tool cost \$15, square insert nonreversible, tool change time 45 s, \$60 machine/labor rate, Taylor's coefficient 0.21 and tool's life 15 minutes for 210 m/min.

2 Calculate the cutting speed for the minimum cost strategy in the turning operation (Exercise 1).

3 Calculate the estimated surface roughness generated in a straight turning operation where a coated carbide insert with 0.4-mm nose radius and 0.22 mm/rev is applied.

4 A drilling operation with a twist drill is conducted in a mild steel workpiece. The tool is a HSS drill with an overhang length of 60 mm and a diameter of 10 mm. Calculate the maximum feed rate to avoid buckling considering a safety factor of 2.

5 A-side milling operation is conducted with a radial depth of cut of 1 mm using a Ø15 mm flat-end mill. Feed rate recommendations are 0.02–0.05 mm/tooth. Considering the thinning effect, calculate the appropriate range of feed rate during the operation.

6 Calculate the power consumption of a face milling operation given the following information: face cutter Ø35 mm (4 inserts), radial depth of cut 27 mm, axial depth of cut 3 mm, cutting speed 150 m/min, feed rate 0.15 mm/tooth. Workpiece material DIN C22 (refer to Appendix 2, A2.3 Cross-reference Materials Table and ISO codes).

7 Calculate the power consumption of a straight turning operation given the following information: distance from workpiece axis 60 mm, constant cutting speed 200 m/min, and feed rate 0.1 mm/rev. Workpiece material AISI 301 (refer to Appendix 2, A2.3 Cross-reference Materials Table and ISO codes).

8 Calculate the power consumption of a drilling operation given the following information: Ø8 mm coated HSS twist drill (118°); no previous hole; wrought aluminum workpiece; cutting speed 140 m/min; and feed rate 0.22 mm/rev.

9 Complete Table 10.9 with the recommended ranges of cutting speed and feed rate. Check the information given in this chapter and Appendix 2 at the end of this book.

Table 10.9 Operations and tools for Exercise 9.

Operation/tool and workpiece material	Cutting speed	Feed rate
Face milling with a 45° face mill and carbide inserts. Low alloyed steel	—	—
End milling operations with a Ø15 mm HSS end mill. Gray cast iron material	—	—
End milling operations with a Ø10 mm coated carbide end mill. Copper alloys	—	—
Drilling operation with an uncoated HSS twist drill (Ø8 mm). Stainless steels	—	—

10 Calculate the appropriate cutting parameters (cutting speed, feed rate, depth of cut) using the cutting data given in Appendix 2 to machine the part shown in Figure 10.17. Assume that the raw material is a prismatic part with dimensions 100 × 100 × 50 mm made of low carbon steel (AISI 1015). The machining process plan is as follows (tools refer to Tables 9.8 and 9.9, Chapter 9):

- Job 1. Machining center DMC70V
  - Subjob 1. Vice
    - Face milling. Tool 1, PP-PL-6P-80, inserts SEKW 1204 AFTN
    - Shoulder milling. Tool 11, PP-ES-3P-25, inserts ADHX 110305 ER

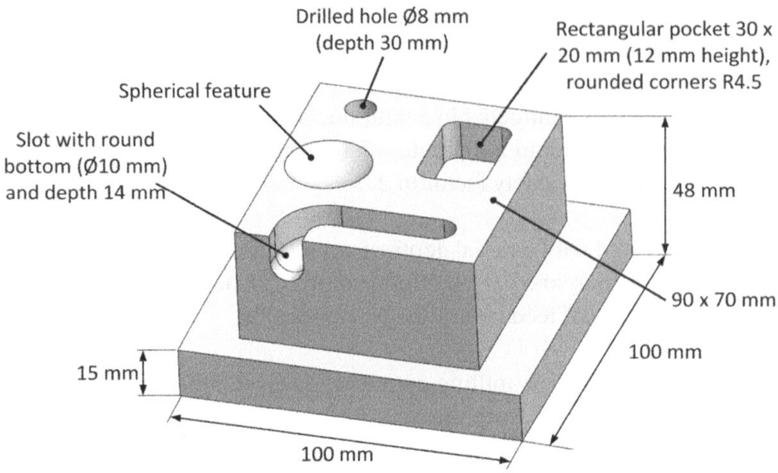

Figure 10.17 Part to be machined. Exercise 10.

- Pocket milling. Tool 77, FR-AC-3F-08-HSS-TiAlN
- Slot milling. Tool 121, FR-CO-2F-10-HSS-TiAlN
- Profile milling. Tool 121, FR-CO-2F-10-HSS-TiAlN
- Spot drilling. Tool 132, TR-BR-2F-6-HSS
- Drilling. Tool 138, TR-BR-2F-8-HSS-TiAlN

References

Abellán-Nebot, J.V., Vila, C., Serrano, J. et al. (2024). *Exercises and Solutions for Machining Operations: Milling, Turning and Drilling*. Independently published ISBN 9798884832541.

Albrecht Präzision GmbH. (2023). Outstanding precision chucks. (2023) Catalogue and technical data. https://www.albrecht-germany.com (accessed May 2023).

Çakīr, O., Yardimeden, A., Ozben, T., and Kilickap, E. (2007). Selection of cutting fluids in machining processes. *Journal of Achievements in Materials and Manufacturing Engineering* 25 (2): 99–102.

Dapra Milling Solutions. (2023). Feed, speed & diameter compensation for ball nose finishing mills. Technical data. https://www.dapra.com (accessed May 2023).

DeVries, W.R. (1991). *Analysis of Material Removal Processes*. Springer Science & Business Media.

Goodno, B.J., and Gere, J.M. (2020). *Mechanics of Materials*. Cengage Learning.

Grzesik, W. (2008). *Advanced Machining Processes of Metallic Materials: Theory, Modelling and Applications*. Elsevier.

Oberg, E., Jones, F.D., Horton, H.L., and Ryffel, H.H. (2004). *Machinery's Handbook 27th Edition-Full Book*. New York: Industrial Press Inc.

Röhm. (2024). *Lathe Chucks and Independent Chucks*. Edition 8. Online Catalogue. https://www.roehm.biz/fileadmin/content/pdf/products/en/PG3_Lathe_chucks_Independent_chucks_en.pdf (accessed Dec 2024).

Sandvik Coromant. (2017). *Training Handbook*. Metal Cutting Technology https://www.sandvik.coromant.com.

Stephenson, D.A., and Agapiou, J.S. (2016). *Metal Cutting Theory and Practice*, 3e. Boca Raton, FL: CRC Press.

TaeguTec LTD. (2023). Technical guide – tooling system. https://www.taegutec.com (accessed May 2023).

11

Geometric Validation

Tolerance Charting

11.1 Introduction

The primary aim of process planning is to create a cost-effective process plan that ensures the production of components within design specifications. The geometric validation of process plans is based on tolerance analysis where it is evaluated the impact of the natural variations of manufacturing processes and raw materials on the resulting machined dimensions. For uncomplicated parts that can be machined directly with most or all of their design dimensions, there is little need for tolerance analysis. In those cases, the design dimensions are obtained using as datum (i.e., locating surface) the surface itself that defines the design dimensions. Consequently, the tolerance of the dimension matches the tolerance of the manufacturing process, and there is no accumulation of tolerances stemming from datum changes. Under these scenarios, geometric validation is ensured while the selected processes and machinery are capable of producing parts within the specified design tolerances.

However, tolerance analysis becomes significantly more challenging when dealing with complex parts, tight design tolerances, frequent datum changes, and a large number of operations required. Under these conditions, tolerances from various operations accumulate, leading to stackups on the design dimensions. Thus, to validate the process plan, both design dimensions and stock removals should be analyzed to ensure machining feasibility and part tolerances. To address this issue, a formal technique called tolerance charting has been applied for decades.

Tolerance charting is widely utilized in the aerospace and automotive industries, particularly for parts with tight tolerances produced through machining processes like turning, milling, and grinding. In these industries, tolerance charting is typically carried out either manually or with the assistance of computer programs. Despite the significant advantages it offers, tolerance charting remains relatively unpopular outside of the automotive and aerospace sectors due to be considered a complex tool.

This chapter presents a methodology for 1D tolerance charting from references Whybrew and Britton (1997), Jeang (1996), and Thumm et al. (2002) based on traditional tolerance charting with a rooted (datum hierarchy) tree technique. Although there are commercial tools for 3D tolerance analysis to ensure geometric product validation within computer-aided (CAD) environments (CETOL 6σ is one of these popular software), these tools are mainly focused on assembly validation. 1D tolerance charting techniques may provide sufficient insights for geometric validation in simple machined products considering machining and fixture tolerances. Tolerance charting is

Manufacturing Process Planning: A Practical Approach for Mechanical Engineering, First Edition.
José V. Abellán-Nebot, Carlos Vila Pastor, and Héctor R. Siller.

Companion website: www.wiley.com/go/Manufacturing_Process_Planning_1e

commonly applied after machining sequencing to validate the geometrical product specifications, and, if specifications are not met, re-sequencing of the machining operations is conducted. Alternatively, tolerance charting can be also applied to calculate the efficiency of different process plans to facilitate process plan selection (Britton et al. 2001).

11.2 Definition of Terms

To construct tolerance charts, the following terms should be defined: design dimensions, preform dimensions, working dimensions, balancing dimensions, resultant dimensions, and stock removal allowances. These terms are defined in the following and – for illustrative purposes – their definition is applied to the example shown in Figure 11.1a, b.

- Design dimensions. The workpiece dimensions indicated in the design drawing or CAD model are known as design dimensions (Figure 11.1a). They may also be referred to as blueprint dimensions. They consist of two parts, a nominal value and a tolerance, which is the allowed deviation from the nominal dimension. It is important to convert dimensions so that the tolerances are in evenly disposed bilateral form to make the calculation easier. For instance, a dimension like $20.0^{+0.3}_{+0.1}$ must be transformed into 20.2 ± 0.1 mm.
- Preform dimensions. The dimensions of the raw material, such as a sawn bar, forging, or casting, are known as preform dimensions. Additionally, they include nominal and tolerance elements. In Figure 11.1a, the preform dimensions are 87 ± 0.8 mm.
- Working dimensions. Working dimensions are those assigned to each machining operation. A nominal dimension and a tolerance are both components of a working dimension. The tolerance is the process tolerance, that is, the variability of the machining operation due to accuracy limitations. As shown in Figure 11.1b, the working dimensions are the dimensions cut at each operation with the corresponding variability. At the first subjob, the nominal working dimensions for face turning and internal face turning are 86 and 71 mm, respectively. Assuming a machining tolerance of 0.02 mm, the working dimensions are 86 ± 0.02 and 71 ± 0.02 mm.
- Balancing dimensions. Working dimensions can be used to derive balancing dimensions, which are distances between workpiece surfaces. The balancing dimension is an intermediate outcome

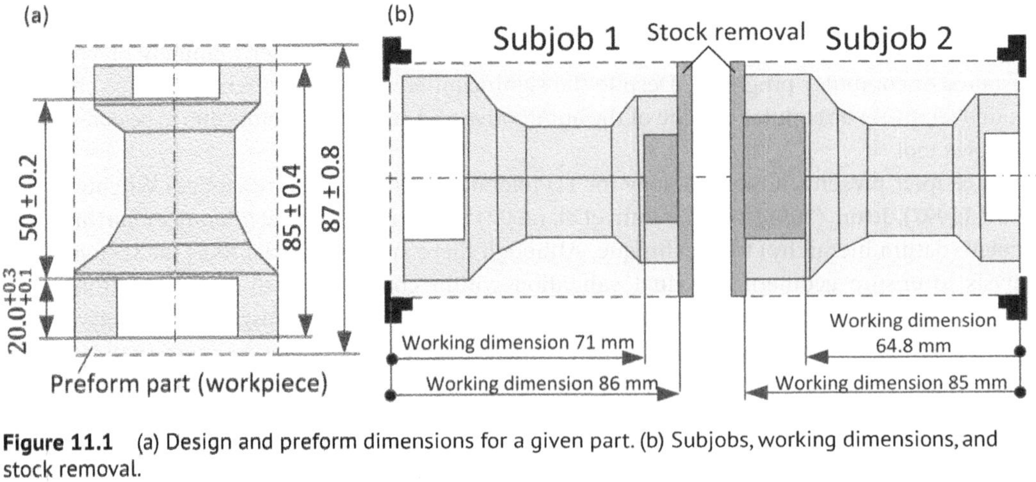

Figure 11.1 (a) Design and preform dimensions for a given part. (b) Subjobs, working dimensions, and stock removal.

dimension if additional processing is to be performed on either one of the two surfaces constituting the dimension. Because balancing dimensions are derived from working dimensions, they also include nominal and tolerance components. For instance, the distance between the two machined surfaces at the first subjob is a balancing dimension with a value of 15 ± 0.04 mm. As can be seen, the result of 0.04 mm comes from a tolerance stackup since the process tolerance is accumulated within the two operations, each one with a process tolerance of ± 0.02 mm.

- Resultant dimensions. A resultant dimension is a special case of a balancing dimension when it results in a design dimension. To ensure that the design dimensions can be met, nominal and tolerance stackups of the resultant dimensions must be evaluated and they should fall within the range of design dimensions. Otherwise, a percentage of the manufactured products would be out of specification. For instance, the distance between the two working dimensions from subjob 2 results in the design dimension that, according to the initial specifications, should be within the value 20.2 ± 0.1 mm. The working dimensions are 64.8 and 85 mm with a process tolerance of ± 0.02 mm, and thus, the resultant dimension from these two working dimensions is $85 \pm 0.02 - (64.8 \pm 0.02) = 20.2 \pm 0.04$ mm, where a tolerance stackup of 0.04 mm arises. The upper and lower limits of the resultant dimension must fall within the range of the design dimensions to be considered acceptable, as occurs in this example.
- Stock removal allowance. The stock removal allowance is the difference between the dimensions before and after machining. In other words, it is the quantity of material left on a surface for that surface's subsequent machining operation and it depends on the type of operation to be conducted, the precision of previous manufacturing operations, etc. As it has been explained, a working dimension has a process tolerance due to machining variability. Therefore, the stock removed in a machining operation will present a tolerance value. Additionally, the machining allowance from previous manufacturing processes (e.g., forming, casting) presents some degree of variability. To define a correct process plan, the stock removal allowance must be checked to be positive (i.e., enough material should exist to conduct machining) to ensure machining feasibility. Otherwise, there might be insufficient stock material to be removed and surface cutting may be not produced, leaving the dimensions out of specifications. Note that the stock removal may present a tolerance stackup depending on the sequence of machining. An example of tolerance stackup in stock removal is shown in Figure 11.1. The raw material has a preform dimension of 87 ± 0.8 mm, and the final length of the part is 85 mm. The machining process is carried out in a lathe, with two subjobs. At the first subjob, a face turning operation is conducted with a working dimension of 86 mm. The process tolerance is ± 0.02 mm; thus, the stock removal is $(87-86) \pm 0.8 \pm 0.02 = 1 \pm 0.82$ mm. In the worst-case scenarios, the machining operation would remove a maximum value of 1.82 mm or a minimum value of 0.18 mm. As can be seen, even in these worst-case scenarios, the machining operation is feasible (nonnegative values). At the second subjob, the left side of the part is machined. Initially, there is a machining allowance of 1 mm (86 – 85 mm = 1 mm). Since the right surface has been previously machined to 86 ± 0.02 mm, the stock removal is $(86 - 85) \pm 0.02 \pm 0.02 = 1 \pm 0.04$ mm. At this subjob, the worst-case scenario is 1.04 and 0.96 mm since there is no effect of the preform dimensional tolerance.

11.3 Steps for Tolerance Charting

A tolerance chart is a combined graphical and tabular record of a machining sequence that facilitates checking of 1D tolerance stacks (Thimm et al. 2002). Its purpose is to assess whether a specific machining sequence is suitable for producing the part with the desired dimensions and tolerances and to ensure that material removal operations are appropriately assigned during the machining process. In simpler terms, the chart is employed to geometrically validate the process plan (Kim and Dong 2005).

Figure 11.2 illustrates the key components of a tolerance chart. At the top center, a depiction of the part is provided. Leader lines, also known as face lines, extend from this depiction to indicate the surfaces, points, and centerlines on the part that will define the locating surface and the machined surfaces. These lines are numbered sequentially from left to right using two-digit numbers (e.g., 01, 02, etc.) and serve as start and end points for tolerance chart symbols (Figure 11.3). The tail of the arrows defined in these chart symbols indicates the locating surface for the operation and the head indicates the machined surface. Within the middle section of the chart, data pertaining to operations, working dimensions, operation symbols, and tolerances related to stock removal are recorded. The chart shows the sequence of machining operations, from top to bottom. At the bottom of the chart, the resultant dimensions are calculated and compared to the design data.

In addition to the tolerance chart, the rooted tree and the stock removal set table are also drawn (Figure 11.4). The rooted tree is a directed graph of working dimensions (i.e., machining operations) with no loops that indicates the datum hierarchy for a machining sequence (Thimm et al. 2002; Whybrew et al. 1990). The nodes represent surfaces and the arcs represent operations. The nodes and arcs can include tolerance information about fixturing tolerance and machining tolerance, respectively. The stock removal set defines a table with dimensional information with respect to a surface reference that is used to calculate the nominal dimensions between surfaces.

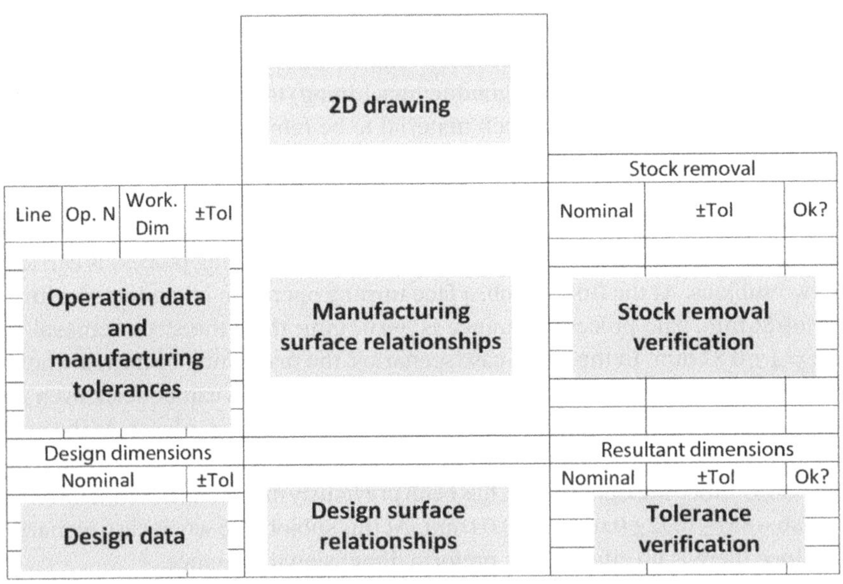

Figure 11.2 Basic features of a tolerance chart.

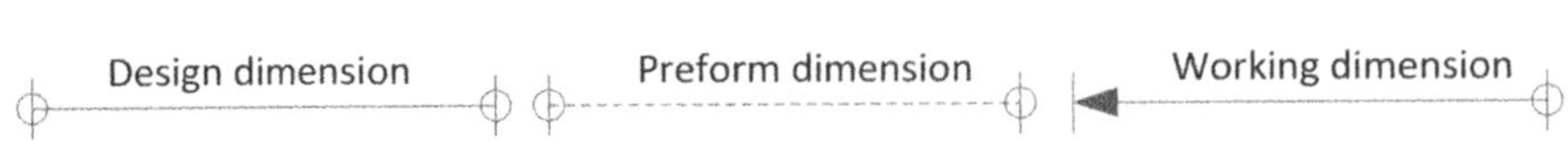

Figure 11.3 Tolerance chart symbols.

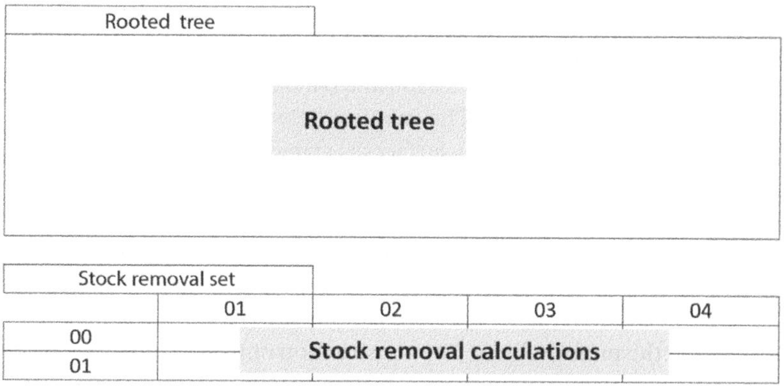

Figure 11.4 Additional features to build the tolerance chart. Rooted tree and stock removal set.

Tolerance charting can be conducted in the following six main steps:

1) Create the rooted tree and start chart preparation.
2) Write design dimensions, process tolerances, and raw surface tolerances.
3) Define a surface reference and write the stock removal set table.
4) Calculate the resultant dimensions and verify that they are within design tolerances.
5) Calculate the stock removal for critical surfaces and verify its positive value.
6) Calculate working dimensions.

To understand the tolerance charting procedure, the process of drawing the chart for the previous example (Figure 11.1) is discussed step by step in the following sections.

11.3.1 Step 1: Create the Rooted Tree and Start Chart Preparation

The initial step in creating the chart involves starting with a blank template chart. Then, the outline of the finished part is drawn at the top of the chart and the face lines are added. Subsequently, the sequence of machining operations is analyzed and the rooted tree graph is created. To draw the rooted tree, the surfaces should be numerated distinguishing the surface state before and after machining. Each surface is assigned a unique four-digit number where the initial two digits designate the stock removal set and the last two digits are related to the number of cuts made on a surface.

From previous example, four workpiece surfaces are identified together with four stock removal sets numbered 01, 02, 03, and 04. The first state of each surface, i.e., the first creation of the surface, adds the digits 00 to its removal set. If the surface is generated after cutting a surface that was previously created, then, the digits 01 are added. Consequently, 0200 means the creation of surface 02, while 0101 denotes the first cut on surface 01, which existed from a prior operation. From the example, it can be seen that surfaces 0100 and 0101 refers to surface 01 from casting and to surface 01 after machining the casting surface, respectively. However, surface 0200 refers to surface 02 after machining because there is no previous casting surface. Therefore, surfaces before and after machining form part of the same stock removal set.

After numeration, the tree graph is built to represent the sequence of machining and the hierarchy of datums. In the rooted tree, each node on the graph denotes either a processed surface, a reference surface, or both, and each link connecting two immediate nodes corresponds to a single machining operation with the corresponding machining tolerance. The tree starts with the preform surfaces drawn at the top. For example, the surfaces 0100 and 0400 are drawn first since they are the casting surfaces available before machining. The relation between preform surfaces is shown in dashed lines, and the tolerance of the preform surfaces is stated in the diagram. For this example, a casting tolerance is indicated as $t_p = \pm 0.8$ mm.

Then, below the preform surfaces, the surfaces that are created in the first subjob are drawn. A link is added between the datum surface from subjob 1 and the surfaces generated. The links are solid lines to differentiate from those related to the preform dimensions with an arrowhead pointing toward the machined surface. Each link in the rooted tree is numbered following the manufacturing sequence and numbering first the preform links. The construction of the tree proceeds from top to bottom, and when a new job/subjob is identified (i.e., a different setup for part location), the rooted tree is extended downward.

After the creation of the rooted tree (Figure 11.5), the left part of the chart shown in Figure 11.2 can be completed. The column "Line" refers to the number of each link of the rooted tree whereas the column "Op. N" refers to the number of the machining operation according to the established machining sequence. The center of the chart is also completed with the same information of the rooted tree, relating with solid or dashed lines the surfaces of the workpiece that is represented at the top of the chart. The chart symbols from Figure 11.3 are used to indicate if the line refers to a preform, working, or design dimensions.

11.3.2 Step 2: Write Design Dimensions, Process Tolerances, and Raw Surface Tolerances

Next, the part drawing or CAD model's part dimensions and tolerances are retrieved, and these values are put at the bottom of the chart. Tolerances related to processes and raw surfaces are known and are explicitly listed in the tolerance chart under column ± Tol.

After steps 1 and 2, the resulting tolerance chart is shown in Figure 11.6.

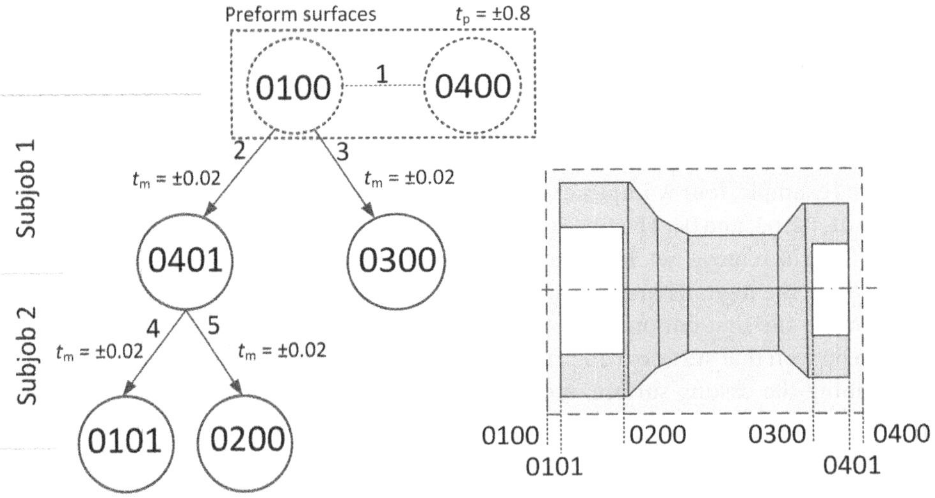

Figure 11.5 Rooted tree for the previous example.

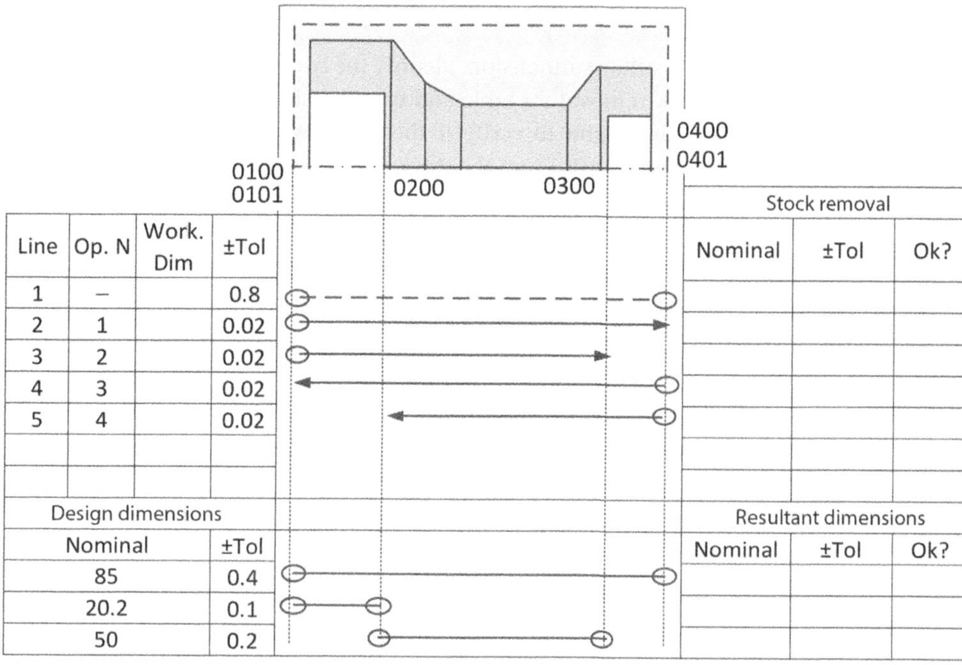

Figure 11.6 Tolerance chart for the example. Steps 1 and 2. Units in mm.

11.3.3 Step 3: Define a Surface Reference and Write the Stock Removal Set Table

The stock removal set table is a useful tool to obtain the working and resultant dimensions in a straightforward way. In this table, the columns indicate the stock removal sets and the rows are the cut number (positions) in the sets. To construct the table, a surface reference is first defined. In this example, the left surface of the machined part, named 0101, is considered the reference, and thus, its value is 0. The raw state of this surface is named 0100 and its value is −1 mm since there is a 1-mm machining allowance from the raw surface. The rest of the dimensions can be obtained knowing that surface 0101 is the reference (i.e., origin for dimensions), the dimensions of all surfaces are known from the design drawing and the machining allowance is 1 mm. Following this approach, the stock removal set is defined as shown in Figure 11.7.

Surface 0200 with respect to 0101 (reference), the dimension is 20.2 mm (design dimension)

Surface 0400 is surface 0401 before machining. As machining allowance from the preform is 1 mm, 0400 with respect to 0101 is 86 mm

Stock removal set

	01	02	03	04
00	−1	20.2	70	86
01	0			85

Surface 0101 is chosen as the reference

Surface 0401 with respect to 0101 (reference), the dimension is 85 mm (design dimension)

Figure 11.7 Stock removal set table for the example. Units in mm.

11.3.4 Step 4: Calculate the Resultant Dimensions and Verify

To calculate the nominal value of a resultant dimension, identify the two surfaces involved. Extract their surface positions from the stock removal set table and take the absolute difference between them. For instance, the resultant dimension to verify if the first design dimension is reached (85 ± 0.4 mm) involves surfaces 0101 and 0401. From the stock removal set (Figure 11.7), the nominal values of 0101 and 0401 are 0 and 85 mm, respectively, which results in a nominal value of $85 - 0 = 85$ mm.

After calculating the nominal value of the resultant dimension directly from the stock removal set, the tolerance of the resultant dimension is evaluated. The following calculation method is used to easily calculate the tolerance stackup for the resultant dimensions (Britton 2000; Whybrew and Britton 1997):

1) Identify one of the two surfaces that define the resultant dimension.
2) Find the path on the rooted tree between this surface and the other surface of the resultant dimension.
3) Add the tolerances of all the operations that contribute to the path to determine the tolerance stacks for the resulting dimension.

This process is repeated for all resultant dimensions. After the tolerance stackup has been calculated, its value should be lower than or equal to the corresponding design tolerance to ensure this specification. Otherwise, the process planner should consider other alternative machining sequences to reduce the tolerance stackup and ensure part quality.

Applying this procedure for the first resultant dimension, the path on the tree is 0401 – 0101. In this case, the tolerance of the resultant dimension is just defined by the tolerance of a machining operation (i.e., there is no stackup) and it is equal to ± 0.02 mm. Thus, the resultant dimension is 85 ± 0.02 mm, which is within the design dimensions, of 85.0 ± 0.4 mm.

However, the second resulting dimension defined by surfaces 0101 and 0200 determines the path 0101-0401-0200 from the rooted tree. The path shows two machining operations, thus, a stackup tolerance of $\pm 0.02 \pm 0.02 = \pm 0.04$ mm is generated. Since design tolerance is ± 0.1 mm for this specification, the sequence is validated.

Surfaces 0200 and 0300 define the third design dimension. Following the path from the rooted tree (see Figure 11.8), lines 2, 3, and 5 yield the resultant tolerance, which is $\pm 0.02 \pm 0.02 \pm 0.02 = \pm 0.06$ mm. This value is entered in the chart, and it is again compared with the tolerance of the design dimension to verify if the specification is met.

11.3.5 Step 5: Calculate the Stock Removal and Verify

The nominal and tolerance of the stock removal values are calculated similar to the resultant dimensions. First, the stock removal set is used to calculate the nominal values, and then, the tolerance stackup is obtained by adding the tolerances identified in the path of the rooted tree. The calculations are listed below:

- Stock removal set 01

The stock removal set 01 refers to the difference between the raw state and the machined state of surface 01 and it is defined by the dimension between 0100 and 0101. The nominal value is 1 mm according to the stock removal set table (Figure 11.7, absolute difference between 0100 and 0101). The tolerance of the stock removal set is obtained by analyzing the path 0100-0101 from the rooted tree. As shown in the rooted tree, the path is 0100-0401-0101 and the tolerance stackup is

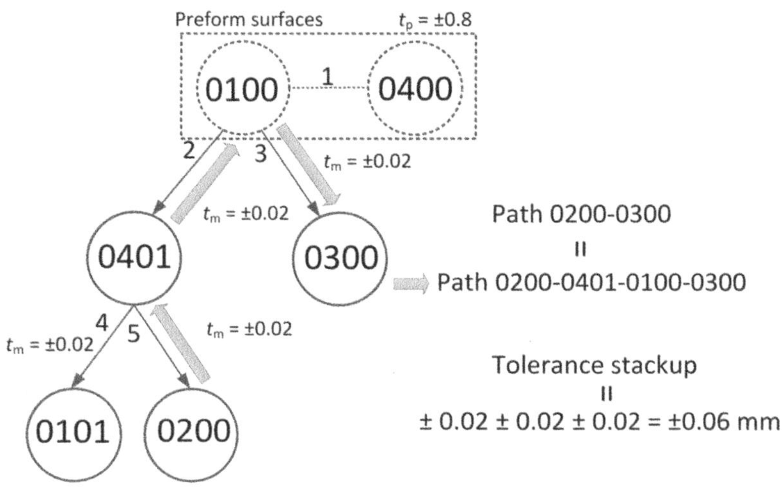

Figure 11.8 Path between 0200 and 0300 surfaces and resulting stackup. Units in mm.

$\pm 0.02 \pm 0.02 = \pm 0.04$ mm. This value is added to the chart below the ±tol column for stock removals.

- Stock removal set 04

The nominal value is 1 mm since this stock removal set is obtained from 0400-0401. The path from the rooted tree is 0400-0100-0401 and lines 1 and 2 contribute to the tolerance stackup. The stackup tolerance is $\pm 0.8 \pm 0.02 = \pm 0.82$ mm since the tolerance of the preform dimension is included in the path.

Finally, sets 02 and 03 contain only one surface state, and thus, there is no stock removal to check. In these cases, the stock removal allowance is set to "solid" in the chart.

At this step, the tolerance stackup values are compared with the stock removal allowances to verify that, even in the worst-case scenario, enough stock material is available to conduct the machining operation. For this example, in both cases, the allowances are greater than the stackups since the nominal stock allowance is 1 mm, and the tolerance stackups are ±0.04 and ±0.82 mm.

11.3.6 Step 6: Calculate Working Dimensions

The nominal values of the working dimensions can be computed once the machining sequence has been validated by verifying the design specifications and the stock removals. Following the same procedure, the following working dimensions are defined:

- Operation 1, line 2 of the chart. Surface 0100-0401. Nominal value from the stock removal set: $85-(-1)=86$ mm.
- Operation 2, line 3 of the chart. Surface 0100-0300. Nominal value from the stock removal set: $70-(-1)=71$ mm.
- Operation 3, line 4 of the chart. Surface 0401-0101. Nominal value from the stock removal set: $85-0=85$ mm.
- Operation 4, line 5 of the chart. Surface 0401-0200. Nominal value from the stock removal set: $85-20.2=64.8$ mm.

After this step-by-step methodology, the final tolerance chart is shown in Figure 11.9.

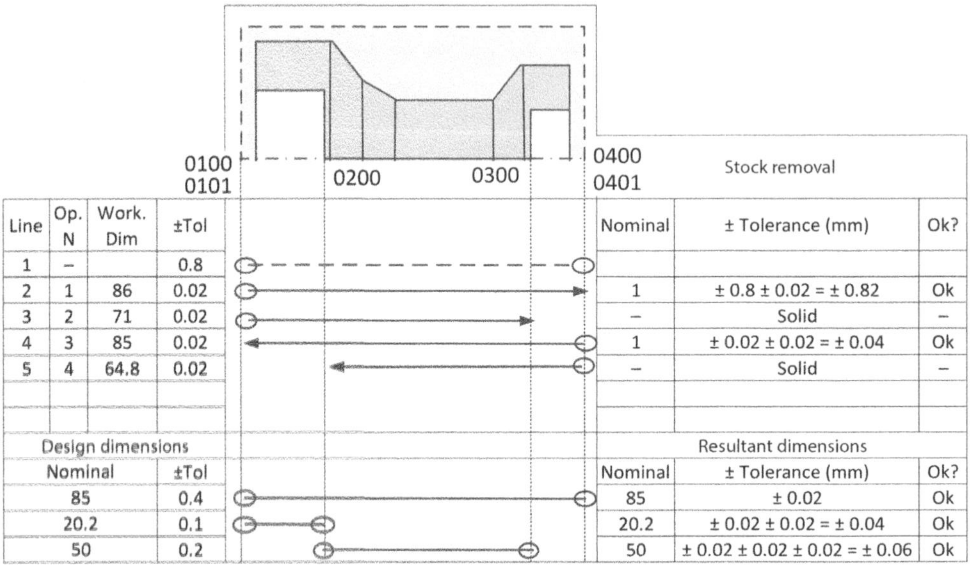

Figure 11.9 Resulting tolerance chart for the example. Units in mm.

11.3.7 Adding Fixturing Tolerances

The methodology previously presented has been reported in several technical documents (Britton 2000; Jeang 1996) and illustrates how to add the machining and preform tolerances (e.g., casting tolerances) for process planning validation. The methodology can also include the fixturing tolerance, which is one of the main factors in tolerance stackup. This tolerance is added to the rooted tree below each surface that is used as a locating surface. To understand the stackup mechanism for fixturing tolerances, consider the dimension 0101-0200 from the previous example. As can be seen, this dimension is obtained under the same subjob, by clamping surface 0401 and machining both 0101 and 0200. If a fixturing error exists, its effect will show up on both machined surfaces, but the dimension defined by the distance between both surfaces will remain unaffected. In other words, fixturing errors do not affect when the resultant distance looked for in the rooted tree is in the same subjob.

Now consider the dimension 0300-0101. In this case, the dimension is between two features machined in different subjobs, and the fixture error will have an impact. According to the rooted tree, the path is 0300-0100-0401-0101. The fixturing error in the first subjob (i.e., when the locating surface is 0100) does not affect the path since the error in the dimension 0300-0100 and the error 0100-0401 is compensated, and only machining errors will influence the path 0300-0100-0401. However, at the second subjob the fixturing error will affect the final tolerance additionally to the machining error. Therefore, the dimension 0300-0101 will present a stackup of three machining tolerances (lines 2, 3, and 4), and the fixturing tolerance at the second subjob (0401) while the fixturing tolerance at the first subjob (0100) is not included. Assuming a fixture tolerance of ± 0.1 mm, the resulting tolerance stackup is $\pm 0.02 \pm 0.02 \pm 0.02 \pm 0.1 = \pm 0.16$ mm. Figure 11.10 illustrates the path studied in the rooted tree and the resulting tolerance stackup extracted from the diagram.

Following this reasoning, the resulting tolerance chart from the previous example is shown in Figure 11.11.

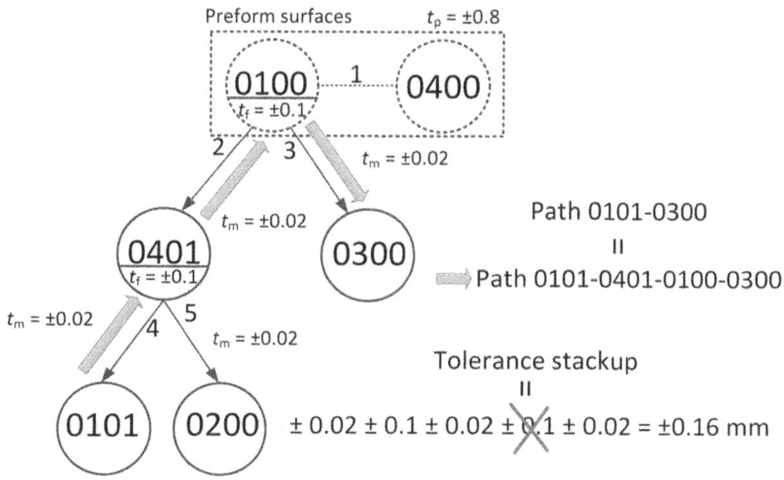

Figure 11.10 Path between 0101 and 0300 surfaces and resulting stackup considering fixturing tolerances added below each locating surface. Note that the fixture tolerance in 0100 is not considered since 0401 and 0300 share the same fixture tolerance. Units in mm.

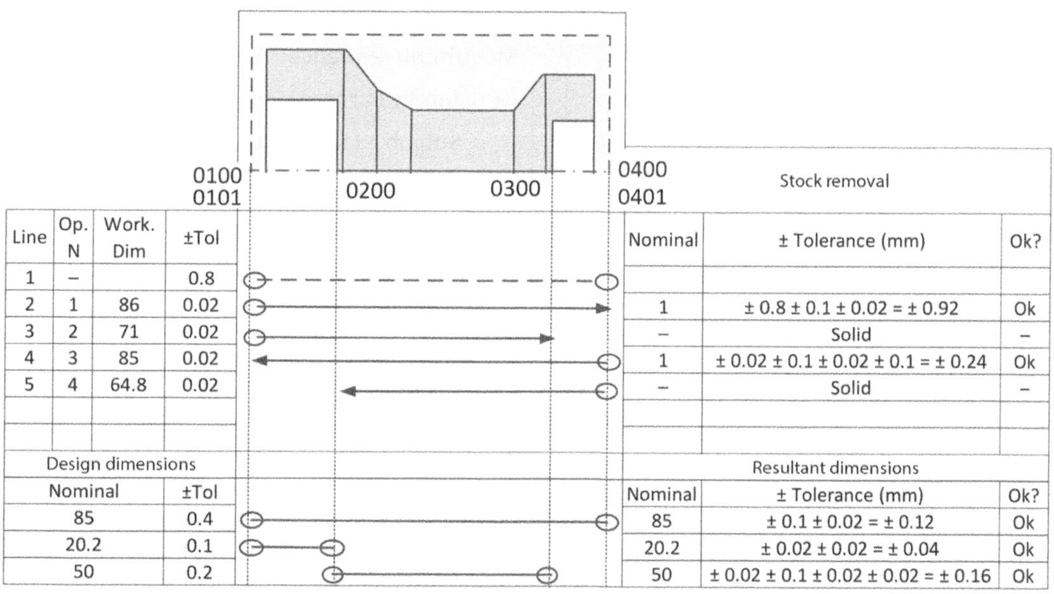

Line	Op. N	Work. Dim	±Tol	Stock removal Nominal	± Tolerance (mm)	Ok?
1	–		0.8			
2	1	86	0.02	1	± 0.8 ± 0.1 ± 0.02 = ± 0.92	Ok
3	2	71	0.02	–	Solid	–
4	3	85	0.02	1	± 0.02 ± 0.1 ± 0.02 ± 0.1 = ± 0.24	Ok
5	4	64.8	0.02	–	Solid	–

Design dimensions Nominal	±Tol	Resultant dimensions Nominal	± Tolerance (mm)	Ok?
85	0.4	85	± 0.1 ± 0.02 = ± 0.12	Ok
20.2	0.1	20.2	± 0.02 ± 0.02 = ± 0.04	Ok
50	0.2	50	± 0.02 ± 0.1 ± 0.02 ± 0.02 = ± 0.16	Ok

Figure 11.11 Resulting tolerance chart for the previous example adding fixturing tolerances. Units in mm.

Questions and Exercises

1 Draw the tolerance chart for the part shown in Figure 11.12. Consider the given rooted tree and a machining tolerance of ±0.05 mm. The part is obtained from casting with a machining allowance of 2 mm and a casting tolerance of ±0.8 mm. Assume fixturing tolerances negligible for this exercise.

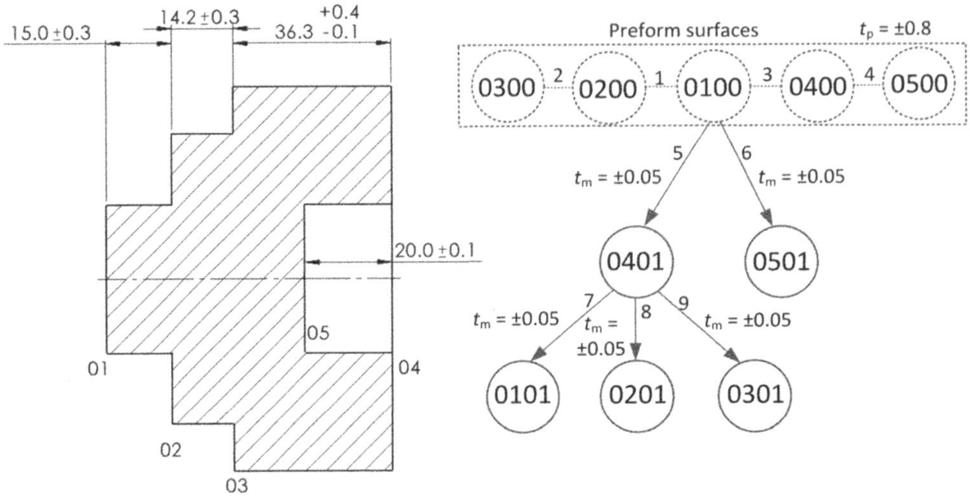

Figure 11.12 Part drawing and rooted tree for Exercise 1. Units in mm.

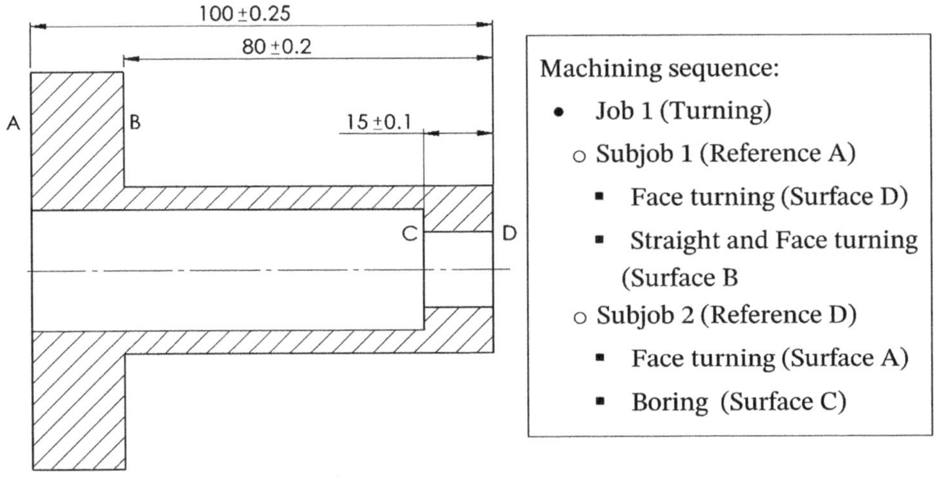

Figure 11.13 Design dimensions and machining sequence for Exercise 2. Units in mm.

2 Verify, using the tolerance charting procedure, if the part specifications shown in Figure 11.13 can be ensured given the process plan provided. The raw material is a round bar obtained from rolling with dimensional tolerances of ±0.5 mm and a machining allowance of 2 mm. The turning operations are assumed to present a variability of ±0.02 mm and fixturing tolerance is assumed ±0.10 mm.

3 Draw the tolerance chart and tolerance tree for the part shown in Figure 11.14. Consider part from rolling with dimensional tolerances of ±0.5 mm and machining allowance of 1 mm; machining tolerance of ±0.02 mm; fixturing tolerance ±0.08 mm.

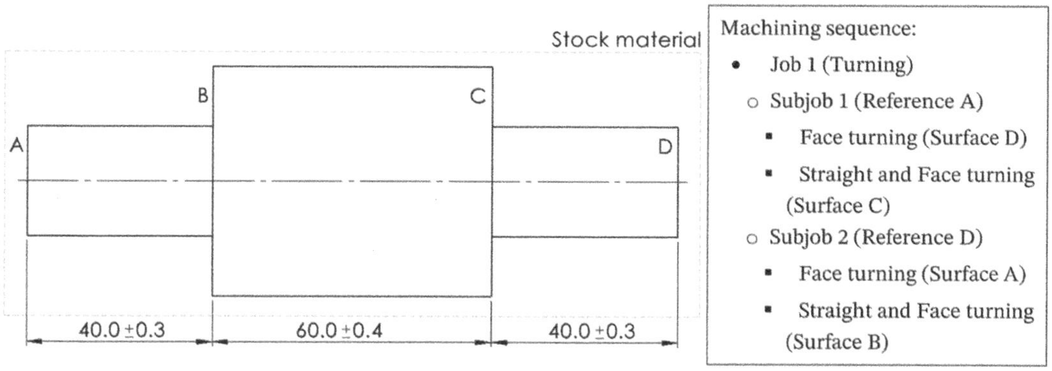

Figure 11.14 Design dimensions and machining sequence for Exercise 3. Units in mm.

References

Britton, G. (2000). Notes on tolerance planning and control, Tampere University of Technology. https://www.academia.edu/1745538/Manufacturing_tolerance_planning_and_control (accessed February 2024).

Britton, G.A., Thimm, G., Fork, S.C., and Jiang, F. (2001). Efficiency measures of process plans for machined parts. *The International Journal of Advanced Manufacturing Technology* 17: 819–824.

Jeang, A. (1996). Tolerance chart balancing for machining process planning. *Quality and Reliability Engineering International* 12 (5): 355–364.

Kim, I. H., and Dong, Z. (2005). Quality measurement of production process plan using tolerance chart. In IEEE International Conference on Automation Science and Engineering, 2005.

Thimm, G., Jiang, F., Beng, T.S., and Britton, G. (2002). A graph representation scheme for process planning of machined parts. *The International Journal of Advanced Manufacturing Technology* 20: 429–438.

Whybrew, K., and Britton, G.A. (1997). Tolerance analysis in manufacturing and tolerance charting. In: *Advanced Tolerancing Techniques* (ed. H.-C. Zhang), 13–41. Wiley.

Whybrew, K., Britton, G.A., Robinson, D.F., and Sermsuti-Anuwat, Y. (1990). A graph-theoretic approach to tolerance charting. *The International Journal of Advanced Manufacturing Technology* 5: 175–183.

12

CNC Programming

12.1 Introduction

There is a wide variety of industrial equipment that works with computer numerical control (CNC), such as bending machines, punching machines, lathes, milling machines, machining centers, electrical discharge machining equipment, welding machines, robots, laser cutters, 3D printing machines, etc. Among all these equipment, machine tools and machining operations are the most common CNC applications.

The main advantages of CNC applied to small and medium production batches in machining are the reduction of total machining time, the increase of manufacturing flexibility (simple changes in the CNC code result in different machining operations), and more accurate movements (i.e., higher precision and accuracy) with respect to manually operated machine tools. Considering that almost all metal components require, at some point, a machining operation, it is easy to understand that CNC machine tools are commonly available in almost all manufacturing companies.

When dealing with process planning in machining, CNC programs should be elaborated and checked using CNC code simulators or computer-aided manufacturing (CAM) programs. In this chapter, the fundamentals of CNC programming to create machining programs are presented. The chapter describes the different coordinate systems that are defined in common machine tools and the basic commands for manual CNC programming according to the International Organization for Standardization (ISO) standard. Basic functions for linear and circular movements, cutting tool compensations, feed and speed programming, or other auxiliary functions are presented employing some examples. The explanation of CNC canned cycles completes the overview of CNC programming. Finally, some guidelines for proper cutting tool trajectories are presented and illustrated with practical examples in milling, turning, drilling, and threading operations. Readers interested in a more detailed explanation of CNC programming may refer to Smid (2003) and Kumar et al. (2020).

12.2 CNC Machine Tools

12.2.1 Nomenclature of Machine Tool Axes

All movements programmed numerically in a machine tool numerical control are referred to as a system of Cartesian axes. The denomination, placement, and orientation of each axis are standardized according to the "ISO 841: 2001 Industrial automation systems and integration, numerical

Manufacturing Process Planning: A Practical Approach for Mechanical Engineering, First Edition.
José V. Abellán-Nebot, Carlos Vila Pastor, and Héctor R. Siller.

Companion website: www.wiley.com/go/Manufacturing_Process_Planning_1e

control of machines, coordinate systems and motion nomenclature" (ISO 2001). The main axes of a machine tool are called *X*, *Y*, and *Z*. These axes are defined as follows:

- *Z*-axis. In machine tools where the workpiece has the rotary cutting motion (for example, on a lathe), the *Z*-axis is parallel to the machine spindle, and a positive movement moves the tool away from the workpiece. In machines where the cutting tool has the rotary cutting motion (such as a machining center), the *Z*-axis is parallel to the tool axis.
- *X*-axis.
  - In lathes, the *X*-axis leads the direction of movement of the tool, and a positive movement moves the tool away from the workpiece.
  - In vertical machining centers, the positive *X*-axis direction is to the right of the programmer when it is in front looking at the machine.
  - In horizontal machining centers, the *X*-axis is parallel to the workbench.

  In all cases, the *X*-axis is perpendicular to the *Z*-axis defined above.
- *Y*-axis. The *Y*-axis is determined from the *X*-axis and the *Z*-axis so that the three form a coordinate system (right-hand rule).

Figure 12.1 shows graphically the definitions of the axes in a CNC lathe and a vertical machining center.

12.2.2 Coordinate Systems in Machine Tools: Reference Systems

Machine tools are systems where the position of cutting tools is defined by CNC programs to create specific trajectories. The CNC programs specify the coordinates of the cutting tool in *X*, *Y*, and *Z* directions with respect to a coordinate system. To understand the definition of the coordinate system in CNC programs, it is important to understand the different coordinate systems that are defined in machine tools and their interrelation. The following coordinate systems can be used to define the reference system in the CNC program: machine coordinate system (MCS), fixture coordinate system (FCS), and workpiece/part coordinate system (WCS) (Fagor Automation 2024a).

- Machine coordinate system (MCS).

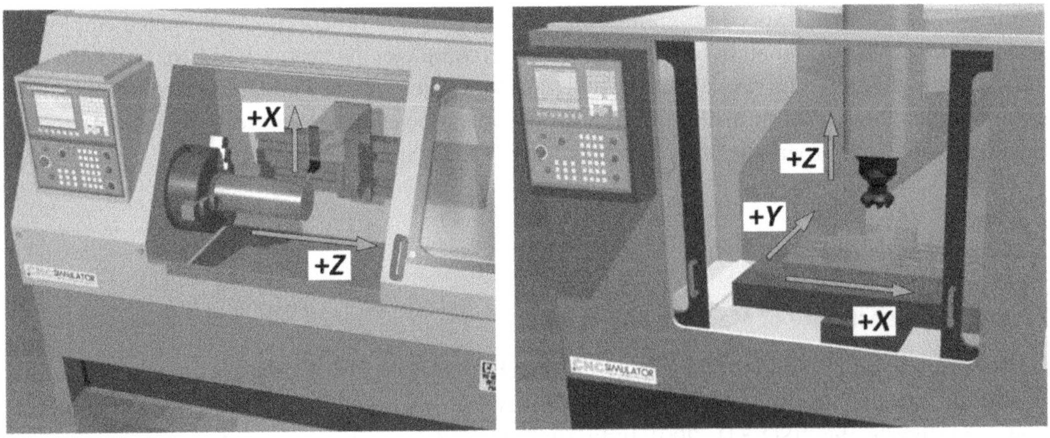

Figure 12.1 Definition of axes *X*, *Y*, and *Z* in a CNC lathe and a vertical machining center.

The MCS is defined by the vendor and specifies the physical movement of the axis of the machine tool following the standard ISO 841: 2001. The origin of this coordinate system is named as machine zero, and at this position, the value of all axes is 0. The physical location of the machine zero depends on the type of machine tool and it is defined by the machine tool manufacturer. In lathes, machine zero is usually placed at the frontal face of the spindle and over the *Z*-axis. In milling machines, the machine zero is usually placed at one extreme point of its machining volume.

- Fixture coordinate system (FCS).

The FCS is defined by the fixture used, and it is defined with respect to the MCS as the "fixture offset." To use the FCS, it must be set in the CNC program to activate it or should be previously defined from the CNC's front panel. When the machine has several fixtures, each one may have its own reference system associated with it. Similar to the previous coordinate system, the origin of this coordinate system is named as fixture zero.

- Workpiece/Part coordinate system (WCS).

To facilitate the programming task, the CNC programmer may define a WCS to which the cutting tool movements are referred. The origin point of this coordinate system is named as part zero, and it should be clearly defined in the machining operation sheet so that the machinist can define the workpiece zero in relationship with the machine zero as the "part offset." Similar to the FCS, to use the WCS, it must be set in the CNC program to activate it or should be previously defined from the CNC's front panel.

Figure 12.2 illustrates the different coordinate systems in CNC lathes and machining centers.

In addition to these coordinate systems, two other reference coordinate systems are defined: the tool coordinate system and the reference or home coordinate system.

- Tool coordinate system (TCS).

The TCS is used to calibrate cutting tools since it defines the position of the cutting tool tip with respect to the MCS. When calibrating the cutting tools, the position of the cutting tool tip with respect to this coordinate system is set. Thus, a proper radius and length compensation during machining can be done (Figure 12.3a, b).

- Reference or home coordinate system (RCS).

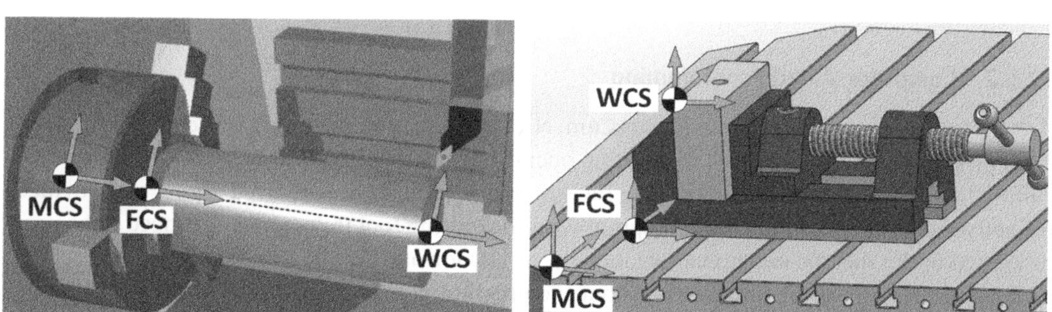

Figure 12.2 Machine tool coordinate system (MCS), fixture coordinate system (FCS), and workpiece coordinate system (WCS) in a lathe and a milling machine.

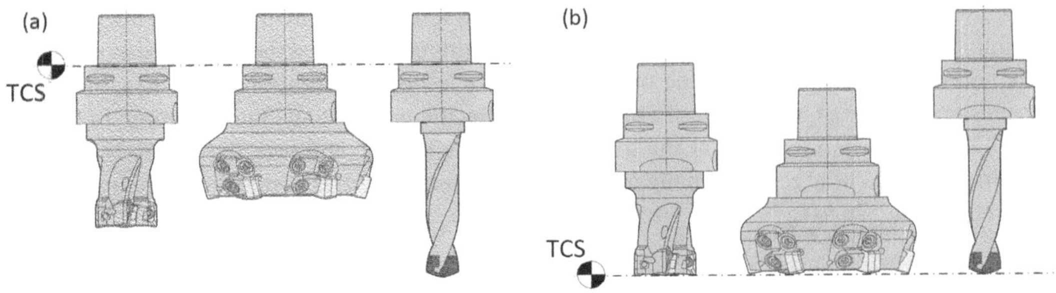

Figure 12.3 Tool coordinate system (a) before and (b) after tool height compensation. *Source:* Fagor Automation (2024a).

When initializing the machine tool, the controller starts to perform a "home search" and the axes move to the machine reference point that the machine manufacturer has previously set with respect to the MCS. When reaching this reference point, the CNC assumes the coordinate values assigned to that point, and the machine tool is ready to operate. The RCS is commonly set to a corner of the working volume to avoid any type of collision when powering up the machine tool and start searching home.

12.3 Programming CNC Machine Tools

A CNC program can be generated in three different ways: manual programming, computer-aided programming, and conversational programming.

12.3.1 Manual Programming

The code is created directly by the programmer, line by line. The program can be done using ISO codes (G-codes) or a specific language compatible with the CNC controller. The manual programming is conducted using specific software to simulate CNC code, where the programmer can check the cutting tool trajectories before launching the program. Figure 12.4 shows an example of G-code simulation using the software CNC simulator.

12.3.2 Computer-Aided Programming

CAM software, such as SolidCam, MasterCam, NX CAM, ProEngineer, etc., is used to generate the CNC program. In this case, each machining operation is defined by the user with its appropriate cutting tool and cutting conditions using the software. When all the required inputs are provided, the software generates a high-level program with the cutting tool trajectories that need to be post-processed in the specific machine tool language or ISO format.

The main benefit of this procedure is that the programmer does not have to define line by line the cutting tool trajectories since this is now the computer task. Furthermore, the computer-aided approach detects cutting tool collisions, quality defects, etc. Figure 12.5 shows an example of CNC simulation and rapid collision detection in SolidWorks CAM. Figure 12.6 shows an example of postprocessing the CNC program for a FANUC controller in SolidWorks CAM.

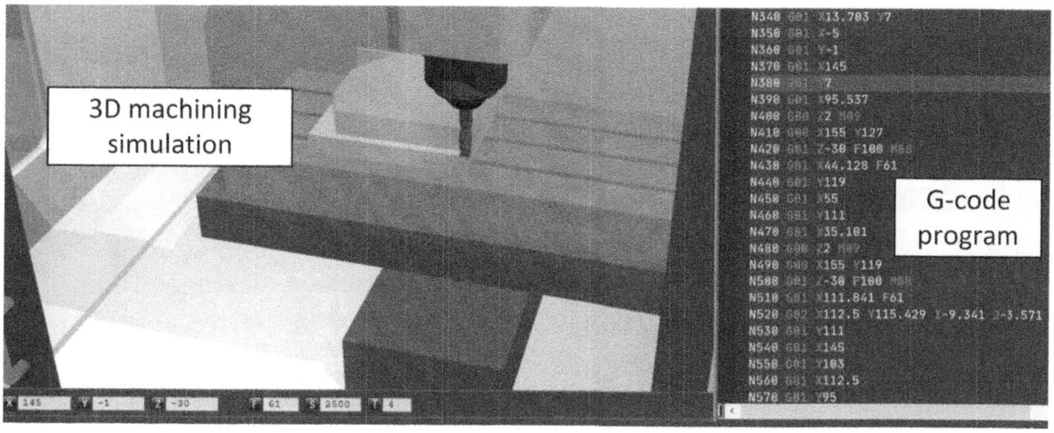

Figure 12.4 Example of manual programming and 3D simulation by the software CNCSimulator.

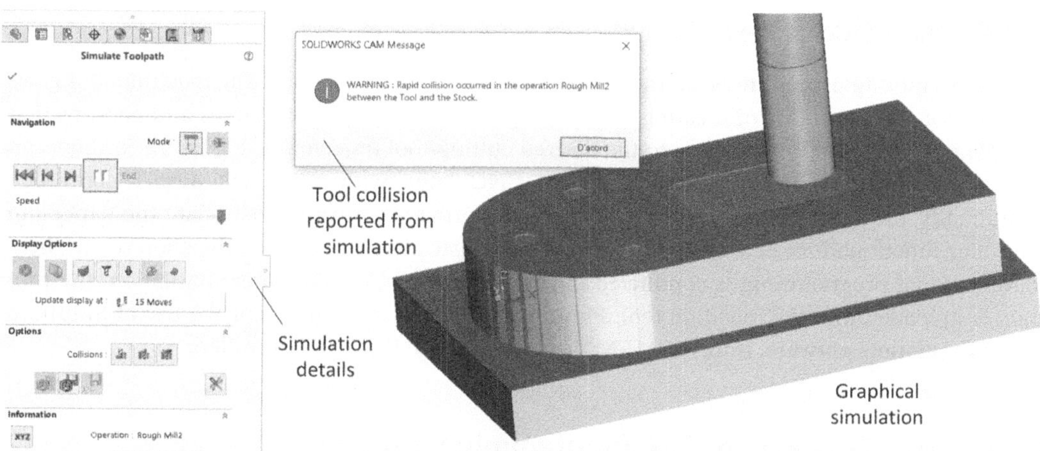

Figure 12.5 Example of computer-aided manufacturing (CAM) with SolidWorks CAM.

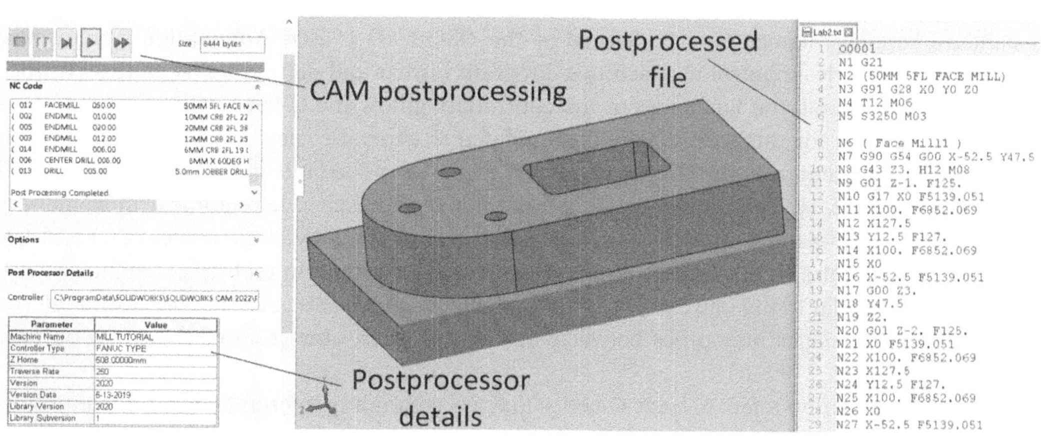

Figure 12.6 Example of postprocessing in CAM.

(a) (b)

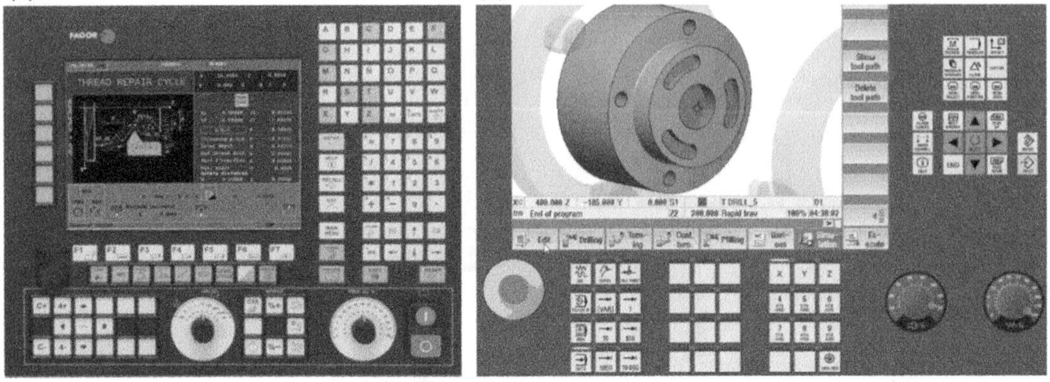

Figure 12.7 Example of conversational programming. (a) FAGOR 8055-T panel and thread cycle. (b) Sinutrain v4.8 programming software.

12.3.3 Conversational Programming

This programming procedure is conducted directly on the machine tool. The machine tool gives the operator the possibility of creating a CNC program using a series of menus and pop-ups that the operator has to fill in according to the desired cutting tool trajectories. Figure 12.7a shows an example of conversational programming of a CNC lathe with a FAGOR 8055-T controller. Additionally, most of CNC controller manufacturers provide software to simulate the panel and controller functionalities at a computer workstation instead of working on the shop floor. Then, conversational programming is conducted in the technical office and, after simulation, the program is uploaded into the machine tool. Figure 12.7b shows an example of the use of Sinutrain v4.8, a simulation software, from SIEMENS to simulate SINUMERIK controllers.

12.4 Basic Concepts of CNC Programming. G-Code

To program CNC machine tools, different programming languages can be applied depending on the type of controller installed in the machine tool. However, almost all controllers can interpret a standard language named G-code defined by the "ISO 6983-1: 2009 Automation systems and integration. Numerical control of machines. Program format and definitions of address words" (ISO 2009). In general, the languages of the controllers follow the G-code with some functions defined specifically for them, especially those related to canned cycles. In some cases, the language may be different from the G-code to make it easier to understand and use in conversational mode. For instance, Heidenhain controllers use a native language where a linear movement is defined by the command L or a circular interpolation is defined by circle center (CC) and DR (direction of rotation). In other cases, programs can also be written in high-level languages with a terminology similar to the one used by other languages, such as $IF, $GOTO, etc. Although the programming languages may differ between CNC controllers, almost all of them can execute standard G-codes.

The following section refers to basic G-codes that are commonly interpretable by all controllers and can be applied to elaborate simple programs. However, it should be noted that more complex parts are typically programmed using canned cycles that require specific codes and parameters depending on the controller or CAM programs, where postprocessing using the correct controller language should be selected.

12.4.1 Program Structure

The CNC program is defined by a sequence of blocks. Each block may contain the following characters:

- *N*. Block number.
- *G*. Preparatory function.
- *X*, *Y*, and *Z*. Axis values.
- *F*. Feed speed.
- *S*. Spindle speed.
- *T*. Tool number.
- *D*. Tool offset number.
- *M*. Auxiliary function.

The block number is defined by the letter *N* with a number. Usually, the blocks are numbered in steps of 10 (N10, N20, N30, ...). A general CNC block may have the following structure:

N... G... X±... Y±-... Z±-... F... S... T... D... M...

The data has no set order; it may be programmed anywhere in the block except the block number, *N*, which must always be programmed at the beginning. At any moment, comments can be added to facilitate program interpretation. Although the way of adding comments may depend on the controller, usually the text within parentheses at the end of a block or the text after the symbols "#" or ";" are accepted by CNC controllers as comments.

It should be noted that, for some functions, the data do not have to be repeatedly written every sentence. These functions are named modal functions, and they are kept active until it is called another function that is incompatible with them or until a stop instruction or reset is reached. In general, when it is called a *G* or *M* function this is activated until another *G* or *M* function that modifies the previous one is called. In other words, it is not necessary to rewrite at every CNC block the same *G* or *M* function. Once that function has been called, it is active in all the following blocks. The same occurs with the words *F*, *S*, and *T*, which are, by definition, modal functions.

12.4.2 List of Preparatory Functions

Preparatory functions are the *G*-codes that identify the type of activities the machine will execute. A program block may contain one or more *G*-codes. Preparatory functions are programmed using the letter *G* with two digits. These functions are usually placed following the block number and are used to determine the geometry of the part and working conditions such as units, planes of work, etc. If, in the same CNC block, two incompatible G functions are called, the CNC keeps the last one. Some general preparatory functions are defined in Table 12.1.

Note that the previous list is common to any type of controller (FAGOR, FANUC, SIEMENS etc.), but *G*-codes about cycles or subroutines may differ among controllers.

12.4.3 Programming Modes

12.4.3.1 Measurement Units. G20/G21 or G70/G71

The coordinates can be indicated in mm or inches. The use of G20/G70 indicates that coordinates are in inches whereas G21/G71 means in millimeters.

12.4.3.2 Absolute and Incremental Programming: G90/G91

Coordinates can be programmed in absolute values (G90) or by incremental steps (G91). Using G90, the coordinates of the point programmed refers to the zero workpiece or, alternatively, to the zero fixture. Using G91, the coordinates of a point programmed refer to the previous programmed point.

Table 12.1 General CNC G-codes.

G00	Positioning at rapid speed	G40	Tool radius compensation off
G01	Linear interpolation	G41	Tool radius compensation (left)
G02	Circular interpolation, clockwise	G42	Tool radius compensation (right)
G03	Circular interpolation, counterclockwise	G43/G44	Tool height compensation on
G04	Dwell	G49	Tool height compensation off
G17	*XY* plane for arc machining	G53	MCS
G18	*ZX* plane for arc machining	G54-59	Work coordinate systems
G19	*ZY* plane for arc machining	G7x-8x	Fixed cycles (depends on controller)
G20/G70	Inch units	G90	Absolute programming
G21/G71	Metric units	G91	Incremental programming
G95	Feed rate per revolution	G94	Feed rate per minute
G97	Constant spindle speed (lathe)	G96	Constant surface speed (lathe)

The format of Cartesian coordinates for turning and milling operations is as follows:

- Lathe: *X*±... *Z*±...
- Milling: *X*±... *Y*±... *Z*±...

Note that polar coordinates can also be used in CNC programs, but they are omitted here for simplicity.

12.4.4 Changes in Reference Systems

A CNC program could be written using the MCS. However, this is not practical since the MCS has an origin that is nowhere near the workpiece. As previously indicated, a workpiece coordinate system (WCS) is defined to facilitate the programming task. Thus, the CNC programs should refer to this change of reference system to make the movements be referred to the WCS and not to the MCS.

This change in the reference system is mainly done by defining the desired WCS and calling it by the corresponding *G*-code, from G54 to G59, which are codes to call up to six different WCSs. The WCSs are defined with respect to the MCS, and their coordinates are stored in the work offset register, a table defined in the CNC controller, and each register is associated with a *G*-code (G54, G55, etc., ...). To illustrate the process, Figure 12.8 shows the work offset register previously defined by the operator to define the WCS and a CNC program where the first WCS is loaded when calling G54.

In some controllers, it is also possible to define a fixture offset register that defines the FCS with respect to the MCS. This is useful since – for large production volumes – the FCS is a permanent position that is not changed, and the dimensions of the resulting parts are with respect to the fixture. However, when dealing with prototypes or single production parts, the WCS can be directly set by touching with a probe several surfaces from the raw part.

12.4.5 Programming of Movements

12.4.5.1 Rapid Movement: G00

Rapid movements are used for initial movements or cutting tool replacement movements where there is no cutting engagement. This is done in a high-speed movement.

When the CNC is turned on, after M02/M30, or after an emergency or reset command, the CNC assumes the G00 code. G00 is modal and incompatible with G01, G02, and G03. After using G00,

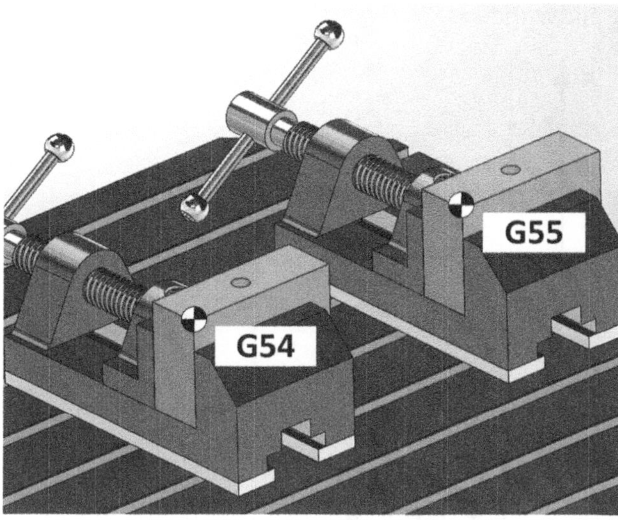

Work offset register			
G-code	*X*-axis	*Y*-axis	*Z*-axis
G54	150	150	200
G55	300	150	200
G56	0	0	0
G57	0	0	0

CNC program

```
N10 T01 D01 M06
N20 G90 G94 G17 G00 X100 Y100 Z400 S2000
M03
N30 G54
N40 X32 Y12.5 Z2
N50 G01 Z-10 F150
N60 G00 Z15
N70 G55
N80 X32 Y12.5 Z2
N90 G01 Z-10 F150
N100 G00 Z15
```

Figure 12.8 Example of changing a reference system in a CNC program and work offset register. Codes G54-G59.

the last feed speed programmed is not erased, so when programming G01, G02, or G03, if the feed speed is not provided, the previous value is used again.

12.4.5.2 Linear Interpolation: G01

G01 moves the cutting tool in a linear trajectory with the feed speed indicated by the code *F*. G01 is modal and incompatible with G00, G02, and G03.

Example: Generate the CNC program in Cartesian and absolute coordinates that define the turning trajectory of Figure 12.9. Consider the program in diameter for *X* coordinates.

Solution:

```
N40   G90 G00 X18 Z0 (rapid movement from the origin)
N45   G01 G95 X0 F0.2 (face turning, feed speed of 0.2 mm/rev)
N50   G00 Z2 (withdrawal in rapid movement)
N55   X15 (tool positioned for turning)
N60   G01 Z-15 (turning Ø15 × 15)
N65   X38.1 Z-35 (contour turning)
N70   X47 (face turning)
N75   G00 X200 Z200 (withdrawal to the origin)
```

Note that the text added within the parentheses can be kept in the program as comments.

12.4.5.3 Circular Interpolation: G02/G03

Functions G02/G03 allow circular trajectories with the feed speed defined in the F parameter. To program a circular trajectory, it is necessary to define the rotation direction (clockwise or counterclockwise), the last point of the circular trajectory, and the position of the center of the trajectory or its radius. For a clockwise trajectory, the G02 code is applied; for a counterclockwise movement, the code to be applied is G03. Figure 12.10 shows the rotation direction according to the working plane (G17, G18, or G19).

The format of a CNC block for a circular movement is as follows:

- Lathe. G02 (G03) *X*±... *Z*±... *I*±... *K*±... *F*...

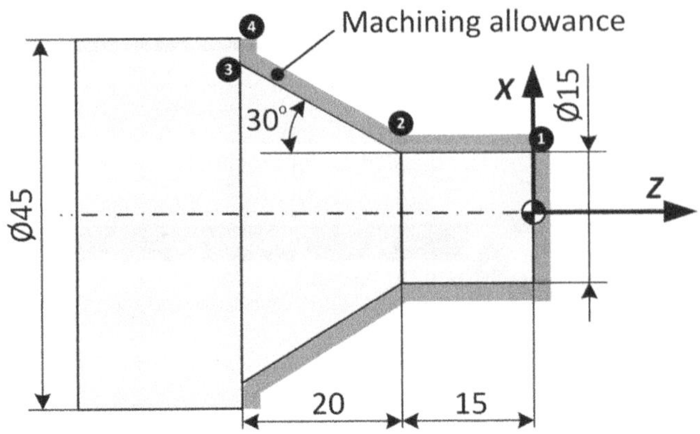

Figure 12.9 Example of linear movements in a lathe. Units in mm.

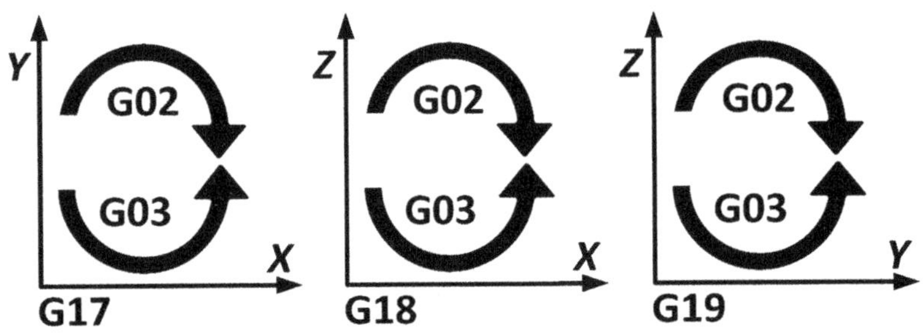

Figure 12.10 Meaning of codes G02 and G03 according to the use of G17, G18, or G19.

In a milling machine, the format depends on the reference plane from which the circular movement is conducted. Thus, the format would be as follows:

- Plane *XY*.G17 G02/G03 *X*±-... *Y*±... *I*±... *J*±... *F*...
- Plane *XZ*.G18 G02/G03 *X*±-... *Z*±... *I*±... *K*±... *F*...
- Plane *YZ*.G19 G02/G03 *Y*±-... *Z*±... *J*±... *K*±... *F*...

If G17-G19 is not specified, the default value is G17 (plane *XY*). The values of *I*, *J*, and *K* define the center of the circular trajectory, where:

- *I*. Distance from the initial point to the center along the *X*-axis.
- *J*. Distance from the initial point to the center along the *Y*-axis.
- *K*. Distance from the initial point to the center along the *Z*-axis.

The values of *I*, *J*, and *K* are programmed with signs, and they should be included even if their value is 0. In a lathe, though the *X*-axis is programmed in diameter, the *I* value is always defined in radius.

Additionally, the circular movement can be programmed using radius instead of the position of the center of the movement. Using this format, the circular movement is defined as:

- Lathe. G02/G03 *X*±... *Z*±... *R*±... *F*...
- Milling machine (XY plane). G02/G03 *X*±... *Y*±... *R*±... *F*...

As can be noted, instead of using *I*, *J*, and *K*, the value of the radius of the circular movement is used. If the circular movement is less than 180°, the radius is programmed with a positive sign; if it is higher than 180°, it is programmed with a negative sign. This format is not allowed to program a complete circle since infinite solutions exist.

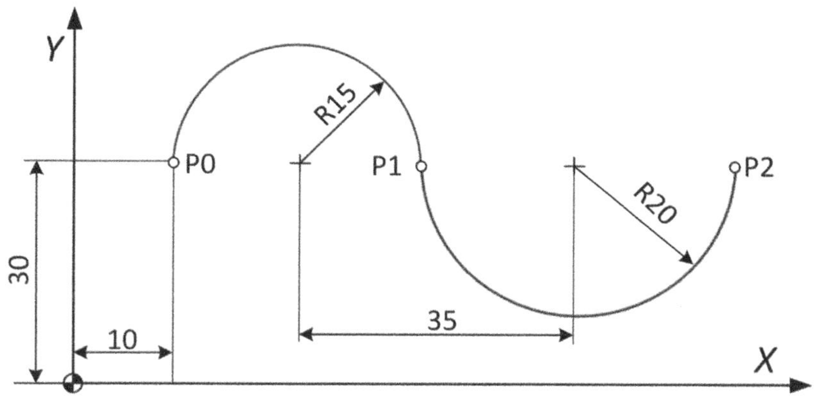

Figure 12.11 Example of circular movements and the use of G02 and G03. Units in mm.

Example. Generate the CNC program in a milling machine to define the trajectory shown in Figure 12.11. The cutting tool is initially at the point X10 Y30.

Solution:

- Absolute coordinates:
  N20 G90 G02 X40 Y30 I15 J0 (arc P0-P1) or N20 G90 G02 X40 Y30 R15
  N25 G03 X80 Y30 I20 J0 (arc P1-P2) or N25 G03 X80 Y30 R20
- Incremental coordinates:
  N30 G91 G02 X30 Y0 I15 J0 (arc P0-P1) or N30 G91 G02 X30 Y0 R15
  N35 G03 X40 Y0 I20 J0 (arc P1-P2) or N35 G03 X40 Y0 R20

12.4.6 Programming of Cutting and Feed Speeds

12.4.6.1 Feed Speed (*F*): G94/G95

The feed speed defined by the letter *F* can be programmed in mm/rev (G95) or in mm/min (G94). Usually, the feed speed is programmed in mm/rev in a lathe and in mm/min in a machining center.

The feed speed is used for linear and circular interpolations (G01, G02, and G03). In rapid movements (G00), the feed speed is the one defined in the machine tool for rapid movements.

12.4.6.2 Spindle Speed (*S*). G96/G97

The spindle speed is defined in rev/min (rpm), and it is denoted by the letter *S*. Unlike milling operations where the spindle speed is programmed in rev/min, turning operations can be programmed in m/min (G96) or in rev/min (G97). With the exception of threading or drilling, the constant cutting speed is recommended, so the G96 is used.

Example. N30 G96 S150 M03 (constant cutting speed value of 150 m/min, spindle rotation clockwise).

12.4.6.3 Formulae

The common procedure in programming cutting conditions is to define the cutting speed and the feed per tooth of the cutting tool, given the cutting tool material and the workpiece material. Knowing the cutting speed, the spindle speed can be obtained as follows:

$$N(\text{rpm}) = \frac{V_c \times 1{,}000}{\times D}$$

where D is the cutting tool diameter or the workpiece diameter if it is a turning operation.

Example. Calculate the spindle speed of a drilling of 15-mm diameter if the cutting tool vendor recommends a cutting speed of 80 m/min.

$$N = \frac{V_c \times 1{,}000}{\times D} = \frac{80 \times 1{,}000}{\times 15} = 1{,}697\,\text{rpm}$$

From cutting data handbooks, the feed rate in mm/rev (f_n) or in mm/tooth (f_z) is given. However, in the CNC program, the feed rate is usually given in mm/min, so the following expression is required.

$$V_f\left(\text{mm / min}\right) = f_z \times z \times N = f_n \times N$$

Example. Calculate the feed rate of an end mill with four teeth, knowing that the recommended feed rate in mm/tooth is 0.023 and the spindle speed is 500 rpm.

$$V_f = f_z \times z \times N = 0.023 \times 4 \times 500 = 46\,\text{mm / min}$$

12.4.7 Cutting Tool Change

In machine tools with an automatic tool changer, a CNC command can change the cutting tool automatically. Although it may depend on the type of controller, the tool change is usually done by applying two codes: the number of the tool to be prepared, the T code, and the execution of the tool change, the M06 code.

The number of the tool is indicated by the code T, and the numbers that follow the code refer to the position of the tool in the physical turret or carousel tool changer. An example of a machining center with an automatic tool changer is as follows:

N10 T01 (the tool changer moves to place the tool 1 ready for the change)
N20 M06 (execution of tool change)

Depending on the CNC controller, the T code may indicate the tool number (position on the tool changer) and the tool offset where the dimensions of the tool after tool calibration are stored. For instance, in FANUC series 0-TC controllers for CNC lathes, the T code is represented as TXXYY, where XX are the digits for the tool number and YY are the digits for tool offset. However, in FAGOR 8060 controllers for CNC lathes, the tool number is indicated by the code TXX (XX, the position on the tool changer) followed by a code DYY where YY represents the tool offset. In both cases, when applied to CNC lathes, the M06 is not needed, and the T code directly executes the tool change. Following is an example from both controllers:

FANUC Series 0-TC. N10 T0101 (tool 01 with offset in register 01 is loaded)
FAGOR 8060. N10 T01 D01 (tool 01 with offset in register 01 is loaded)

12.4.8 Tool Offsets

As shown previously, to identify a tool, the position of the tool in the turret/carousel tool changer (Figure 12.12a), and the register number of the tool offset is needed. The operator may manually add the information presented in this register, or it may be automatically filled by the CNC after executing a tool calibration program where the tool's dimensions are measured and stored in the

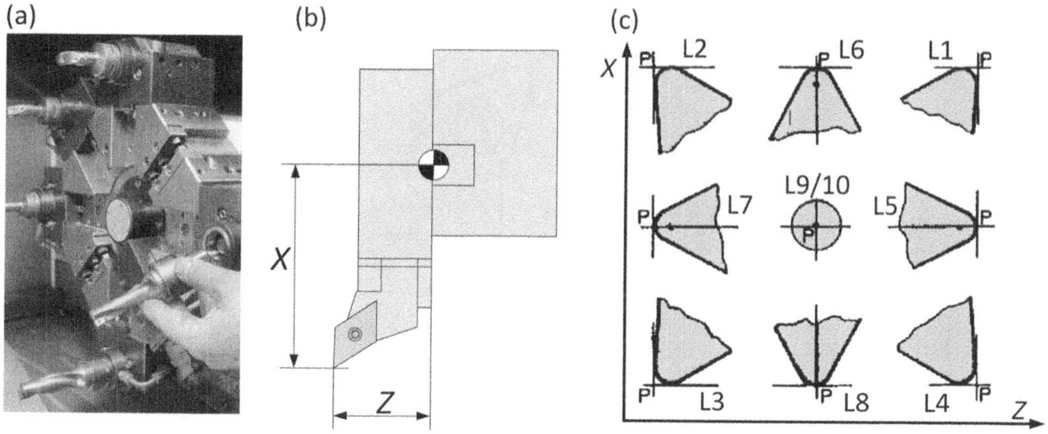

Figure 12.12 (a) Positions of the cutting tools in a lathe turret; (b) X, Z values used to compensate the position of the cutting tool; and (c) Shape factor of cutting tools in a lathe.

corresponding register. In a lathe, the register for the tool offsets (correctors) typically has the following information:

- *X*. Length of the cutting tool along *X*-axis (Figure 12.12b).
- *Z*. Length of the cutting tool along *Z*-axis (Figure 12.12b).
- *R*. Tool nose radius.
- *L*. Code related to the tool tip orientation (Figure 12.12c).
- Other values related to compensate the tool wear of the cutting tool.

In a milling machine, the register has the following information:

- *R*: Cutting tool radius.
- *L*: Length of the cutting tool.
- Other values related to compensate the tool wear of the cutting tool.

12.4.9 Radius Compensations (G40, G41, and G42)

12.4.9.1 Compensation in Turning

When programming nonparallel lines to the lathe axis, the radius of the tool nose prevents the machined part from having the same profile as the design part (Figure 12.13). To overcome this problem, the programmer can use the functions G41/G42 that make the controller compensate for the radius of the tool nose and, therefore, keep the machined profile as the one programmed.

To use the *G*-codes G41 and G42, it is mandatory to have previously defined the cutting tool geometry, such as the shape of the cutting tool and the value of the tool nose radius, in the corresponding register of tool offsets. G41 corresponds to a left compensation, whereas G42 refers to a right compensation, as shown in Figure 12.14. To deactivate the compensation, the *G*-code G40 should be used.

12.4.9.2 Compensation in Milling

When programming cutting tool movements in milling, the coordinates programmed refer to the center of the cutting tool. When milling the contour of a part, the CNC coordinates of the center of the cutting tool do not correspond to the part coordinates since there is an offset with the value of the cutting tool radius. In milling, the cutting tool radius compensation is conducted using the codes G41 or G42 according to the position of the cutter with respect to the direction of the machining operation.

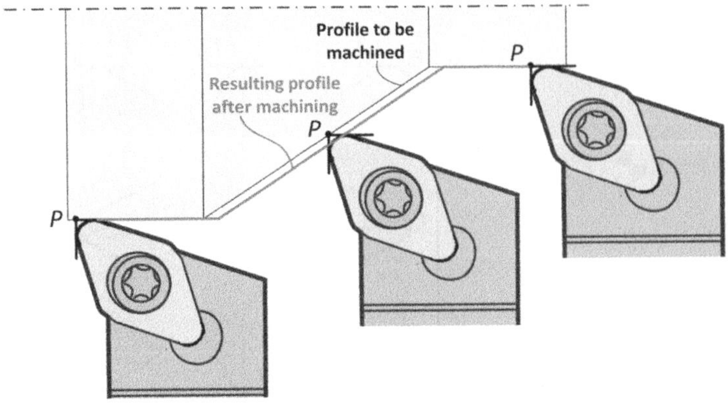

Figure 12.13 Machining error due to tool nose radius in contour turning operations.

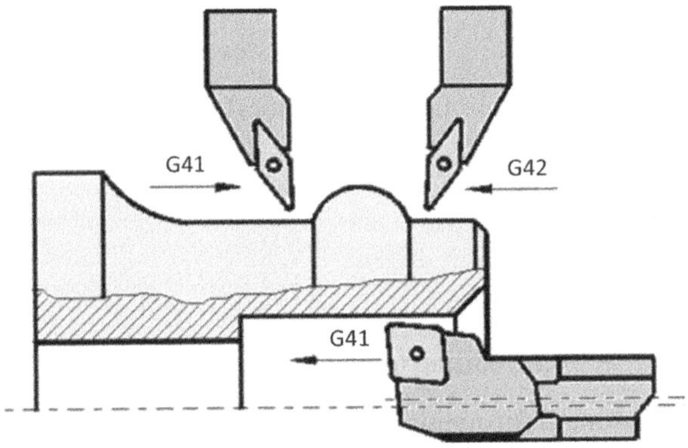

Figure 12.14 Compensation functions in a lathe. G41 and G42.

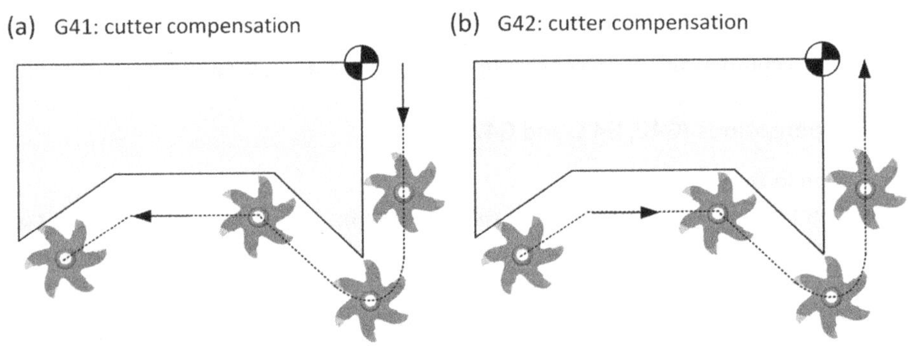

Figure 12.15 Compensation of cutting tool radius on (a) G41 or (b) G42.

Figure 12.15 shows an example of left and right radius compensation in milling. The G40 code deactivates the compensation. Similar to lathe operations, to compensate for the radius, the cutting tool geometry should be defined in the corresponding register of the cutting tool table of the CNC.

Figure 12.16 shows an end milling operation with radius compensation. At the last movement, before deactivating the compensation, the cutting tool is perpendicular to the last point defined in the trajectory. When deactivating the radius compensation (G40), the cutting tool moves horizontally, and part of the machined profile may be spoiled. To avoid this kind of problem, the activation and deactivation of the compensations should be done at a safe distance from the final geometry of the part. An example of a correct deactivation of radius compensation is shown in Figure 12.17.

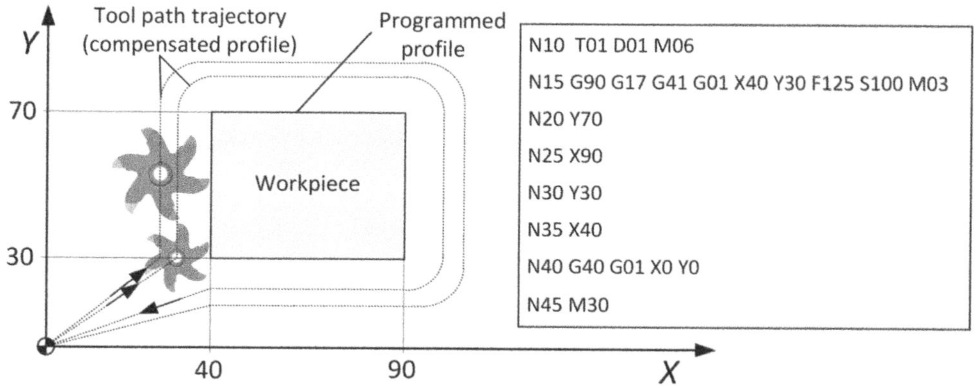

Figure 12.16 Special caution should be paid when the radius compensation is removed.

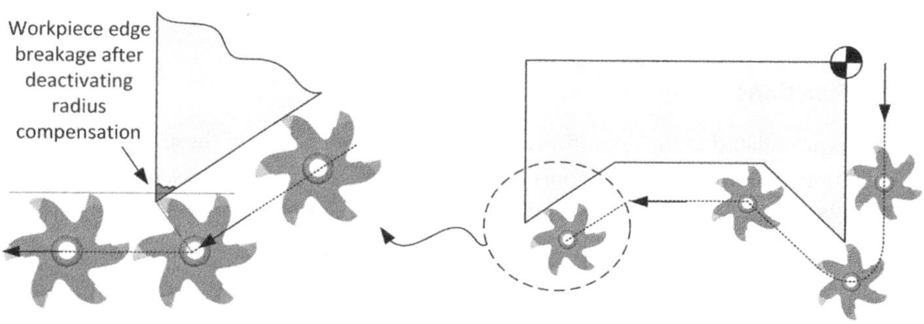

Figure 12.17 Example of cutting tool radius compensation in milling. Note that the same program can be used for different tool diameters thanks to radius compensation G-codes.

12.4.10 Length Compensations (G43, G44, and G49)

In some controllers (e.g., FAGOR 8060), tool length compensation is activated when selecting a tool offset DXX, where XX is the tool offset number, and there is no specific command for height compensation. To cancel this compensation, a D0 value is added to the program. An example of length compensation is given in the following:

FAGOR 8060. N10 T05 D04 M06 (tool 5 is loaded with the height compensation indicated in register 4)

...

N50 T04 D0 M06 (the height compensation from register 4 is canceled. The CNC assumes D1 as height compensation.)

In other controllers (e.g., FANUC 0-TC), the codes G43 and G44, followed by a Z movement and an HXX code, which indicates the tool height offset to be loaded from the register, activate the tool length compensation. In these cases, G43 is a positive compensation (the offset is added), and G44 is a negative compensation (the offset is subtracted). The code G49 cancels the tool length compensation.

FANUC controller. N80 G43 Z1 H1 (tool length offset number 1 is added, and the tool moves to $Z+1$)

....

N200 G49 Z100 (height compensation is canceled)

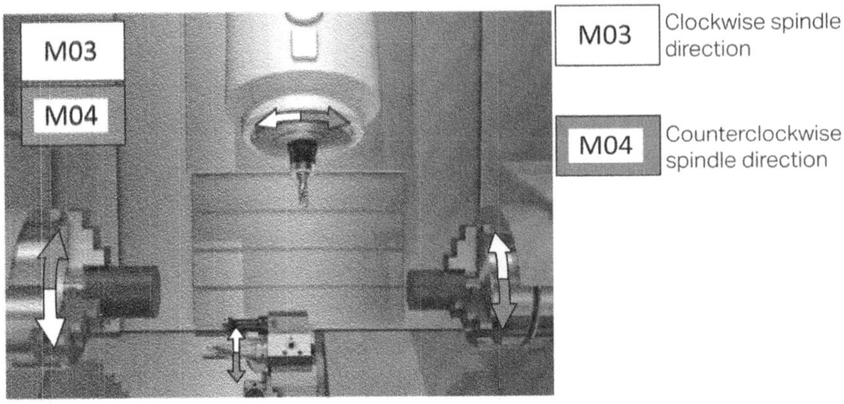

Figure 12.18 Spindle rotation and commands M03 and M04 for different spindle orientations. *Source:* Sandvik Coromant (2017)/Sandvik Coromant.

12.4.11 Auxiliary Functions

The auxiliary functions are related to the operation of the machine tool. Some of these operations are spindle rotation activation, coolant activation, tool change, etc. Similar to *G* functions, more than one auxiliary function can be called in the same program block. Some of the auxiliary functions are:

- M00. Program stop. When the CNC reads block M00, the program code is paused. It also turns off the coolant. If the run button is pressed, the program continues to the next block of the program.
- M02. End of program. This code indicates the end of the program and performs a function of "general reset" of the CNC (initial conditions).
- M03. Start the spindle clockwise (clockwise, Figure 12.18).
- M04. Start the spindle counterclockwise (counterclockwise, Figure 12.18).
- M05. Spindle stop.
- M06. Tool change.
- M08. Coolant on.
- M09. Coolant off.
- M30. End of the program and back to the beginning of the program. Identical to M02 except that the CNC returns to the starting block program.

12.5 CNC Canned Cycles

CNC programs can be created by using the ISO codes shown earlier. However, even a simple part may require a large number of CNC blocks when using simple *G*-codes such as G00, G01, G02, and G03. For instance, a part with a rectangular pocket may require several tens of lines of code. To simplify this tedious job, all CNC controllers have predefined CNC programs that can be easily configured to create common machining operations such as rough straight turning, pocket milling, drilling, etc. These predefined programs are named CNC canned cycles and are usually defined in the *G*-codes G7X-G8X or higher, and their definition depends on each CNC controller.

To illustrate the use of CNC canned cycles, following are two cycles from the controller CNC FAGOR 8065 T/M for turning and milling extracted from FAGOR programming manuals (Fagor Automation 2024b; Fagor Automation 2024c).

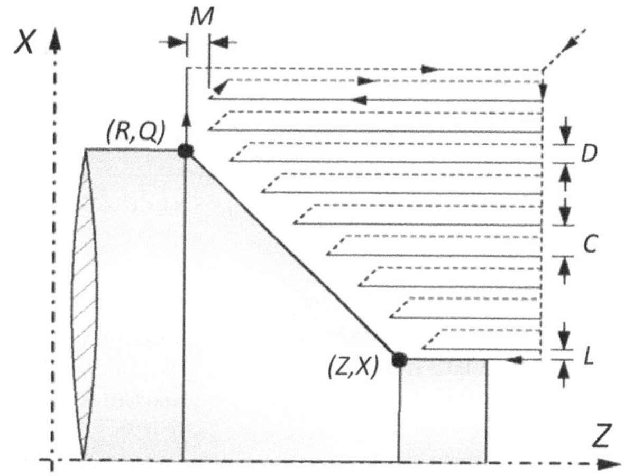

Figure 12.19 Turning canned cycle with straight sections. *Source:* Fagor Automation (2024b)/Fagor Automation.

12.5.1 G81 Turning Canned Cycle with Straight Sections

This canned cycle turns the programmed section, maintaining the indicated pass between the successive turning passes. The canned cycle is called in the CNC program in Cartesian coordinates as follows:

G81*X*...*Z*...*Q*...*R*...*C*...*D*...*L*...*M*...*F*...*H*...

where the parameters required are (Figure 12.19) listed as follows:

X. This defines the *X* coordinate of the initial point of the profile. It must be programmed in absolute values and in the active units (radius or diameter).
Z. This defines the *Z* coordinate of the initial point of the profile. It must be programmed in absolute values.
Q. This defines the *X* coordinate of the final point of the profile. It must be programmed in absolute values and the active units (radius or diameter).
R. This defines the *Z* coordinate of the final point of the profile.
C. This defines the turning pass, which is given by a positive radius value. The whole turning operation is carried out with the same pass, which is the same or smaller than the programmed one "*C*."
D. This defines the safety distance to withdraw the tool after each pass.
L. This defines the finishing stock along *X*, and it is given in radius. If not programmed, a value of 0 is assumed.
M. This defines the finishing stock along *Z*. If not programmed, a value of 0 is assumed.
F. This defines the feed rate for the final roughing pass. If it is not programmed or programmed with a value of 0, there is no final roughing pass.
H. This defines the feed rate for the finishing pass. If not programmed or programmed with a value of 0, it is considered that no finishing pass is desired.

Table 12.2 shows an example of this canned cycle, with the final part and the corresponding CNC program.

12.5.2 G87. Rectangular Pocket Canned Cycle

This canned cycle mills the programmed section, maintaining the indicated pass between the successive milling passes. The canned cycle is called in the CNC program in Cartesian coordinates as follows:

G87*Z*...*I*...*D*...*A*...*J*...*K*...*M*...*Q*...*B*...*C*...*L*...*H*...*V*...

Table 12.2 Example of G81 turning canned cycle and resulting trajectory from FAGOR CNC simulator.

G81 canned cycle with straight sections. Exterior profile	**G81 canned cycle with straight sections. Internal profile**
N10 T2 D01 M06 N20 G95 G96 F0.2 S180 M04 N30 G00 X90 Z20 N40 G01 G42 X85 Z5 N50 G81 X18 Z0 Q78 R-75 C2 D2 L0.3 M0.1 F0.2 H0.3 N60 G00 G40 X20 Z0 N70 G01 X-0.8 N80 Z5 N90 G00 Z10 N100 M30	N10 T8 D1 M06 N20 G95 G96 F0.2 S180 M04 N30 G00 X90 Z5 N40 G01 G41 X18 Z5 N50 G81 X78 Z0 Q18 R-75 C2 D2 L0.3 M0.1 F0.2 H0.3 N60 G01 X85 Z5 N70 Z0 N80 X76 N90 G00 G40 Z5 N100 Z10 N110 M30
Tool path trajectories 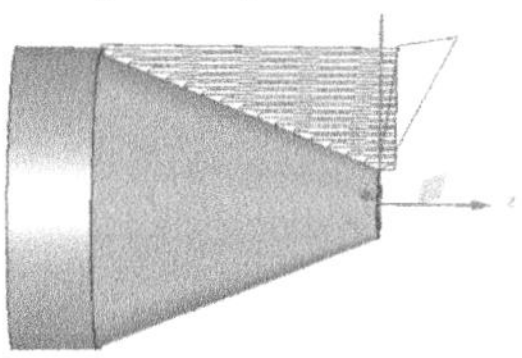	Tool path trajectories 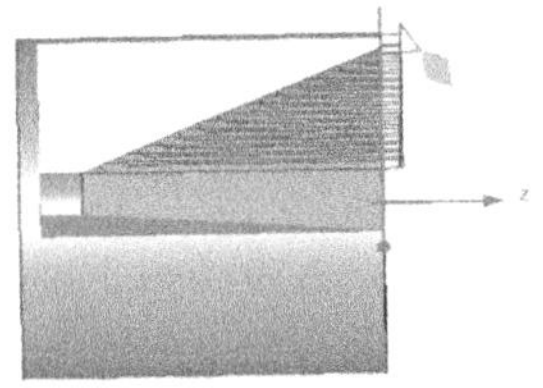

Source: Fagor Automation (2020a).

where the parameters required are (Figure 12.20) listed as follows:

Z. Reference plane. In G90, the coordinate is referred to as part zero. In G91, coordinate referred to the starting plane (*Zi*).
I. Pocket depth. In G90, the coordinate is referred to as part zero. In G91, the coordinate referred to the reference plane (*Z*).
D. Distance between the reference plane and the part surface. If not programmed, it assumes a value of 1 mm.
A. Angle, in degrees, between the pocket and the abscissa axis. If not programmed, it assumes a value of 0.
J. Half the length of the pocket. The sign indicates the pocket machining direction: (*J*+) clockwise, (*J*–) counterclockwise.
K. Half the width of the pocket.
M. Type of corner. (0) square, (1) rounded, (2) chamfered. If not programmed, it assumes a value of 0.
Q. Rounding radius or chamfer size.
B. Depth of pass.
C. Milling pass or width.
L. Finishing pass. If not programmed or programmed with a 0 value, it does not run the finishing pass.
H. Feed rate for the finishing pass. If not programmed or programmed with a 0 value, it is carried out at the roughing feed rate.
V. Tool penetrating feed rate. If not programmed or programmed with a 0 value, it is carried out at 50% of the feed rate in the plane.

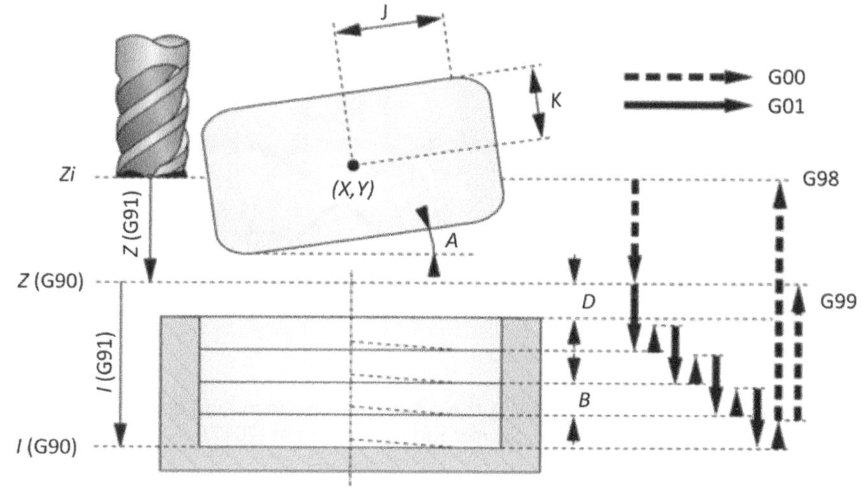

Figure 12.20 Rectangular pocket canned cycle. *Source:* Fagor Automation (2024c)/Fagor Automation.

Table 12.3 Example of G87 pocket canned cycle and resulting trajectory from FAGOR CNC simulator.

G87. Rectangular pocket canned cycle	
CNC program	**Tool path trajectories**
N10 T7 D1 M06 N20 G00 G90 X0 Y0 Z25 F350 S1200 M03 M08 M41 N30 G99 X60 Y35 N40 G87 Z2 I-20 D2 A15 J-40 K20 M1 Q10 B5 C5 L1 H280 V100 N50 X200 Y135 N60 G98 X350 Y235 N70 M30 *Three pockets are machined with center points X60 Y35, X200 Y135 and X350 Y235	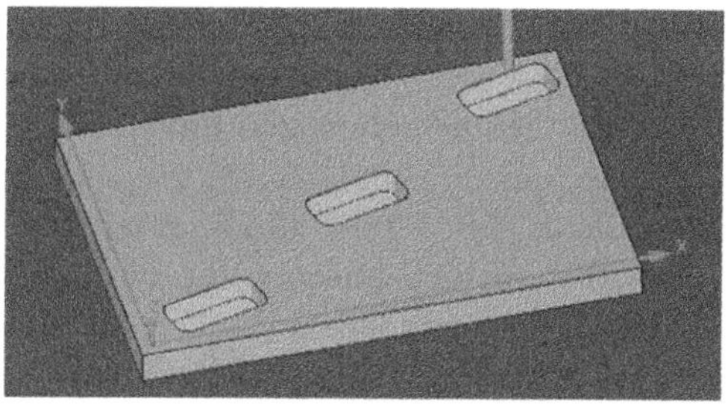

Source: Adapted from Fagor Automation (2020b).

Table 12.3 shows an example of this canned cycle, with the final part and the corresponding CNC program.

12.6 Machining Methods/Cutting Strategies

The following considerations should be taken into account when defining cutting tool trajectories in general machining operations.

12.6.1 Milling

a) Cutter engagement. If multiple passes are necessary, the radial depth of cut should be approximately 3/4 the cutter diameter rather than the full diameter at each pass, as this helps to ensure good chip formation and an appropriate cutting-edge load (Figure 12.21a). Additionally, steer clear of centering cutter placement, which could cause vibrations (Figure 12.21b).

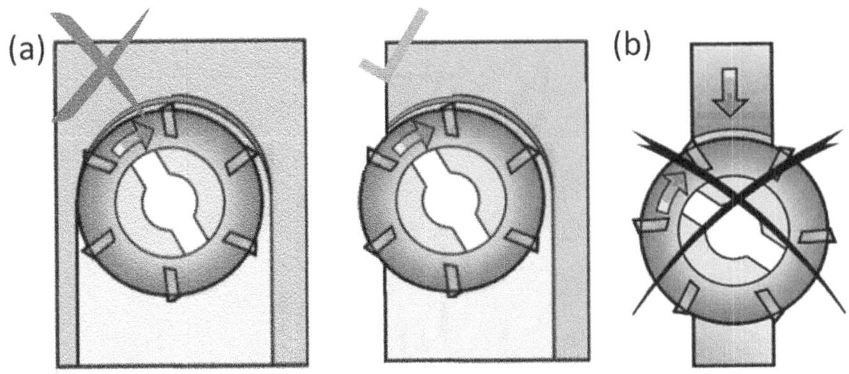

Figure 12.21 (a) Radial passes should be approximately ¾ the cutter diameter in milling operations. (b) Avoid positioning cutters centered on the workpiece. *Source:* Sandvik Coromant (2017).

b) Tool path entries/exits. Avoid entries and exits through tool path programming (Figure 12.22). Keeping the cutter constantly engaged is a better strategy to avoid frequently entering and leaving the workpiece, which might result in chatter tendencies or adverse pressures on the cutting edge.

c) Climb versus conventional milling. In climb milling (Figure 12.23a), also known as down milling, the workpiece is fed in the same direction as the cutter's rotation within the cutting area. This causes the chip thickness to gradually decrease from the beginning of the cut until it reaches zero at the end of the cut. Conversely, in conventional milling (Figure 12.23b), also referred to as up milling, the workpiece is fed in the opposite direction to the cutter's rotation within the cutting area. In this case, the chip thickness begins at zero and increases as the cut progresses. This necessitates the insertion of the cutter into the material, leading to friction, elevated temperatures, and often contact with a work-hardened surface left behind by the previous insert. Additionally, these forces can cause the workpiece to lift from the machine tool table. In contrast, in climb milling, the insert initiates the cut with a substantial chip thickness, which prevents the burnishing effect, reduces heat generation, and minimizes work-hardening tendencies. The significant chip thickness is advantageous, as it causes the cutting forces to pull the workpiece into the cutter, maintaining the insert's position within the cut. Therefore, climb milling is the preferred method whenever the machine tool, fixturing, and workpiece permit it.

d) Cutting entry from outside. Whenever possible, the cutting operation should enter the workpiece from one side and not with a vertical entry. In cases where this is not possible, such as pocketing operations, the cutter should enter the workpiece following a ramp or a helicoidal trajectory, and the feed rate should be reduced to 1/3 of the normal feed rate. If the cutter presents a central edge, the entry can be conducted vertically but applying the same feed rate reduction.

e) Minimum one-tooth engagement. The cutter must have sufficient teeth density to keep at least one cutting tooth engaged during machining. Failure to do this could cause severe pounding, which can lead to chipped cutting edges, a damaged cutter, and excess wear on the machine. Thus, the radial depth of cut and cutter pitch should be chosen always to maintain more than one insert in the cut.

f) Roughing/semi-finishing/finishing. Large cavity surfaces (e.g., mold cavities) are sequentially machined through three operation types: roughing, semi-roughing/semi-finishing, and finishing (Figure 12.24). Roughing operations seek to maximize material removal rate and reduce machining time. Semi-finishing operations seek to keep a constant machining stock allowance

Figure 12.22 Keep cutter constantly engaged. *Source:* Sandvik Coromant (2023)/Sandvik Coromant.

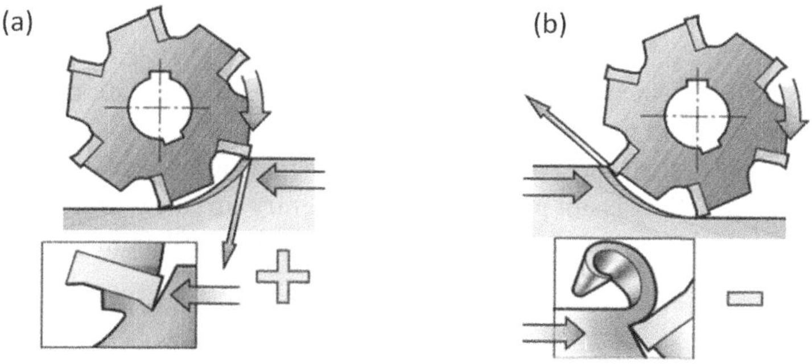

Figure 12.23 (a) Climb milling (down milling); (b) conventional milling (up milling). *Source:* Sandvik Coromant (2017).

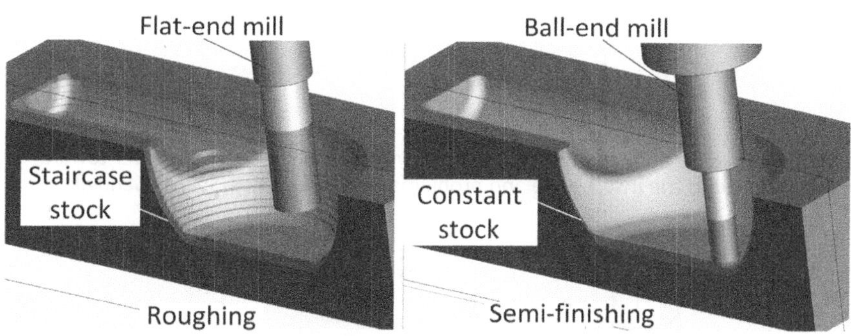

Figure 12.24 Profile milling for mold cavities. Example of roughing and semi-finishing operations. Using flat-end mills generates more pronounced staircase stocks.

for finishing operations to ensure that finishing operations can be conducted without important depth of cut changes increasing geometrical part quality.

g) Round insert cutters. In roughing and semi-finishing operations for profile milling, long edge cutters or shoulder face/end mills will leave a staircase stock that needs to be removed. As a result, cutting forces and tool deflections vary. The result is an uneven stock for finishing, which will affect the final shape's geometric precision. Utilizing round insert cutters or ball-end mills is a preferable strategy because they produce smooth transitions between passes, leave less stock for the profiling operation, and produce components of higher quality.

12.6.2 Turning

When vertical cuts are required, the cut should be from outer diameter toward inner diameter to avoid damage of the insert edge and problems with wrap chips. If face turning is required in a workpiece, this is conducted at the end after the straight turning cuts are done (Figure 12.25).

12.6.3 Drilling

When drilling nonflat surfaces (Figure 12.26), the following rules should be considered:

- Convex surfaces. They are possible to drill if the radius is larger than four times the drill diameter and the hole is perpendicular to the radius. The feed should be reduced to half the normal rate when entering.
- Concave surfaces. They can be drilled if the radius is larger than 15 times the drill diameter. When entering, the feed should be reduced to a third of the normal rate.
- Inclined surface. If inclination is within the range 5–10°, start by performing a centering operation. If inclination is larger than 10°, drilling is not possible unless the entry surface is prepared. In any case, the feed should be reduced to 1/3 of normal feed rate until cutting full diameter.

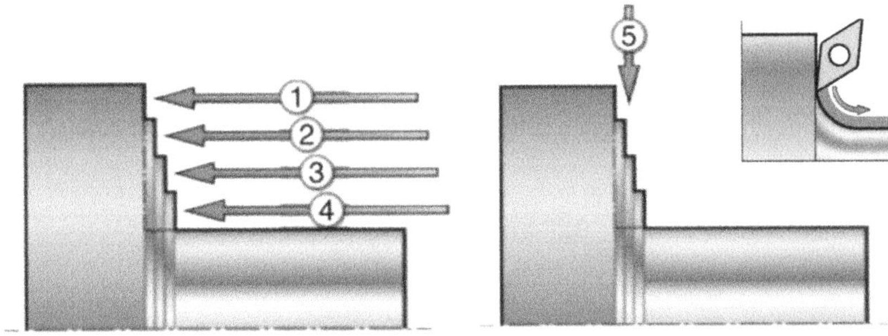

Figure 12.25 The final cut (5) shall be done in one vertical cut starting from the outer diameter toward the inner diameter to avoid damage to the insert edge and problems with wrap chips. *Source:* Sandvik Coromant (2017).

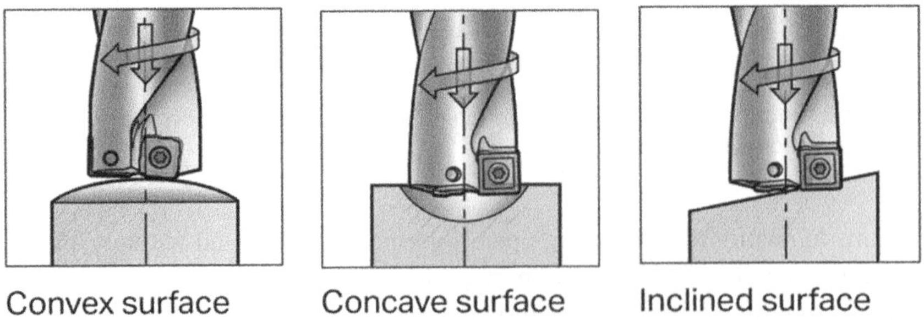

Figure 12.26 Specific strategies should be applied when drilling nonflat surfaces. At least, the feed rate should be reduced to a third of the normal rate. *Source:* Courtesy of Sandvik Coromant (2017).

When drilling holes larger than three times the diameter, a peck drilling strategy is recommended. Peck drilling (i.e., drill a short distance, withdraw the tool, and then drill again) reduces chip packing and prevents drilling problems. The deeper the hole, the more frequently the drill must be retracted to be effective. The depth for each pecking is usually one or two times the drill diameter. Figure 12.27 shows a peck drilling example with four retractions.

For steeped holes that are machined by drilling two holes, the first drilling operation should be the larger diameter and then the smaller one. Under this strategy, the large diameter assures the centering of the drill point and avoids the chipping of cutting edges. The smaller drill should have a drill point angle equal to or smaller than the large drill to ensure that the point hits the workpiece first (Figure 12.28).

In deep-hole applications where coolant cannot be effectively applied, feeds and speeds should be reduced up to 50% to reduce the excessive heat generated. Finally, when drilling through holes, an additional depth is provided. It is recommended that an additional depth of 1.27 mm (0.05 inches) be provided for general through holes.

12.6.4 Threading

In turning threading, the thread is machined after threading the workpiece several times. By dividing the full cutting depth of the thread into smaller cuts, the sensitive nose radius of the insert is not overloaded. The number of passes required depends on the pitch of the thread. Vendors provide

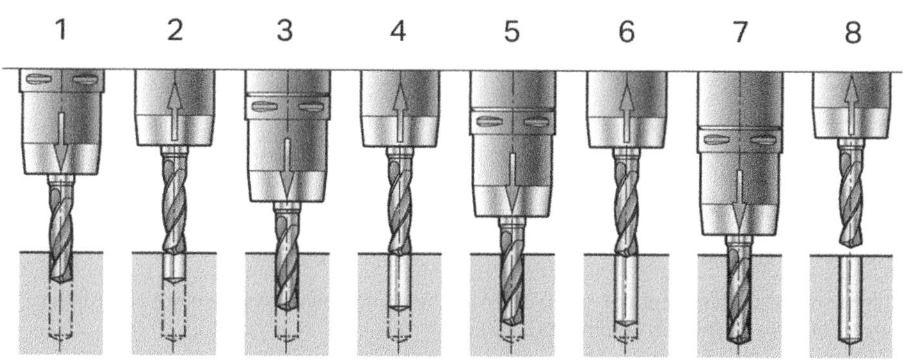

Figure 12.27 After each drilling cycle, retract the drill from the hole to ensure that no chips are stuck onto the drill. *Source:* Courtesy of Sandvik Coromant (2017).

Figure 12.28 Steps for drilling a hole with different diameter sections. *Source:* Courtesy of Sandvik Coromant (2017).

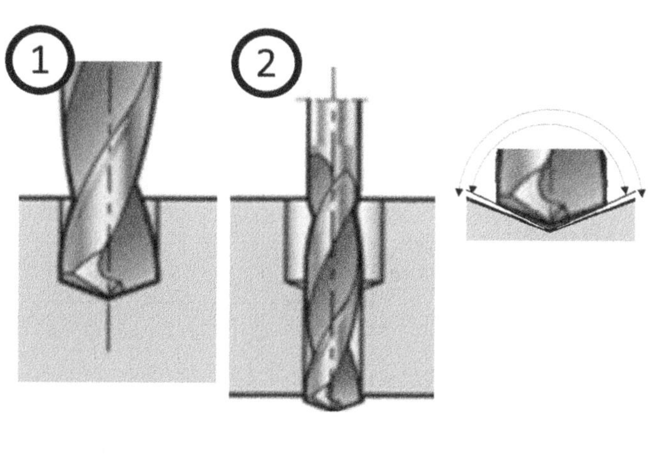

tables to calculate the number of passes and the depth of cut at each pass to set up the threading operation. For illustrative purposes, consider an external thread M60×1.5 to be machined. According to Figure 12.29, the profile depth of the thread for a 1.5-pitch metric thread is 0.94 mm, which is machined by 6 passes. The first cutting passes threads 0.22 mm, the second one 0.21 mm, etc., and the sixth one 0.08 mm to reach the total depth of cut of 0.94 mm.

A general strategy for threading with machine taps is shown in Figure 12.30 and is as follows:

- Enter feed rate according to thread pitch (or 1–2% lower) and set spindle to starting point with 0.08-mm clearance.
- Start spindle forward with right-hand rotation until reaching desired depth.
- Stop feed and rotation and reverse to the starting point.

No. of infeeds (nap)	Pitch, mm ← Reduce cutting speed 0.5	0.75	1.0	1.25	1.5	1.75	2.0	2.5	3.0	3.5	4.0	4.5	5.0	5.5	6.0
	Radial infeed per pass, mm														
1	0.11	0.17	0.19	0.20	0.22	0.22	0.25	0.27	0.28	0.34	0.34	0.37	0.41	0.43	0.46
2	0.09	0.15	0.16	0.17	0.21	0.21	o.24	0.24	0.26	0.31	0.32	0.34	0.39	0.40	0.43
3	0.07	0.11	0.13	0.14	0.17	0.17	0.18	0.20	0.21	0.25	0.25	0.28	0.32	0.32	0.35
4	0.07	0.07	0.11	0.11	0.14	0.14	0.16	0.17	0.18	0.21	0.22	0.24	0.27	0.27	0.30
5	0.34	0.50	0.08	0.10	0.12	0.12	0.14	0.15	0.16	0.18	0.19	0.22	0.24	0.24	0.27
6			0.67	0.08	0.08	0.10	0.12	0.13	0.14	0.17	0.17	0.20	0.22	0.22	0.24
7				0.80	0.94	0.10	0.11	0.12	0.13	0.15	0.16	0.18	0.20	0.20	0.22
8						0.08	0.08	0.11	0.12	0.14	0.15	0.17	0.19	0.19	0.21
9						1.14	1.28	0.11	0.12	0.14	0.14	0.16	0.18	0.18	0.20
10								0.08	0.11	0.12	0.13	0.15	0.17	0.17	0.19
11								1.58	0.10	0.11	0.12	0.14	0.16	0.16	0.18
12									0.08	0.08	0.12	0.13	0.15	0.15	0.16
13									1.89	2.20	0.11	0.12	0.12	0.13	0.15
14											0.08	0.10	0.10	0.13	0.14
14											2.50	2.80	3.12	0.12	0.12
16														0.10	0.10
														3.41	3.72

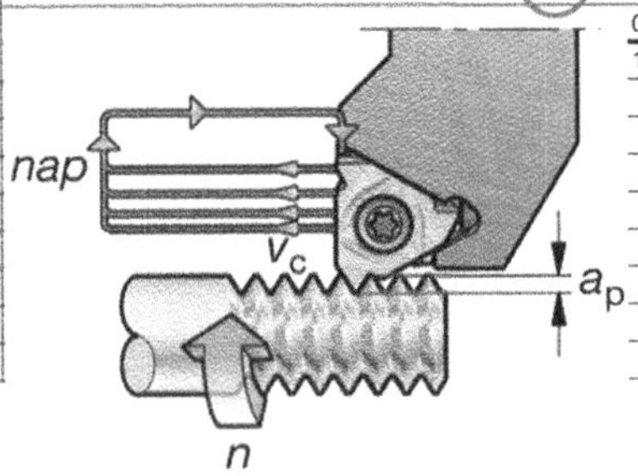

Figure 12.29 Number of passes and depth of cut in external threading operations (turning) according to pitch (in mm). *Source:* Courtesy of Sandvik Coromant (2017).

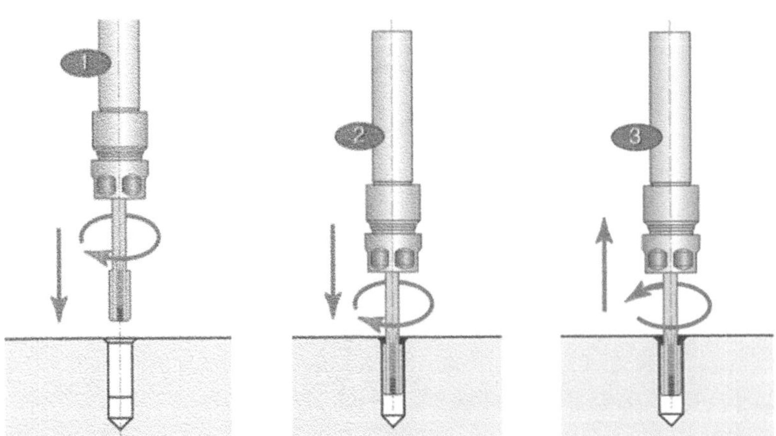

Figure 12.30 Steps for threading with machine taps. *Source:* Courtesy of TaeguTec Ltd.

Questions and Exercises

1 Write the CNC program to mill the workpiece shown in Figure 12.31. Use absolute coordinates and tool radius compensation. Consider that the tool after a tool change is at the position X100 Y100 Z100. Assume: cutting speed 75 m/min, feed rate 0.05 mm/tooth, tool number 1 and tool offset 1 (diameter 15 mm and four teeth) for external profile, and tool number 2 and tool offset 2 (diameter 8 mm and three teeth) for the inner profile. The workpiece presents the same geometry with a 2-mm machining allowance and 20-mm thickness.

2 Write the CNC program to turn the workpiece shown in Figure 12.32. Consider that the initial workpiece is a bar of 40-mm diameter, and a straight turning operation for roughing and a

Figure 12.31 Machined part for Exercise 1. Units in mm.

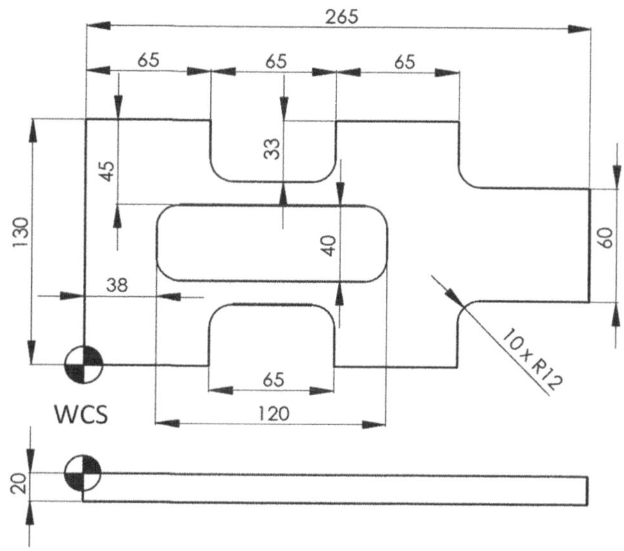

Figure 12.32 Machined part for Exercise 2. Units in mm.

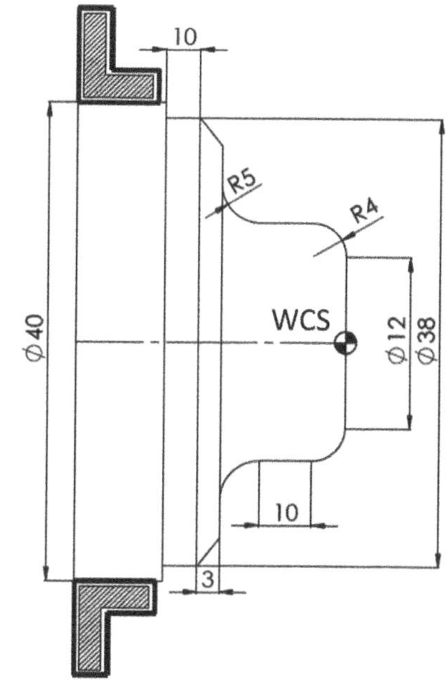

contour turning operation for finishing is required. Consider that the CNC lathe is configured to use *X* coordinates as diameter. Assume the following: cutting speed, 180 m/min; feed rate, 0.15 mm/rev; and maximum depth of cut, 4 mm. The tool for straight turning is tool number 1, tool offset 1, and the tool for contour turning is tool number 2 and tool offset 2.

3 Write the CNC program to turn the workpiece shown in Figure 12.33. Consider that the initial workpiece is a bar of 25-mm diameter and a first face turning operation of 1 mm depth of cut is conducted. Then, a straight turning operation to reduce the diameter up to 24 mm is carried out, and a sequence of straight turning operations for roughing the part up to 10-mm diameter is conducted later. Consider that the CNC lathe is configured to use *X* coordinates as diameter. Assume: cutting speed 180 m/min, feed rate 0.15 mm/rev, maximum depth of cut 3 mm. The tool for both face turning and straight turning operations is tool number 1 and tool offset 2.

4 Write the CNC programs for Exercises 2 and 3, but using CNC cycles.

5 Write the CNC program to machine the workpiece shown in Figure 12.34, which requires a pocketing operation and eight drilling operations. Pocketing is conducted using an end mill

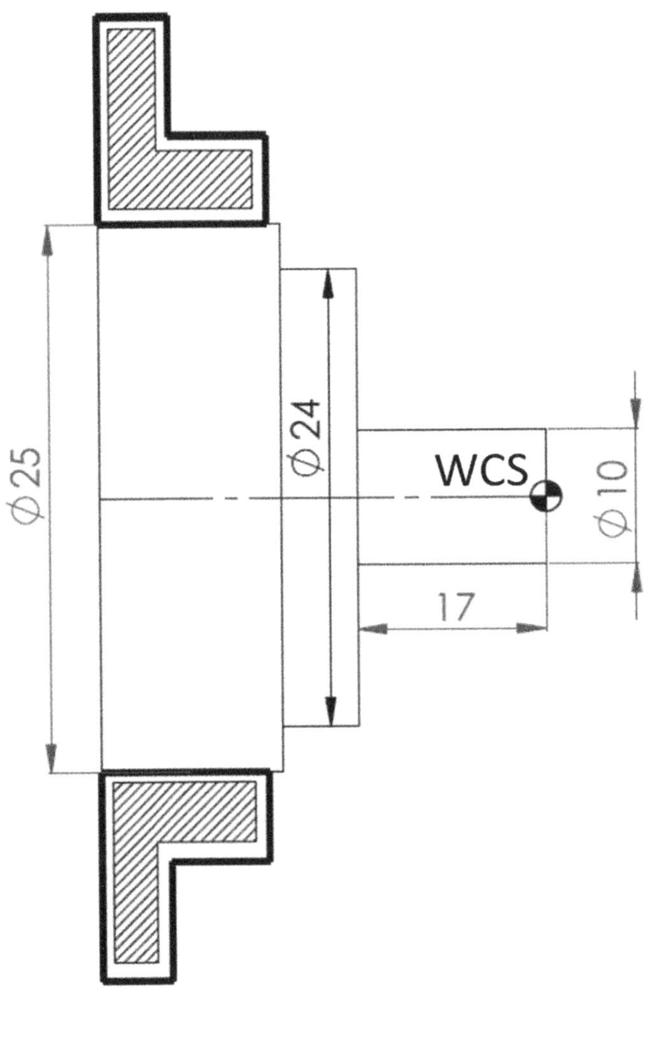

Figure 12.33 Machined part for Exercise 3. Units in mm.

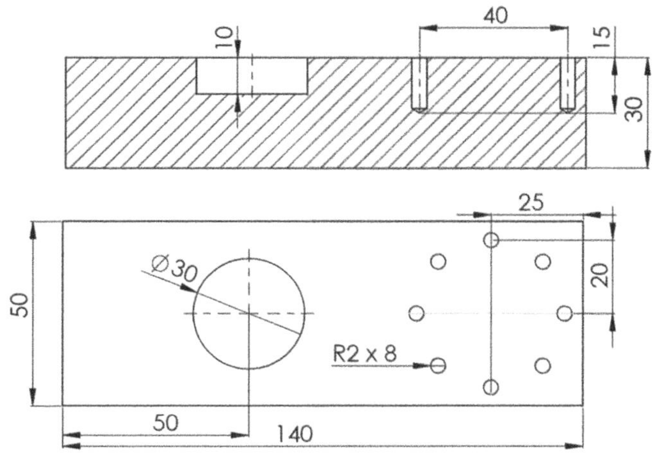

Figure 12.34 Machined part for Exercise 5. Units in mm.

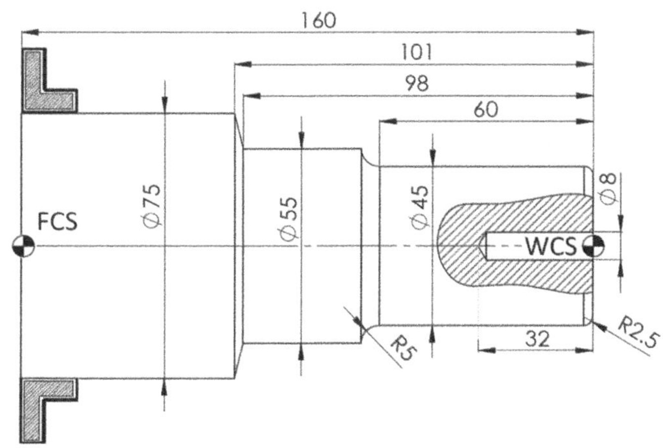

Figure 12.35 Machined part for Exercise 6. Units in mm.

tool of 18-mm diameter (tool number 5 and tool offset 5), and the cutting conditions are 70 m/min and 0.4 mm/rev with a 5-mm maximum axial depth of cut. During the penetration of the pocketing operation, the feed rate is decreased to 1/4. For the drilling operation, a twist drill of 4-mm diameter is used with the cutting conditions of 50 m/min and 0.075 mm/rev. The twist drill is tool number 6 with tool offset 6. Note that there is no WCS defined in the drawing and, first of all, you have to clearly define one.

6 Write the CNC program to turn the workpiece shown in Figure 12.35. Consider that the initial workpiece is a bar of 75-mm diameter and 162-mm length. Assume that the FCS and the WCS are defined as shown in Figure 12.35 (you can choose any of them to write the CNC program). The machining process plan defined by the process planner is as follows:

- Job. CNC lathe
  - Subjob 1. Three-jaw chuck
    - OP 1. Face turning, tool number 3 (tool offset 3), cutting speed 175 m/min, feed rate 0.1 mm/rev, coolant on.
    - OP 2. Straight turning, tool number 4 (tool offset 4), cutting speed 150 m/min, feed rate 0.25 mm/rev, coolant on, maximum depth of cut recommended 2.5 mm.
    - OP 3. Contour turning, tool number 5 (tool offset 5), cutting speed 175 m/min, feed rate 0.1 mm/rev, coolant on.
    - OP 4. Drilling, tool number 8 (tool offset 8), spindle speed 750 rpm, feed rate 30 mm/min, coolant on.

References

Fagor Automation. (2020a). Teoría de programación CNC 8060/65 TORNO (Ref. 2009V2). From Fagor Automation (Spanish).

Fagor Automation. (2020b). Teoría de la programación CNC 8060/65 FRESA (Ref. 2005). From Fagor Automation (Spanish).

Fagor Automation. (2024a). CNC 8060/8065 programming manual (Ref. 1402). https://www.fagorautomation.com (accessed from February 2024).

Fagor Automation. (2024b). 8058/8060/8065 canned cycles (T model). Ref. 2102. https://www.fagorautomation.com (accessed from February 2024).

Fagor Automation. (2024c). 8058/8060/8065 canned cycles (M model). Ref. 2102. https://www.fagorautomation.com (accessed from February 2024).

ISO. (2001). *ISO 841: 2001 Industrial Automation Systems and Integration, Numerical Control of Machines, Coordinate Systems and Motion Nomenclature*. Geneva, Switzerland: International Organization for Standardization.

ISO. (2009). *ISO 6983-1: 2009 Automation Systems and Integration. Numerical Control of Machines. Program Format and Definitions of Address Words*. Geneva, Switzerland: International Organization for Standardization.

Kumar, K., Ranjan, C., and Davim, J.P. (2020). *CNC programming for machining*. Springer.

Sandvik Coromant. (2017). Training handbook. Metal cutting technology. https://www.sandvik.coromant.com (accessed from May 2023).

Sandvik Coromant. (2023). Face milling. Knowledge & services. https://www.sandvik.coromant.com (accessed from May 2023).

Smid, P. (2003). *CNC programming handbook: a comprehensive guide to practical CNC programming*. Industrial Press Inc.

Part IV

Process Planning in Inspection and Costing

13

Inspection Process Planning

13.1 Introduction

An inspection plan is required prior to the production process to study the manufacturing process capability and validate the current manufacturing process plan. In many contracts between customers and manufacturers, the manufacturer is demanded to provide documentation related to a complete characterization of the first produced parts to ensure process capabilities and to deliver the first units to the customer after a complete inspection.

Furthermore, at the production level, statistical process control (SPC) is usually implemented based on inspection plans to find when the process is statistically out of control and rapidly act on the process to avoid nonconformities. Additionally, inspection plans based on acceptable quality level (AQL) sampling are also required to accept or reject incoming raw materials lots or manufactured batches after sampling procedures and according to pre-established customer–manufacturer agreements.

Although quality engineering departments may conduct inspection planning, process planners may have an active role in it. For instance, process planners may aid in defining inspection locations, identifying corrective actions for nonconforming parts, defining process capabilities, and proposing inspection methods.

In this chapter, different inspection plans are presented, distinguishing inspection plans related to the part and process approval stage when new products are being produced, plans for process control, or plans for decisions related to accepting or rejecting a batch of goods. In all cases, the inspection plan requires the proper selection of gauges and inspection instruments, the procedure for each measurement, the acceptable limits for product conformity, etc.

13.2 Inspection Plans for Product and Process Validation

In the world of regulated industries, engineers are required to provide evidence that the manufacturing processes are in control and capable of meeting design requirements. In the automotive, aerospace, or defense industry, a so-called production part approval process (PPAP) is a mandatory process to assess the manufacturability of the supplier to fulfill customer quality requirements. The PPAP is included in the advanced product quality planning (APQP) process initially developed for the automotive industry to introduce a new product in the market or incorporate changes in the

Manufacturing Process Planning: A Practical Approach for Mechanical Engineering, First Edition.
José V. Abellán-Nebot, Carlos Vila Pastor, and Héctor R. Siller.

Companion website: www.wiley.com/go/Manufacturing_Process_Planning_1e

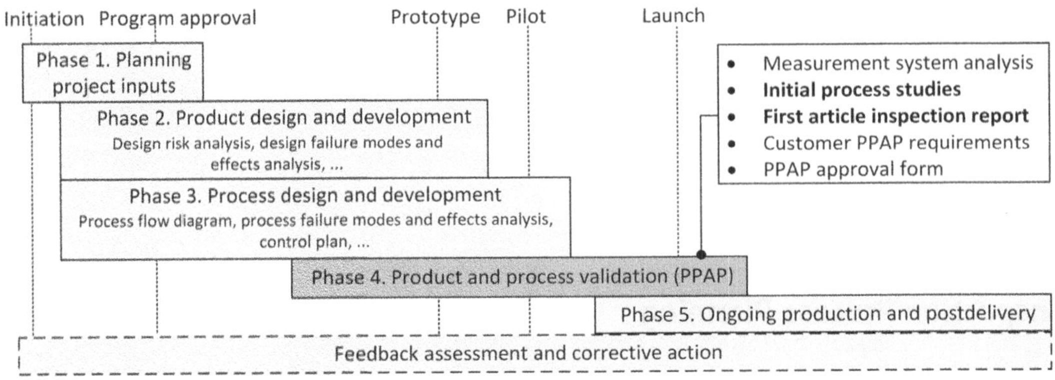

Figure 13.1 Phases of APQP and some activities at the PPAP phase.

product after its release. Other industries also require similar processes to validate their manufacturing procedures prior to production.

Although the details of APQP and PPAP are out of the scope of this chapter, Figure 13.1 shows an overview of the phases in the APQP process and some key activities related to the PPAP phase. Two of these activities are related to initial inspection reports and initial process studies, critical tasks to ensure product and process validation. As can be seen, the main idea of these processes is to prove that engineering designs and specifications are ensured during the manufacturing process prior to production. For this purpose, inspection plans are deployed at the first runs of the manufacturing process prior to production when the product and process validation phase is carried out. Product validation is conducted through a complete inspection during the first produced units, reporting documents such as first article inspection (FAI) reports that should prove product specifications. Process validation is conducted through process capability studies to demonstrate and document that the manufacturing process consistently produces products of the desired quality.

13.2.1 Product Validation. First Article Inspection (FAI)

The FAI is a comprehensive verification process of a product before mass production. By an FAI, manufacturers can assure, prior to volume production, that product specifications are being met, and reengineering will not be necessary at an inconvenient point of time in the future. Therefore, FAI is particularly important for new product developments or changes in the production process, where the parts are being made under this manufacturing process for the first time.

A basic FAI typically includes an annotated drawing that identifies all the part's characteristics, along with a listing of the characteristic requirements and actual results. Therefore, the creation of the FAI involves identifying the characteristics (often called ballooning) and documenting the requirements. Figures 13.2 and 13.3 show an example of an FAI report.

For industries such as aerospace, the AS9102 standard establishes the baseline requirements for performing and documenting the FAI (SAE 2000). This standard is based on the following three distinct forms (Luke 2021):

- AS9102 Form 1. Part Number Accountability: Form 1 specifies the part for the FAI, including any associated subassemblies or detailed parts.

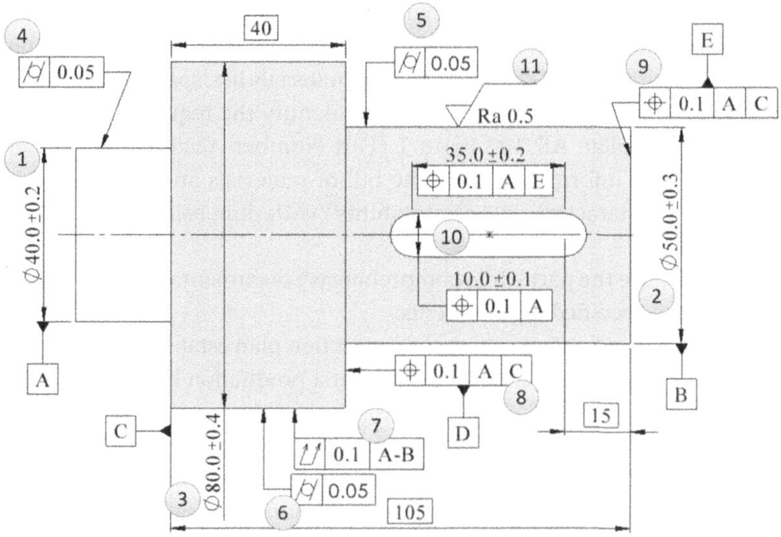

Figure 13.2 First article inspection (FAI). Example. Balloons. Units in mm.

Part number: PRT_321A
Part name: Shaft

QA inspector: J.V. Abellán
QA inspection date: 25/3/2023
Units: mm
Inspection frequency: 3

#	Dimension	Conformity zone	Measurement method	Measurement uncertainty	Part 1	Part 2	Part 3	Decision	Observations
1	Ø40 ± 0.2	[40.18, 39.82]	Caliper	±0.02	40.02	40.05	40.00	Pass	---
2	Ø50 ± 0.3	[50.28, 49.72]	Caliper	±0.02	50.20	50.15	50.22	Pass	---
3	Ø80 ± 0.4	[80.38, 79.62]	Caliper	±0.02	79.88	79.82	79.90	Pass	---
4	⌭ 0.05	[0, 0.04]	Between centers and dial indicator	±0.01	0.02	0.02	0.03	Pass	---
5	⌭ 0.05	[0, 0.04]	Between centers and dial indicator	±0.01	0.01	0.02	0.01	Pass	---
6	⌭ 0.05	[0, 0.04]	Between centers and dial indicator	±0.01	0.03	0.04	0.03	Pass	---
7	⌰ 0.1 A-B	[0, 0.09]	Between centers and dial indicator	±0.01	0.05	0.07	0.04	Pass	---
8	⌖ 0.1 A C	[39.96, 40.04]	Height gauge	±0.01	39.99	40.01	40.01	Pass	---
9	⌖ 0.1 A C	[104.96, 105.04]	Height gauge	±0.01	104.98	105	105.00	Pass	---
10	⌖ 0.1 A E ⌖ 0.1 A	Pass	Inspection gauge A	-	Pass	Pass	Pass	Pass	---
11	√Ra 0.5	Ra < 0.5 μm	Profilometer	-	0.28	0.35	0.25	Pass	---

Figure 13.3 Example of first article inspection (FAI) report. Units in mm.

- AS9102 Form 2. Product Accountability: Form 2 is dedicated to documenting all material and process specifications, as well as any special processes and functional testing defined as design requirements.
- AS9102 Form 3. Characteristic Accountability: Form 3 focuses on capturing all product characteristics, such as dimensions, tolerances, and notes. An inspection drawing or model is required, with all inspection characteristics (defined using CAD annotations) clearly marked with unique inspection balloon numbers.

To perform an FAI in accordance with AS9102, the following steps are followed (Luke 2021):

Step 1. Preplan the FAI. Gather the necessary documentation for the FAI, including engineering drawings, digital product definition datasets, bill of materials/raw materials list, specifications, etc.

Step 2. Create an inspection plan. Balloon the part drawing to identify the requirements to be inspected with balloon numbers. Complete AS9102 Form 1 (Part Number Accountability) and Form 2 (Product Accountability) using information from the bill of materials and the specifications list. Complete AS9102 Form 3 (Characteristic Accountability) with dimensions, tolerances, notes, and inspection tools.

Step 3. Manufacture the part. Fabricate the part while comprehensive documentation of processing and inspection results to meet the specified requirements.

Step 4. Collect data. Inspect the part in accordance with the inspection plan established in Step 2. Data are collected from a random selection of products from the first production batch. A sample size of 1–10 is typically applied.

Step 5. Finalize reporting. Record measurement result data in AS9102 Form 3.

It is worth noting that during the FAI, any out-of-specification conditions encountered must be assessed for their significance, followed by implementing corrective actions. Typically, this involves making dimensional changes to the part print or modifying the tooling. Once the FAI is approved by the customer, the process validation phase starts.

An FAI may be required under the following scenarios:

- When the part is manufactured for the first time.
- When a design or manufacturing process change takes place.
- After a customer request.
- When the factory changes the location.
- When restarting a production line that has not been used for a long period of time.

13.2.2 Process Validation. Process Capability Studies

Process validation shows that the process is reliable, capable of producing parts with the desired dimensions, and stable. Process validation is typically carried out by performing three successive runs of the manufacturing process. Throughout each run, samples should be taken at regular intervals, sequentially labeled, and submitted to metrology for measurement and compliance. If all measurements are within the conformity zone, previously defined by the upper specification limit (USL) and the lower specification limit (LSL), the process is said to be capable. Before analyzing process capability, the process should be checked to be stable. A process is stable or in control when the process behaves consistently over time, that is, the process variation is random and is not influenced by special circumstances.

Both process capability and process stability can be demonstrated using statistical tools such as control charts (Khlebnikova 2012). Control charts are used to detect process variation, and they can show if the process is capable and if it is in statistical control (stable). Figure 13.4 shows four different types of processes according to their stability and capability. Each graph represents the normal distribution of a specific dimension and its evolution over time within the manufacturing process.

To check process stability, different samples are inspected. All samples should be within the control limits of a control chart, which means that process variation is random and there is no special cause of variation. To check the stability of the process, all measured data should follow a normal distribution. A graphical way to prove it is to build a control chart with the upper control limit (UCL) and the lower control limit (LCL) to cover 99.73% of the normal distribution (i.e., $\pm 3\sigma$, where σ is the standard deviation of the data). If all the samples are within the control

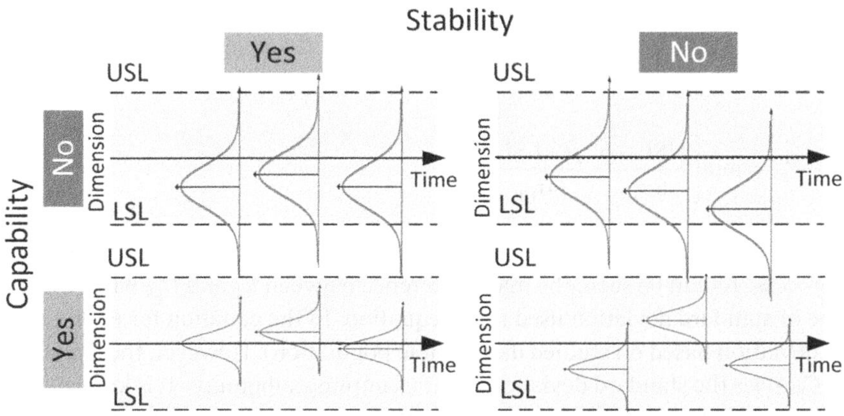

Figure 13.4 Example of four types of processes according to their stability and capability characteristics.

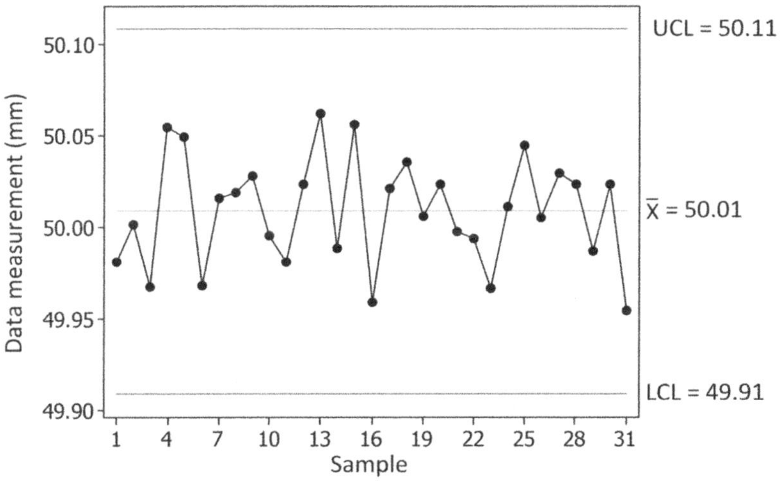

Figure 13.5 Example of process stability analysis under a control chart.

limits, the process can be assumed to be stable, as shown in Figure 13.5. Otherwise, the process is unstable or out of control and a search of the special variation sources should be conducted. It should be noted that for a better analysis of process stability, data trends, and other issues may be checked in the measured data. Specific details about process stability analysis can be found in Oakland and Oakland (2007), Qiu (2013) and Stamatis (2002).

To quantify process capability, several indices are defined. Essentially, these indices measure the degree of variation within the process in comparison to the specified requirements. Typically, short-term indices such as C_p and C_{pk} or long-term indices such as P_p and P_{pk} are employed to assess process performance. C_{pk} aims to answer whether the current production sample complies with the specifications, while P_{pk} addresses the question of whether the process can consistently meet specifications over the long term. Mathematically, these indices are obtained as follows (Arcidiacono and Nuzzi 2017):

$$C_p = \frac{\text{USL} - \text{LSL}}{6\sigma_{\text{ST}}}, C_{pk} = \min\left[\frac{\text{USL} - \mu}{6\sigma_{\text{ST}}}, \frac{\mu - \text{LSL}}{6\sigma_{\text{ST}}}\right]$$

$$P_p = \frac{\text{USL} - \text{LSL}}{6\sigma_{\text{LT}}}, P_{pk} = \min\left[\frac{\text{USL} - \mu}{6\sigma_{\text{LT}}}, \frac{\mu - \text{LSL}}{6\sigma_{\text{LT}}}\right]$$

where μ is the mean value of the samples, and σ_{ST} and σ_{LT} are the short-term and long-term standard deviations of the process. As can be seen, the main difference between C_p and C_{pk} with respect to P_p and P_{pk} is the type of standard deviation used in the equation. In the equation for P_p and P_{pk}, it is used the standard deviation based on studied data (whole population). However, the equation for calculating C_p and C_{pk} uses the standard deviation within sampling subgroups. It is known that when the process has a normal distribution and is under statistical control, P_p is essentially close to C_p and P_{pk} to C_{pk} because the short-term and long-term standard deviations are more or less similar. It is important to emphasize that evaluating process capability can only be carried out once the process is under statistical control.

As per the definition, a C_{pk} value of 1.0 signifies that approximately 99.73% of the produced parts fall within the specified limits. Practically, only about 3 out of every 1,000 parts deviate from the specified requirements. A C_{pk} of 1.33 corresponds to 99.994% of the products meeting the specifications. At the same time, a C_{pk} of 2.0, known as a 6σ process, indicates that 99.99966% of the products conform to the specifications, resulting in just 3.4 defective parts out of 1 million. It should be remarked that many companies nowadays require their suppliers to achieve C_{pk} indices of 1.33 or even 2.0 for production approval, otherwise they may ask for 100% inspection.

13.3 Inspection Plans in Production

Inspection plans are commonly deployed at three stages in manufacturing companies:

- Incoming parts. Raw materials and outsourced components are inspected to confirm that they are conformed to specifications.
- During production. Inspection is conducted while production is underway.
- Preshipment inspection. Manufactured goods may be inspected to detect defects before shipping.

At these stages, different strategies to inspect and control the process can be conducted, such as SPC and acceptance sampling. The SPC can be defined as the application of statistical techniques to monitor and control a manufacturing process. Unlike SPC, the purpose of acceptance sampling is to determine the decision related to a lot of goods in terms of accept, reject, or screen. Therefore, the focus is the decision of accepting or rejecting a manufactured or purchased lot and letting it be shipped or to returning the lot to the supplier.

13.3.1 Statistical Process Control

To monitor and control the manufacturing process, SPC techniques are commonly used. The manufacturing process is monitored by sampling the resulting part and representing the result of the inspection in a chart named the control chart (Figure 13.5). This chart presents UCL and LCL that, if the resulting inspection is outside these limits, means that the process is statistically out of control and a significant cause of variation exists. Under SPC, the inspection plan indicates the characteristics to be inspected, the number of subsamples to average each inspection point in the chart, the frequency of inspection, and the limits and patterns that, if reached, may indicate the existence of significant causes of variations. Depending on the process capability (C_{pk} and P_{pk} values) and how critical are the inspected features, the features to be inspected will present different subsample sizes and inspection frequencies. The details of how to build a control chart are more related to quality engineering, and it is out of the scope of this chapter. Interesting books about SPC and how to define the sampling frequency, number of samples, etc., can be found in Oakland and Oakland (2007), Qiu (2013) and Stamatis (2002).

13.3.2 Inspection Sampling for Reception/Shipment. Acceptance Sampling

Three main inspection strategies can be found to accept or reject a lot of goods: 100% inspection, acceptance sampling, and no inspection. The inspection of 100% of the production that is produced in massive quantities at low cost can be too costly and unrealistic (products such as bolts, nuts, etc.). However, 100% may be the correct strategy for expensive products such as life-supporting products. In general, costs are the major factor to consider while deciding the inspection strategy. A measure used is the break-even point (BEP). The BEP, also known as break-even quality, is the percentage of defects at which the benefits gained from fewer defects being released are on the same level as the costs associated with performing a number of 100% inspections. The break-even point is calculated as (Taylor 1993):

$$\text{BEP}(\%) = 100 \times \frac{k_1}{k_2}$$

The unit cost of executing the 100% inspection is denoted by k_1, and the loss suffered by the customer and others as a result of the release of a single defective item is denoted by k_2. It is better to release lots that have no faults (zero defects), as 100% examining such lots is a waste of time and resources. The expense of a 100% inspection, however, is minimal compared to the effects of the defects it eliminates. Thus, lots with high levels of defects should undergo it.

Example.

A new supplier delivers subassemblies of electronic panels to a car assembly company in quantities of 1,000 pieces. At receiving, testing an electronic panel subassembly costs \$0.10, and the correct assembly operation results in a \$1.5 added value. The cost of stopping the assembly line to replace a damaged electronic panel that has been assembled on a car is \$100. Only 0.05–0.1% of the electronic panel subassemblies are anticipated to be faulty. Determine the process's break-even point.

Solution:

To graphically represent the value if the lot is inspected 100% and the value of the lot if no inspection is conducted, two points per each inspection strategy are calculated considering a different percentage of defects, which is denoted as d.

Value if inspected 100%:

- If $d = 0\%$: value $= 1{,}000 \times 1.5 - 1000 \times 0.1 = \$1{,}400$
- If $d = 2\%$: value $= (1{,}000 \times 0.98) \times 1.5 - 1{,}000 \times 0.1 = \$1{,}370$

Value if no inspection:

- If $d = 0\%$: value $= 1000 \times 1.5 = \$1{,}500$
- If $d = 2\%$: value $= (1000 \times 0.98) \times 1.5 - (1{,}000 \times 0.02) \times 100 = -\$530$

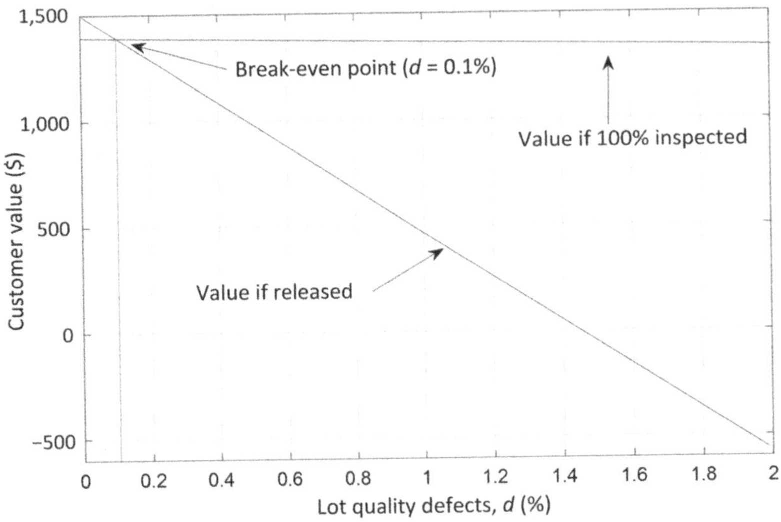

Figure 13.6 Break-even quality of the example.

Figure 13.6 graphically shows the values according to the different percentages of defects in a lot. According to the figure, the BEP is in 0.1% of defects. For a percentage of defects higher than this value, a 100% inspection may provide higher customer value.

Mathematically, the BEP is straightforwardly calculated by the expression shown above, which results in the following:

$$\text{BEP} = 100 \times \frac{k_1}{k_2} = 100 \times \frac{0.1}{100} = 0.1\%$$

Instead of conducting either 100% inspection or no inspection at all, an alternative approach is the utilization of the acceptance sampling strategy, which involves inspecting a specific number of parts and making a decision based on the number of defects detected within the sample. Acceptance sampling is a method of inspection applied to lots/batches of items produced in large quantities, and it can be used for both attribute-based (count-based) and variable-based (measurement-based) inspections. Typically, these lots represent incoming purchased items or finished products awaiting shipment to warehouses or customers. The primary objective of acceptance sampling is to determine whether a lot meets preestablished standards or specifications for critical characteristics of the item. If a lot meets these standards, it is considered passed or accepted; if not, it is rejected. Rejected lots may undergo 100% inspection, or if they were purchased, they may be returned to the supplier for credit or replacement, particularly when destructive testing is involved.

Buyers or consumers may deem lots containing a small percentage of defective items as "acceptable," especially when the cost associated with a few defects is minimal. Typically, this percentage falls within the range of 0.01–2%, which is referred to as the acceptable quality level (AQL). The choice of the AQL should be determined based on the criticality of the characteristic being inspected. For example, critical part specifications that impact product functionality would warrant a lower AQL value compared to minor defects that only affect the appearance of the part.

If AQLs for a product are seen as the point at which the quality is considered acceptable, the formula previously mentioned can be employed to establish AQLs, with the value of k_2 depending on the severity of the defect. As a general rule, being d the estimated proportion of defectives in the lot, the inspection strategy to be conducted is defined as (Krajewski et al. 2010, Supplement G):

- If $d \approx$ BEP, use acceptance sampling.
- If $d >$ BEP, use 100% inspection.
- If $d <$ BEP but d is variable, use acceptance sampling.
- If $d <$ BEP but d is stable, do not inspect.

The sample plan is a crucial component of acceptance sampling. Given the lot size, N, the sampling plan defines the sample size, n, and the acceptance/rejection criteria based on the number of defects in the samples. To understand the procedure for building sampling plans, the following terms should be defined: the operating characteristic (OC) curve, the AQL, the lot tolerance percent defective (LTPD), the consumer risk, and the producer risk.

13.3.2.1 Operating Characteristic Curve

The OC curve, which is depicted in Figure 13.7, displays the probabilities of accepting lots based on their proportion of defectives. In this case, a lot containing 0.5% defectives would have an acceptance chance of 0.86.

Because some low-quality lots will inevitably be accepted and some high-quality lots will inevitably be rejected, a sampling plan cannot completely distinguish between good and bad lots. Only 100% examination of each lot would lead to perfect discrimination, so the lot will only be accepted if the proportion of defects is below the value unambiguously agreed upon by the customer and supplier. The OC curve's shape influences sampling plan's capacity to discriminate; the steeper the curve, the more discriminating the sampling plan is (see Figure 13.8). As a general rule, the curve is steeper, and the discrimination performance is better when larger sample sizes are applied.

13.3.2.2 Acceptable Quality Level

When the cost of flaws is minimal, buyers (or customers) are typically ready to accept lots that contain tiny percentages of defectives as "good." The AQL is a percentage that is typically set between 0.01% and 2%. Based on how crucial the feature under inspection is, the AQL should be set. The features to be inspected are divided into three categories depending on their criticality, that

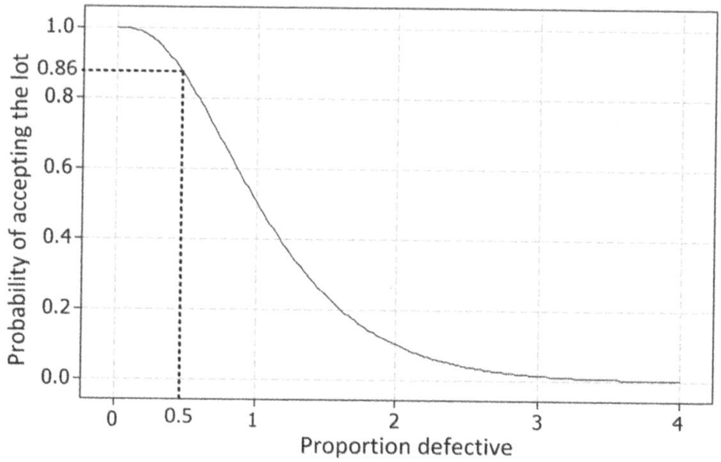

Figure 13.7 A typical OC curve for a specific sampling plan.

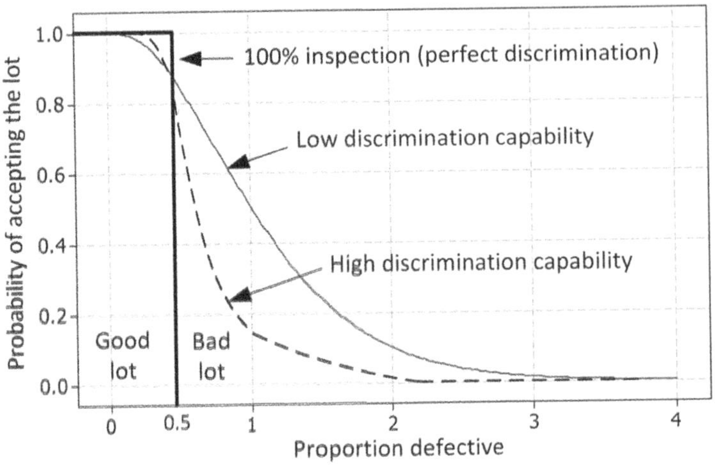

Figure 13.8 The steeper the OC curve, the more discriminating the sampling plan.

is critical faults, major defects, and minor defects. The type of feature determines the AQL values, resulting in higher AQL values when less critical characteristics are inspected. For instance, an airhead with Bluetooth connectivity may have 0.01% critical flaws (e.g., Bluetooth communication problems), 0.05% major defects (e.g., defects on wire welds), and 2% minor defects (e.g., surface scratches).

13.3.2.3 Lot Tolerance Percent Defective

Certain lots that include more defects than the AQL will be accepted due to the inherent uncertainty of sampling. The consumer's tolerance for the proportion of defective items in acceptable lots is named the LTPD. Similar to AQL, LTPD should be determined by the degree of criticality of the characteristic under inspection; the greater the degree of criticality, the lower the LTPD.

13.3.2.4 Consumer's and Producer's Risk

As previously mentioned, sampling plans do not accurately distinguish between excellent and bad lots. The consumer's risk, also known as beta (β) error or Type II error, is the likelihood that a bad lot with flaws equal to the LTPD will be accepted. The producer's risk sometimes referred to as the alpha (α) error or Type I error, is the likelihood that a good lot with defects equivalent to the AQL will be rejected. An example of the OC curve with the AQL, LTPD, producer's risk, and consumer's risk is shown in Figure 13.9. Keep in mind that the likelihood of accepting a lot with AQL quality is $1 - \alpha$, and the likelihood of accepting a lot with LTPD quality is β.

13.3.2.5 Determining the Sampling Plan

Various types of sampling plans can be outlined (Heizer and Render 2005):

- Single sampling. A single sampling plan is defined by two parameters: the number of items to be sampled (n) and a predetermined acceptable threshold for the number of defects (c). If the number of defects in the lot is equal to or lower than the acceptance threshold, the entire batch is accepted. If the number of defects exceeds this threshold, the entire lot is rejected or subjected to 100% inspection.
- Double sampling. In cases where a lot of items may be exceptionally good or bad, it is possible to reach a quality conclusion by taking a smaller sample compared to what would be used in a

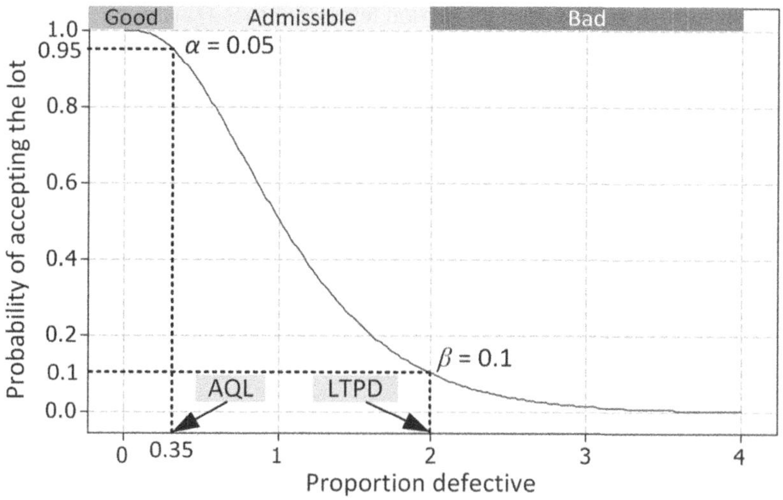

Figure 13.9 An OC curve with the AQL, LTPD, producer's risk α, and consumer's risk β.

single sampling plan. If the number of defects in this smaller sample (with size n_1) is equal to or below a specified lower limit (c_1), the lot can be accepted. Conversely, if the number of defects surpasses an upper limit (c_2), the entire lot is rejected. However, if the number of defects in the n_1 sample falls between c_1 and c_2, a second sample (with size n_2) is drawn. The combined results of both samples determine whether the lot should be accepted or rejected.

- Sequential sampling. Another approach involves inspecting smaller samples sequentially until a definitive decision can be reached. In sequential sampling, units are randomly selected from a lot and tested one by one, with the cumulative count of inspected items and defects recorded over time.

For single sampling plans, determining the sample size n and acceptance number c is equivalent to determining the sampling plan's OC curve. One method to define the single sampling plan is by using the US military standard MIL-STD-105E, currently obsolete and transferred to ANSI/ASQ Z1.4 and ISO 2859 (ANSI 2003, ISO 1999). Although replaced by ASQ Z1.4, many companies still use the notation 105E in their inspection documents. ASQ Z1.4 can be used to inspect AQLs around 0.01% or higher and is not intended to inspect products with quality levels in the parts per million range. ASQ Z1.4 uses the AQL point, the lot size N, the level of inspection desired, and the type of inspection required (which in turn depends on the history of the inspection results) to determine n and c.

There are three primary or general inspection levels (I, II, and III) and four special inspection levels (S-1 to S-4). These general inspection levels are defined as follows (ANSI 2003):

- Level I. Suitable for evaluating features of minor significance. Minor defects are unlikely to render the part unusable for its intended purpose. This level offers lower customer protection but comes with reduced inspection costs.
- Level II. Appropriate for assessing features of major importance. Major defects may lead to part failure when used but do not pose a hazardous condition. Level II involves normal inspection costs and offers standard customer protection. It is the most commonly used default inspection level.
- Level III. Employed when inspecting features that could potentially result in critical defects. Critical defects can lead to hazardous conditions. This inspection level provides the highest level of customer protection but comes with higher inspection costs.

In contrast to the general inspection levels, special inspection levels (S-1 to S-4) are applied when the inspection may damage the item being examined, prove costly, or consume a significant amount of time. The key distinction among the four special inspection levels is the sample size they generate. S-1 yields the smallest sample size, while S-4 yields the largest among the special levels. However, all four levels result in smaller sample sizes compared to primary levels.

Regarding the type of inspection, three categories are recognized:

- Normal inspection. This is employed to ensure a high likelihood of acceptance when the process quality exceeds the AQL and there are no indications that the process's quality is unacceptable.
- Reduced inspection. Reduced inspection means sampling fewer parts during the inspection. To initiate a reduced inspection, ten or more consecutive lots must pass inspection, with the total number of defects from those ten lots falling below a specified threshold in the standard. Additionally, production must be consistent.
- Tightened inspection. If two out of five lots are rejected, a tightened inspection is necessary. Tightened inspection involves sampling more parts during the inspection. To revert to normal inspection, five consecutive lots with tightened inspection sample sizes must pass inspection.

The values of *n* (sample size) and *c* (acceptance number) required to construct the sampling plan's OC curve are derived from tables provided in ASQ Z1.4 and ISO 2859 (see Appendix 3). Alternatively, online resources, such as http://www.sqconline.com/mil-std-105.html, can be utilized to define the sampling plan according to the lot size, the AQL, the chosen inspection level, and the type of inspection.

The sampling plan must be presented together with the engineering drawing or as a note on the drawing itself, as shown in the example given in Figure 13.10. In the last case, a simple coding can be

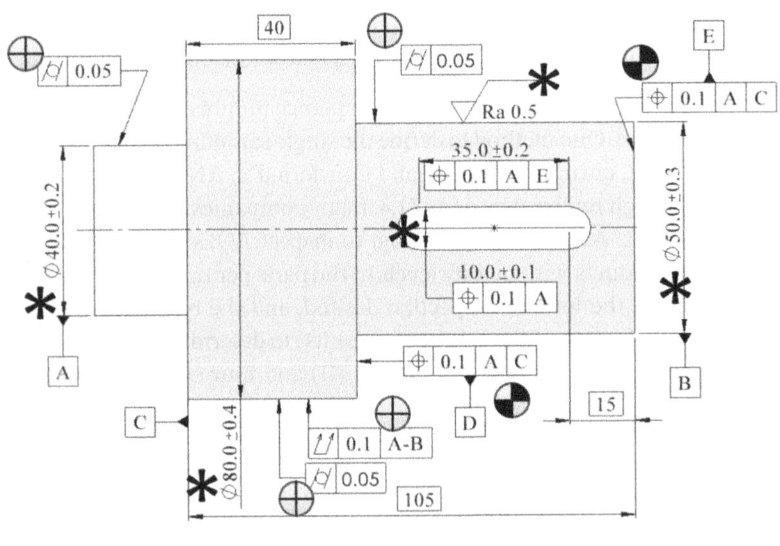

		AQL values					
		Critical dim. ⊕		Major dim.		Minor dim. *	
Lot size	Sample size	Acc.	Rej.	Acc.	Rej.	Acc.	Rej.
100 – 500	32	0	1	1	2	3	4
501 – 1,000	80	1	2	2	3	7	8
1,001 – 5,000	95	2	3	3	4	10	11

Figure 13.10 AQL is shown on a drawing in the form of a note. Units in mm.

used. For instance, the symbols ⊕, ◕, and ★ can be used to define critical, major, and minor dimensions, respectively. A chart explaining the sampling procedure should answer any additional questions.

Given the values of n and c, the sampling plan's OC curve can be obtained. The graph is commonly obtained using statistical software or the same online tools described above. If the lot size N is very large as compared to the sample size n, then the distribution of the number of defectives, d, in a random sample of n items approximately follows a binomial distribution with parameters n and p, where p is the fraction of defectives per lot. Mathematically, the probability of acceptance of a lot with a number of defectives less or equal to c for a process with p fraction of defectives per lot is (NIST/SEMATECH 2014, Chapter 6):

$$P_a = P(d \leq c) = \sum_{d=0}^{c} \frac{n!}{d! \times (n-d)!} \times p^d \times (1-p)^{n-d}$$

Practical Example

Assume the lot size N is 400 and the AQL is 1%. The inspection type is considered normal with an inspection level II. Find the acceptance number c and the sample size n for a single sampling plan. What is the likelihood of accepting the lot if there are 5% defects?

Solution 1: using online tools

In http://www.sqconline.com/mil-std-105.html it is chosen the right range for N, AQL, inspection level, and type of inspection. Then, the "Submit" button is clicked (Figure 13.11).

Figure 13.12 gives the results for the single and double sampling procedures. As it is shown, the acceptance number of defects, c, is equal to 1, and the sample size, n, is equal to 50. Furthermore, according to the OC curve, the probability of accepting the lot if the number of defects is 5% ($p = 0.05$) is around 0.3 (30%).

Solution 2: using AQL Tables (Appendix 3)

The same solution is reached using the AQL Tables from Appendix 3. Using the tables in this appendix, the sample size code is H for a lot size of 400 (range between 281 and 500) and an Inspection Level II. According to the sample size code H, the sample size is defined as 50 and the decision for an AQL 1% is Accept if defects are 1 or less, and Reject if defects are 2 or more.

The probability of accepting the lot (i.e., the number of defects in the sample is 1 or 0) if the number of defects is 5% is defined as

$$P_a = P(d \leq 1) = \sum_{d=0}^{1} \frac{50!}{d! \times (50-d)!} \times (0.05)^d \times (1-0.05)^{50-d} = 0.279$$

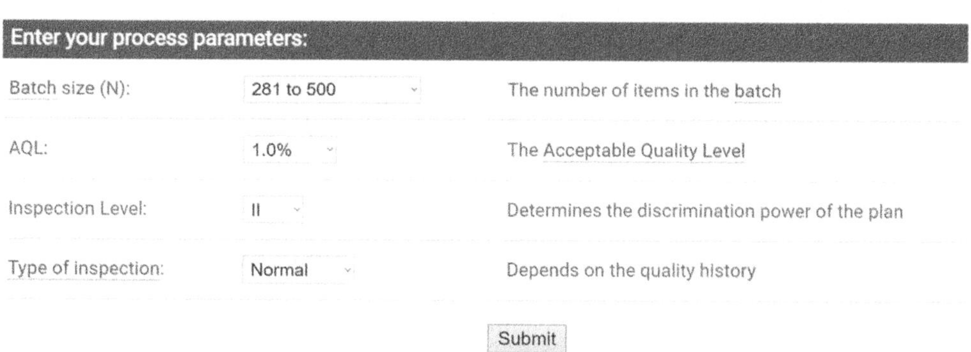

Figure 13.11 Input data for sampling plan.

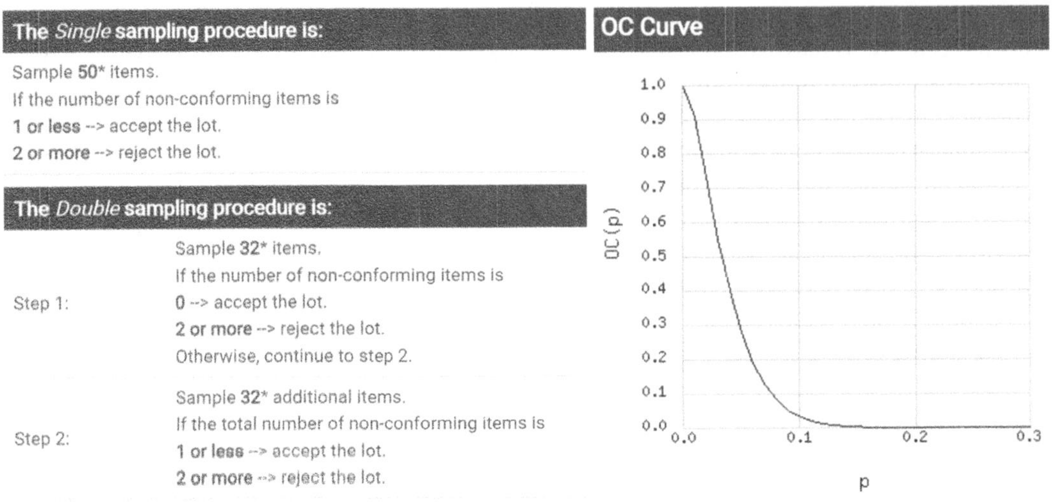

Figure 13.12 Results of the single and double sampling plan.

13.4 Features to be Inspected

As previously shown, when a new product is launched, the first units produced may require a total inspection through an FAI to validate the product. Process validation may also be required to validate the process by estimating the C_{pk} and P_{pk} values. When dealing with product inspection, the features to be included in the inspection plans must be defined. If there exists a prior inspection plan for a similar part, it can serve as a reference for adapting it into the new plan. If this is not the case, process planners or quality engineers should evaluate the functioning of the part and carefully analyze which characteristics of its features need inspection to ensure the part operates as intended. It is important to note that in certain cases, measuring a feature may not be necessary. For instance:

a) When the dimension is under the control of a proven die or mold.
b) When an attribute can be confirmed using methods other than measurement, such as comparing strength to diameter or water leak tests.
c) When a dimension holds no significance for the functionality of the workpiece and is known to be inconsequential in the batch or part being tested.
d) When a feature can be accepted or rejected at a lower cost.

Note that depending on the criticality of the feature to be inspected and the corresponding manufacturing process capability, the sampling size and frequency adopted in the inspection plans may vary notably.

13.5 Selection of Measurement Instruments for Inspection

The main goal of inspection is to inspect the required features with the highest possible level of accuracy while ensuring that it is not consumed more inspector time and machine resources than the part justifies. To achieve this, the inspection plan should consider the following:

1) The total quantity of parts being produced.
2) The availability of appropriate measurement instruments.
3) The time required for setups using the available metrology equipment.
4) The significance of time required for inspecting a particular part in relation to its importance.
5) The investment cost associated with acquiring and using measurement instruments that are currently unavailable.
6) Whether personnel training is a factor in implementing this inspection plan.
7) The potential sources of measurement uncertainties.
8) The acceptable probabilities of accepting a defective feature or rejecting a conforming one.

As it can be seen, a detailed study of the cost of the inspection equipment and procedure should be conducted for an optimal inspection process plan. However, if data about measurement time or equipment cost are not available, a simple procedure for a cost-effective inspection plan can be obtained by selecting the measurement instruments with low precision while still ensuring a good conformity zone. The main reason behind this idea is that a more precise instrument means, generally, a more expensive one. For these cases, the methodology shown below for selection of inspection equipment can be followed:

- Step 1. From the equipment available, select the ones that are adequate for the type of measurement to be conducted, that is, select the instruments according to the measurement to be done, such as internal length, external length, depths, weight, etc.
- Step 2. From the previous selection, select the instruments that are adequate according to the range of measurement.
- Step 3. From the previous selection, select the instruments with a ratio T/2U between 3 and 10, where T is the tolerance of the dimension to be measured and U is the uncertainty of the instrument used. The number 3 is used to avoid rejecting too many good parts due to an excessive uncertainty measurement, and the number 10 is used to avoid the use of too precise instruments (i.e., too expensive instruments), although many times instead of 10, the use of 20 or more is applied.

Since the measurement instrument is not perfect and has some uncertainty during its use, the decision to consider a good or bad part according to the measurements is made based on the conformity zone. Figure 13.13 shows the conformity zone for a part with USL and LSL, depending on the inspection uncertainty U. Mathematically, the conformity zone is defined as (LSL + U, USL − U).

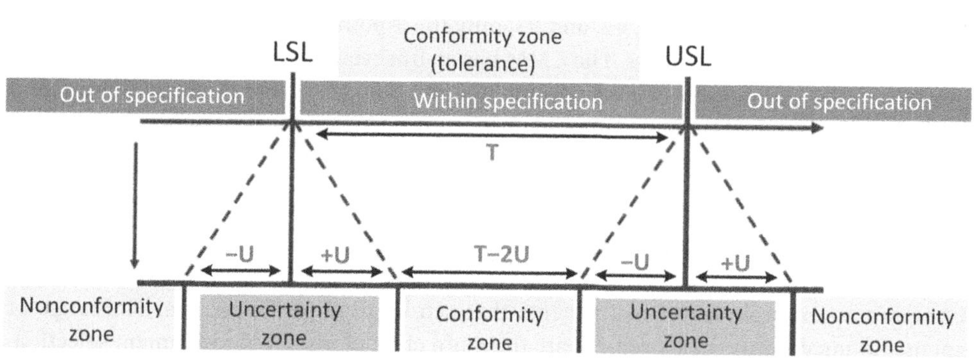

Figure 13.13 Representation of the conformity zone and uncertainty zone in the inspection process.

Table 13.1 Table of instruments.

Instrument no.	Name	Measurement range (mm)	Uncertainty (mm)
1	Coordinate machine measurement (CMM)	0–2,000	0.001
2	Digital vernier	0–100	0.02
3	3-Point internal micrometer	25–50	0.005
4	Vernier depth gauge	0–130	0.02
5	2-Point internal micrometer	50–100	0.01

The uncertainty of an instrument is usually referred to as the expanded uncertainty, which is obtained by multiplying the combined standard uncertainty by a coverage factor, k (JCGM 2008). A value of coverage factor of 2 ($k = 2$) means a 95.4% confidence in the reported conformity zone.

Practical Example

A precision hole with specification $Ø27^{+0.023}_{-0.012}$ should be inspected in the quality department. Table 13.1 shows the available measurement instruments. Select the most adequate instrument.

Solution: The most adequate instrument is selected following the previous methodology:

- Step 1. Discard instruments that do not fit the type of measurement. Instrument #4 is discarded since the vernier depth gauge is only used for measuring depth.
- Step 2. Discard instruments that do not fit the range of the measurement. Instrument #5 is discarded since the specification to be inspected, $Ø27^{+0.023}_{-0.012}$, is not in the range of the instrument, $50-100$ mm.
- Step 3. Discard instruments that do not fit the ratio tolerance/uncertainty. The range considered as adequate is

$$3 \leq \frac{T}{2U} \leq 10$$

Therefore, the uncertainty limits for this example are

$$U_{max} \leq \frac{T}{6} = \frac{0.035}{6} = 0.0058 \text{ mm}, \; U_{min} \geq \frac{T}{20} = \frac{0.035}{20} = 0.0017 \text{ mm}$$

From the remaining instruments, #1, #2, and #3, only the 3-point internal micrometer presents an uncertainty that is within these limits. The CMM has an uncertainty of 0.001 mm, which means that the precision is too high for this inspection, and thus, it is considered too expensive. The digital vernier, with an uncertainty of 0.02 mm, presents a too-low precision for the application. A complete guide for instrument selection and practical examples can be found in Vila et al. (2023).

13.6 Example Data: Measurement Instrument Specifications

Table 13.2 shows an example list of instruments that can be used for inspection. The range of measurement and uncertainty measurement are the main characteristics of instrument selection. Other information, such as the last date of calibration and calibration sheet results, may be provided.

Table 13.2 Example of list of instruments for inspection.

Instrument no.	Name	Brand	Range of measurement	Measurement uncertainty[a]	Date of calibration	Calibration sheet
1	Caliper	MITUTOYO	0–150 mm	0.15 mm	1-3-23	C1-01
2	Caliper	INPRESA	0–150 mm	0.2 mm	1-3-23	C1-02
3	Caliper	TESA	0–300 mm	0.1 mm	1-3-23	C1-03
4	Digital caliper	TESA	0–180 mm	0.02 mm	1-3-23	C1-04
5	Micrometer	MITUTOYO	0–25 mm	0.02 mm	1-3-23	C1-05
6	Micrometer	MITUTOYO	25–50 mm	0.02 mm	1-3-23	C1-06
7	Micrometer	MITUTOYO	50–75 mm	0.02 mm	1-3-23	C1-07
8	Micrometer	MITUTOYO	75–100 mm	0.02 mm	1-3-23	C1-08
9	Digital micrometer	TESA	0–25 mm	0.004 mm	6-6-22	C1-09
10	Digital micrometer	TESA	25–50 mm	0.004 mm	6-6-22	C1-10
11	Digital micrometer	TESA	50–75 mm	0.004 mm	6-6-22	C1-11
12	Digital micrometer	TESA	75–100 mm	0.004 mm	6-6-22	C1-12
13	2-point internal micrometer	ETALON	50–170 mm	0.03 mm	6-6-22	C1-13
14	3-point internal micrometer	TESA	6–12 mm	0.004 mm	6-6-22	C1-14
15	3-point internal micrometer	TESA	11–20 mm	0.008 mm	6-6-22	C1-15
16	Vernier depth gauge	ETALON	0–200 mm	0.1 mm	6-6-22	C1-16
17	Vernier depth gauge	ALCA	0–160 mm	0.15 mm	6-6-22	C1-17
18	Universal bevel protractor	EZQUERRA	0–90°	10’	15-2-23	C1-18
19	Dial indicator	MITUTOYO	0–10 mm	0.03 mm	15-2-23	C1-19
20	Dial indicator	DIAL-GAUGE	0–10 mm	0.03 mm	15-2-23	C1-20
21	Dial indicator (horizontal)	MITUTOYO	0–1 mm	0.03 mm	15-2-23	C1-21

[a] Coverage factor of $k = 2$ (95% confidence interval).

Questions and Exercises

1 Explain the purpose of the first article inspection (FAI) report.

2 Explain the meaning of process capability and process stability.

3 To conduct a process validation in a manufacturing line, the following inspection data are obtained related to the size of a dimensional feature (mm):

Day 1: 50.02, 50.04, 49.97, 50.01, 49.96, 49.94, 49.98, 50.00, 50.02, 50.03, 49.95, 50.00.
Day 2: 50.02, 50.03, 49.95, 50.00, 49.97, 49.99, 50.02, 50.07, 50.04, 50.03, 50.04, 49.96.

The specification of the feature is Ø50 ± 0.15. Analyze the stability and capability of the process.

4 Determine the sampling plan to be conducted at the reception point of raw materials in a company if the lot size of the purchase is 3,500 units and the AQL value is 3%. Since the defects are considered minor, assume a normal type of inspection with level I.

5 Determine the sampling plan to be conducted before shipping in a company. The manufacturer/supplier agreement defines an AQL value of 2% and an LTPD value of 5%. The lot size is 9,000 units. Assume a single sampling plan with a normal type of inspection at level II. What is the probability of shipping a lot that exceeds the LTPD value?

6 A manufacturer is shipping a lot of 1,500 electrical devices to the customer and wants to inspect the product before shipping. For each electrical device, the inspection requires powering on, checking connectivity, and other actions that take 5 minutes. Considering an hourly rate for inspection of $15/hour, calculate the cost and time needed for the inspection sampling and compare it for a 100% inspection strategy. Assume a single sampling plan with the normal type of inspection with level II and an AQL value of 2.5%.

7 What is the meaning of a ratio $T/2U > 10$ in relation to the conformity zone?

8 What is the meaning of a ratio $T/2U < 3$ in relation to the conformity zone?

9 A hole with a dimensional specification of Ø18 ± 0.15 is being inspected with a three-point internal micrometer. The measurement range of the instrument is 15–25 mm, and the measurement uncertainty is 0.01 mm. Calculate the conformity zone of the inspection process. Is the instrument adequate for the inspection?

10 Given the part shown in Figure 13.14 and the measurement instruments from Table 13.3, select the most appropriate instrument for inspecting the part.

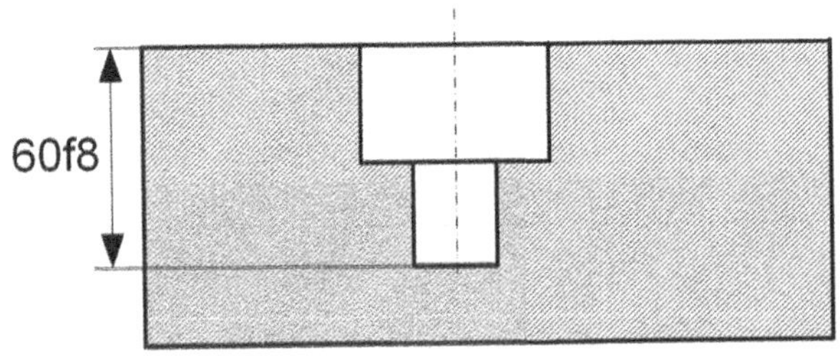

Figure 13.14 Dimension to be inspected. Exercise 10.

Table 13.3 List of instruments for inspection. Exercise 10.

Instrument no.	Name	Brand	Range of measurement (mm)	Measurement uncertainty (mm)
1	Caliper	MITUTOYO	0–150	0.15
4	Digital caliper	TESA	0–180	0.02
8	Micrometer	MITUTOYO	75–100	0.02
13	2-Point internal micrometer	ETALON	50–170	0.03
15	3-Point internal micrometer	TESA	50–75	0.008
16	Vernier depth gauge	ETALON	0–200	0.10

References

ANSI. (2003). *ANSI/ASQ Z1.4-2003 (R2018): Sampling Procedures and Tables for Inspection by Attributes*. American National Standards Institute (ANSI) and American Society for Quality (ASQ).

Arcidiacono, G., and Nuzzi, S. (2017). A review of the fundamentals on process capability, process performance, and process sigma, and an introduction to process sigma split. *International Journal of Applied Engineering Research* 12 (14): 4556–4570.

Heizer, J., and Render, B. (2005). *Principles of Operations Management. Tutorial Acceptance Sampling*. Prentice Hall.

ISO. (1999). *ISO 2859-1:1999. Sampling Procedures for Inspection by Attributes—Part 1: Sampling Schemes Indexed by Acceptance Quality Limit (AQL) for Lot-by-Lot Inspection*. Geneva, Switzerland: International Organization for Standardization.

JCGM. (2008). *JCGM 100:2008 Evaluation of Measurement Data – Guide to the Expression of Uncertainty in Measurement*. Geneva, Switzerland: International Organization for Standardization.

Khlebnikova, E. (2012). Statistical tools for process qualification. *Journal of Validation Technology* 18 (2): 60.

Krajewski, L.J., Ritzman, L.P., and Malhotra, M.K. (2010). *Operations Management: Processes and Supply Chains*. New Jersey: Pearson.

Luke, M. (2021). How to simplify your AS9102 first article inspection process. Quality Magazine 60, number 7. https://digitaledition.qualitymag.com (accessed May 2023).

NIST/SEMATECH. (2014). e-Handbook of statistical methods. https://www.itl.nist.gov/div898/handbook (accessed April 2024).

Oakland, J., and Oakland, J.S. (2007). *Statistical Process Control*. Routledge.

Qiu, P. (2013). *Introduction to Statistical Process Control*. CRC Press.

SAE. (2000). *SAE AS9102. Aerospace First Article Inspection Requirement*. SAE International.

Stamatis, D.H. (2002). *Six Sigma and Beyond: Statistical Process Control*, vol. 4. CRC Press.

Taylor, W.A. (1993). *Classifying Defects and Selecting AQLs*, 101. FDC Control, Food Drug & Cosmetic Division ASQC.

Vila, C., Abellán-Nebot, J.V., and Bruscas, G.M. (2023). *Especificaciones Geométricas de Productos y Procedimientos Metrológicos*. Paraninfo (Spanish): Problemas Resueltos ISBN: 9788428338851.

14

Product Cost Estimation

14.1 Introduction

Almost any industrial inquiry related to the manufacturing process, such as floor space, delivery deadlines, lead time, work in progress, and more relies on process planning as a fundamental source of data. For instance, consider the introduction of a new product within a company. The finance department seeks to determine its manufacturing cost, and subsequently, the selling price. To provide a reasonably accurate answer, the bill of materials (product structure) for the product must be studied, listing all the necessary components and their quantities for a single product. Based on the components to be made and bought, manufacturing cost estimation is then conducted to analyze the economic feasibility of a product. It typically requires considering costs related to material, tooling, equipment operation, labor costs, etc. Given a fair manufacturing cost estimation, finance personnel can translate all this data into cost figures for a final customer quotation or selling price definition.

Additionally, to determine the selling price of a product, product cost estimation can also be conducted to determine how much can be invested in equipment, evaluate alternative designs of a product, conduct make-or-buy decisions, etc. It should be noted that product cost estimating is generally carried out by engineers or technical personnel before the actual production of a product. Cost estimation should be differentiated from cost accounting, which is usually carried out by accounting personnel with the issue of the order for production of a product, and it gives the actual cost of the product based on the data related to actual expenditures of a product.

In this chapter, a global view of different costs that can be found in process planning is reviewed. Standard terms such as direct and indirect costs, prime costs, and profits are defined, and general figures are provided to facilitate a straightforward cost estimation. Given the definition of all different costs in a manufacturing company, the chapter reviews the cost terms to estimate the manufacturing cost of three manufacturing processes: sand casting, machining, and three-dimensional (3D) printing. For the first two processes, the methodology followed is mainly based on reference handbooks about costing (Boothroyd et al. 2002). The figures related to specific costs, such as labor rates, materials' price, etc., have been obtained from the US Bureau of Labor Statistics, Occupational Employment and Wage Statistics (US Bureau of Labor Statistics 2023a), and other websites. In some cases, the costs have been kept from reference (Boothroyd et al. 2002) but adding a cost

Manufacturing Process Planning: A Practical Approach for Mechanical Engineering, First Edition.
José V. Abellán-Nebot, Carlos Vila Pastor, and Héctor R. Siller.

Companion website: www.wiley.com/go/Manufacturing_Process_Planning_1e

increase of 71% with respect to the given values, which is the average price rise from 2002 to 2023 reported by the US Bureau of Labor Statistics (US Bureau of Labor Statistics 2023b). The chapter also includes some computer-aided cost estimators and web-based tools useful for early cost estimations in process planning.

14.2 Elements of Costs

Different costs throughout the company influence a finished product's selling price. These include the cost of production, cost of design, research and development, and overheads, the latter usually including the cost of marketing, sales, customer support, and administration. Figure 14.1 depicts them, illustrating how much each contributes to the ultimate selling price using common figures in the industry (Scallan 2003). The specific numbers may vary significantly between different organizations, depending on the product and its market context. For instance, companies producing highly intricate or cutting-edge products are more likely to allocate a larger proportion of their expenses to research, development, and product design.

However, the primary focus here lies in the costs related to manufacturing, which typically constitute the most substantial portion of the final selling price. These manufacturing costs can be broken down into four primary components: direct labor, indirect labor, parts and material, and energy, equipment, and depreciation. Figure 14.1 shows these four primary components along with their typical percentage allocations. According to this figure, approximately 20% of the overall cost can be attributed to parts and materials (Scallan 2003). The remaining costs are primarily associated with labor and machinery.

These elements of cost can be grouped into different categories to define the final cost of the part. Given the final cost of the part and adding the corresponding profit, the selling price is derived. The first category that is defined is the prime cost, which is composed of the direct costs. Direct costs refer to the expenses incurred by the organization that can be directly linked to the production of a particular item. These direct costs encompass three key elements: direct labor costs (e.g., welders, machinists, etc.), direct material costs, and any other expenses directly associated with the production of a product (e.g., hiring special tools and equipment). Therefore, the prime cost is derived from the direct costs, and it can be defined as follows:

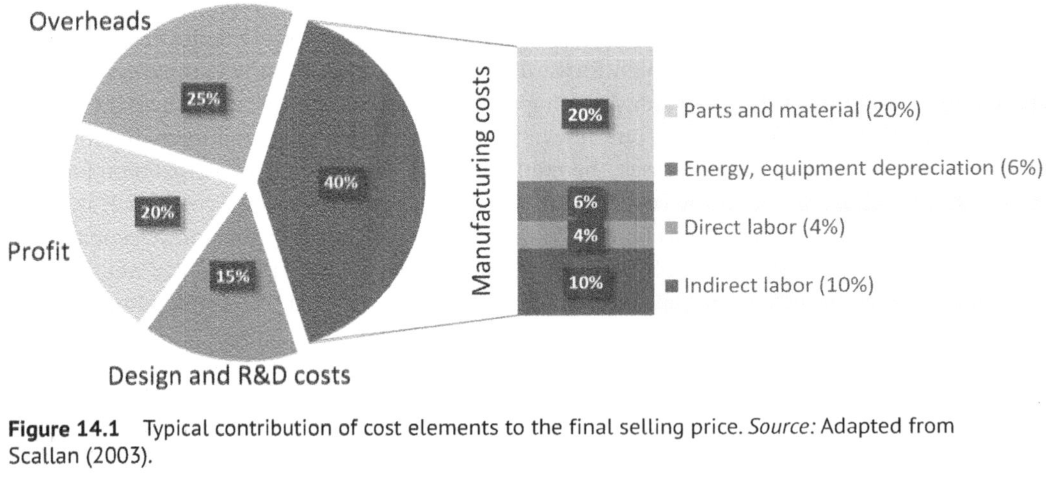

Figure 14.1 Typical contribution of cost elements to the final selling price. *Source:* Adapted from Scallan (2003).

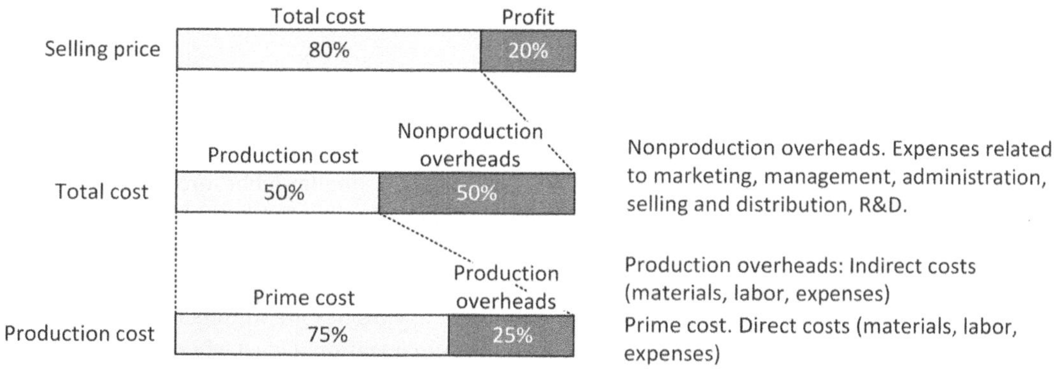

Figure 14.2 Definition of the selling price based on different cost terms and common figures in manufacturing companies. *Source:* Adapted from Scallan (2003).

$$\text{Prime costs} = \text{Direct labour} + \text{Direct material} + \text{Direct expenses}$$

On the contrary, indirect manufacturing costs are those that can only indirectly be attributed to the production of a specific item. In general, these costs include expenses other than the direct costs related to the manufacturing process. They may encompass indirect materials, indirect labor (e.g., supervisors, maintenance operators, etc.), and indirect overhead expenses. All these costs are also referred to as production overheads.

The second category of cost is the production costs, which include both prime costs and indirect manufacturing costs (i.e., production overheads). Thus, the production cost is defined as follows:

$$\text{Production cost} = \text{Prime cost} + \text{Production overheads}$$

Other overhead expenses, such as those related to marketing, management, administrative expenses, selling and distribution expenses, and research and development, should also be included to determine the overall cost of creating a particular product. Therefore, the last category is the total cost, which includes these additional overheads, and it is defined as follows:

$$\text{Total cost} = \text{Production cost} + \text{Nonproduction overheads}$$

Given the total cost and the profit of the company, the final selling price of the product can be defined. The final difference between the selling price and the total cost is the profit; hence,

$$\text{Profit} = \text{Selling price} - \text{Total cost}$$

Figure 14.2 shows the different types of cost in a product and its tentative selling price. Common percentages related to industrial practices are shown to understand the importance of each cost category (Scallan 2003).

14.3 Make-or-Buy Decision

When a company deals with the production of a new product, various components listed in the bill of materials necessitate a decision: should they be manufactured in-house or outsourced? To facilitate this decision, popularly known as a "make-or-buy decision," a cost estimate may be required for comparison purposes, although other factors such as company strategy, vendor quality, and vendor delivery reliability come into play.

A make decision entails securing suitable production equipment, qualified personnel, raw materials, sufficient workspace, oversight, adherence to design standards, and the management of indirect and hidden costs such as maintenance, taxes, insurance, and managerial attention. Conversely, a buy decision involves lower initial investments, a smaller labor force, reduced handling, lower building costs for facilities, diminished overhead expenses, lower taxes and insurance, less supervision, and fewer challenges related to employee-management relationships. However, a make decision brings control of the production and generally lowers the cost per part.

Companies may opt to procure parts, subassemblies, or assemblies for various reasons, and the rationale behind these decisions varies depending on the organizational level where the decision is made. Two levels may be distinguished, the tactical level and the operational level. The tactical level is given when the make-or-buy decision is due to a strategic position of the company in the market, control of the manufacturing knowledge behind the product, control of delivery deadlines to reduce risks in the supply chain, safeguarding intellectual property, reduction of the labor force, etc. On the contrary, the operational level is given when the make-or-buy decision is based solely on existing equipment, resources, skills, and capacity, and the cost is the main decision driver. Figure 14.3 shows common buy-and-make decisions in terms of economic factors and risks. Strengthening relationships with reliable suppliers as partnerships is also another strategy to mitigate the risks and take advantage of supplier's lower costs.

At the operational level, the items evaluated for a make-or-buy decision can be categorized into one of three groups:

- Must-buy items. These items cannot be feasibly manufactured based on current capabilities.
- Must-make items. These items involve processes and operations critical for ensuring product integrity and quality. For instance, a product that is prone to frequent design changes may complicate third-party manufacturing.
- Items that can be either make or buy. These items require thorough analysis to reach a decision, often hinging on cost considerations.

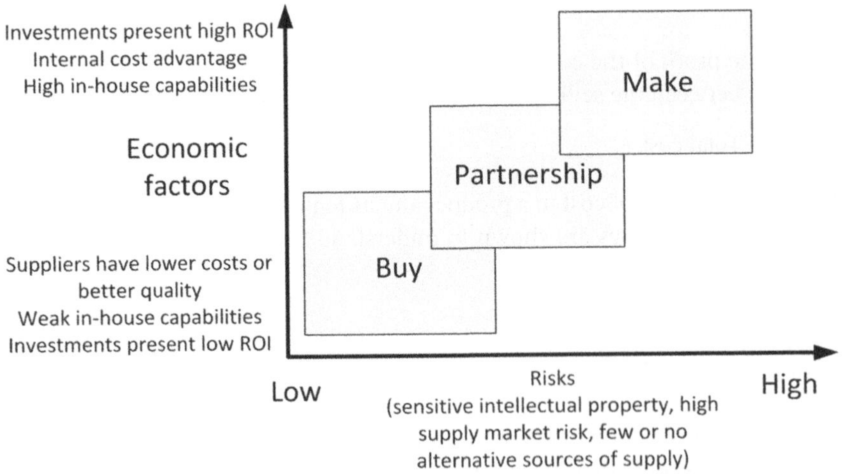

Figure 14.3 Overview of buy-and-make decision according to economic factors and risk.

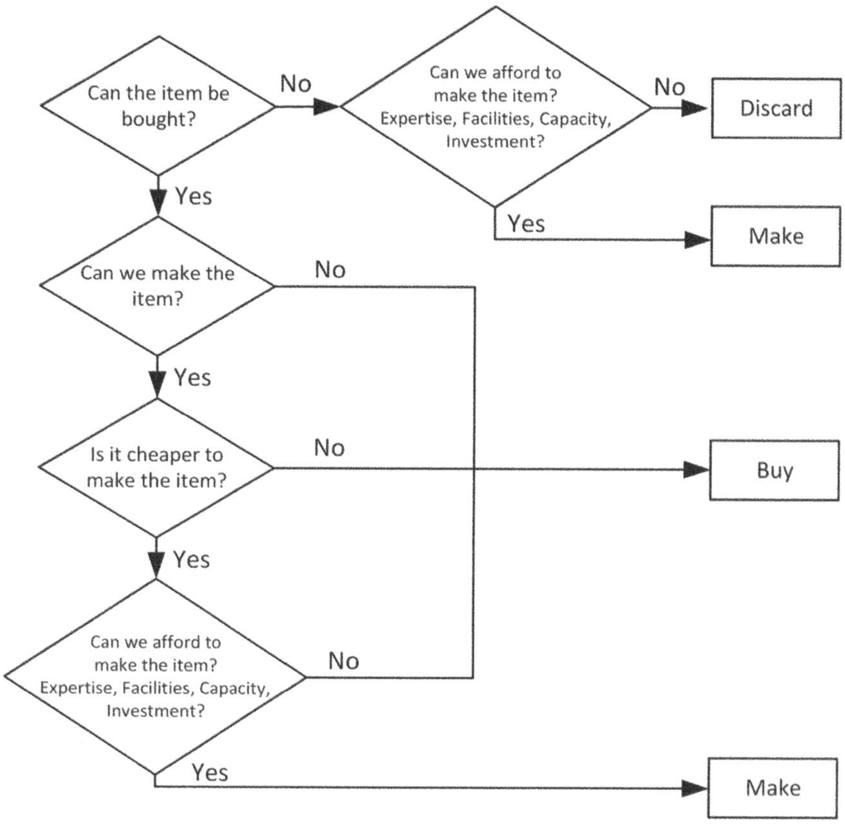

Figure 14.4 Flow chart for "make-or-buy" decision process at operational level. *Source:* Adapted from Scallan (2003).

As a general guideline, a flow chart adapted from Scallan (2003) with the steps for a make-or-buy decision is presented in Figure 14.4. Four fundamental questions guide the make-or-buy decision. The initial inquiry focuses on whether the item can be procured externally. If so, the subsequent question is, "Can the item be produced internally?" Typically, if the answer is negative, the part will be procured unless the decision is being made at the tactical level. If in-house production is feasible, the next step is to compare the costs of making versus buying. Assumptions are often required regarding factors such as expected volume, demand patterns, waste, inventory levels, and fluctuations in material prices. If making is more cost-effective, it is analyzed if the required investment in equipment and resources is affordable, considering the potential alternative uses of capital and the cost of external funding if necessary.

14.4 Pricing

The pricing activity of a company may be conducted under two different approaches:

- Top–down approach. The price is defined according to similar products in the market. In other words, the price is what the customers are ready to pay, and it is not related to the actual manufacturing costs. Therefore, the price is defined by the market, and the profit is calculated depending on the total manufacturing cost. In this case, the price may also allocate the maximum costs to make the product economically viable. This situation may be common in exclusive or traditionally high-priced markets.

- Bottom–up approach. The price is defined based on the total manufacturing cost of the product. After estimating the total cost of the product, a percentage of profit is added to set the final price of the product. This profit percentage depends on the type of product, market characteristics, etc., but a good profit margin for manufacturing companies is usually within the range of 10–20%.

The final price in the market will also depend on the distributors/resellers. If the company relies on distributors/resellers, they often make a typical 60% markup, which increases notably the final price of the product to the consumer. The markup value depends on the type of product. Some consumer products go through long, complex, and expensive sales chains to get to the end user, leading to high markup values.

To illustrate the company's profit and final price of the product, consider a company with a total cost of $50 per product and a final price of the product of $65. In this case, the gross margin (profit) obtained is as follows:

$$\text{Gross margin} = \frac{\text{Revenue} - \text{Total cost}}{\text{Revenue}} \times 100 = \frac{65 - 50}{65} \times 100 = 23\%$$

However, if the same company applies the same gross margin percentage and uses a distributor/reseller to sell its products with a 60% markup, the final price of the product is:

$$\text{Retail price} = (\text{Price to sell to distributor}) \times (\text{Markup} + 1)$$

$$\text{Retail price} = 65 \times 1.6 = \$104$$

Despite the large margins applied by distributors/resellers, they make it easier for products to sell since they handle shipping, orders and returns, and have direct contact with customers.

Pricing is critical when evaluating a new product development. At this point, it is necessary to consider the capital investment for new equipment or other requirements must be considered in setting the final product price. In these cases, the analysis of the return on investment (ROI) is needed to estimate the profitability of the new product.

A basic calculation of ROI is given as follows:

$$\text{ROI}(\%) = \frac{(\text{Unit price} - \text{Unit total cost}) \times \text{Volume of sales}}{\text{Capital investment}} \times 100\%$$

which depends on the final price since it establishes the company's profit. ROIs higher than 7–10% are considered adequate for continuing the product development process. Note that more complex ROI expressions to consider the multi-year effect can be applied, but previous expressions can give a fast overview of the profitability of the investment for a new product release.

14.5 Sand-Casting Cost Estimation (C_{tsc})

Sand-casting cost estimation can be conducted by analyzing three main cost factors: tooling cost ($C_{tooling}$), processing cost (C_{proc}), and metal cost (C_{mat}). These three cost terms can be fairly estimated for a sand-casting process given the material, the equipment to be used, pattern characteristics, production rate, labor cost, etc. The following equation is used for a generic sand-casting cost estimation (C_{tsc}):

$$C_{tsc}(\$) = C_{tooling} + C_{proc} + C_{mat}$$

Each cost term is described in detail in the following sections and some cost tables are provided for a straightforward cost estimation.

14.5.1 Tooling Cost ($C_{tooling}$)

Tooling cost takes into account tooling cost for both molds and cores. Tooling cost related to molds refers to the cost of mold sand (C_{msd}) and the cost of the pattern (C_{pm}) that is needed to create the molds. Tooling cost related to cores refers to the cost of core sand (C_{csd}) which is usually more expensive than the ones needed for molds, and the cost of the core boxes (C_{box}) to fabricate the cores. Therefore, assuming that cores are needed, the tooling cost is defined as:

$$C_{tooling}(\$) = C_{msd} + C_{csd} + \frac{C_{pm} + C_{box}}{N_{tooling}}$$

Note that tooling costs related to pattern and core box fabrication are shared by all parts made under this tooling ($N_{tooling}$). Each cost previously shown can be estimated as discussed in the following sections.

14.5.1.1 Mold Sand and Core Sand Cost (C_{msd}, C_{csd})

To estimate the cost of the molding sand required for building the mold, the volumes of the mold (cope and drag) and their cavities (i.e., sprue, runner, risers, and pattern cavity) should be calculated. Since the mold sand cost has a relatively small impact on the cost of producing a casting (often less than 3% of the total cost), a rough estimation can be straightforwardly applied where the cost of mold sand is estimated as \$0.03/kg of metal poured. This empirical relationship was presented in a survey of American foundries considering recycling and other factors and it is usually employed in some reference books (Boothroyd et al. 2002). Thus, the cost of mold sand per part can be evaluated as follows:

$$C_{msd}(\$) = 0.03 \times W_p$$

where W_p is the weight of the poured casting. At early stages, the sprue, runners, risers, etc., may be not defined, so the estimation of W_p can be roughly done by increasing the weight of the final cast part by 20–40%.

To estimate the cost of core sand, the total weight of the sand needed is straightforwardly obtained since the geometry of the cores is known. Depending on how delicate the core configuration is, and the sort of core-making technique being employed, scrap rates for cores can range from 4% to 20% throughout processing. A scrap rate of 8% will give a reliable estimate in most situations. Therefore, the cost of the core sand for producing a casting can be estimated as:

$$C_{csd}(\$) = \rho_{cs} \times V_c \times \frac{C_{cs}}{1 - \frac{S_c}{100}}$$

where ρ_{cs} is the density of core sand (approximately 1,387 kg/m³); V_c is the volume of core (m³); S_c is the core scrap rate (%); and C_{cs} is the cost of core sand (approximately \$0.15/kg). Note that sand for core fabrication is more expensive than mold sand because stronger cores must be produced using finer sand.

14.5.1.2 Pattern and Core Box Cost (C_{pm}, C_{box})

The plates with the pattern impressions are often made of iron castings for higher volume production. If the mold's parting surface is flat, the top and bottom plate surfaces are machined in accordance with that shape, and flask pins and bushings are then inserted. Cast parts that do not have a level parting line require more intricate mounting plate castings. These, in turn, necessitate more involved machining, which can substantially raise tooling costs.

The pattern and core box costs depend on the material cost and the machining cost to generate the required imprints. The material cost is straightforwardly evaluated, knowing the volume of the pattern plate or the core box and the material cost per kg. Thus, the material cost for the pattern plate is expressed as follows:

$$C_{\text{Mpm}}(\$) = C_{\text{rpm}} \times \rho_{\text{pm}} \times V_{\text{pm}}$$

where C_{rpm} is the raw material cost per kg, ρ_{pm} is the density of the material used for the pattern plate and V_{pm} is the volume of the plate. Table 14.1 shows some reference values for typical materials used in pattern plates. A similar equation is applied when dealing with the cost of the core box:

$$C_{\text{Mbox}}(\$) = C_{\text{rbox}} \times \rho_{\text{box}} \times V_{\text{box}}$$

where the terms refer to the core box characteristics.

Calculating the cost of machining is the next step after determining the material cost of the pattern plate or the core box. Machining cost depends on the complexity of the surface to be generated and the surface roughness requirement that applies. For sand-casting processes, the surface roughness requirements are not critical, and common machining capabilities are enough without any increase in cost. To evaluate surface complexity, the number of surface patches is counted. A surface patch refers to a region with a constant or smoothly varying curvature that can be machined under the same machining operation without lifting the tool. Two surface patches coming together create an abrupt shift in slope or curvature. The machining cost will be a function of the number of different patches identified. In Boothroyd et al. (2002), the machining cost for a set of pattern impressions is estimated as:

$$C_{\text{pi}}(\$) = R_{\text{t}} \times \left(0.313 \times N_{\text{sp}}^{1.27} + 0.085 \times A_{\text{p}}^{1.2}\right) \times N_{\text{pi}}^{0.7}$$

Table 14.1 Tool life and relative costs.

Pattern or core box material	Density (kg/m³)	Cost (\$/kg)	Relative machining cost factor, R_{tf}	Tool life per cavity (# of uses)
Wood	500–900	2–10	0.25	2,500
Plastic	900–1,300	0.5–4	0.40	5,000
Cast iron	6,920	1.5–2.5	1.00	150,000
Stainless steel	7,750	2.5–6.0	1.18	180,000

Source: Adapted from Boothroyd et al. (2002)

where R_t is the toolmaking cost rate ($/hour, typical values are $70–80/hour); N_{sp} is the number of different surface patches; N_{pi} is the number of identical impressions; and A_p the projected area of the impressions (cm^2).

The previous cost applies when patterns or core boxes are made of cast iron, a material that is commonly applied to make high-production tools. For different materials, such as wood or plastic for low-volume production or stainless steel for high production volumes, a relative cost factor, R_{tf}, can be used to adjust the cost. Table 14.1 shows relative machining cost factors for different tooling materials and a typical number of cycles that a tooling can be used before needing to be replaced.

Given the material cost and the machining cost, the cost for both pattern plates and core boxes are obtained as follows:

$$C_{pm}(\$) = C_{Mpm} + R_{tf} \times C_{pi}$$

$$C_{box}(\$) = C_{Mbox} + R_{tf} \times C_{pi}$$

Other references present alternative methods to estimate tooling costs in sand casting. For instance, in Chougule and Ravi (2006), the tooling cost is obtained based on a regression equation where the casting volume and factors related to complexity and accuracy are considered.

For illustrative purposes, Table 14.2 shows common pattern costs from a Chinese manufacturer, Liaoning Borui Machinery Co. Ltd.

14.5.2 Processing Cost (C_{proc})

The cost of creating the cores (C_{core}), the cost of mold processing (C_{mp}), and the cost of cleaning the castings (C_{cl}) make up the three primary categories of processing cost. Thus, the processing cost is evaluated as:

$$C_{proc}(\$) = C_{core} + C_{mp} + C_{cl}$$

Each cost previously shown can be estimated as follows.

Table 14.2 Common pattern cost from a Chinese manufacturer.

	Very simple shape		Normal complex shape		Very complex shape	
Casting process	**Small size ($)**	**Large size ($)**	**Small size ($)**	**Large size ($)**	**Small size ($)**	**Large size ($)**
Manual green sand casting (Resin pattern)	800	1,200	1,200	1,600	—	—
Manual green sand casting (Aluminum pattern)	1,000	1,500	1,500	2,500	3,000	5,000
Automatic molding casting (Iron pattern)	1,500	2,000	2,000	3,000	3,500	5,000

Source: Liaoning Borui Machinery Co. Ltd. (https://www.iron-foundry.com/mould-pattern-cost-cast-iron-sand-castings-steel-castings.html)

14.5.2.1 Cost of Producing Cores (C_{core})

The cost of producing a core depends on a wide range of factors. For instance, some cores need to be coated with refractory material, which requires special tools and labor to coat the cores and bake them to dry the coating. To facilitate and get a rough estimation of core cost, it is evaluated the weight of cores that are expected to be produced at the shop floor every hour and divided this value by the total number of workers that work on it. This gives the production rate for the shop floor that is expressed in core kilograms per worker-hour. An example of this rate is given in Boothroyd et al. (2002), where the production rate for cores was determined to be around 53 kg/worker-hour from a survey of several big foundries. Therefore, the cost of processing a particular core is given (Boothroyd et al. 2002) as follows:

$$C_{\text{core}}(\$) = \rho_{cs} \times V_c \times \frac{R_{\text{cm}}}{\left[P_{\text{cm}} \times \left(1 - \frac{S_c}{100}\right) \times \left(\frac{P_{ff}}{100}\right)\right]}$$

where ρ_{cs} is the density of core sand (kg/m^3); V_c is the volume of core (m^3); P_{ff} is the plant efficiency ([actual production time/total available time] × 100); R_{cm} is the worker rate for core making ($/hour, typical values are $20–40/hour); S_c is the core scrap rate; and P_{cm} is the core production rate (kg/hour). The scrap rate depends on the type of core-making method and the core's design, and it usually ranges from 4% for a straightforward core to 20% for a delicate one, and a scrap rate of 8% is suitable for early costing situations. Plant efficiency depends on the plant itself, but efficiencies of 60–85% can be reasonable.

14.5.2.2 Cost of Producing Molds (C_{mp})

The cost of producing molds is determined by the number of castings produced per hour, the number of workers needed to support the line, and the labor rate per worker. The cost per casting related to mold production is determined by (Boothroyd et al. 2002) the following equation:

$$C_{\text{mp}}(\$) = N_{\text{mw}} \times \frac{R_{\text{mp}}}{N_c \times P_{\text{mp}} \times \left(1 - \frac{S_{\text{m}}}{100}\right) \times \left(\frac{P_{\text{ff}}}{100}\right)}$$

where N_{mw} is the number of line workers; R_{mp} is the labor rate for molding line production ($/h, reasonable values are $130–200/hour); N_c is the number of mold cavities (castings per mold); P_{mp} is the molding line production rate (molds/hour); S_m is the casting scrap rate (%); and P_{ff} is the plant efficiency (%). The scrap rate may depend on the geometry of the item and the type of metal being poured. As presented in Boothroyd et al. (2002), a figure of 2% may be appropriate for early cost estimation.

14.5.2.3 Cost of Cleaning Castings (C_{cl})

The castings' removal from the mold is the last process on the mold line. Typically, a ram is used to drive the sand mold out of the flask and shatter the mold. The castings are brought to the cleaning section while the mold sand is recycled.

The size and geometry of the part play a major role in how much it will cost to clean a casting. From the literature, the overall labor time required to clean castings – including the removal of sand, the removal of runner systems and flash, and shot blasting – has been proven to be roughly equal to the square root of the casting weight (Boothroyd et al. 2002). Thus, the following empirical expression is used for estimating the cleaning time per casting:

$$T_{cl}(h) = 0.024 \times W_p^{0.44}$$

and the cleaning cost is then given by

$$C_{cl}(\$) = R_{cl} \times T_{cl}$$

where T_{cl} is the total cleaning time (hour) and R_{cl} is the worker labor rate ($/hour). Labor rates of $20–30/hour are reasonable values for small automated foundries in cleaning operations.

14.5.3 Metal Cost (C_{mat})

The metal cost must consider the following terms: raw material (C_{rm}), energy cost related to melting (C_{en}), amortization cost of furnace equipment (C_{fk}), and the labor cost for activities related to furnace and pouring (C_{lk}). The total of these expenses is frequently referred to as "at the spout" metal cost and can be expressed in $/kg as (Boothroyd et al. 2002) follows:

$$C_{ms}(\$/kg) = C_{rm} + C_{en} + C_{fk} + C_{lk}$$

To consider the metal loss in melting and pouring, the total weight to be cast is increased by a generic factor of loss. Additionally, the wasted metal in the feed system and risers is reutilized, and thus, the material cost is adjusted considering its scrap value. Therefore, the final metal cost is estimated as follows:

$$C_{mat}(\$) = \left(C_{ms} \times W_p - C_{cv} \times \left(W_p - W_{fc}\right)\right) \times f_{loss}$$

where W_p is the weight of poured metal (kg); W_{fc} is the weight of the finished casting without feeder and risers (kg); C_{cv} is the metal scrap value ($/kg); and f_{loss} is the metal loss factor in melting and pouring that typical ranges from 1.02 to 1.14. The weight of metal poured at early stages may be hard to estimate since the sprue, runners, risers, etc., may be not defined. If this information is not provided, a rough estimation for raw material weight can be done by increasing the weight of the final cast part by 20–40%.

The different cost terms associated with the metal cost is discussed in the following.

14.5.3.1 Cost of Raw Material (C_{rm})

The cost of raw material in $/kg is straightforwardly obtained from databases and vendors' quotations. Table 14.3 shows common raw material costs and their scrap values for casting.

14.5.3.2 Cost of Melting Energy (C_{en})

The raw material should be heated to reach the required pouring temperature. The furnace energy cost is estimated (Boothroyd et al. 2002) as follows:

$$C_{en}(\$/\text{kg}) = E_{ct} \times \frac{M_{me}}{F_{ff}}$$

where E_{ct} is the cost of energy [$/(kWh)]; M_{me} is the minimum melt energy of the material (kWh/kg); F_{ff} is the furnace efficiency. Values of 0.16 $/kWh for E_{ct} and 45–75% for F_{ff} can be used

Table 14.3 Properties and costs of commonly used sand-casting alloys.[a]

Generic alloy	Density (kg/m^3)	Minimum melt energy (kW·h/kg)	Cost ($/kg)	Scrap value ($/kg)
Ductile iron	7,100	0.391	2–3.5	0.5–0.9
Gray iron	6,920	0.390	1.5–2.5	0.5–0.8
Low carbon steel	7,830	0.393	0.6–1.0	0.1–0.2
High carbon steel	7,830	0.393	1.0–2.5	0.2–0.4
Austenitic stainless steels	7,750	0.405	2.5–5.0	0.4–0.8
Ferritic stainless steels	7,750	0.405	3.0–6.0	0.5–0.9
Aluminum 6061	2,710	0.326	2.0–4.0	0.5–1
Aluminum 7075	2,710	0.326	7.0	1.5
Magnesium alloys	1,800	0.332	4.00–10.0	1–2.5
Bronze	8,800	0.185	5.0–15.0	1.5–4

[a] Data gathered from different databases and website sources.

for typical foundry operations with electric arc furnaces. M_{me} values for a range of typical casting metals are given in Table 14.3.

14.5.3.3 Cost of Equipment (C_{fk})

A fixed furnace cost is typically added to consider the cost of the equipment, that is, the amortization cost. This cost is related to the shop-floor equipment and can be estimated as follows:

$$C_{fk}\left(\$/\text{kg}\right) = \frac{P_{fc}}{A_y \times n_{life}}$$

where P_{fc} is the price of the furnace equipment ($), A_y is the expected amount of kg heated per year (kg/year), and n_{life} is the effective life of the equipment (years). Alternative, common depreciation rates in casting machinery are 5–10%, and the cost of equipment can be rewritten as follows:

$$C_{fk}\left(\$/\text{kg}\right) = d_r \times \frac{P_{fc}}{A_y}$$

where d_r is the depreciation rate. For instance, an electric arc furnace with a 500-kg capacity may cost around $230,000. Assuming a depreciation cost of 10%, 100 days of work per year, and 2 castings per day, the resulting cost is $0.23/kg.

14.5.3.4 Cost of Labor (C_{lk})

The labor expense for running the furnace, expressed as a labor cost per kilogram of metal poured (C_{lk}), must also be included in the metal cost. A foundry can estimate this on a shift-by-shift basis by dividing the cost of furnace workers for one shift by the weight of metal poured during that shift. For a medium-sized automated foundry, the furnace labor cost is typically calculated at $0.034/kg (Boothroyd et al. 2002).

14.5.4 Practical Example

Analyze the sand-casting cost for casting 2,500 disks of ductile iron with dimensions of Ø200 mm and a width of 40 mm with a hole of Ø50 mm. Assume: 5-mm machining allowance; manual casting process; a production rate of 20 molds/hour with 2 workers in the molding line; pattern plate dimensions 30 × 30 × 4 cm. Assume any reasonable value for other required data.

The sand-casting cost is estimated as follows:

$$C_{\text{tsc}} = C_{\text{tooling}} + C_{\text{proc}} + C_{\text{mat}}$$

The three terms of cost are individually calculated as follows:

- Tooling cost (C_{tooling}):
  The tooling cost is

$$C_{\text{tooling}} = C_{\text{msd}} + C_{\text{csd}} + \frac{C_{\text{pm}} + C_{\text{box}}}{N_{\text{tooling}}}$$

where

- Mold sand cost:

$$C_{\text{msd}} = 0.03 \times W_{\text{p}}$$

where W_{p} is the weight of the poured casting, which is estimated as 1.3 times the final cast weight to include approximately the weight of feeders, runners, and sprue. The volume of the finished casting (V_{fc}) and its weight (W_{fc}) are as follows:

$$V_{\text{fc}} = \left(\quad \times \left((0.1+0.005)^2 - (0.025-0.005)^2\right)\right) \times (0.04 + 2 \times 0.005) = 1.67 \times 10^{-3}\,\text{m}^3$$

$$W_{\text{fc}} = \rho \times V_{\text{fc}} = 7{,}100 \times 1.67 \times 10^{-3} = 11.86\,\text{kg}$$

Therefore, the mold sand cost is

$$C_{\text{msd}} = 0.03 \times (1.3 \times W_{\text{fc}}) = 0.03 \times 1.3 \times 11.86 = \$0.46$$

- Core sand cost:

Assuming a core print of 10 mm per side, the core volume is

$$V_{\text{c}} = \left(\quad \times 0.02^2 \times (0.05 + 0.02)\right) = 2.8 \times \quad \times 10^{-5}\,\text{m}^3$$

Therefore, the core sand cost is

$$C_{\text{csd}} = \rho_{\text{cs}} \times V_{\text{c}} \times \frac{C_{\text{cs}}}{1 - \frac{S_{\text{c}}}{100}} = 1{,}387 \times 2.8 \times \quad \times 10^{-5} \times \frac{0.15}{1 - \frac{8}{100}} = \$0.019$$

- Pattern plate cost:

Since the number of parts to be cast is 2,500, a pattern plate made of wood may be reasonable. A plate of dimensions 300 × 300 mm with a thickness of 40 mm can fit the imprints for the required pattern. Assuming a cost of \$5/kg and a density of 600 kg/m^3, the material cost of the pattern plate is calculated as follows:

$$C_{\text{Mpm}} = C_{\text{rpm}} \times \rho_{\text{pm}} \times V_{\text{pm}} = 5 \times 600 \times 0.3 \times 0.3 \times 0.04 = \$10.8$$

The cost of machining the pattern is

$$C_{\text{pi}} = R_{\text{t}} \times \left(0.313 \times N_{\text{sp}}^{1.27} + 0.085 \times A_{\text{p}}^{1.2}\right)$$

$$C_{\text{pi}} = 70 \times \left(0.313 \times 4^{1.27} + 0.085 \times \left(\pi \times \frac{21^2}{4}\right)^{1.2}\right) = 70 \times (1.82 + 94.8) = \$6,763$$

where R_{t} is assumed as \$70/h; N_{sp} is the number of surface patches which is four (Figure 14.5); and A_{p} is the projected area or the pattern imprints. Note that N_{pi} is not included in the above expression since there are no identical impressions.

Thus, the pattern plate cost is

$$C_{\text{pm}} = C_{\text{Mpm}} + R_{\text{tf}} \times C_{\text{pi}} = 10.8 + 0.25 \times 6,763 = \$1,701$$

where R_{tf} is 0.25 (Table 14.1) since the material of the plate is wood.

- Core box cost:

  For the core box, the dimensions of the cores are Ø40 mm and length 70 mm. A core box of dimensions 90 × 60 × 60 mm can be enough. Assuming cast iron material, the cost of the core box is:

$$C_{\text{Mpm}} = C_{\text{rpm}} \times \rho_{\text{pm}} \times V_{\text{pm}} = 2 \times 6,920 \times 0.09 \times 0.06 \times 0.06 = \$4.5$$

The machining cost is:

$$C_{\text{pi}} = 70 \times \left(0.313 \times 3^{1.27} + 0.085 \times (28)^{1.2}\right) = 70 \times (1.26 + 4.63) = \$412$$

since the number of patches is 3 (Figure 14.5) and the projected area is 7 × 4 cm. The total cost of the core box is

$$C_{\text{box}} = C_{\text{Mbox}} + R_{\text{tf}} \times C_{\text{pi}} = 4.5 + 1 \times 412 = \$416.5$$

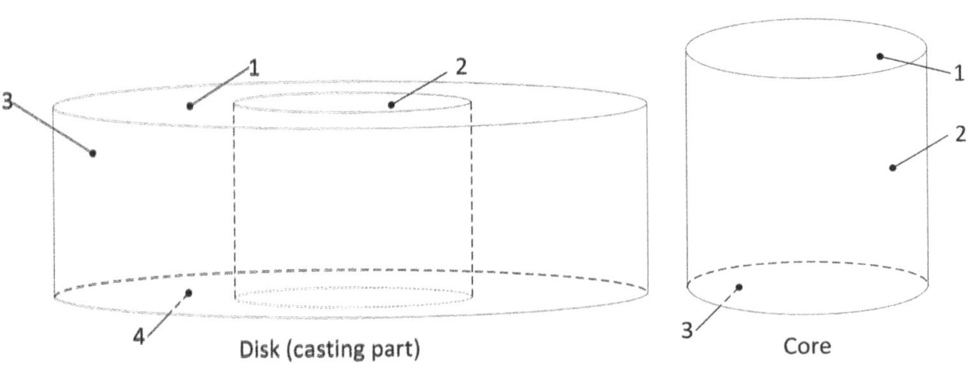

Figure 14.5 Number of patches of the casting part and the core for pattern and core box machining cost estimation.

As a result, the final tooling cost is computed as follows:

$$C_{\text{tooling}} = C_{\text{msd}} + C_{\text{csd}} + \frac{C_{\text{pm}} + C_{\text{box}}}{N_{\text{tooling}}}$$

$$C_{\text{tooling}} = 0.46 + 0.019 + \frac{1{,}701}{2{,}500} + \frac{416.5}{150{,}000} = \$1.16$$

Note that N_{tooling} varies for pattern plates (wood material) and core boxes (cast iron material).

- Processing cost (C_{proc}):
The processing cost is expressed as

$$C_{\text{proc}} = C_{\text{core}} + C_{\text{mp}} + C_{\text{cl}}$$

where

- Processing cost of cores:
The cost of processing a particular core is

$$C_{\text{core}} = \rho_{\text{cs}} \times V_{\text{c}} \times \frac{R_{\text{cm}}}{\left[P_{\text{cm}} \times \left(1 - \frac{S_{\text{c}}}{100}\right) \times \left(\frac{P_{\text{ff}}}{100}\right)\right]}$$

$$C_{\text{core}} = 1{,}387 \times 2.8 \times \quad \times 10^{-5} \times \frac{60}{\left[10 \times \left(1 - \frac{8}{100}\right) \times \left(\frac{85}{100}\right)\right]} = \$0.94$$

Assuming R_{cm} to be \$60/h; S_{c} of 8%; P_{ff} of 85%; and P_{cm} of 10 kg/worker-hour.

- Processing cost of molds:
The mold line processing cost per casting is

$$C_{\text{mp}} = N_{\text{mw}} \times \frac{R_{\text{mp}}}{N_{\text{c}} \times P_{\text{mp}} \times \left(1 - \frac{S_{\text{m}}}{100}\right) \times \left(\frac{P_{\text{ff}}}{100}\right)}$$

$$C_{\text{mp}} = 2 \times \frac{130}{1 \times 20 \times \left(1 - \frac{2}{100}\right) \times \left(\frac{85}{100}\right)} = \$15.6$$

Assuming R_{mp} to be 130 \$/h; S_{m} of 2%, P_{ff} of 85%; one mold cavity per mold (N_{c}).

- Processing cost of cleaning:
The cleaning labor time is

$$T_{\text{cl}} = 0.024 \times W_{\text{p}}^{0.44} = 0.024 \times (1.3 \times 11.86)^{0.44} = 0.08\,\text{h}$$

and the cleaning cost is then given by

$$C_{\text{cl}} = R_{\text{cl}} \times T_{\text{cl}} = 25 \times 0.08 = \$2$$

where the worker rate for cleaning the room is assumed to be $25/hour.
As a conclusion, the total processing cost per casting is

$$C_{\text{proc}} = C_{\text{core}} + C_{\text{mp}} + C_{\text{cl}} = 0.94 + 15.6 + 2 = \$18.54$$

- Metal cost (C_{mat}).
The metal cost is given by the equation:

$$C_{\text{mat}} = C_{\text{ms}} \times W_{\text{p}} \times f_{\text{loss}}$$

where the metal cost in $/kg, C_{ms}, is defined as

$$C_{\text{ms}} = C_{\text{rm}} + C_{\text{en}} + C_{\text{fk}} + C_{\text{lk}}$$

These four terms are estimated as follows:

– Raw material cost:
The material is ductile iron, then

$$C_{\text{rm}} = \$3 / \text{kg}$$

– Melting energy:
The cost related to the energy for melting the metal is

$$C_{\text{en}} = E_{\text{ct}} \times \frac{M_{\text{me}}}{F_{\text{ff}}} = 0.16 \times \frac{0.391}{0.6} = \$0.10 / \text{kg}$$

assuming E_{ct} to be $0.16/kWh and F_{ff} to be 60%.

– Equipment cost:
To consider the machinery amortization, a depreciation value of 5% is assumed. The price of the furnace is assumed to be $200,000 with an average production volume per year of 25,000 kg.

$$C_{\text{fk}} = d_{\text{r}} \times \frac{P_{\text{fc}}}{A_{\text{y}}} = 0.05 \times \frac{200{,}000}{25{,}000} = \$0.4 / \text{kg}$$

– Labor cost:
It is assumed a labor cost of $0.034/kg.

$$C_{\text{lk}} = \$0.034 / \text{kg}$$

As a result, the total metal cost per kg is

$$C_{\text{ms}} = C_{\text{rm}} + C_{\text{en}} + C_{\text{fk}} + C_{\text{lk}} = 3 + 0.1 + 0.4 + 0.034 = \$3.53 / \text{kg}$$

Considering a metal loss factor of 1.1, the weight of the final cast part, the weight of the metal poured and the scrap value of ductile iron (assumed to be $0.8/kg), the total metal cost is

$$\begin{aligned} C_{\text{mat}} &= \left(C_{\text{ms}} \times W_{\text{p}} - C_{\text{cv}} \times \left(W_{\text{p}} - W_{\text{fc}}\right)\right) \times f_{\text{loss}} \\ &= \left(3.53 \times (1.3 \times 11.86) - 0.8 \times (1.3 \times 11.86 - 11.86)\right) \times 1.1 \\ &= \$56.9 \end{aligned}$$

Finally, adding all three cost terms, the final sand-casting cost per part is

$$C_{\mathrm{tsc}} = C_{\mathrm{tooling}} + C_{\mathrm{proc}} + C_{\mathrm{mat}} = 1.16 + 18.54 + 56.9 = \$76.6$$

14.6 Machining Cost Estimation (C_{tmach})

The total machining cost of a part can be estimated (Boothroyd et al. 2002) as follows:

$$C_{\mathrm{tmach}}(\$) = C_{\mathrm{mat}} + C_{\mathrm{setup}} + C_{\mathrm{non-pr}} + C_{\mathrm{mach}} + C_{\mathrm{tooling}}$$

where C_{mat} is the material cost, C_{setup} is the setup cost, $C_{non\text{-}pr}$ is the nonproductive cost, C_{mach} is the machining cost, and $C_{tooling}$ is the cost related to tooling considering both cutting tools and fixtures. How to estimate all these cost terms is explained in detail in the following sections.

14.6.1 Material Cost (C_{mat})

The material cost is frequently the most crucial element in the overall cost of a machined component, and it frequently implies more than half of the overall cost. Material cost is straightforwardly calculated as

$$C_{\mathrm{mat}}(\$) = C_{\mathrm{rm}} \times W_{\mathrm{p}}$$

where C_{rm} is the raw material cost in \$/kg, and W_p is the weight of the preform to be machined. Table 14.4 lists several materials' densities and typical costs in dollars per kilogram in their most common basic shapes. The statistics in Table 14.4 can be used as a general reference, but process planners can get more precise quotations from material suppliers.

14.6.2 Setup Cost (C_{setup})

In machining, the main setup cost is related to the preparation time of machine tools and cutting tools (e.g., computer numerical control [CNC] programming, definition of coordinate systems,

Table 14.4 Densities and approximate costs in \$/kg for various metals under different raw shapes.[a]

		Price per kg (\$/kg)				
Material	**Density (kg/m^3)**	**Bar**	**Rod**	**Sheet**	**Plate**	**Tube**
Carbon steel	7,830	1.5–2.4	1.75–2.65	1.1–1.75	1.1–2.0	1.75–3.3
Stainless steel	7,750	5.5–11.0	5.5–11.0	3.3–6.6	5.50–9.90	5.5–11.0
Magnesium alloys	1,800	13–33	11–26	11–26	13–33	15–33
Aluminum Alloys	2,770	3.30–6.60	3.30–6.60	1.75–3.30	1.75–3.30	3.30–6.60
Brass	8,580	6.6–13.2	6.6–13.2	6.6–13.2	6.6–13.2	5.50–9.90
Titanium alloys	4,510	40–77	35–70	31–60	22–55	44–110

[a] Values from different databases and website resources.

Table 14.5 Setup time for different machine tools and setup time for each additional tool.

Machine tool	Basic setup time (h)	Additional setup per tool (h)
Manual turret lathe	1.2	0.2
CNC turret lathe	0.5	0.15
Milling machine	1.5	—
Drilling machine	1.0	—
Broaching machine	0.6	—
Grinding machine	0.6	—
CNC machining center	0.7	0.05

Source: Adapted from Boothroyd et al. (2002).

checking CNC programs, tools and fixture calibrations, etc.). Table 14.5 shows typical setup times for basic machine tools and cutting tool preparation (Boothroyd et al. 2002). To get the setup time per machined part, the overall setup time must be divided by the batch size. Therefore, the setup cost is defined as follows:

$$C_{\text{setup}}(\$) = \frac{\sum\left(t_{\text{setup}} \times R_{\text{setup}}\right)}{N_{\text{batch}}}$$

where N_{batch} is the number of parts per batch, R_{setup} the worker labor rate in setup operations ($/hour), and t_{setup} the time required for each setup operation (hour). Labor rates of $20–30/h can be considered as a reference value.

For CNC machines, CNC programs should be created and checked in the technical office with CNC simulators or computer-aided manufacturing (CAM) software. Although the time spent to create these programs highly depends on many factors (complexity of the part, production volume to optimize tool paths, etc.), a typical programming time of 2–5 hours can be a reasonable estimate. This cost is shared among all the parts produced under the same CNC program. Note that the worker labor rate can be notably increased for CAM operations in order to amortize the cost of software licenses.

14.6.3 Nonproductive Cost ($C_{\text{non-pr}}$)

Nonproductive costs are related to machine loading and unloading ($C_{\text{ld-uld}}$), handling parts between machines (C_{hn}), and nonmachining costs ($C_{\text{non-mach}}$). The nonproductive costs are estimated as

$$C_{\text{non-pr}}(\$) = C_{\text{ld-uld}} + C_{\text{hn}} + C_{\text{non-mach}}$$

Some guidelines to estimate these types of nonproductive costs are presented in the following.

14.6.3.1 Machine Loading and Unloading ($C_{\text{ld-uld}}$)

Every time a workpiece is inserted into (and then removed from) a machine tool, nonproductive costs are incurred. The cost associated with this activity depends on the time required and the labor cost, and it can be calculated as

$$C_{\text{ld-uld}}(\$) = t_{\text{ld-uld}} \times R_{\text{ld-uld}}$$

Table 14.6 Loading and unloading time (in seconds) versus workpiece weight.

Workholding	Workpiece weight			
	0–5 kg	5–14 kg	14–25 kg	Crane
Collet	15	21	—	—
Three-jaw chuck	23	32	53	—
Four-jaw chuck	41	50	71	—
Between centers with carrier	40	58	98	250
Clamp on table	34	40	59	270
Vice	19	25	40	175
Rotary table	36	45	73	280
V-blocks	30	36	78	365

Source: Adapted from Boothroyd et al. (2002).

where $t_{\text{ld-uld}}$ is the time of loading and unloading per part (hour), and $R_{\text{ld-uld}}$ is the labor rate for these activities ($/hour). Table 14.6 shows common loading and unloading times depending on the weight of the workpiece and the workholding system applied (Boothroyd et al. 2002).

14.6.3.2 Handling between Machines (C_{hn})

The cost of transporting batches of partially machined workpieces between machines should also be considered. At the shop floor, forklift trucks are usually employed to transfer stacks of pallets of workpieces across the facility. Depending on the type of forklift and its use, common cost rates are $2–5/h ($R_{\text{hn}}$). Knowing the distance (l_{travel}) and the speed of the truck (v_{travel}, common speeds during transportation are 8–10 km/h), the travel time and its associated cost are estimated. Thus, the cost of handling between machines is defined as

$$C_{hn}(\$) = R_{\text{hn}} \times \frac{l_{\text{travel}}}{v_{\text{travel}}}$$

14.6.3.3 Nonmachining Movements ($C_{\text{non-mach}}$)

Machining and nonmachining movements should be estimated for cost purposes. Nonmachining movements are included in the nonproductive cost. These movements are related to cutting engagement, tool position, withdrawal movements, etc. The nonmachining costs can be estimated as

$$C_{\text{non-mach}}(\$) = C_{\text{mr}} \times T_{\text{non-mach}}$$

where $T_{\text{non-mach}}$ is the time related to nonmachining movements during cutting tool trajectory (hour), and C_{mr} is the machine's shop rate ($/hour). Table 14.7 lists some time components for these nonmachining movements. For most machining shop floors, the machine's shop rate ranges $60–90/h.

Table 14.7 Time to tool's engage/disengage and tool change time (nonproductive times).

Machine tool	Time to tool's engage and disengage (s)	Tool change time (s)
Manual turret lathe	9	25
CNC turret lathe	1.5	1.5–2.5
Milling machine	30	120
Drilling machine	9	20 (keyed-drill chuck)
Broaching machine	13	—
Grinding machine	19	—
CNC machining center	8	1 (multiple arms) 2–5 (swing arm) 45 (manual)

Source: Boothroyd et al. (2002), Polgar (1996), and Stephenson and Agapiou (2016).

14.6.4 Machining Cost (C_{mach})

The machining cost depends on the time the machine tool is working and the machine's shop rate. Machining time per part is evaluated as

$$T_{\text{mach}}(\text{hour}) = \frac{1}{60} \times \frac{L}{f_n \times N}$$

where L is the length to be machined (mm), f_n is the feed rate (mm/rev), and N is the spindle speed (rpm). The spindle speed and the cutting speed are related through the equation:

$$N(\text{rpm}) = \frac{V_c \times 1{,}000}{\times D}$$

where V_c is the cutting speed (m/min) and D refers to the diameter of the cutting tool (mm) in milling/drilling operations or to the workpiece cutting diameter in turning operations. Knowing the machining time, the machining cost is defined as

$$C_{\text{mach}}(\$) = C_{\text{mr}} \times T_{\text{mach}}$$

where C_{mr} is the machine's shop rate ($/hour). Typical rates are $60–90/h.

14.6.5 Tooling Costs ($C_{tooling}$)

In machining, tooling cost is related to both cutting tools and workholding/fixturing equipment. Considering both terms, the tooling costs are defined as follows:

$$C_{\text{tooling}}(\$) = C_{\text{cutting-tools}} + C_{\text{fixturing}}$$

The specific expressions to estimate both terms are discussed in the following.

14.6.5.1 Cutting Tool Costs ($C_{\text{cutting-tools}}$)

Cutting tool costs include the costs of the cutters due to the natural wearing out process (C_{cutter}), and the cost of tool setting every time a cutting tool replacement is required (C_{setting}). Therefore, the cost related to cutting tools per machined part is calculated as follows:

$$C_{\text{cutting-tools}}(\$) = C_{\text{cutter}} + C_{\text{setting}}$$

The cutter cost depends on the type of tool. On the one hand, for cutting tools based on cutting inserts, the insert costs are calculated as follows:

$$C_{\text{cutter}}(\$) = \left[C_{\text{insert-edge}} \times \frac{T_{\text{m}}}{T_{\text{life}}}\right] \times z_{\text{inserts}}$$

where $C_{\text{insert-edge}}$ is the cost per cutting edge of the tool, z_{inserts} is the number of inserts in the indexable cutter, T_{m} is the machining time to machine a part, and T_{life} is the tool's life (15 and 20 minutes are common tool life values for cutting tools). The cost per cutting edge of the tool is defined as follows:

$$C_{\text{insert-edge}}(\$/\text{edge}) = \frac{C_{\text{insert}}}{n_{\text{edges}}}$$

where C_{insert} is the total cost of the insert, and n_{edges} is the number of available cutting edges, which depends on the geometry of the insert and whether the insert is reversible or not. Table 14.8 shows common insert geometries and their corresponding number of cutting edges.

On the other hand, for regrindable tools such as high-speed steel (HSS) tools (Figure 14.6), the cost of regrinding and their uses must also be included. In those cases, the cost per cutter is calculated as (Juneja 2003):

$$C_{\text{cutter}}(\$) = \left[\frac{C_{\text{solid}}}{n_{\text{reground}}} + C_{\text{regrinding}}\right] \times \frac{T_{\text{m}}}{T_{\text{life}}}$$

where C_{solid} is the cost of the solid cutter, n_{reground} is the number of times the tool can be reground, and $C_{\text{regrinding}}$ is the cost of regrinding. Typically, most of the tools can be reground at different times (e.g., most of the end mills can be reground at least three times), and the cost of re-sharpening a tool is about 30% of the cost of the new one.

The cost of the insert or the solid cutter depends on the cutting tool material, shape, type of coating, etc. Table 14.9 shows different cutting tools and purchase prices from Hoffmann GmbH Quality Tools, a worldwide supplier of tools and manufacturing equipment.

Another important cost related to the cutting tools is the setting cost. For instance, cutting inserts have to be loosened, repositioned by bringing another cutting edge into the cutting position, and then clamping the screw tight. This is conducted by a machining operator, and its associated cost is referred to as setting cost. The tool setting cost is estimated (Sandvik Coromant 2017) as follows:

$$C_{\text{setting}}(\$) = R_{\text{setting}} \times \frac{T_m}{T_{\text{life}}} \times T_{\text{setting}}$$

where R_{setting} is the machine's shop rate (\$/hour) for this operation, T_{m} is the machining cutting time (minutes), T_{life} is the cutting tool's life (minutes), and T_{setting} is the time required for setting the

Table 14.8 Number of edges for different insert shapes. Reversible and nonreversible inserts.

Insert shape	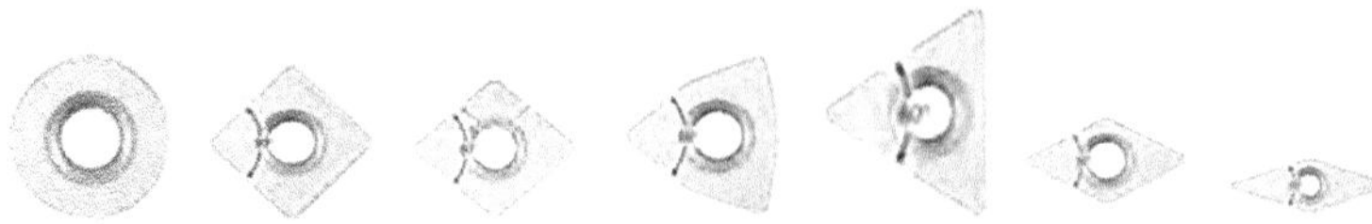						
ISO (first letter)	R	S	C	W	T	D	V
N° of edges (reversible)	8[a]	8	4	6	6	4	4
N° of edges (nonreversible)	4[a]	4	2	3	3	2	2

[a] Depending on a_p.
Source: Courtesy of Sandvik Coromant (2017).

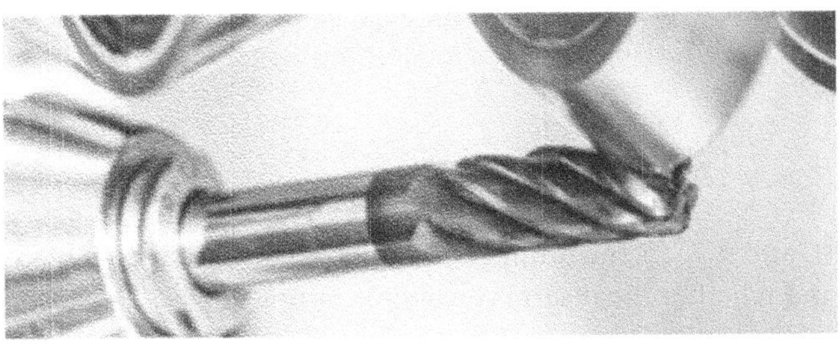

Figure 14.6 Regrinding operation. *Source:* Courtesy of Hoffmann GmbH Quality Tools.

Table 14.9 Cost of common cutting inserts, tools, and tool holders.[a]

Tool type	**Tool type and size (mm)**	**Cost ($)**	**Tool type and size (mm)**	**Cost ($)**
Inserts	Coated carbides	5–11	Ceramic inserts	15–25
	PCD inserts	80–150	CBN inserts	40–90
Solid tools (milling)	HSS Ø10, $z = 3$ uncoated	15	HSS Ø10, $z = 3$ coated TiAlN	37
	Carbide Ø10, $z = 2$, coated TiAlN	41	Carbide Ø20, $z = 2$, coated TiAlN	165
Drills	Twist drill HSS Ø10, Type N uncoated	5.5	Twist drill HSS Ø10, Type N coated TiN	6.5
	Twist drill carbide Ø10, Type N uncoated	44	Twist drill carbide Ø10, Type N coated TiN	51
Tool holders (milling)	Face milling cutter 45°, Ø63, $z = 5$	437	Shoulder mill 90°, Ø20, $z = 2$	142
Tool holders (turning)	DDJN 93°tool holder, inserts 15 mm	85	A.PDUN 93°, boring tool holder, inserts 15 mm	185
Tool holders (drilling)	Tool holder DIN 238, tools 0–10 mm	82	—	—

[a] The value of z refers to the number of teeth.
Source: Data from vendor Hoffmann GmbH Quality Tools (https://www.hoffmann-group.com).

tool (hour). Setting time can range from 5 minutes for manual machine tools to 1 minute for CNC lathes and machining centers.

14.6.5.2 Fixturing Cost ($C_{fixturing}$)

Workholding devices were reviewed in Chapter 7, and they can be basically grouped into three main categories: permanent fixturing, modular fixturing, and general-purpose workholding devices. The estimation of fixturing cost is based on the following expression:

$$C_{\text{fixturing}}(\$) = C_{\text{run}} + \frac{C_{\text{setup}}}{N_{\text{batch}}} + \frac{C_{\text{ini}}}{N_{\text{life}}}$$

The initial fixturing cost (C_{ini}) is estimated by calculating the design expenses and the cost of all fixture components, including the cost of assembling. This cost is shared among all the expected parts produced under this workholding/fixturing device (N_{life}). Additional costs are related to the setup process (C_{setup}) for every batch operation such as inspection, calibration, cleaning, and greasing operations. This cost is shared among the parts expected to be produced in a batch (N_{batch}). Finally, the workholding/fixturing device may present a running cost (C_{run}) related to loading and unloading activities (previously included in nonproductive costs, but they can instead be added here), power actuators, etc. Very often, the running cost includes the machining and nonproductive costs since the workholding selection may influence cutting tool trajectories, cutting parameters, etc.

14.6.6 Practical Example

Analyze the machining cost per part to manufacture a batch of 500 cylindrical parts in a CNC lathe. The operations required are a face turning (cutting length of 15 mm) and a straight turning operation (cutting length of 40 mm). The part is loaded onto a 3-jaw chuck, and the same cutting tool is used for both turning operations. The workpiece is a low carbon steel cylindrical bar of Ø30- and 60-mm in length. Assume a cutting speed of 150 m/min, a feed rate of 0.2 mm/rev for straight turning, and a feed rate of 100 mm/min for face turning. The cutting tool is a square, nonreversible plain carbide insert. Assume any reasonable value for any additional information needed for cost estimation.

The final machining cost is estimated by the following expression:

$$C_{\text{tmach}} = C_{\text{mat}} + C_{\text{setup}} + C_{\text{non-pr}} + C_{\text{mach}} + C_{\text{tooling}}$$

Each of the cost terms shown above is calculated as follows:

- Material cost (C_{mat}).
  The material cost is

$$C_{\text{mat}} = C_{\text{rm}} \times W_{\text{p}} = 2.0 \times 0.33 = \$0.66$$

Assuming the low carbon steel price of \$2/kg. The weight of the part is calculated as follows:

$$W_{\text{p}} = 7.83 \times 10^3 \times \left(\quad \times \frac{0.03^2}{4} \times 0.06 \right) = 0.33\,\text{kg}$$

- Setup cost (C_{setup}).
  According to Table 14.5, machine tool and cutting tool setup times are estimated as 0.5 and 0.15 hours, respectively. Additionally, the part is very simple, so a CNC programming of 15 minutes is added as a reasonable setup time. Assuming a working labor rate (R_{setup}) of \$40/h, the setup cost is as follows:

$$C_{setup} = \frac{\Sigma\left(t_{setup} \times R_{setup}\right)}{N_{batch}} = \frac{\left(0.5+0.15+0.25\right)\times 40}{500} = \$0.072$$

- Nonproductive cost ($C_{non\text{-}pr}$).
  The nonproductive cost is estimated as

$$C_{non-pr} = C_{ld-uld} + C_{hn} + C_{non-mach}$$

where

- Machine loading and unloading:
  According to Table 14.6, the loading and unloading time for a part weight of 0.33 kg and a 3-jaw chuck workholding is 23 seconds. Thus, the cost of handling the part in the CNC lathe is

$$C_{ld-uld} = t_{ld-uld} \times R_{ld-uld} = \frac{23}{3{,}600}\times 30 = \$0.19$$

assuming a labor rate of \$30/hour.

- Handling between machines:
  There is no other machining station, so this cost is 0.

$$C_{hn} = \$0$$

- Nonmachining movements:
  The nonmachining time is assumed to be 1.5 seconds to engage the tool per operation (Table 14.7). Thus, the nonmachining cost is

$$C_{non-mach} = C_{mr} \times T_{non-mach} = 70\times\frac{1.5\times 2}{3{,}600} = \$0.058$$

assuming machine's shop rate of \$70/hour and 1.5 seconds for the engage time in both machining operations.

As a result, the nonproductive cost is

$$C_{non-pr} = C_{ld-uld} + C_{hn} + C_{non-mach} = 0.19+0+0.058 = \$0.25$$

- Machining cost (C_{mach})
  The machining time per part for straight turning is

$$T_{mach1} = \frac{L}{f_n \times N} = \frac{40}{0.2\times 1{,}591} = 0.126\,\text{minute}$$

since the spindle speed is

$$N = \frac{1{,}000 \times V_c}{\times D} = \frac{(1{,}000 \times 150)}{\times 30} = 1{,}591\,\text{rpm}$$

For the face turning operation, the machining time is

$$T_{\text{mach}2} = \frac{L}{f_n \times N} = \frac{15}{100} = 0.15\,\text{minute}$$

Thus, the machining cost is

$$C_{\text{mach}} = C_{\text{mr}} \times \Sigma T_{\text{mach}} = 70 \times \frac{0.15 + 0.126}{60} = \$0.322$$

assuming a shop-floor rate of $70/hour.

- Tooling cost (C_{tooling})
  The tooling cost is estimated as

$$C_{\text{tooling}} = C_{\text{cutting-tools}} + C_{\text{fixturing}}$$

where

– Cutting tool costs:
Cutting tool costs include cutter and setting costs:

$$C_{\text{cutting-tools}} = C_{\text{cutter}} + C_{\text{setting}}$$

The cutting tool is a square nonreversible plain carbide insert; thus, the cutter cost is

$$C_{\text{cutter}} = \left[C_{\text{insert-edge}} \times \frac{T_{\text{m}}}{T_{\text{life}}}\right] \times z_{\text{inserts}} = \left[1.5 \times \frac{0.15 + 0.126}{15}\right] \times 1 = \$0.028$$

assuming 15-minutes tool's life and the insert cost per edge is

$$C_{\text{insert-edge}} = \frac{C_{\text{insert}}}{n_{\text{edges}}} = \frac{6}{4} = \$1.5\,/\,\text{edge}$$

Note that the inserts are square and nonreversible (four cutting edges), and the cost of each insert is assumed to be $6 (Table 14.9).

The tool setting cost is

$$C_{\text{setting}} = R_{\text{setting}} \times \frac{T_{\text{m}}}{T_{\text{life}}} \times T_{\text{setting}} = 20 \times \frac{0.15 + 0.126}{15} \times \frac{2}{60} = \$0.012$$

assuming $20/hour rate for setting and 2 minutes time for cutting tool setting.

As a result, the cutting tool cost per part is

$$C_{\text{cutting-tools}} = C_{\text{cutter}} + C_{\text{setting}} = 0.028 + 0.012 = \$0.040$$

- Fixturing costs:

  In relation to fixturing cost, the setup is assumed to be already included in the machine tool setup cost (C_{setup}), so only the depreciation of the chuck during machining is included. Although hard to estimate, it is assumed that 50,000 parts can be manufactured within the chuck's working life, and a 3-jaw chuck cost is assumed to be $1,100. Therefore,

$$C_{fixturing} = \frac{C_{ini}}{N_{life}} = \frac{1,100}{50,000} = \$0.022$$

As a result, the total tooling cost is

$$C_{tooling} = C_{cutting-tools} + C_{fixturing} = 0.040 + 0.022 = \$0.062$$

Finally, the total machining cost per part is obtained by adding all previous terms

$$C_{tmach} = C_{mat} + C_{setup} + C_{non-pr} + C_{mach} + C_{tooling}$$

$$C_{tmach} = 0.66 + 0.072 + 0.25 + 0.322 + 0.062 = \$1.37$$

14.7 3D Printing Cost Estimation (C_{3Dpr})

To estimate the cost of 3D printed parts, three main terms should be evaluated: material cost (C_{mat}), operating cost (C_{op}), and postprocessing cost (C_{post}). Therefore, the 3D printing cost is estimated as

$$C_{3Dpr}(\$) = C_{mat} + C_{op} + C_{post}$$

Each cost from the previous equation can be estimated as discussed in the following.

14.7.1 Material Cost (C_{mat})

The material cost highly depends on the 3D printing technique applied. For instance, in fused deposition modeling (FDM), also named as fused filament fabrication (FFF), the cost of common materials such as Polylactic acid (PLA), Acrylonitrile butadiene styrene (ABS), and Polyethylene terephthalate glycol (PETG) is around \$20–35/kg. However, for metallic filaments composed of 90% by weight of sinterable metallic powders and 10% binder polymers, the cost may range \$150–400/kg. When selective laser sintering (SLS) technology is applied, materials are in the form of powder, which notably increases their cost with respect to their filament equivalents. A sintered powder of polyamide (PA12) costs around \$120/kg to \$170/kg, which means 20–30 times higher than the same material in the form of pellets for plastic injection molding and 2.5 times higher than the material in filament form for FDM. Other technologies, such as stereolithography (SLA), use photopolymer resins with a cost of around \$50–200/kg or higher, depending on the type of resins. More advanced technologies, such as direct melting laser sintering (DMLS), uses metal powders that may cost \$100–500/kg, depending on the metal alloy.

The material cost in 3D printing is straightforwardly estimated given the 3D part design. The software used for slicing and creating the 3D printing layers provides information about the kilograms of material needed to build the part, including supports if required. Therefore, the material cost is calculated as follows:

$$C_{mat}(\$) = C_{rm} \times W_p$$

Table 14.10 Material cost for different materials and technologies in 3D printing.[a]

Technology	Material	Cost ($/kg)	Material	Cost ($/kg)
FDM	PLA	20	TPU	55
	ABS	22	PA12	70
	PETG	26	ASA	33
	316L (90% metal, 10% binder)	180	Inconel 718 (86% metal, 14% PLA)	600
SLS	PA12	170	TPU	190
SLA	Standard resin	35	Biocompatible resin for orthodontics	350
	Rigid resin	210		
DMLS	316L	100	Inconel 718	300

[a] Data from different vendors.

where C_{rm} is the raw material cost of the filament per kilogram (Table 14.10), and W_p is the weight of the printed part. If the weight of the printed part is not obtained from the software that creates the layers for printing, it can be estimated considering the volume of the part and the density of the material. However, in this case, it should be noted that 3D printing parts are printed under an infill parameter that will define the resulting density of the part. The required infill for the part should be estimated to determine the total weight of the part after printing. Common infill values are 15–30% for parts that do not require mechanical strength.

14.7.2 Operating Cost (C_{op})

The operating cost is the result of adding the costs related to manpower (C_{man}), electricity (C_{elec}), and the cost of the equipment (C_{eq}). The operating cost is then defined as follows:

$$C_{op}(\$) = C_{man} + C_{elec} + C_{eq}$$

Each of these costs is estimated as discussed in the following.

14.7.2.1 Cost of manpower (C_{man})

The use of software for slicing the part and the cost of preparing the printer and setting the correct printing values should be considered here. The time required for preparing the 3D printer and the slicing process may vary depending on the technology used, but in general the time is not high, and 0.5–2 hours may be reasonable. The 3D printing process is automated, so the time of printing does not require the operator's action. The cost of manpower can be defined as follows:

$$C_{man}(\$) = R_{3Dpr} \times T_{setup}$$

where T_{setup} is the time for preparing the 3D printer and the slicing process (hour), and R_{3Dpr} is the labor rate ($/hour). A labor rate of $20–30/h can be assumed for personnel working on the 3D printers, although these figures can increase notably depending on the 3D printer type and technology used. Note that this cost related to setup is shared within the production volume if more than one part is printed.

14.7.2.2 Electricity Cost (C_{elec})

The electrical power consumption for most 3D printers of FDM technology ranges from 0.1 to 0.5 kW. Other technologies based on power lasers, such as SLS technologies, can range from 1 to 5 kW, especially for large industrial printers (Gebhardt and Hötter 2016). When dealing with metal additive manufacturing, technologies such as DMLS can reach up to 20 kW of power consumption in the case of large-scale printers.

In general, the electricity cost is very low in 3D printing technologies based on polymers, and it is often overlooked. For instance, given a common electricity cost of \$0.16/kWh with a printer of 500 W, the cost per hour is \$0.08. The electricity cost for any technology can be derived as follows:

$$C_{\text{elec}}(\$) = R_{\text{elec}} \times P_{\text{print}} \times T_{\text{print}}$$

where R_{elec} is the electricity cost per kWh (e.g., \$0.16/kWh), P_{print} is the power consumption of the 3D printer in kW, and T_{print} is the printing time (hour).

14.7.2.3 Cost of the equipment (C_{eq})

When calculating the cost of using any equipment, the most common practice is to use amortization or depreciation costs. The way to calculate amortization is to divide the price of the equipment (i.e., the 3D printer) by the number of hours that the machine is going to be used until it is amortized. This amortization cost is then added to the total manufacturing cost to consider the equipment cost due to the natural "wearing out" process. Therefore, the cost of the equipment as the amortization cost can be written as follows:

$$C_{\text{eq}}(\$) = \frac{P_{\text{3Dpr}}}{T_{\text{total}}} \times T_{\text{print}}$$

where P_{3Dpr} is the purchase cost of the 3D printing equipment, T_{total} is the total time (hour) that the equipment is expected to be used along its lifecycle, and T_{print} is the printing time (hour) required for the 3D printing part under evaluation.

14.7.3 Postprocessing Cost (C_{post})

Postprocessing of 3D printed parts is a labor-intensive operation, and it is almost required for any 3D printed part. Costs related to polishing, painting, removal of supports or excess powder, ultraviolet (UV) curing, etc., are included here. In many cases the cost of postprocessing may be higher than the rest of the total costs. A National Institute of Standards and Technology (NIST) study found that postprocessing expenditures account for 4–13% of overall production costs in 3D printing parts (Thomas and Gilbert 2014). To estimate this cost, consider the hourly rate and the time required for postprocessing:

$$C_{\text{post}}(\$) = T_{\text{post}} \times R_{\text{post}}$$

where T_{post} is the total postprocessing time (hour) and R_{post} the worker labor rate (\$/hour). In an early estimation cost, a 10% increase in the overall production cost is reasonable.

Note that additional costs may be included if metal 3D printing parts are considered. For instance, metal FDM parts require a debinding and sintering process after printing, which increases the cost notably. Metallic filament vendors offer these services at approximately \$80–100/kg.

14.7.4 Practical Example

Analyze the 3D printing cost for a part made of an ABS part printed using the FDM technology. The information from the slicing system gives a material weight of 150 g and a printing time of 16 hours 30 minutes. After printing, a postprocessing operation is required to remove supports and improve surface roughness, which is estimated to last 1 hour. Estimate the manufacturing cost and assume reasonable values for labor rates and other needed terms.

The 3D printing cost of the part is estimated by the following equation:

$$C_{3\mathrm{Dpr}} = C_{\mathrm{mat}} + C_{\mathrm{op}} + C_{\mathrm{post}}$$

where the three cost terms are calculated as follows:

- Material cost.

  The material cost of ABS filament, according to Table 14.10, is \$22/kg. Then, the material cost is

$$C_{\mathrm{mat}} = C_{\mathrm{rm}} \times W_{\mathrm{p}} = 22 \times 0.15 = \$3.3$$

- Operating cost.

  The operating cost is estimated as

$$C_{\mathrm{op}} = C_{\mathrm{man}} + C_{\mathrm{elec}} + C_{\mathrm{eq}}$$

where these three terms are calculated as follows.

- Cost of manpower.

  The manpower cost related to the setup process is estimated. A fixed 30-minute setup time (printer setup and slicing process) and a labor rate of \$20/hour are assumed reasonable for polymer FDM printers. Therefore,

$$C_{\mathrm{man}} = R_{3\mathrm{Dpr}} \times T_{\mathrm{setup}} = 20 \times 0.5 = \$10$$

- Electricity cost.

  The electricity cost is evaluated knowing the printing time, the 3D printing power consumption and the electricity cost. Assuming a middle-range printer for polymer FDM technology, 400 W can be a reasonable power consumption with an electricity cost of \$0.15/kWh. Therefore,

$$C_{\mathrm{elec}} = R_{\mathrm{elec}} \times P_{\mathrm{print}} \times T_{\mathrm{print}} = 0.15 \times 0.4 \times 16.5 = \$0.99$$

- Equipment cost.

  The equipment cost is estimated as:

$$C_{\mathrm{eq}} = \frac{P_{3\mathrm{Dpr}}}{T_{\mathrm{total}}} \times T_{\mathrm{print}}$$

If a middle-range printer for polymer FDM technology is assumed, a purchase cost of around $5,000 can be reasonable. Assuming a useful lifespan of 7 years and an operating time per year of around 1,000 hours, the amortization cost is

$$C_{eq} = \frac{5{,}000}{7 \times 1{,}000} \times 16.5 = \$11.8$$

As a result, the operation cost is

$$C_{op} = C_{man} + C_{elec} + C_{eq} = 10 + 0.99 + 11.8 = \$22.79$$

- Postprocessing cost.
  As postprocessing cost, 1 hour of postprocessing at a labor rate of $20/hour is considered. Therefore,

$$C_{post} = T_{post} \times R_{post} = 1 \times 20 = \$20$$

Finally, the total 3D printing cost is estimated as:

$$C_{3Dpr} = C_{mat} + C_{op} + C_{post} = 3.3 + 22.79 + 20 = \$46.09$$

14.8 Computer-Aided Cost Estimators

Many different software can help designers with an early estimation of manufacturing costs. These tools let designers modify and check product designs in a very efficient way. For instance, most of CAD software have integrated tools to estimate machining costs, casting costs, sheet metal costs, 3D printing costs, etc. The estimation is based on templates that can be easily edited to include shop-floor characteristics, including labor rates, overhead rates, plant efficiency, etc. Figure 14.7 shows an example of cost estimation using SolidWorks Costing for machining parts. There are other specialized cost-estimating software packages designed to handle comprehensive manufacturing cost calculations. Examples include aPriori, Costimator, DFM Concurrent Costing, CostWorks, etc.

14.9 Web-Based Tools for Cost Estimation

Nowadays, many websites from the industry obtain good manufacturing cost estimations using some basic geometry data, labor cost, and some other information from the shop floor. Although many of the equations that evaluate the final cost are not visible to the user, their results can be used as a good starting point for early cost estimations in different product designs and process plans. In fact, many of these websites use similar mathematical relationships to the ones shown in this chapter. Figure 14.8 shows an example of a popular website for rapid cost estimation in plastic injection molding, die casting, sand-casting, and machining processes. Other popular websites are listed in the following:

- Die casting cost estimation (price calculator) for die castings (China Sourcing): http://www.sourcing-cn.com/diecasting/
- Cost estimation for welding processes (ESAB company): https://esab.com/us/nam_en/support/tools/calculators/quick-weld-productivity-analyzer/

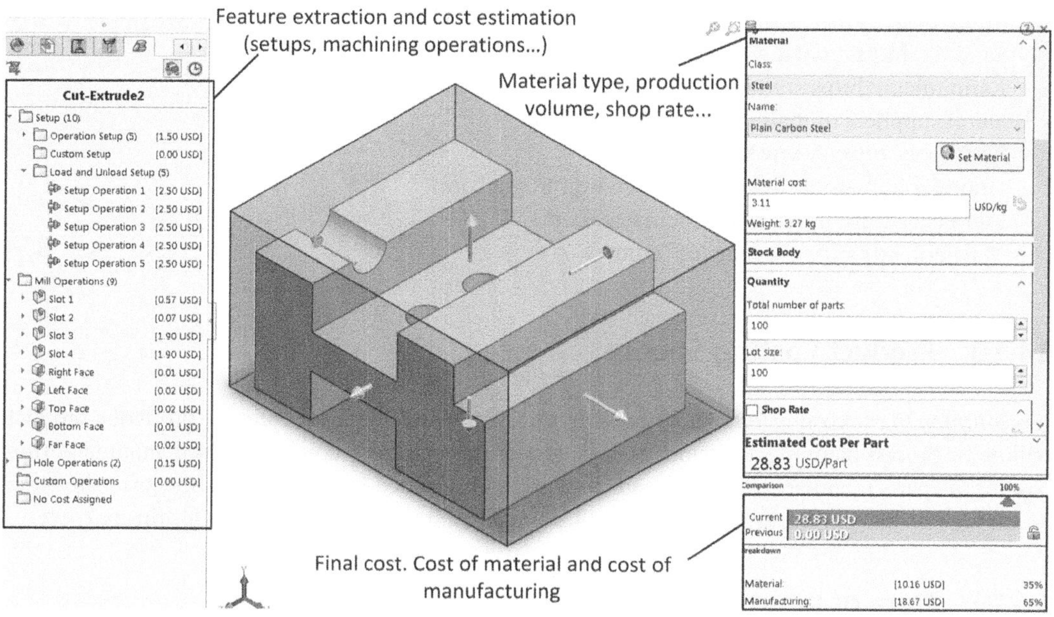

Figure 14.7 Example of cost estimation using SolidWorks Costing. *Source:* SolidWorks Costing (2024).

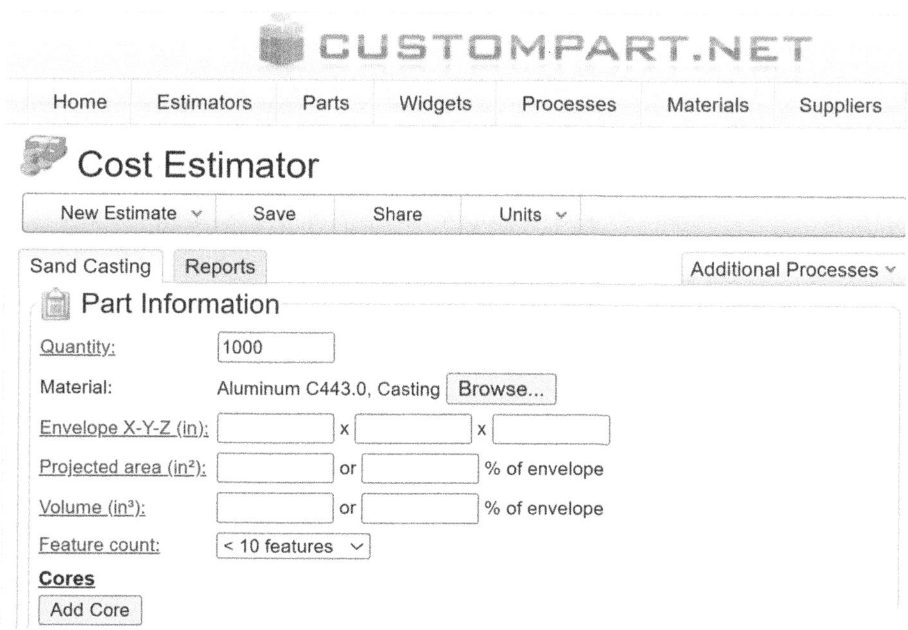

Figure 14.8 Online cost estimator website for plastic injection molding, die casting, sand-casting, and machining estimations. *Source:* CustomPartNet http://www.custompartnet.com/last accessed on June 28, 2024.

- Cost estimation for injection molding (BASF company): https://pmtools-na.basf.com/quickcost/

Besides specific websites for manufacturing cost estimation, many websites are available from vendors to estimate the costs of cutting tools, fixture devices, and any other component required for the machining process. Some interesting websites of worldwide vendors are as follows:

- Cutting tools for machining
  - Sandvik. https://www.sandvik.coromant.com
  - Kennametal. https://www.kennametal.com
- General suppliers of manufacturing equipment
  - KBC tools. http://www.kbctools.com/
  - Grainger industrial supply. http://www.grainger.com/
  - McMaster-Carr. http://www.mcmaster.com/
  - Maritool. https://www.maritool.com/

14.10 Product Costing Sheets

It is common to use costing templates or costing sheets to estimate the cost and the profit of a product within the process planning. Decisions about make-or-buy, source of bought material, manufacturing cost for the parts manufactured, overhead cost, etc., are indicated in these costing sheets to show the economic feasibility of the product. Figures 14.9 and 14.10 show simple templates for this purpose.

14.11 Data of Interest: Wage Statistics for Primary Metal Manufacturing Industries

Table 14.11 shows some salary wage values according to the US Bureau of Labor Statistics for primary metal manufacturing industries (US Bureau of Labor Statistics 2023a). The salary wage shown depends on the standard occupational classification (SOC) and the specific occupation.

Bill of materials

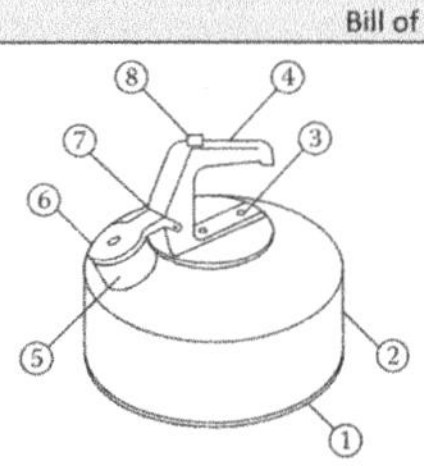

No.	Part name	Quantity
1	Base	1
2	Body	1
3	Screw	2
4	Handle	1
5	Spout	1
6	Upper lid part	1
7	Roll pin	1
8	Lid lever	1

Document description

Document	Manufacturing cost evaluation		
Version	1	Data	25-3-23
Product manager	J.V Abellán-Nebot		
Product name	Teakettle model 100		

Product stock code	Product description	Component stock code	Component description	Component material	Unit of measurement	Quantity	Yield	Make/ buy	Purchase price ($)	Source	Manufacturing cost ($)
		Bas-1	Base	3/64" sheet aluminium 3033 alloy (AL3003)	Units	1	100%	Make	-	-	0.15
		Bod-1	Body	3/64" AL3003	Units	1	100%	Make	-	-	0.18
		Scr-5	Machine screw	Low carbon plated steel	Units	2	95%	Buy	0.008	Lone star screw company (LSS)	-
Teak-100	Teakettle model 100	Han-3	Handle	Thermosetting phenolic compound mineral (Bakelite)	Units	1	100%	Buy	0.1	Plastic eng. corp (PEC)	-
		Spo-2	Spout	3/64" AL3003	Units	1	95%	Make	-	-	0.04
		UppL-1	Upper lid part	Bakelite	Units	1	100%	Buy	0.01	PEC	-
		Roll-2	Roll pin	Stainless steel spring	Units	1	90%	Buy	0.006	Schindre's machine works	-
		LidL-1	Lid lever	Bakelite	Units	1	100%	Buy	0.008	LSS	-
									0.132		0.37
									Total component cost		0.502

Figure 14.9 Template for estimating the cost components of a product. *Source:* Teakettle model adapted from Sule (2008).

Product stock code	Product description	Total cost				Selling price ($)	Profit ($)
		Production cost		Overheads			
		Manufacturing product cost ($)	Assembly product cost ($)	Office expenses ($)	Selling expenses ($)		
Teak-100	Teakettle model 100	0.502	0.2	0.32	0.1	1.5	0.378

Figure 14.10 Template for estimating the final profit of a product according to materials, manufacturing, and overhead costs.

Table 14.11 Salary wage for different occupations in primary metal manufacturing industries.

Occupation	Salary wage ($/h)	Occupations	Salary wage ($/h)
Managers (SOC 11-0000)			
Operations manager	67	Facilities managers	53
Marketing managers	66	Industrial production managers	53
Sales manager	69	Purchasing managers	56
Human resources managers	58	Engineering managers	64
Engineering occupations (SOC 17-0000)			
Industrial engineers	41	Engineering technologists and technicians	30
Electrical engineers	44	Materials engineers	41
Mechanical engineers	40	Mechanical drafters	25
Office and administrative support occupations (SOC 43-0000)			
Shipping, receiving clerks	21	Administrative assistants	21
Maintenance and repair occupations (SOC 49-0000)			
Maintenance workers	26	Electrical repairers	30
Production occupations (SOC 51-0000)			
Supervisors of production	32	Inspectors, samplers	22
Machine tool operators	21	Production workers (general)	21
Metal workers	20	CNC operators	21
Grinding, lapping, polishing operators	20	Tool and die makers	24
Welders, solders, brazers	24	Chemical equipment operators	26
Coating, painting, spraying operators	19	Packaging and filling operators	15
Assemblers and fabricators	19	Foundry operators	20
Transportation and material moving occupations (SOC 53-0000)			
Motor vehicle operators	21	Crane and tower operators	22

Source: US Bureau of Labor Statistics (2023a).

Questions and Exercises

1 Explain the differences between the production cost and the prime cost.

2 The total cost of a part is estimated as $55. What is its tentative selling price if a profit margin of 10% is desired?

3 For the previous exercise, if a distributor with a 60% markup is used, what is the final selling price to the costumer?

4 In a company, the prime cost of manufacturing a component is $11. Assuming typical figures of cost in manufacturing companies (refer to Figure 14.2), estimate the total cost and a reasonable selling price of the product.

5 Indicate in which situations a buy decision is preferred:
- Standard components such as bolts, screws, springs. . . .
- A component that can be made in a company with a similar cost of a bought component.
- A component that requires specific machinery but cannot be bought (not a standard component).
- A component that is critical for product performance and provides a competitive advantage with respect to other competitors.

6 Calculate the manufacturing cost of the part shown in Figure 14.11 produced by sand casting. Consider the following process characteristics: number of parts to be cast is 1,000; gray cast iron material; 3 mm machining allowance; manual casting process; a production rate of 12 molds/hour with 1 worker in the molding line; a pattern plate of 30 × 30 cm (plate made of wood). Assume reasonable values for any additional information needed.

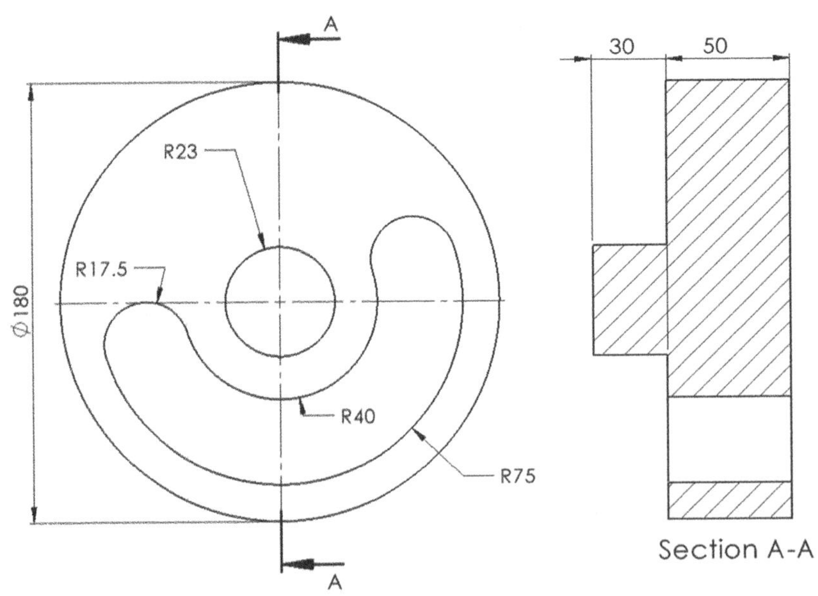

Figure 14.11 Casting part for Exercise 6. Units in mm.

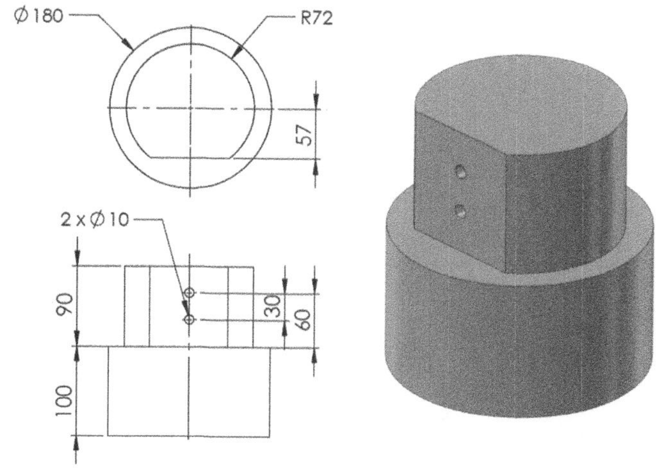

Figure 14.12 Casting part for Exercise 7. Units in mm. *Note:* Hole depth is 20 mm.

7 Calculate the manufacturing cost of the part shown in Figure 14.12 produced by machining in lots of 1,500 parts. Consider the following process characteristics: raw material in the form of cylindrical bars of Ø185 mm and 200 mm in length; low carbon steel as raw material; CNC turning and CNC machining center with a labor rate of $50/hour and $75/hour, respectively. The process planning is

- Job 1. CNC lathe
  - Subjob 1.
- OP 1. Face turning. Tool T1, nonreversible coated carbide tool, ISO shape C. Cutting conditions: 140 m/min, 0.2 mm/rev, 2.5 mm depth of cut, two passes.
- OP 2. Straight turning. Tool T2, nonreversible coated carbide tool, ISO shape S. Cutting conditions: 170 m/min, 0.15 mm/rev, 2.5 mm depth of cut, one pass.
  - Subjob 2.
- OP 1. Face turning. Tool T1, nonreversible coated carbide tool, ISO shape C. Cutting conditions: 140 m/min, 0.2 mm/rev, 2.5 mm depth of cut, two passes.
- OP 2. Straight turning. Tool T2, nonreversible coated carbide tool, ISO shape S. Cutting conditions: 170 m/min, 0.15 mm/rev, 5.125 mm depth of cut, four passes.
- Job 2. CNC machining center.
  - Subjob 1.
- OP 1. End milling. Tool T3, uncoated HSS flat end mill of 10 mm diameter (four teeth). Cutting conditions: 60 m/min, 0.03 mm/tooth, 5 mm axial depth of cut and 7.5 mm radial depth of cut, three axial passes.
- OP 2. Drilling. Tool T4, uncoated twist drill of 10 mm diameter. Cutting conditions: 25 m/min, 0.05 mm/rev.

Note: all cutting tools are expected to last 15 minutes. Assume reasonable values for any additional information needed (tool cost, load/unload times, etc.).

8 Calculate the manufacturing cost of the part shown in Figure 14.13 produced by 3D printing. Consider the following process characteristics: ABS material; total weight per part 78 g; acetone postprocessing bath for 15 minutes; printing time 10 hours 15 minutes; setup time 15 minutes. Consider labor rate $30/h; unattended printing process where electricity cost is $0.15/kWh; ABS cost material $20/kg.

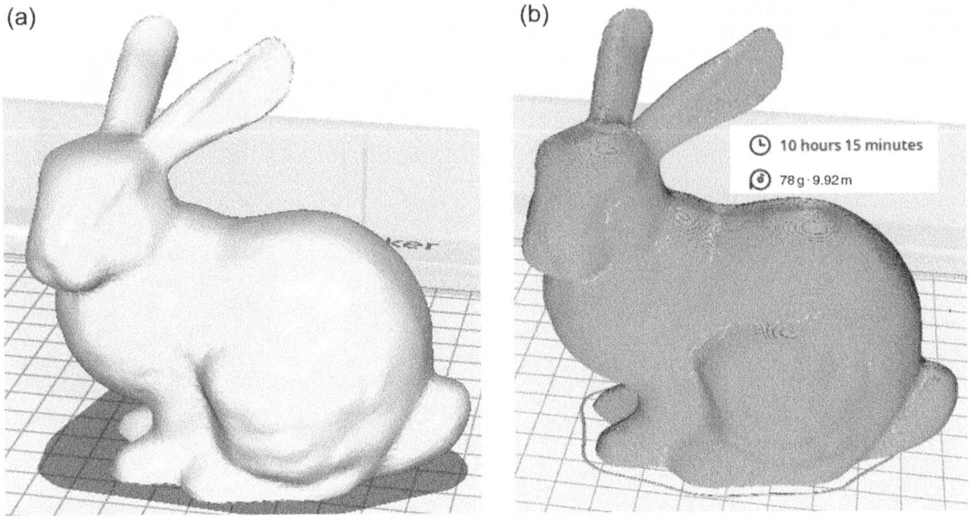

Figure 14.13 Casting part for Exercise 8. Part (a) and printing time, grams, and layers in Cura software (b).

References

Boothroyd, G., Dewhurst, P., and Knight, W. (2002). *Product Design for Manufacture and Assembly*. New York: Marcel Dekker.

Chougule, R.G., and Ravi, B. (2006). Casting cost estimation in an integrated product and process design environment. *International Journal of Computer Integrated Manufacturing* 19 (7): 676–688.

Gebhardt, A., and Hötter, J.S. (2016). *Additive Manufacturing: 3D Printing for Prototyping and Manufacturing*. Carl Hanser Verlag GmbH Co KG.

Juneja, B.L. (2003). *Fundamentals of Metal Cutting and Machine Tools*. New Delhi, India: New Age International.

Polgar, K.C. (1996). Simplified Time Estimation Booklet for Basic Machining Operations. Massachusetts Institute of Technology (Doctoral dissertation, MS Thesis. *Department of Mechanical Engineering*, Cambridge, MA, USA).

Sandvik Coromant. (2017). *Training Handbook*. Metal Cutting Technology www.sandvik.coromant.com.

Scallan, P. (2003). *Process Planning: the Design/Manufacture Interface*. Butterworth-Heinemann.

SolidWorks Costing. (2024). SolidWorks costing overview. Internet source. https://www.cati.com/3d-cad/features/solidworks-costing/ (accessed February 2024).

Stephenson, D.A., and Agapiou, J.S. (2016). *Metal Cutting Theory and Practice*, 3e. Boca Raton, FL: CRC Press.

Sule, D.R. (2008). *Manufacturing Facilities: Location, Planning, and Design*. CRC Press.

Thomas, D.S., and Gilbert, S.W. (2014). *Costs and Cost Effectiveness of Additive Manufacturing*, vol. 1176, 12. NIST Special Publication.

US Bureau of Labor Statistics (2023a). Occupational employment and wage statistics. https://www.bls.gov/oes/current/naics4_331500.htm (accessed October 2023).

US Bureau of Labor Statistics (2023b). CPI inflator calculator. https://www.bls.gov/data/inflation_calculator.htm (accessed May 2023).

Appendix 1: ISO Dimensional Tolerances

Manufacturing Process Planning: A Practical Approach for Mechanical Engineering, First Edition.
José V. Abellán-Nebot, Carlos Vila Pastor, and Héctor R. Siller.

Companion website: www.wiley.com/go/Manufacturing_Process_Planning_1e

Table A1.1 Values of standard tolerances grades.

Nominal Size (mm)		Standard Tolerance Grades																			
		IT01	IT0	IT1	IT2	IT3	IT4	IT5	IT6	IT7	IT8	IT9	IT10	IT11	IT12	IT13	IT14	IT15	IT16	IT17	IT18
		Standard Tolerance Values																			
Above	Up to and including	µm													mm						
–	3	0.3	0.5	0.8	1.2	2	3	4	6	10	14	25	40	60	0.1	0.14	0.25	0.4	0.6	1	1.4
> 3	6	0.4	0.6	1	1.5	2.5	4	5	8	12	18	30	48	75	0.12	0.18	0.3	0.48	0.75	1.2	1.8
> 6	10	0.4	0.6	1	1.5	2.5	4	6	9	15	22	36	58	90	0.15	0.22	0.36	0.58	0.9	1.5	2.2
> 10	18	0.5	0.8	1.2	2	3	5	8	11	18	27	43	70	110	0.18	0.27	0.43	0.7	1.1	1.8	2.7
> 18	30	0.6	1	1.5	2.5	4	6	9	13	21	33	52	84	130	0.21	0.33	0.52	0.84	1.3	2.1	3.3
> 30	50	0.6	1	1.5	2.5	4	7	11	16	25	39	62	100	160	0.25	0.39	0.62	1	1.6	2.5	3.9
> 50	80	0.8	1.2	2	3	5	8	13	19	30	46	74	120	190	0.3	0.46	0.74	1.2	1.9	3	4.6
> 80	120	1	1.5	2.5	4	6	10	15	22	35	54	87	140	220	0.35	0.54	0.87	1.4	2.2	3.5	5.4
> 120	180	1.2	2	3.5	5	8	12	18	25	40	63	100	160	250	0.4	0.63	1.1	1.6	2.5	4	6.3
> 180	250	2	3	4.5	7	10	14	20	29	46	72	115	185	290	0.46	0.72	1.15	1.85	2.9	4.6	7.2
> 250	315	2.5	4	6	8	12	16	23	32	52	81	130	210	320	0.52	0.81	1.3	2.1	3.2	5.2	8.1
> 315	400	3	5	7	9	13	18	25	36	57	89	140	230	360	0.57	0.89	1.4	2.3	3.6	5.7	8.9
> 400	500	4	6	8	10	15	20	27	40	63	97	155	250	400	0.63	0.97	1.55	2.5	4	6.3	9.7
> 500	630			9	11	16	22	32	44	70	110	175	280	440	0.7	1.1	1.75	2.8	4.4	7	11
> 630	800			10	13	18	25	36	50	80	125	200	320	500	0.8	1.25	2	3.2	5	8	12.5
> 800	1,000			11	15	21	28	40	56	90	140	230	360	560	0.9	1.4	2.3	3.6	5.6	9	14
> 1,000	1,250			13	18	24	33	47	66	105	165	260	420	660	1.05	1.65	2.6	4.2	6.6	10.5	16.5
> 1,250	1,600			15	21	29	39	55	78	125	195	310	500	780	1.25	1.95	3.1	5	7.8	12.5	19.5
> 1,600	2,000			18	25	35	46	65	92	150	230	370	600	920	1.5	2.3	3.7	6	9.2	15	23
> 2,000	2,500			22	30	41	55	78	110	175	280	440	700	1,100	1.75	2.8	4.4	7	11	17.5	28
> 2,500	3,150			26	36	50	68	96	135	210	350	540	860	1,350	2.1	3.3	5.4	8.6	13.5	21	33

Source: Table 1 from ISO 286–1:2010 Geometrical product specifications (GPS) — ISO code system for tolerances on linear sizes — Part 1: Basis of tolerances, deviations and fits.

Table A1.2 Fundamental deviation values A to M for internal features (in μm).

Nominal Size mm		Fundamental Deviation Values																		
		Lower Limit Deviation, EI												*Upper Limit Deviation, ES*						
		All Standard Tolerance Grades												IT6	IT7	IT8	Up to and including IT8	Above IT8	Up to and including IT8	Above IT8
Above	Up to and including	A[a]	B[a]	C	CD	D	E	EF	F	FG	G	H	JS	J			K		M[b]	
–	3	+270	+140	+60	+34	+20	+14	+10	+6	+4	+2	0	Deviations = ± ITn/2, where n is the standard tolerance grade number	+2	+4	+6	0	0	−2 +D	−2
3	6	+270	+140	+70	+46	+30	+20	+14	+10	+6	+4	0		+5	+6	+10	−1 +D		−4 +D	−4
6	10	+280	+150	+80	+56	+40	+25	+18	+13	+8	+5	0		+5	+8	+12	−1 +D		−6 +D	−6
10	14	+290	+150	+95	+70	+50	+32	+23	+16	+10	+6	0		+6	+10	+15	−1 +D		−7 +D	−7
14	18	+290	+150	+95	+70	+50	+32	+23	+16	+10	+6	0		+6	+10	+15	−1 +D		−7 +D	−7
18	24	+300	+160	+110	+85	+65	+40	+28	+20	+12	+7	0		+8	+12	+20	−2 +D		−8 +D	−8
24	30	+300	+160	+110	+85	+65	+40	+28	+20	+12	+7	0		+8	+12	+20	−2 +D		−8 +D	−8
30	40	+310	+170	+120	+100	+80	+50	+35	+25	+15	+9	0		+10	+14	+24	−2 +D		−9 +D	−9
40	50	+320	+180	+130	+100	+80	+50	+35	+25	+15	+9	0		+10	+14	+24	−2 +D		−9 +D	−9
50	65	+340	+190	+140		+100	+60		+30		+10	0		+13	+18	+28	−2 +D		−11 +D	−11
65	80	+360	+200	+150		+100	+60		+30		+10	0		+13	+18	+28	−2 +D		−11 +D	−11
80	100	+380	+220	+170		+120	+72		+36		+12	0		+16	+22	+34	−3 +D		−13 +D	−13
100	120	+410	+240	+180		+120	+72		+36		+12	0		+16	+22	+34	−3 +D		−13 +D	−13
120	140	+460	+260	+200		+145	+85		+43		+14	0		+18	+26	+41	−3 +D		−15 +D	−15
140	160	+520	+280	+210		+145	+85		+43		+14	0		+18	+26	+41	−3 +D		−15 +D	−15
160	180	+580	+310	+230		+145	+85		+43		+14	0		+18	+26	+41	−3 +D		−15 +D	−15

(*Continued*)

Table A1.2 (Continued)

Nominal Size mm		Fundamental Deviation Values																		
		Lower Limit Deviation, EI												*Upper Limit Deviation, ES*						
		All Standard Tolerance Grades												IT6	IT7	IT8	Up to and including IT8	Above IT8	Up to and including IT8	Above IT8
Above	Up to and including	A[a]	B[a]	C	CD	D	E	EF	F	FG	G	H	JS	J			K		M[b]	
180	200	+660	+340	+240		+170	+100		+50		+15	0		+22	+30	+47	−4 +D		−17 +D	−17
200	225	+740	+380	+260		+170	+100		+50		+15	0		+22	+30	+47	−4 +D		−17 +D	−17
225	250	+820	+420	+280		+170	+100		+50		+15	0		+22	+30	+47	−4 +D		−17 +D	−17
250	280	+920	+460	+300		+190	+110		+56		+17	0		+25	+36	+55	−4 +D		−20 +D	−20
280	315	+1,050	+540	+330		+190	+110		+56		+17	0		+25	+36	+55	−4 +D		−20 +D	−20
315	355	+1,200	+600	+360		+210	+125		+62		+18	0		+29	+39	+60	−4 +D		−21 +D	−21
355	400	+1,350	+680	+400		+210	+125		+62		+18	0		+29	+39	+60	−4 +D		−21 +D	−21
400	450	+1,500	+760	+440		+230	+135		+68		+20	0		+33	+43	+66	−5 +D		−23 +D	−23
450	500	+1,650	+840	+480		+230	+135		+68		+20	0		+33	+43	+66	−5 +D		−23 +D	−23
500	560					+260	+145		+76		+22	0					0		−26	
560	630					+260	+145		+76		+22	0					0		−26	
630	710					+290	+160		+80		+24	0					0		−30	
710	800					+290	+160		+80		+24	0					0		−30	
800	900					+320	+170		+86		+26	0					0		−34	
900	1,000					+320	+170		+86		+26	0					0		−34	

1,000	1,120	+350	+195	+98	+28	0	0	−40
1,120	1,250	+350	+195	+98	+28	0	0	−40
1,250	1,400	+390	+220	+110	+30	0	0	−48
1,400	1,600	+390	+220	+110	+30	0	0	−48
1,600	1,800	+430	+240	+120	+32	0	0	−58
1,800	2,000	+430	+240	+120	+32	0	0	−58
2,000	2,240	+480	+260	+130	+34	0	0	−68
2,240	2,500	+480	+260	+130	+34	0	0	−68
2,500	2,800	+520	+290	+145	+38	0	0	−76
2,800	3,150	+520	+290	+145	+38	0	0	−76

[a] Fundamental deviations A and B shall not be used for nominal sizes ≤ 1 mm.

[b] Special case: for tolerance class M6 in the range above 250 mm up to 315 mm, ES = −9 μm (instead of −11 μm according to the calculation).

Source: Table 2 from ISO 286-1:2010. Geometrical product specifications (GPS) — ISO code system for tolerances on linear sizes — Part 1: Basis of tolerances, deviations and fits.

Table A1.3 Fundamental deviation values N to ZC for internal features (in μm).

Nominal Size mm		Fundamental Deviation Values																				
		Upper Limit Deviation, ES																Values for D				
		Up to and including IT8	Above IT8	Up to and including IT7	Standard tolerance grades above IT7													Standard tolerance grades				
Above	Up to and including	N[a]		P to ZC	P	R	S	T	U	V	X	Y	Z	ZA	ZB	ZC	IT3	IT4	IT5	IT6	IT7	IT8
–	3	−4 +D	−4	Values as for standard tolerance grades above IT7 incresed by D	−6	−10	−14		−18		−20		−26	−32	−40	−60	0	0	0	0	0	0
3	6	−8 +D	0		−12	−15	−19		−23		−28		−35	−42	−50	−80	1	1.5	1	3	4	6
6	10	−10 +D	0		−15	−19	−23		−28		−34		−42	−52	−67	−97	1	1.5	2	3	6	7
10	14	−12 +D	0		−18	−23	−28		−33		−40		−50	−64	−90	−130	1	2	3	3	7	9
14	18	−12 +D	0		−18	−23	−28		−33	−39	−45		−60	−77	−108	−150	1	2	3	3	7	9
18	24	−15 +D	0		−22	−28	−35		−41	−47	−54	−63	−73	−98	−136	−188	1.5	2	3	4	8	12
24	30	−15 +D	0		−22	−28	−35	−41	−48	−55	−64	−75	−88	−118	−160	−218	1.5	2	3	4	8	12
30	40	−17 +D	0		−26	−34	−43	−48	−60	−68	−80	−94	−112	−148	−200	−274	1.5	3	4	5	9	14
40	50	−17 +D	0		−26	−34	−43	−54	−70	−81	−97	−114	−136	−180	−242	−325	1.5	3	4	5	9	14
50	65	−20 +D	0		−32	−41	−53	−66	−87	−102	−122	−144	−172	−226	−300	−405	2	3	5	6	11	16
65	80	−20 +D	0		−32	−43	−59	−75	−102	−120	−146	−174	−210	−274	−360	−480	2	3	5	6	11	16
80	100	−23 +D	0		−37	−51	−71	−91	−124	−146	−178	−214	−258	−335	−445	−580	2	4	5	7	13	19
100	120	−23 +D	0		−37	−54	−79	−104	−144	−172	−210	−254	−310	−400	−525	−690	2	4	5	7	13	19
120	140	−27 +D	0		−43	−63	−92	−122	−170	−202	−248	−300	−365	−470	−620	−800	3	4	6	7	15	23
140	160	−27 +D	0		−43	−65	−100	−134	−190	−228	−280	−340	−415	−535	−700	−900	3	4	6	7	15	23
160	180	−27 +D	0		−43	−68	−108	−146	−210	−252	−310	−380	−465	−600	−780	−1,000	3	4	6	7	15	23
180	200	−31 +D	0		−50	−77	−122	−166	−236	−284	−350	−425	−520	−670	−880	−1,150	3	4	6	9	17	26
200	225	−31 +D	0		−50	−80	−130	−180	−258	−310	−385	−470	−575	−740	−960	−1,250	3	4	6	9	17	26
225	250	−31 +D	0		−50	−84	−140	−196	−284	−340	−425	−520	−640	−820	−1,050	−1,350	3	4	6	9	17	26

250	280	−34 +D	0	−56	−94	−158	−218	−315	−385	−475	−580	−710	−920	−1,200	−1,550	4	4	7	9	20	29
280	315	−34 +D	0	−56	−98	−170	−240	−350	−425	−525	−650	−790	−1,000	−1,300	−1,700	4	4	7	9	20	29
315	355	−37 +D	0	−62	−108	−190	−268	−390	−475	−590	−730	−900	−1,150	−1,500	−1,900	4	5	7	11	21	32
355	400	−37 +D	0	−62	−114	−208	−294	−435	−530	−660	−820	−1,000	−1,300	−1,650	−2,100	4	5	7	11	21	32
400	450	−40 +D	0	−68	−126	−232	−330	−490	−595	−740	−920	−1,100	−1,450	−1,850	−2,400	5	5	7	13	23	34
450	500	−40 +D	0	−68	−132	−252	−360	−540	−660	−820	−1,000	−1,250	−1,600	−2,100	−2,600	5	5	7	13	23	34
500	560	−44		−78	−150	−280	−400	−600													
560	630	−44		−78	−151	−310	−450	−660													
630	710	−50		−88	−175	−340	−500	−740													
710	800	−50		−88	−185	−380	−560	−840													
800	900	−56		−100	−210	−430	−620	−940													
900	1,000	−56		−100	−220	−470	−680	−1,050													
1,000	1,120	−66		−120	−250	−520	−780	−1,150													
1,120	1,250	−66		−120	−260	−580	−840	−1,300													
1,250	1,400	−78		−140	−300	−640	−960	−1,450													
1,400	1,600	−78		−140	−330	−720	−1,050	−1,600													
1,600	1,800	−92		−170	−370	−820	−1,200	−1,850													
1,800	2,000	−92		−170	−400	−920	−1,350	−2,000													
2,000	2,240	−110		−195	−440	−1,000	−1,500	−2,300													
2,240	2,500	−110		−195	−460	−1,100	−1,650	−2,500													
2,500	2,800	−135		−240	−550	−1,250	−1,900	−2,900													
2,800	3,150	−135		−240	−580	−1,400	−2,100	−3,200													

[a] Fundamental deviations N for standard tolerance grades above IT8 shall not be used for nominal sizes ≤ 1 mm.

Source: Table 3 from ISO 286-1:2010. Geometrical product specifications (GPS) — ISO code system for tolerances on linear sizes — Part 1: Basis of tolerances, deviations and fits.

Table A1.4 Fundamental deviation values for external features a to j (in μm).

Nominal Size mm		Fundamental Deviation Values														
		Upper limit deviation, es												Lower deviation, ei		
		All standard tolerance grades												IT5 and IT6	IT7	IT8
Above	Up to and including	a[a]	b[a]	c	cd	d	e	ef	f	fg	g	h	js	j		
–	3	−270	−140	−60	−34	−20	−14	−10	−6	−4	−2	0	Deviations = ± ITn/2, where n is the standard tolerance grade number	−2	−4	−6
3	6	−270	−140	−70	−46	−30	−20	−14	−10	−6	−4	0		−2	−4	
6	10	−280	−150	−80	−56	−40	−25	−18	−13	−8	−5	0		−2	−5	
10	14	− 290	− 150	− 95	− 70	− 50	− 32	− 23	− 16	− 10	− 6	0		− 3	− 6	
14	18	− 290	− 150	− 95	− 70	− 50	− 32	− 23	− 16	− 10	− 6	0		− 3	− 6	
18	24	− 300	− 160	− 110	− 85	− 65	− 40	− 28	− 20	− 12	− 7	0		− 4	− 8	
24	30	− 300	− 160	− 110	− 85	− 65	− 40	− 28	− 20	− 12	− 7	0		− 4	− 8	
30	40	−310	−170	−120	− 100	− 80	− 50	− 35	− 25	− 15	− 9	0		− 5	− 10	
40	50	−320	−180	−130		− 80	− 50		− 25		− 9	0		− 5	− 10	
50	65	−340	−190	−140		− 100	− 60		− 30		− 10	0		− 7	− 12	
65	80	−360	−200	−150		− 100	− 60		− 30		− 10	0		− 7	− 12	
80	100	−380	−220	−170		− 120	− 72		− 36		− 12	0		− 9	− 15	
100	120	−410	−240	−180		− 120	− 72		− 36		− 12	0		− 9	− 15	
120	140	−460	−260	−200		− 145	− 85		− 43		− 14	0		− 11	− 18	
140	160	−520	−280	−210		− 145	− 85		− 43		− 14	0		− 11	− 18	
160	180	−580	−310	−230		− 145	− 85		− 43		− 14	0		− 11	− 18	
180	200	−660	−340	−240		− 170	− 100		− 50		− 15	0		− 13	− 21	
200	225	−740	−380	−260		− 170	− 100		− 50		− 15	0		− 13	− 21	
225	250	−820	−420	−280		− 170	− 100		− 50		− 15	0		− 13	− 21	

250	280	-920	-460	-300		- 190	- 110		- 56		- 17	0		- 16	- 26
280	315	-1,050	-540	-330		- 190	- 110		- 56		- 17	0		- 16	- 26
315	355	-1,200	-600	-360		- 210	- 125		- 62		- 18	0		- 18	- 28
355	400	-1,350	-680	-400		- 210	- 125		- 62		- 18	0		- 18	- 28
400	450	-1,500	-760	-440		- 230	- 135		- 68		- 20	0		- 20	- 32
450	500	-1,650	-840	-480		- 230	- 135		- 68		- 20	0			
500	560					- 260	- 145		- 76		- 22	0			
560	630					- 260	- 145		- 76		- 22	0			
630	710					- 290	- 160		- 80		- 24	0			
710	800					- 290	- 160		- 80		- 24	0			
800	900					- 320	- 170		- 86		- 26	0			
900	1,000					- 320	- 170		- 86		- 26	0			
1,000	1,120					- 350	- 195		- 98		- 28	0			
1,120	1,250					- 350	- 195		- 98		- 28	0			
1,250	1,400					- 390	- 220		- 110		- 30	0			
1,400	1,600					- 390	- 220		- 110		- 30	0			
1,600	1,800					- 430	- 240		- 120		- 32	0			
1,800	2,000					- 430	- 240		- 120		- 32	0			
2,000	2,240					- 480	- 260		- 130		- 34	0			
2,240	2,500					- 480	- 260		- 130		- 34	0			
2,500	2,800					- 520	- 290		- 145		- 38	0			
2,800	3,150					- 520	- 290		- 145		- 38	0			

[a] Fundamental deviations a and b should not be used for nominal sizes ≤ 1 mm.

Source: Table 4 from ISO 286–1:2010. Geometrical product specifications (GPS) — ISO code system for tolerances on linear sizes — Part 1: Basis of tolerances, deviations and fits.

Table A1.5 Fundamental deviation values for external features k to zc (in µm).

Nominal Size mm		Fundamental Deviation Values															
		Lower limit deviation, ei															
		IT4 to IT7	Up to and including IT3 and above IT7	All standard tolerance grades													
Above	Up to and including	k		m	n	p	r	s	t	u	v	x	y	z	za	zb	zc
–	3	0	0	+2	+4	+6	+10	+14		+18		+20		+26	+32	+40	+60
3	6	+1	0	+4	+8	+12	+15	+19		+23		+28		+35	+42	+50	+80
6	10	+1	0	+6	+10	+15	+19	+23		+28		+34		+42	+52	+67	+97
10	14	+1	0	+7	+12	+18	+23	+28		+33		+40		+50	+64	+90	+130
14	18	+1	0	+7	+12	+18	+23	+28		+33	+39	+45		+60	+77	+108	+150
18	24	+2	0	+8	+15	+22	+28	+35		+41	+47	+54	+63	+73	+98	+136	+188
24	30	+2	0	+8	+15	+22	+28	+35	+41	+48	+55	+64	+75	+88	+118	+160	+218
30	40	+2	0	+9	+17	+26	+34	+43	+48	+60	+68	+80	+94	+112	+148	+200	+274
40	50	+2	0	+9	+17	+26	+34	+43	+54	+70	+81	+97	+114	+136	+180	+242	+325
50	65	+2	0	+11	+20	+32	+41	+53	+66	+87	+102	+122	+144	+172	+226	+300	+405
65	80	+2	0	+11	+20	+32	+43	+59	+75	+102	+120	+146	+174	+210	+274	+360	+480
80	100	+3	0	+13	+23	+37	+51	+71	+91	+124	+146	+178	+214	+258	+335	+445	+580
100	120	+3	0	+13	+23	+37	+54	+79	+104	+144	+172	+210	+254	+310	+400	+525	+690
120	140	+3	0	+15	+27	+43	+63	+92	+122	+170	+202	+248	+300	+365	+470	+620	+800
140	160	+3	0	+15	+27	+43	+65	+100	+134	+190	+228	+280	+340	+415	+535	+700	+900
160	180	+3	0	+15	+27	+43	+68	+108	+146	+210	+252	+310	+380	+465	+600	+780	+1,000

180	200	+4	0	+17	+31	+50	+77	+122	+166	+236	+284	+350	+425	+520	+670	+880	+1,150
200	225	+4	0	+17	+31	+50	+80	+130	+180	+258	+310	+385	+470	+575	+740	+960	+1,250
225	250	+4	0	+17	+31	+50	+84	+140	+196	+284	+340	+425	+520	+640	+820	+1,050	+1,350
250	280	+4	0	+20	+34	+56	+94	+158	+218	+315	+385	+475	+580	+710	+920	+1,200	+1,550
280	315	+4	0	+20	+34	+56	+98	+170	+240	+350	+425	+525	+650	+790	+1,000	+1,300	+1,700
315	355	+5	0	+21	+37	+62	+108	+190	+268	+390	+475	+590	+730	+900	+1,150	+1,500	+1,900
355	400	+5	0	+21	+37	+62	+114	+208	+294	+435	+530	+660	+820	+1,000	+1,300	+1,650	+2,100
400	450	+5	0	+23	+40	+68	+126	+232	+330	+490	+595	+740	+920	+1,100	+1,450	+1,850	+2,400
450	500	+5	0	+23	+40	+68	+132	+252	+360	+540	+660	+820	+1,000	+1,250	+1,600	+2,100	+2,600
500	560	0	0	+26	+44	+78	+150	+280	+400	+600							
560	630	0	0	+26	+44	+78	+155	+310	+450	+660							
630	710	0	0	+30	+50	+88	+175	+340	+500	+740							
710	800	0	0	+30	+50	+88	+185	+380	+560	+840							
800	900	0	0	+34	+56	+100	+210	+430	+620	+940							
900	1,000	0	0	+34	+56	+100	+220	+470	+680	+1,050							
1,000	1,120	0	0	+40	+66	+120	+250	+520	+780	+1,150							
1,120	1,250	0	0	+40	+66	+120	+260	+580	+840	+1,300							
1,250	1,400	0	0	+48	+78	+140	+300	+640	+960	+1,450							
1,400	1,600	0	0	+48	+78	+140	+330	+720	+1,050	+1,600							
1,600	1,800	0	0	+58	+92	+170	+370	+820	+1,200	+1,850							
1,800	2,000	0	0	+58	+92	+170	+400	+920	+1,350	+2,000							
2,000	2,240	0	0	+68	+110	+195	+440	+1,000	+1,500	+2,300							
2,240	2,500	0	0	+68	+110	+195	+460	+1,100	+1,650	+2,500							
2,500	2,800	0	0	+76	+135	+240	+550	+1,250	+1,900	+2,900							
2,800	3,150	0	0	+76	+135	+240	+580	+1,400	+2,100	+3,200							

Source: Table 5 from ISO 286-1:2010. Geometrical product specifications (GPS) — ISO code system for tolerances on linear sizes — Part 1: Basis of tolerances, deviations and fits.

Appendix 2: Cutting Data

This appendix provides information about cutting data in turning, milling, and drilling operations. The following information is given along with the appendix:

- Recommended cutting data for milling/turning/drilling. Ranges for cutting speed (V_c) and feed rate (f_z or f_n). *Source*: different cutting tool vendors (Walter Titex, Fette, Sandvik, Kennametal).
- Specific cutting force to estimate cutting forces according to Chapter 10.
- Cross-reference material table and ISO codes to identify a specific workpiece material according to the standards DIN, AISI/SAE, UNE, AFNOR, UNI and JIS.

The data provided in this appendix should be used as a starting point for machining. A more adequate cutting data can be obtained considering the information provided by the cutting tool vendor of the specific cutting tools employed and even after conducting some experimental tests.

A2.1 Recommended Cutting Data

A2.1.1 Turning Cutting Data. Carbide Inserts.

Manufacturing Process Planning: A Practical Approach for Mechanical Engineering, First Edition.
José V. Abellán-Nebot, Carlos Vila Pastor, and Héctor R. Siller.

Companion website: www.wiley.com/go/Manufacturing_Process_Planning_1e

Table A2.1 External and internal turning (straight turning, face turning, and profiling). Coated carbide inserts.

			Coated carbide (higher wear resistance)		Coated carbide (higher toughness)	
	Group material	**Material**	**Vc (m/min)**	**fn (mm/rev)**	**Vc (m/min)**	**fn (mm/rev)**
External and internal turning (straight turning, face turning, profiling)	**P**	Non-alloyed steel, HB 150	650–530–420	0.05–0.1–0.2	165–120–85	0.1–0.4–0.8
		Low-alloy steel, HB 180	530–450–360	0.05–0.1–0.2	155–110–70	0.1–0.4–0.8
		High-alloy steel, HB 200	395–330–250	0.05–0.1–0.2	145–100–65	0.1–0.4–0.8
	M	Ferritic / Martensitic Stainless steels, HB 200	290–240	0.1–0.2	130–110–90	0.2–0.4–0.6
		Austenitic stainless steels, HB 180	–	–	100–90–75	0.2–0.4–0.6
	K	Gray cast iron, HB 220	760–650–540	0.1–0.25–0.4	140–115–95	0.1–0.3–0.5
	S	Ni- or Co-based alloys,HB 300	400–260	0.1–0.2	40–30–20	0.1–0.3–0.5
		Titanium Alloys, UTS 950 MPa	85–70–55	0.1–0.2–0.3	65–55–45	0.1–0.3–0.5
	N	Aluminum alloys, HB 75	2000	0.05–0.4	770	0.15–0.8
		Aluminum Si 13-15%, HB 130	1550	0.05–0.4	300	0.15–0.8
		Copper alloys, HB 110	500	0.05–0.4	200	0.15–0.8
	H	Hardened Steels, HRC 55	295–225–185	0.05–0.15–0.25	140–115–95	0.1–0.25–0.4

Table A2.2 External and internal turning (grooving, parting, and threading). Coated carbide inserts.

			Grooving / Parting				Threading*	
			Coated carbide (higher wear resistance)		Coated carbide (higher toughness)		Coated carbide	
	Group material	Material	Vc (m/min)	fn (mm/rev)	Vc (m/min)	fn (mm/rev)	Vc (m/min)	nap
External and internal turning (grooving, parting and threading)	**P**	Non-alloyed steel, HB 150	330–140	0.05–0.5	220–155	0.05–0.5	205–160	See Table A2.3
		Low-alloy steel, HB 180	290–135	0.05–0.5	205–145	0.05–0.5	135–115	
		High-alloy steel, HB 200	260–115	0.05–0.5	130–100	0.05–0.5	120–105	
	M	Ferritic / Martensitic Stainless steels, HB 200	400–175	0.05–0.5	195–135	0.05–0.5	145–90	
		Austenitic stainless steels, HB 180	435–190	0.05–0.5	160–110	0.05–0.5	145–75	
	K	Gray cast iron, HB 220	250–120	0.05–0.5	140–70	0.05–0.5	130–80	
	S	Ni- or Co-based alloys, HB 300	90–60	0.05–0.3	30–13	0.05–0.3	25–19	
		Titanium Alloys, UTS 950 MPa	–	–	65–37	0.05–0.3	65–50	
	N	Aluminum alloys, HB 75	2100	0.05–0.5	1500	0.05–0.8	500–425	
		Aluminum Si 13-15%, HB 130	1600	0.05–0.5	400	0.05–0.8	270–210	
		Copper alloys, HB 110	600	0.05–0.5	350	0.05–0.8	460–370	
	H	Hardened Steels, HRC 60	125–120	0.05–0.1	–	–	120–45	

Note: nap: number of cutting passes.
* Values for metric threads with 60° and pitch 1.25 mm

Table A2.3 Number of infeeds (nap), infeed per pass, and total depth of thread. Coated carbide inserts.

	Pitch (mm)													
	0.50	0.75	1.00	1.25	1.50	1.75	2.0	2.5	3.0	3.5	4.0	4.5	5.0	5.5
Number of infeeds (nap)	Radial infeed per pass (mm)													
1	0.11	0.17	0.19	0.20	0.22	0.22	0.25	0.27	0.28	0.34	0.34	0.37	0.41	0.43
2	0.09	0.15	0.16	0.17	0.21	0.21	0.24	0.24	0.26	0.31	0.32	0.34	0.39	0.40
3	0.07	0.11	0.13	0.14	0.17	0.17	0.18	0.20	0.21	0.25	0.25	0.28	0.32	0.32
4	0.07	0.07	0.11	0.11	0.14	0.14	0.16	0.17	0.18	0.21	0.22	0.24	0.27	0.27
5	**0.34**	**0.50**	0.08	0.10	0.12	0.12	0.14	0.15	0.16	0.18	0.19	0.22	0.24	0.24
6			**0.67**	0.08	0.08	0.10	0.12	0.13	0.14	0.17	0.17	0.20	0.22	0.22
7				**0.80**	**0.94**	0.10	0.11	0.12	0.13	0.15	0.16	0.18	0.20	0.20
8						0.08	0.08	0.11	0.12	0.14	0.15	0.17	0.19	0.19
9						**1.14**	**1.28**	0.11	0.12	0.14	0.14	0.16	0.18	0.18
10								0.08	0.11	0.12	0.13	0.15	0.17	0.17
11								**1.58**	0.10	0.11	0.12	0.14	0.16	0.16
12									0.08	0.08	0.12	0.13	0.15	0.15
13									**1.89**	**2.20**	0.11	0.12	0.12	0.13
14											0.08	0.08	0.10	0.13
15											**2.50**	**2.80**	**3.12**	0.12
16														0.10
														3.41

A2.1.2 Milling Cutting Data

A2.1.2.1 Cutters with Carbide Cutting Inserts

Table A2.4 Milling operations with large engagement and coated carbide inserts.

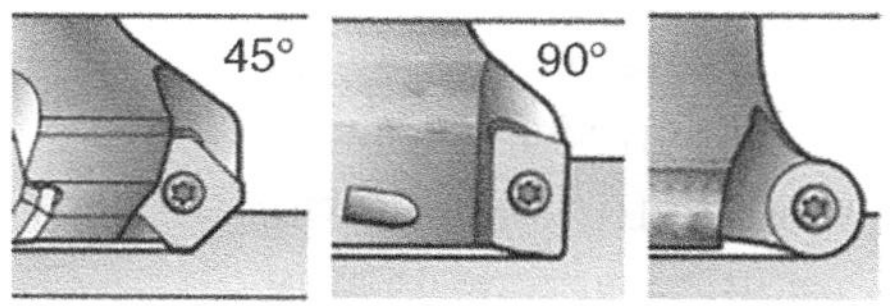

Large engagement

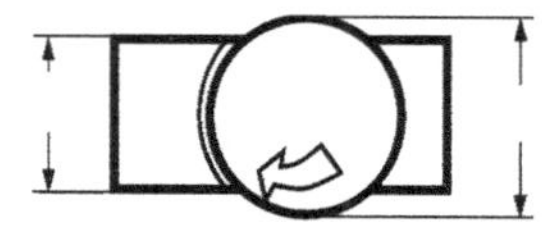

Material group	Material	Coated carbide (higher wear resistance)		Coated carbide (higher toughness)	
		Vc (m/min)	fz (mm/tooth)	Vc (m/min)	fz (mm/tooth)
P	Non-alloyed steel, HB 150	385–350–315	0.1–0.15–0.2	240–205–150	0.1–0.2–0.4
	Low-alloy steel, HB 180	300–275–245	0.1–0.15–0.2	185–160–120	0.1–0.2–0.4
	High-alloy steel, HB 200	230–205–185	0.1–0.15–0.2	140–120–90	0.1–0.2–0.4
M	Ferritic / Martensitic Stainless steels, HB 200	285–255–230	0.1–0.15–0.2	185–160–115	0.1–0.2–0.4
	Austenitic stainless steels, HB 200	265–240–215	0.1–0.15–0.2	170–150–110	0.1–0.2–0.4
K	Gray cast iron, HB 180	850–720–620	0.1–0.15–0.2	290–240–195	0.1–0.2–0.3
S	Ni- or Co-based alloys, HB 300	50–20	0.1–0.2	50–20	0.1–0.25
	Titanium Alloys, UTS 950 MPa	65–60–55	0.1–0.15–0.2	65–60–55	0.1–0.15–0.25
N	Aluminum alloys, HB 75	1,900–1,750–1,600	0.1–0.15–0.2	940–870–810	0.1–0.15–0.2
	Aluminum Si 13-15%, HB 130	760–700–650	0.1–0.15–0.2	380–350–325	0.1–0.15–0.2
	Copper alloys, HB 110	940–870–810	0.1–0.15–0.2	470–435–405	0.1–0.15–0.2
H	Hardened Steels, HRC 59	160–140–115	0.07–0.12–0.2	40–36–29	0.07–0.12–0.2

Table A2.5 Milling operations with small engagement and coated carbide inserts

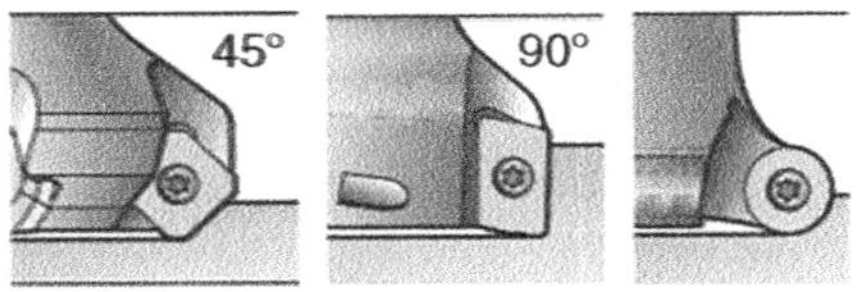

Small engagement

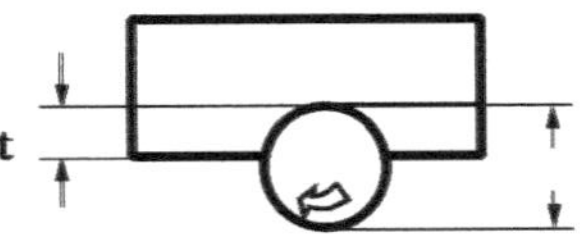

Material group	Material	Coated carbide (higher wear resistance)		Coated carbide (higher toughness)	
		Vc (m/min)	fz (mm/tooth)	Vc (m/min)	fz (mm/tooth)
P	Non-alloyed steel, HB 150	450–440–430	0.1–0.15–0.2	270–260–250	0.1–0.2–0.3
	Low-alloy steel, HB 180	350–345–335	0.1–0.15–0.2	210–205–195	0.1–0.2–0.3
	High-alloy steel, HB 200	265–260–255	0.1–0.15–0.2	160–155–150	0.1–0.2–0.3
M	Ferritic / Martensitic Stainless steels, HB 200	340–335–325	0.1–0.15–0.2	210–200–195	0.1–0.2–0.3
	Austenitic stainless steels, HB 200	320–310–300	0.1–0.15–0.2	195–185–180	0.1–0.2–0.3
K	Gray cast iron, HB 180	1,100–1,050–1,000	0.1–0.15–0.2	340–325–310	0.1–0.2–0.3
S	Ni- or Co-based alloys, HB 300	60–30	0.05–0.2	60–30	0.05–0.15–0.25
	Titanium Alloys, UTS 950 MPa	80–75–75	0.05–0.15–0.2	75–70–70	0.05–0.15–0.25
N	Aluminum alloys, HB 75	2,100–2,100–2,050	0.1–0.15–0.2	850–830–820	0.1–0.15–0.2
	Aluminum Si 13-15%, HB 130	850–840–820	0.1–0.15–0.2	425–420–410	0.1–0.15–0.2
	Copper alloys, HB 110	1,050	0.1–0.15–0.2	530–520–510	0.1–0.15–0.2
H	Hardened Steels, HRC 59	190–180–175	0.07–0.12–0.2	45	0.07–0.12–0.2

A2.1.2.2 HSS End Mill Tools (Coated and Uncoated)

Table A2.6 Large engagement milling operations with HSS end mill tools (coated and uncoated). For small engagement, apply a thinning factor.

Large engagement. Axial depth of cut ap = 1 × D (D = Tool diameter)						
End Mills						
		Uncoated HSS	Coated HSS (TiAlN...)	Feed (mm/tooth)		
Material group	Material	Vc (m/min)		Ø 3 a 10	Ø 10 a 20	Ø 20 a 40
P	Non-alloyed steel, HB 150	20–45	40–60	0.006–0.015	0.01–0.03	0.05–0.09
	Low-alloy steel, HB 180	15–25	30–50	0.005–0.013	0.008–0.027	0.04–0.08
	High-alloy steel, HB 200	12–20	20–40	0.004–0.010	0.007–0.025	0.03–0.07
M	Austenitic stainless steels, HB 180	12	18	0.004–0.010	0.007–0.025	0.03–0.07
K	Gray cast iron, HB 220	30	40	0.005–0.013	0.008–0.027	0.04–0.08
N	Aluminum alloys, HB 75	120	160	0.006–0.015	0.01–0.03	0.05–0.09
S	Heat resistant super alloys	–	–	–	–	–
H	Hardened steels, HRC 60	–	–	–	–	–

Note: Cutting conditions for finishing operations. For roughing, feed rate can be increased × 3.

A2.1.2.3 Carbide End Mill Tools (Uncoated)

Table A2.7 General milling operations with uncoated carbide end mill tools.

Material Group	Vc (m/min)	Feed (mm/tooth) for different end mill diameters (mm)									
		3	4	5	6	8	10	12	16	20	25
Low Carbon Steels	70–125	0.01	0.02	0.02	0.03	0.04	0.06	0.08	0.10	0.15	0.18
Medium Carbon Steels	35–90	0.02	0.020	0.03	0.04	0.04	0.05	0.05	0.08	0.10	0.15
Stainless Steel Soft (300 series)	60–125	0.01	0.02	0.02	0.03	0.04	0.04	0.05	0.08	0.10	0.15
Cast Iron (Soft)	70–180	0.025	0.030	0.04	0.05	0.06	0.07	0.08	0.12	0.15	0.20
Aluminium Low <8% Silicon	600–900	0.01	0.025	0.04	0.05	0.07	0.08	0.10	0.12	0.15	0.20
Aluminium High >8% Silicon	200–400	0.01	0.025	0.03	0.04	0.06	0.07	0.08	0.10	0.12	0.15
Brass and Bronze	70–120	0.03	0.030	0.04	0.05	0.06	0.07	0.08	0.09	0.10	0.13
Copper and Copper Alloys	120–250	0.025	0.030	0.04	0.05	0.06	0.07	0.08	0.09	0.10	0.15
Magnesium/Magnesium Alloys	300–460	0.025	0.030	0.04	0.05	0.07	0.08	0.10	0.13	0.15	0.25
Titanium (Soft)	15–125	0.01	0.02	0.02	0.03	0.03	0.04	0.05	0.08	0.10	0.15
Monel/High Nickel Steels	60–80	0.01	0.02	0.02	0.03	0.03	0.04	0.05	0.06	0.076	0.10
High Temperature Alloys	8–30	0.01	0.01	0.02	0.02	0.02	0.02	0.03	0.03	0.035	0.05
Hardened Steels	9–30	0.005	0.007	0.01	0.01	0.02	0.02	0.03	0.04	0.05	0.075
Plastics	240–490	0.038	0.050	0.07	0.08	0.11	0.13	0.15	0.20	0.250	0.380

A2.1.2.4 Carbide End Mill Tools (Coated)

Table A2.8 General milling operations with coated carbide end mill tools.

					Feed (mm/tooth) for different end mill diameters (mm)									
Material Group	**TiN**	**TiCN**	**TiAlN**	**Vc (m/min)**	**3**	**4**	**5**	**6**	**8**	**10**	**12**	**16**	**20**	**25**
Low Carbon Steels	3	1	2	100–240	0.02	0.03	0.03	0.05	0.05	0.06	0.07	0.08	0.08	0.10
Medium Carbon Steels	3	2	1	60–100	0.15	0.02	0.04	0.04	0.04	0.05	0.05	0.06	0.10	0.10
High-alloy Steels	1	2	3	30–100	0.02	0.02	0.03	0.03	0.04	0.05	0.05	0.06	0.06	0.08
Stainless Steel Soft (300 series)	3	2	1	60–120	0.02	0.03	0.03	0.04	0.04	0.05	0.05	0.06	0.06	0.08
Cast Iron (Soft)	3	1	2	100–225	0.03	0.03	0.04	0.05	0.06	0.07	0.07	0.08	0.09	0.10
Aluminium Low <8% Silicon	3	1	2	max avail.	0.02	0.03	0.04	0.06	0.07	0.10	0.15	0.15	0.20	0.20
Aluminium High >8% Silicon	3	1	2	200–600	0.02	0.02	0.03	0.05	0.06	0.10	0.10	0.10	0.15	0.15
Brass and Bronze	1	2	3	80–365	0.02	0.02	0.03	0.04	0.05	0.07	0.08	0.09	0.10	0.10
Copper and Copper Alloys	1	2	3	130–330	0.03	0.04	0.04	0.05	0.06	0.07	0.09	0.09	0.10	0.10
Titanium (Soft)	3	2	1	60–105	0.01	0.02	0.02	0.04	0.04	0.05	0.05	0.06	0.06	0.07
Monel/High Nickel Steels	3	2	1	45–75	0.01	0.03	0.03	0.04	0.04	0.05	0.05	0.06	0.08	0.10
High Temperature Alloys	3	2	1	15–35	0.01	0.02	0.02	0.03	0.04	0.04	0.04	0.05	0.06	0.07
Hardened Steels	3	2	1	30–60	0.01	0.01	0.02	0.02	0.03	0.03	0.04	0.04	0.05	0.06

Note: First Choice (1), Second Choice (2), Third Choice (3).

A2.1.3 Drilling/Reaming/Threading Cutting Data

A2.1.3.1 Center Drills

Table A2.9 Center drills HSS. Cutting speeds and curve for feed rate.

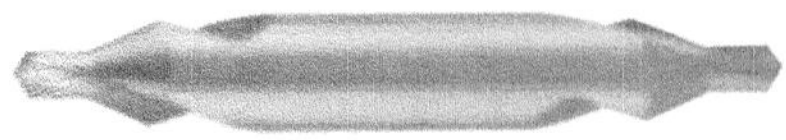

Center drills HSS uncoated (Form A)					
Material Group	**Vc (m/min)**	**Feed Curve**	**Material Group**	**Vc (m/min)**	**Feed Curve**
Non-alloyed steels	28–36	8	Aluminum Si<5%	40–60	12
Low-alloy steels	14–25	7	Aluminum Si 5-10%	28–40	10
High-alloy steels	7–10	5	Copper alloys	40–60	10
Stainless steels	6–8	3	Magnesium alloys	28	10
Cast iron	20–30	10	Hardened Steels	–	–
Ni- or Co-based alloys	4.5	3	Thermoplastics	38	12
Titanium Alloys	7–9	4			

Table A2.10 Feed curves for HSS center drills.

	Feed (mm/rev) for center drill diameter (mm)						
Feed Curve	**1**	**2**	**4**	**6**	**8**	**10**	**12**
3	0.01	0.02	0.04	0.055	0.063	0.071	0.077
4	0.013	0.027	0.053	0.073	0.084	0.094	0.1
5	0.017	0.033	0.067	0.091	0.11	0.12	0.13
7	0.023	0.047	0.093	0.13	0.15	0.16	0.18
8	0.027	0.053	0.11	0.15	0.17	0.19	0.21
10	0.033	0.067	0.13	0.18	0.21	0.24	0.26
12	0.04	0.08	0.16	0.22	0.25	0.28	0.31

A2.1.3.2 Spot Drills

Table A2.11 HSS spot drills. Cutting speed and feed curves.

Spot drill (uncoated HSS)					
Material Group	**Vc (m/min)**	**Feed Curve**	**Material Group**	**Vc (m/min)**	**Feed Curve**
Non-alloyed steels	21–32	9	Aluminum Si<5%	40–60	16
Low-alloy steels	11–30	8	Aluminum Si 5-10%	30–40	12
High-alloy steels	8–21	7	Copper alloys	40–65	12
Stainless steels	5–8	5	Magnesium alloys	30	12
Cast iron	18–30	12	Hardened Steels	–	
Ni- or Co-based alloys	6	3	Thermoplastics	40	12
Titanium Alloys	6–10	4			

Table A2.12 Feed curves for HSS spot drills.

	Feed (mm/rev) for diameter									
Feed Curve	**1**	**2**	**4**	**6**	**8**	**10**	**12**	**15**	**20**	**25**
3	0.01	0.02	0.04	0.055	0.063	0.071	0.077	0.087	0.1	0.11
4	0.013	0.027	0.053	0.073	0.084	0.094	0.1	0.12	0.13	0.15
5	0.017	0.033	0.067	0.091	0.11	0.12	0.13	0.14	0.17	0.19
7	0.023	0.047	0.093	0.13	0.15	0.16	0.18	0.2	0.23	0.26
8	0.027	0.053	0.11	0.15	0.17	0.19	0.21	0.23	0.27	0.3
9	0.03	0.06	0.12	0.16	0.19	0.21	0.23	0.26	0.3	0.34
12	0.04	0.08	0.16	0.22	0.25	0.28	0.31	0.35	0.4	0.45
16	0.053	0.11	0.21	0.29	0.34	0.38	0.41	0.46	0.53	0.6

A2.1.3.3 Twist Drills (Uncoated and Coated HSS Drills)

Table A2.13 HSS twist drills (uncoated and coated drills). Cutting speed and feed curves.

		Cutting Speed (m/min)	
Material Group	**Feed Curve**	**Uncoated HSS drill**	**Coated HSS drill**
Non-alloyed steels	6	20–32	35–50
Low-alloy steels	5	16–25	30–40
High-alloy steels	3	15–19	18–30
Stainless Steels	4	14–18	18–40
Cast iron	7	20–30	30–40
Aluminum Si<5%	7	70–140	140–180
Copper alloys	5	45–90	90–170
Aluminum Si 5-10%	7	30–60	60–150
Magnesium	6	30–100	100–160
Nickel Alloys < 850 Mpa	3	4–8	8–21
Titanium Alloys	2	4–8	8–21
Nickel Alloys > 850 Mpa	2	3–6	6–16
Thermoplastics	6	–	28

Table A2.14 Feed curves for HSS twist drills (uncoated and coated drills).

	Feed Column Number								
	1	**2**	**3**	**4**	**5**	**6**	**7**	**8**	**9**
Drill Diameter (mm)	Feed (mm/rev)								
0.50	0.004	0.006	0.007	0.008	0.010	0.012	0.014	0.016	0.019
1.00	0.006	0.008	0.012	0.014	0.016	0.018	0.020	0.023	0.025
2.00	0.020	0.025	0.032	0.040	0.050	0.063	0.080	0.100	0.125
2.50	0.025	0.032	0.040	0.050	0.063	0.080	0.100	0.125	0.160
3.15	0.032	0.040	0.050	0.063	0.080	0.100	0.125	0.160	0.160
4.00	0.040	0.050	0.063	0.080	0.100	0.125	0.160	0.200	0.200
5.00	0.040	0.050	0.063	0.080	0.100	0.125	0.160	0.200	0.250
6.30	0.050	0.063	0.080	0.100	0.125	0.160	0.200	0.250	0.315
8.00	0.063	0.080	0.100	0.125	0.160	0.200	0.250	0.315	0.315
10.00	0.080	0.100	0.125	0.160	0.200	0.250	0.315	0.400	0.400
12.50	0.080	0.100	0.125	0.160	0.200	0.250	0.315	0.400	0.500
16.00	0.100	0.125	0.160	0.200	0.250	0.315	0.400	0.500	0.630
20.00	0.125	0.160	0.200	0.250	0.315	0.400	0.500	0.630	0.630
25.00	0.160	0.200	0.250	0.315	0.400	0.500	0.630	0.800	0.800

A2.1.3.4 Twist Drills (Uncoated and Coated Carbide Drills)

Table A2.15 Carbide twist drills (uncoated and coated drills). Cutting speed and feed curves.

Material Group	Feed Curve	Cutting Speed (m/min)	
		Uncoated carbide drill	Coated carbide drill
Non-alloyed steels	6	40–50	50–70
Low-alloy steels	6	30–40	40–50
High-alloy steels	4	30–40	40–50
Stainless Steels	5	20–30	30–40
Cast iron	7	30–50	40–60
Aluminum Si<5%	9	200	250
Copper alloys	6	45–90	90–170
Aluminum Si 5-10%	8	90–160	160–220
Magnesium	9	180	230
Nickel Alloys < 850 Mpa	3	8–18	10–20
Titanium Alloys	2	14–25	20–30
Nickel Alloys > 850 Mpa	3	6–10	12–15
Thermoplastics	12	40	75

Table A2.16 Feed curves for carbide twist drills (uncoated and coated drills).

Feed Curve	Feed (mm/rev) for diameter									
	1	2	4	6	8	10	12	15	20	25
2	0.007	0.013	0.027	0.037	0.042	0.047	0.052	0.058	0.067	0.075
3	0.01	0.02	0.04	0.055	0.063	0.071	0.077	0.087	0.1	0.11
4	0.013	0.027	0.053	0.073	0.084	0.094	0.1	0.12	0.13	0.15
5	0.017	0.033	0.067	0.091	0.11	0.12	0.13	0.14	0.17	0.19
6	0.02	0.04	0.08	0.11	0.13	0.14	0.15	0.17	0.2	0.22
7	0.023	0.047	0.093	0.13	0.15	0.16	0.18	0.2	0.23	0.26
8	0.027	0.053	0.11	0.15	0.17	0.19	0.21	0.23	0.27	0.3
9	0.03	0.06	0.12	0.16	0.19	0.21	0.23	0.26	0.3	0.34
12	0.04	0.08	0.16	0.22	0.25	0.28	0.31	0.35	0.4	0.45

A2.1.3.5 Reamers

Table A2.17 Reamers (uncoated HSS-E and uncoated solid carbide tools). Cutting speed and feed curves.

	Uncoated HSS-E Reamer		Uncoated Solid Carbide Reamer	
Material Group	**Cutting Speed (m/min)**	**Feed Curve**	**Cutting Speed (m/min)**	**Feed Curve**
Non-alloyed steels	13–14	8	35–45	8
Low-alloy steels	10–14	8	20–40	8
High-alloy steels	4	8	15–25	8
Stainless Steels	–	–	6–12	8
Cast iron	8–13	8	30–45	8
Aluminum Si<5%	13–28	10	80–115	8
Aluminum Si 5-10%	13–20	10	65–75	8
Copper alloys	17–30	10	70–80	8
Magnesium alloys	13	10	80	8
Ni- or Co-based alloys	–	–	6–12	8
Titanium Alloys	–	–	6–12	8
Thermoplastics	18	8	40	8

Table A2.18 Feed curves for reamers.

	Feed (mm/rev) for diameter									
Feed Curve	**1**	**2**	**4**	**6**	**8**	**10**	**12**	**15**	**20**	**25**
8	0.05	0.07	0.11	0.14	0.16	0.18	0.2	0.23	0.27	0.31
10	0.06	0.09	0.13	0.17	0.2	0.23	0.25	0.29	0.34	0.39

A2.1.3.6 Tap Drills

Table A2.19 Tap drilling. Cutting speed and feed rate calculation.

		Tapping (rpm)							
Material Group	**Vc (m/min)**	**M4**	**M5**	**M6**	**M8**	**M10**	**M11**	**M12**	**M16**
Low carbon steel	17	1,282	1,106	841	674	561	481	420	336
Medium carbon steel	12	932	804	612	490	408	350	306	244
High carbon steel	6	466	402	306	245	204	175	153	122
Low-alloy steel	6	466	402	306	245	204	175	153	122
High-alloy steel	3	233	201	153	122	102	87	76	61
Stainless steels (series 300)	6	466	402	306	245	204	175	153	122
Gray cast iron	22	1,631	1,498	1,070	857	713	612	535	428
Aluminum cast alloys	18	1,398	1,207	917	735	611	525	459	367
Zinc die castings	25	1,864	1,609	1,223	980	815	700	611	489
Copper	18	1,398	1,207	917	735	611	525	459	367
Brass free machining	18	1,398	1,207	917	735	611	525	459	367
Cast bronze	15	1,165	1,006	764	612	510	437	382	306
Nickel alloys	3	233	201	153	122	102	87	76	61
Titanium alloys	3	233	201	153	122	102	87	76	61
Plastic (thermoplastics)	25	1,864	1,609	1,223	980	815	700	611	489

Note 1: For cutting taps in blind holes, reduce RPM by 25% to 50%.
Note 2: Tap feed rate calculation. Example: M6 x 1 at 1,000 rpm. Vf = 1,000 rpm x 1 mm/rev = 1,000 mm/min.

A2.2 Specific Cutting Force

Table A2.20 Specific cutting force for different materials. Elaborated from different data of cutting tool vendors.

ISO	Material Group	Observations	Specific Cutting Force Kc1 (N/mm^2)	Hardness Brinell (HB)	mc
P	Steel Unalloyed	C = 0.10 -0.25%	1,500	125	0.25
		C = 0.25-0.55%	1,500	150	0.25
		C=0.55—0.80%	1,700	170	0.25
	Low-alloyed (alloying elements <5%)	Non-hardened	1,700	175	0.25
		Hardened and tempered	2,000	300	0.25
	High-alloyed (alloying elements >5%)	Annealed	1,950	200	0.25
		Hardened tool steel	2,150	200	0.25
			2,900	300	0.25
			3,100	380	0.25
	Castings	Unalloyed	1,400	150	0.25
		Low-alloy (alloying elements < 5 %]	1,600	200	0.25
		High-alloy (alloying elements > 5%)	1,950	200	0.25
M	Ferritic/martensitic stainless steel	Non—hardened	1,800	200	0.21
		PH-hardened	2,800	330	0.21
		Hardened	2,300	330	0.21
	Austenitic stainless steel	Non-hardened	2,000	200	0.21
		PH-hardened	2,800	330	0.21
	Austenitic-ferritic	Non-weldable > 0.05%C	2,000	230	0.21
	(Duplex)	Weldable <0.05%C	2,400	260	0.21
	Ferritic/martensitic Stainless steel cast	Non-hardened	1,800	200	0.25
		PH-hardened	2,500	330	0.25
		Hardened	2,100	330	0.25
	Austenitic stainless steel cast	Austenitic	1,800	200	0.25
		PH-hardened	2,500	330	0.25
	Austenitic-ferritic (Duplex)	Non-weldable > 0.05%C	1,800	230	0.25
		Weldable <0.05%C	2,200	260	0.25
K	Malleable cast iron	Ferritic (short chipping)	800	130	0.28
		Pearlitic (long chipping)	900	230	0.28
	Gray cast iron	Low tensile strength	900	180	0.28
		High tensile strength	1,100	245	0.28
	Nodular cast iron	Ferritic	900	160	0.28
		Pearlitic	1,350	250	0.28

Table A2.21 (Continuation) Specific cutting force for different materials. Elaborated from different data of cutting tool vendors.

ISO	Material Group	Observations	Specific Cutting Force Kc1 (N/mm^2)	Hardness Brinell (HB)	mc
N	Aluminium alloys	Wrought or wrought and coldworked, non-aging	400	60	0.25
		Wrought or wrought and aged	650	100	0.25
		Cast, non-aging	600	75	0.25
		Cast or cast and aged	700	90	0.25
		Al>99%	350	30	0.25
		Cast, 13-15% Si	700	130	0.25
		Cast, 16-22% Si	700	130	0.25
	Copper and copper alloys	Free cutting alloys, >1 % Pb	550	110	0.25
		Brass, leaded bronzes, <1% Pb	550	90	0.25
		Bronze and non-leaded copper	1,350	100	0.25
	Magnesium alloys	Magnesium alloys	390	<450[1)]	0.19
S	Heat-resistance super alloys. Iron base	Annealed or solution treated	2,400	200	0.25
		Aged or solution treated and aged	2,500	280	0.25
	Heat-resistant super alloys. Nickel base	Annealed or solution treated	2,650	250	0.25
		Aged or solution treated and aged	2,900	350	0.25
		Cast or cast and aged	3,000	320	0.25
	Heat-resistant super alloys. Cobalt base	Annealed or solution treated	2,700	200	0.25
		Solution treated and aged	3,000	300	0.25
		Cast or cast and aged	3,100	320	0.25
	Titanium alloys	Commercial pure (99.5% Ti)	1,300	400[1)]	0.23
		α, near or and $\alpha+\beta$ alloys, annealed	1,400	950[1)]	0.23
		$\alpha+\beta$ alloys in aged cond., β alloys, annealed or aged	1,400	1050[1)]	0.23
H	Extra hard steel	Hardened and tempered	4,200	59 HRC	0.25
	Chilled cast iron	Cast or cast and aged	2,200	400	0.28
O	Thermoplastics	Without abrasive fillers	150	–	0.2
		Fiber-reinforced	300	–	0.3

[1)] Ultimate tensile strength (Mpa)

A2.3 Cross-Reference Materials Table and ISO codes

Table A2.22 Cross-reference materials table. Elaborated from different data of cutting tool vendors.

	Standard					
ISO	**DIN**	**AISI / SAE**	**UNE**	**AFNOR**	**UNI**	**JIS**
Unalloyed steels						
P	C15	1015	F.111	CC12	C15C16	–
	C22	1020	F.112	CC20	C20C21	–
	C35	1035	F.113	CC35	C35	–
	C45	1045	F.114	CC45	C45	–
	36MN5	1335	36Mn5	40M5	–	SMn438(H)
	C45E	1045	C45K	XC42	C45	S45C
	C55	1055	C55K	–	C55	–
	C101E	1095	F-5117	XC 100	–	–
	C101u	W 1	F-5118	Y105	C36KU	SK 3
	C105W1	W210	F515	Y120	C120KU	SUP4
Low-alloy steel						
P	16Mo3	A204Gr.A	16Mo3	15D3	16Mo3KW	–
	14Ni6	A350LF5	15Ni6	16N6	14Ni6	–
	21 NiCrMo2	8620	20NiCrMo2	20NCD2	20NiCrMo2	SNCM220(H)
	13CrMo4-5	A182 F11;F12	14CrMo45	15CD3.5	14CrMo4 5	–
	10CrMo9 10	A182 F.22	TU.H	12CD9, 10	12CrMo9, 10	–
	14NiCr10	3415	15NiCr11	14NC11	16NiCr11	SNC415(H)
	16MnCr5	51 15	16MnCr5	16MC5	16MnCr5	–
	34CrMo4	4137:4135	34CrMo4	35CD4	35CrMo4	SCM432
	41 CrMo4	4140;4142	42CrMo4	42CD4TS	41 CrMo4	SCM 440

	42CrMo4	4140	42CrMo4	42CD4	42CrMo4	SCM 440(H)
	50NiCr13	L6	F-528	55NCV6	–	–
	36CrNiMo4	9840	35NiCrMo4	40NCD3	38NiCrMo4(KB)	
	34Cr4	5132	35Cr4	32C4	34Cr4(KB)	SCr430(H)
	41Cr4	5140	42Cr4	42C4	41Cr4	SCr440(H)
	51CrV4	6150	51CrV4	50CV4	50CrV4	SUP10
	100Cr6	L3	100Cr6	Y100C6	–	–
	25CrMo4	4130	AM26CrMo4	25CD4	25CrMo4 (KB)	SCM420
High-alloy steel						
P	X210Cr12	D3	X210Cr12	Z200C12	X210Cr12KU	SKD1
	X40CrMoV5 1	H13	X40CrMoV5	Z40CDV5	X40CrMoV511KU	SKD61
	X100CrMoV5 1	A2	X100CrMoV5	Z100CDV5	X100CrMoV51KU	SKD12
	X30WCrV9 3	H21	X30WCrV9	Z30WCV9	X30WCrV9 3KU	SKD5
	HS 6-5-2-5	M 35	F-5613	6-5-2-5	HS 6-5-2-5	SKH 55
Ferritic/martensitic stainless steel						
M	X10CrAL13	405	F.311	Z10C13	X10CrA112	SUS405
	X10CrAL18	430	F3113	Z10CAS18	X8Cr17	SUS430
	X6Cr13	403	F.3110	Z6C13	X6Cr13	SUS403
	X6Cr17	430	F3113	Z8C17	X8Cr17	SUS430
	X12CrS13	416	F-3411	Z11CF13	X12 CrS 13	SUS 416
	X19CrNi17-2	431	F.3427	Z15CNi6.02	X16CrNi16	SUS431
	X14CrMoS17	430F	F.3117	Z10CF17	X10CrS17	SUS430F

(*Continued*)

Table A2.22 (Continued)

	Standard					
ISO	**DIN**	**AISI / SAE**	**UNE**	**AFNOR**	**UNI**	**JIS**
Austenitic stainless steel						
M	X6CrNiNb18-10	347	F3552 F3524	Z6CNNb18.10	X6CrNiNb18 11	SUS347
	X6CrNiMoTi17-12-2	316Ti	F3535	Z6NDT17.12	X6CrNiMoTi17 12	–
	X2CrNiMoN17-11-2	308	F8414	Z1NCDU25.20	–	SCS17
	X9CrNi18-8	301	F3517	Z12CN17.07	X12CrNi17 07	SUS301
	X5CrNi18-10	304	F3551	Z6CN18.09	X5CrNi18 10	SUS304
	X6CrNiTi18-10	321	F3553 F3523	Z6CNT18.10	X6CrNiTi18 11	SUS321
	X8CrNiS18-9	303	F3508	Z10CNF 18.09	X10CrNiS 18.09	SUS303
	X8CrNi25-21	310S	F.331	Z12CN25 20	X6CrNi25 20	SUH310

Table A2.23 (Continuation) Cross-reference materials table. Elaborated from different data of cutting tool vendors.

	Standard					
ISO	**DIN**	**AISI / SAE**	**UNE**	**AFNOR**	**UNI**	**JIS**
Hardened materials						
H	X100CrMo13	440A	–	–	–	C4BS
	X110CrMoV15	610	–	–	–	AC4A
	X65CrMo14	0-2	–	–	–	AC4A
Malleable cast iron						
K	EN-GJMB350-10	32510	–	MN 35-10	–	FCMW330
	EN-GJMB450-6	40010	–	Mn 450	GMN 45	FCMW370
	EN-GJMB550-4	50005	–	MP 50-5	GMN 55	FCMP490
	EN-GJMB650-2	A220-70003	–	Mn 650-3	GMN 65	FCMP590
	EN-GJMB700-2	A220-80002	–	Mn700-2	GMN 70	FCMP690
Gray cast iron						
K	EN-GJL-150	No 25 B	FG15	Ft 15 D	G 15	FC150
	EN-GJL-250	No 35 B	FG 25	Ft 25 D	G 25	FC250
	EN-GJL-300	No 45 B	FG 30	Ft 30 D	G 30	FC300
	EN-GJL-350	No 50 B	FG 35	Ft 35 D	G 35	FC350
Nodular cast iron						
K	EN-GJS-400-15	60-40-18	FGE 38-17	FCS 400-12	GS 370-17	FCD400
	EN-GJS-800-7	80-55-06	FGE 50-7	FGS 500-7	GS 500	FCD500
	EN-GJS-700-2	100-70-03	FGS 70-2	FGS 700-2	GS 700-2	FCD700
Aluminum alloys						
N	G-AISi9MgWA	SC64D	–	A-S7G	–	C4BS
	Al99.5	1050A	L-3051	A5	4507	A1050
	G-AlSi10Mg	A13600	L-2560	A-S10G	3051	JIS AC4 4 (AL 4V)
	G-AlMg5	GD-AlSI12	–	A-SU12	–	AC4A
	G-AlSi12	A413.2	L-2520	A-S13	4514	AC3A
Copper alloys						
N	CuZn15	C23000	–	CuZn15	–	C2300
	CuZn37	C27400	–	CuZn37	–	C2720
	G-CuAl11Ni	B-148-52	–	CuAl11Ni5Fe	G-CuAl11FeNi4	–
Magnesium alloys						
N	MgAl3Zn	52	–	G-A3Z1	–	–
	G-MgAl 8 Zn 1	AZ81	–	AZ81	AZ81 hp	AZ81 hp

(*Continued*)

Table A2.23 (Continued)

	Standard					
ISO	**DIN**	**AISI / SAE**	**UNE**	**AFNOR**	**UNI**	**JIS**
Nickel based alloys						
S	S-NiCr13A16MoNb	5391	–	NC12AD	–	–
	NiCo15Cr10MoAlTi	AMS 5397	–	-	–	–
	NiCr19Co11MoTi	AMS 5399	–	NC19KDT	–	–
	NiCr19Fe19NbMo	AMS 5544	–	NC20K14	–	–
	NiCr22Mo9Nb	5666	–	NC22FeDNB	–	–
Cobalt based alloys						
S	CoCr22W14Ni	5772	–	KC22WN	–	–
Titanium alloys						
S	TiAl5Sn2.5	UNS R54520	–	T-A5E	–	–
	TiAl6V4	UNS R56401	–	T-A6V	–	–
Trade names (nickel Based Alloys)						
S	Haynes 600					
	Hastelloy C					
	Incoloy 825					
	Inconel 600					
	Monel 400					
	Inconel 700					
	Inconel 718					
	Waspaloy					
Trade names (Cobalt Based Alloys)						
S	Air Resist 213					
	Jetalloy 209					

Appendix 3: AQL Tables

Table A3.1 and Table A3.2 show the AQL Tables for determining sampling plans. The tables apply to single sampling plan for normal inspection according to "ANSI/ASQ Z1.4-2003 (R2018): Sampling Procedures and Tables for Inspection by Attributes" and "ISO 2859-1:1999 (1999). Sampling procedures for inspection by attributes — Part 1: Sampling schemes indexed by acceptance quality limit (AQL) for lot-by-lot inspection."

Table A3.1 AQL Tables: sample size code letter for single sampling and normal inspection.

SINGLE SAMPLING PLAN FOR NORMAL INSPECTION ANSI/ASQ STANDARD Z1.4-2003							
	SAMPLE SIZE CODE LETTER						
	General Inspection Levels			Special Inspection Levels			
Lot Size	I	II	III	S1	S2	S3	S4
2 to 8	A	A	B	A	A	A	A
9 to 15	A	B	C	A	A	A	A
16 to 25	B	C	D	A	A	B	B
26 to 50	C	D	E	A	B	B	C
51 to 90	C	E	F	B	B	C	C
91 to 150	D	F	G	B	B	C	D
151 to 280	E	G	H	B	C	D	E
281 to 500	F	H	J	B	C	D	E
501 to 1,200	G	J	K	C	C	E	F
1,201 to 3,200	H	K	L	C	D	E	G
3,201 to 10,000	J	L	M	C	D	F	G
10,001 to 35,000	K	M	N	C	D	F	H
35,001 to 150,000	L	N	P	D	E	G	J
150,001 to 500,000	M	P	Q	D	E	G	J
500,001 and over	N	Q	R	D	E	H	K

Manufacturing Process Planning: A Practical Approach for Mechanical Engineering, First Edition.
José V. Abellán-Nebot, Carlos Vila Pastor, and Héctor R. Siller.

Companion website: www.wiley.com/go/Manufacturing_Process_Planning_1e

Table A3.2 AQL Tables: AQL values according to size code letter, sample size and AQL level for normal inspection type.

Sample Size Code Letter	Sample Size	Acceptable Quality Levels (Normal Inspection)																									
		0.065		0.1		0.15		0.25		0.4		0.65		1		1.5		2.5		4		6.5		10		15	
		Ac	Re	Ac	Re	Ac	Re	Ac	Re	Ac	Re	Ac	Re	Ac	Re	Ac	Re	Ac	Re	Ac	Re	Ac	Re	Ac	Re	Ac	Re
A	2	↓		↓		↓		↓		↓		↓		↓		↓		↓		↓		0	1	↑		↓	
B	3	↓		↓		↓		↓		↓		↓		↓		↓		↓		0	1	↑		↓		1	2
C	5	↓		↓		↓		↓		↓		↓		↓		↓		0	1	↑		↓		1	2	2	3
D	8	↓		↓		↓		↓		↓		↓		↓		0	1	↑		↓		1	2	2	3	3	4
E	13	↓		↓		↓		↓		↓		↓		0	1	↑		↓		1	2	2	3	3	4	5	6
F	20	↓		↓		↓		↓		↓		0	1	↑		↓		1	2	2	3	3	4	5	6	7	8
G	32	↓		↓		↓		↓		0	1	↑		↓		1	2	2	3	3	4	5	6	7	8	10	10
H	50	↓		↓		↓		0	1	↑		↓		1	2	2	3	3	4	5	6	7	8	10	10	14	15
J	80	↓		↓		0	1	↑		↓		1	2	2	3	3	4	5	6	7	8	10	10	14	15	21	22
K	125	↓		0	1	↑		↓		1	2	2	3	3	4	5	6	7	8	10	10	14	15	21	22	↑	
L	200	0	1	↑		↓		1	2	2	3	3	4	5	6	7	8	10	10	14	15	21	22	↑		↑	
M	315	↑		↓		1	2	2	3	3	4	5	6	7	8	10	10	14	15	21	22	↑		↑		↑	
N	500	↓		1	2	2	3	3	4	5	6	7	8	10	10	14	15	21	22	↑		↑		↑		↑	
P	800	1	2	2	3	3	4	5	6	7	8	10	10	14	15	21	22	↑		↑		↑		↑		↑	
R	1,250	2	3	3	4	5	6	7	8	10	10	14	15	21	22	↑		↑		↑		↑		↑		↑	
S	2,000	3	4	5	6	7	8	10	11	14	15	21	22	↑		↑		↑		↑		↑		↑		↑	

Ac = Acceptance number; Re = Rejection number

↓ Use first sampling plan below arrow. If sample size equals or exceeds lot or batch, carry out 100% inspection.

↑ Use first sampling plan above arrow.

Source: ANSI/ASQ Z1.4-2003 (R2018): Sampling Procedures and Tables for Inspection by Attributes.

Index

Manufacturing Process Planning: A Practical Approach for Mechanical Engineering, First Edition.
José V. Abellán-Nebot, Carlos Vila Pastor, and Héctor R. Siller.

Companion website: www.wiley.com/go/Manufacturing_Process_Planning_1e

d

Printed and bound by CPI Group (UK) Ltd, Croydon, CR0 4YY
17/09/2026
14964335-0005